中国科学院科学出版基金资助出版

《国外数学名著系列》（影印版）专家委员会

(按姓氏笔画排序)

丁伟岳　王　元　文　兰　石钟慈　冯克勤　严加安
李邦河　李大潜　张伟平　张继平　杨　乐　姜伯驹
郭　雷

项目策划

向安全　林　鹏　王春香　吕　虹　范庆奎　王　璐

执行编辑

范庆奎

国外数学名著系列(影印版) 37

Methods of Homological Algebra

Second Edition

同调代数方法

(第二版)

Sergei I. Gelfand　　Yuri I. Manin

科学出版社
北京

图字：01-2008-5397

Sergei I. Gelfand, Yuri I. Manin: Methods of Homological Algebra(Second Edition)
© Springer-Verlag Berlin Heidelberg 1997, 2003

This reprint has been authorized by Springer-Verlag (Berlin/Heidelberg/New York) for sale in the People's Republic of China only and not for export therefrom

本书英文影印版由德国施普林格出版公司授权出版。未经出版者书面许可，不得以任何方式复制或抄袭本书的任何部分。本书仅限在中华人民共和国销售，不得出口。版权所有，翻印必究。

图书在版编目(CIP)数据

同调代数方法：第2版=Methods of Homological Algebra (Second Edition) /(德)盖尔范德(Gelfand S. I.)等著. —影印版. —北京：科学出版社，2009

(国外数学名著系列；37)

ISBN 978-7-03-023481-0

I.同… II.盖… III.同调代数-英文 IV. O154

中国版本图书馆 CIP 数据核字(2008) 第 186202 号

责任编辑：范庆奎/责任印制：赵 博/封面设计：黄华斌

科 学 出 版 社 出版
北京东黄城根北街 16 号
邮政编码：100717
http://www.sciencep.com

三河市春园印刷有限公司印刷
科学出版社发行 各地新华书店经销
*
2009 年 1 月第 一 版 开本：B5(720×1000)
2024 年 4 月第六次印刷 印张：24 3/4
字数：469 000
定价：168.00 元
(如有印装质量问题，我社负责调换)

《国外数学名著系列》(影印版)序

要使我国的数学事业更好地发展起来，需要数学家淡泊名利并付出更艰苦地努力。另一方面，我们也要从客观上为数学家创造更有利的发展数学事业的外部环境，这主要是加强对数学事业的支持与投资力度，使数学家有较好的工作与生活条件，其中也包括改善与加强数学的出版工作。

从出版方面来讲，除了较好较快地出版我们自己的成果外，引进国外的先进出版物无疑也是十分重要与必不可少的。从数学来说，施普林格(Springer)出版社至今仍然是世界上最具权威的出版社。科学出版社影印一批他们出版的好的新书，使我国广大数学家能以较低的价格购买，特别是在边远地区工作的数学家能普遍见到这些书，无疑是对推动我国数学的科研与教学十分有益的事。

这次科学出版社购买了版权，一次影印了 23 本施普林格出版社出版的数学书，就是一件好事，也是值得继续做下去的事情。大体上分一下，这 23 本书中，包括基础数学书 5 本，应用数学书 6 本与计算数学书 12 本，其中有些书也具有交叉性质。这些书都是很新的，2000 年以后出版的占绝大部分，共计 16 本，其余的也是 1990 年以后出版的。这些书可以使读者较快地了解数学某方面的前沿，例如基础数学中的数论、代数与拓扑三本，都是由该领域大数学家编著的"数学百科全书"的分册。对从事这方面研究的数学家了解该领域的前沿与全貌很有帮助。按照学科的特点，基础数学类的书以"经典"为主，应用和计算数学类的书以"前沿"为主。这些书的作者多数是国际知名的大数学家，例如《拓扑学》一书的作者诺维科夫是俄罗斯科学院的院士，曾获"菲尔兹奖"和"沃尔夫数学奖"。这些大数学家的著作无疑将会对我国的科研人员起到非常好的指导作用。

当然，23 本书只能涵盖数学的一部分，所以，这项工作还应该继续做下去。更进一步，有些读者面较广的好书还应该翻译成中文出版，使之有更大的读者群。

总之，我对科学出版社影印施普林格出版社的部分数学著作这一举措表示热烈的支持，并盼望这一工作取得更大的成绩。

王 元

2005 年 12 月 3 日

Preface to the Second Edition

In the second edition of our book, many of the typos, errors, and imprecise statements that found their way into the first edition, are corrected.

The authors are extremely grateful to Professor Sergei Merkulov who took upon himself the difficult task of carefully reading the text and incorporating appropriate changes. Without his efforts, most of the errors in the first edition would probably have remained unnoticed. Of course, the authors retain full responsibility for any mistakes that may still remain in the text.

September 2002   S. Gelfand, Yu. Manin

Foreword

> ... utinam intelligere possim rationacinationes pulcherrimas quae e propositione concisa DE QUADRATUM NIHILO EXAEQUARI fluunt.
>
> (... if I could only understand the beautiful consequence following from the concise proposition $d^2 = 0$.)
>
> From Henri Cartan Laudatio on receiving the degree of Doctor Honoris Causa, Oxford University, 1980

1

Homological algebra first arose as a language for describing topological properties of geometrical objects. The emergence of a new language is always an important event in the development of mathematics: Euclidean plane and spatial geometry, Cartesian analytic geometry, the formalization of Newton's fluents and fluxions by Leibniz and later by Lagrange start the series to which homological algebra can be added. As with every successful language, homological algebra quickly realized its tendencies for self-development. As with every successful mathematical language, it rapidly began to expand its semantics, that is, to describe things that it was not originally designed to describe. The computation of the index of an elliptic operator, exact estimates for the number of solutions of congruences modulo a prime, the theory of hyperfunctions, anomalies in quantum field theory – these are only some of the contemporary applications of homological ideas.

The history of homological algebra can be divided into three periods. The first one starts in the 1940's with the classical works of Eilenberg and MacLane, D.K. Faddeev, and R. Baer and ends with the appearance in 1956 of the fundamental monograph "Homological Algebra" by Cartan and Eilenberg which has lost none of its significance up to the present day.

A. Grothendieck's long paper "Sur quelques points d'algèbre homologique" published in 1957 (its appearance had been delayed three years) marks the starting point of the second period, which was dominated by the influence of Grothendieck and his school of algebraic geometry.

The third period, which extends up to the present time, is marked by the ever-increasing use of derived categories and triangulated categories. The basic technique was developed in the thesis of Grothendieck's student J.-L. Verdier in 1963, but was slow in spreading beyond the confines of algebraic geometry. Only in the last fifteen years has the situation changed. First in the work of M. Sato and his school on microlocal analysis, then in the theory of D-modules and perverse sheaves with applications to representation theory, derived categories started to be used as the most suitable instrument.

We now try to characterize these three periods, although we should apologize to the reader for our subjective evaluation and judgment and for the incompleteness of the material: of course, many important developments do not fit into our rigid scheme.

The book by Cartan and Eilenberg contains essentially all the constructions of homological algebra that constitute its computational tools, namely standard resolutions and spectral sequences. No less important, it contains an axiomatic definition of derived functors of additive functors on the category of modules over a ring.

It was this idea that determined the contours of the second period. The logic of the internal development of analytic and algebraic geometry led to the formulation of the notion of a sheaf and to the realization of the idea that the natural argument of a homology theory is a pair consisting of a space with a sheaf on it, rather than just a space (or a space and a coefficient group). Here the fundamental contribution of H. Cartan's seminar and J.-P. Serre's paper "Faisceaux algébriques cohérents" should be mentioned. Grothendieck's paper of 1957 quoted above stresses the analogy between pairs (space, sheaf of abelian groups on it) and pairs (ring, module over it) from the homological point of view and emphasizes the idea that sheaf cohomology should be defined as the derived functor of global sections.

The break with the axiomatic homology and cohomology theory of Eilenberg and Steenrod is in that now an abelian object (a sheaf), rather than a non-abelian one (a space), serves as a variable argument in a cohomology theory. More precisely, a homology or a cohomology theory with fixed coefficients according to Eilenberg and Steenrod is a graded functor from the category of topological spaces into abelian groups that satisfies certain axioms by which it is uniquely determined. The most important of these axioms are the specification of the homology (or cohomology) of the point, and the exact sequence associated with the "excision axiom". The cohomology theory of a fixed topological space according to Grothendieck is a graded functor from the category of sheaves of abelian groups on this space into abelian groups, also satisfying a number of axioms by which it is uniquely determined. The most important of these are the specification of zero-dimensional cohomology as global sections and the exact sequence associated with a short exact sequence of sheaves.

The development of this idea led to a very far-reaching generalization of basic notions of algebraic geometry – Grothendieck topologies and topoi. The essence of this generalization is that since the cohomological properties of a space are completely determined by the category of sheaves over it, it is these categories that should be the primary objects of study in topology, rather than topological spaces themselves. After a suitable axiomatization of the properties of such categories we arrive at the notion of a topos. The development of these abstract ideas was motivated by a very concrete problem – the famous conjectures of A. Weil on the number of solutions of congruences modulo a prime. The very statements of these conjectures include the assumption about the existence of a certain cohomology theory of algebraic varieties in characteristic $p > 0$, which would allow us to apply to this situation the Lefschetz fixed point formula; a cohomology theory of this type was provided by the cohomology of the etale topos constructed by A. Grothendieck and developed by his students.

The main product of the homological algebra of this period was the computation and properties of various derived functors $R^p F$, where F is the functor of global sections, of direct image, of tensor product and so on. These derived functors arise as the cohomology of complexes of the form $F(I^{\cdot})$, where $I^{\cdot}$ are resolutions consisting of injective, projective, flat, or some other objects suitably adapted to F. The choice of a resolution is highly non-unique, but $R^p F$ does not depend on this choice.

In the course of time it came to be understood that one should study all complexes, rather that just resolutions $I^{\cdot}$ (and complexes obtained by applying functors to these resolutions), but modulo a quite complicated equivalence relation, which identifies certain complexes having the same cohomology.

The final version of this equivalence relation seems still not to be completely understood. However, a working definition which has proved its worth was formulated in Verdier's thesis of 1963. The categories of complexes obtained in this way are called derived categories, and axiomatization of their properties leads to the notion of triangularted categories.

It seems to us that the main feature of the third period of homological algebra is the development of a special kind of "thinking in terms of complexes" as opposed to the "thinking in terms of objects and their cohomological invariants" that was typical for the first two periods. Perhaps this appears most vividly in the theory of perverse sheaves; it was shown that the cohomological properties of topological manifolds extend to a substantial degree to spaces with singularities, if we take as coefficients not sheaves but special complexes of sheaves (as objects of the corresponding derived category). The conormal complexes of Grothendieck and Illusie and the dualizing complexes of Grothendieck and Verdier can be considered as earlier constructions of the same kind.

explanations provided us with live examples of thinking in complexes.

J.-P. Serre, J.N. Bernstein and M.M. Kapranov have read the manuscript and made a series of very useful comments.

2

This book is intended as an introductory textbook on the technique of derived categories. Up to now, as far as we know, a mathematician willing to learn this subject has had to turn either to the two original sources, the abstract of Verdier's thesis and the notes of Hartshorne's seminar, or to the oral tradition, in those mathematical centers where it still has been maintained.

Thus the central part of the book is Chaps. III–IV, and the reader with even a slight acquaintance with abelian categories and functors can start directly from Chap. III.

Chapter II is directed to the reader who has hardly had anything to do with categories, and we have tried to make clear the intuitive meaning of standard categorical constructions, and to give examples of "thinking in categories". The main practical aim of this chapter is an introduction to abelian categories.

Finally, Chaps. I and V resulted from our attempt (which had cost us a lot of trouble) to separate off homological algebra from algebraic topology, without burning the bridge between them. Triangulated spaces and simplicial sets are perhaps the most direct methods of describing topology in terms of algebra, and we decided to start the book with an introduction to simplicial methods. On the other hand, algebraic topology is unthinkable without homotopy theory, and the book ends with a treatment of the foundations of homotopic algebra in Chap. V.

We worked on this book with the disquieting feeling that the development of homological algebra is currently in a state of flux, and that the basic definitions and constructions of the theory of triangulated categories, despite their widespread use, are of only preliminary nature (this applies even more to homotopic algebra). There is no doubt that similar thoughts have occurred to the founders of the theory, and to everyone who has seriously worked with it; the absence of a monographic exposition is one of the symptoms.

Nevertheless, this period has already lasted twenty years; papers whose main results cannot even be stated in the old language are multiplying; the need for a textbook is growing. We therefore present this book to the benevolent judgment of the reader.

3

The plan of the book evolved gradually over several years when the authors were running seminars in the Mathematics Department of Moscow University, and were in contact with members of the "Homological Algebra Fan Club". A.A. Beilinson, M.M. Kapranov, V.V. Schechtman, whose papers and explanations provided us with live examples of thinking in complexes.

J.-P. Serre, J.N. Bernstein and M.M. Kapranov have read the manuscript and made a series of very useful comments.

V.E. Govorov very kindly to provided us with an extensive card index of works on homological algebra.

We are grateful to all them, and also to V.A. Ginzburg, R. MacPherson, S.M. Khoroshkin and B.L. Tsygan.

Our debt to the founding fathers of the subject, whose books, papers and ideas we have used and have been inspired by, should be obvious from the contents.

Moscow, 1988   S.I. Gelfand, Yu.I. Manin

Reference Guide

1. General References. Five main sources for the classical homological algebra are books by Cartan – Eilenberg [1], MacLane [1], Hilton – Stammbach [1], Bourbaki [1] and the large paper by Grothendieck [1]. Simplicial methods are presented in Gabriel – Zisman [1] and in May [1], sheaves in Godement [1], Bredon [1], Golovin [1], Iversen [1]. Topoi are discussed, in particular, in Goldblatt [1] and Johnstone [1]. Among the books on cohomology of various algebraic structures we mention Brown [1], Serre [8], Guichardet [1], Fuks [2]. A large list of books on algebraic topology contains, among others, Eilenberg – Steenrod [1], Hilton – Wiley [1], Spanier [1], Dold [1], Massey [2], Boardman – Vogt [1], Fuks [1], Dubrovin – Novikov – Fomenko [1], Bott – Tu [1].

Modern algebraic geometry is an ample source of homological algebra of various kind. Here we must mention the pioneering paper by Serre [3] and the publications of Grothendieck and his school: Grothendieck – Dieudonné [1] (especially Chaps. 0 and III) and [2], Grothendieck et al. [SGA] (especially 4, 4 1/2, 6), Artin [1], Hartshorne [1], Berthelot [1], Deligne [1], [2]. Among several textbooks on this subject we mention Hartshorne [2] and Milne [1].

The history of the homological algebra has yet to be written; we can recommend to the interested reader the paper by Grey [1], the corresponding parts from Dieudonné [1] and reminiscences of Grothendieck [5].

2. Topics We Have Not Considered in the Book.

a) Noncommutative cohomology. Some problems in group theory and topology lead to cohomology with non-commutative coefficients. A systematic theory exists only in low dimensions (≤ 2 or ≤ 3). An excellent exposition for the case of group cohomology based on the paper by Dedecker [1] is Serre [7]. Most commonly used is 1-cohomology, or torsors. About intermediate "state of the art" see Giraud [1].

b) Derivatives of non-additive functors. First constructions of derivatives of non-additive functors, such as the symmetric or the exterior power of a module, were suggested by Dold – Puppe [1]. Their technique was developed further by Illusie [1] who applied such functors in certain algebraic geometry situations. Crucial in the construction of these functors are simplicial methods. In Feigin – Tsygan [1], [2] the additive K-theory is interpreted as the derivation of the functor that associates to each ring its quotient by the commutant.

c) Continuous cohomology. Functional analysis and infinite-dimensional geometry produce some cohomology-like constructions in various categories of algebraic structures with topology, such as linear topological spaces, Banach algebras, Lie groups, etc. However, most of these categories (and the most important ones) are non-abelian, and the standard technique does not work. Usually in definition and computations the authors exploit some specific classes of complexes. See Helemski [1], Guichardet [1], Borel – Wallach [1], Johnson [1].

d) Products and duality. Some odds and ends the reader can find in various parts of the book, but a satisfactory general theory in the framework of homological algebra presumably does not exist. See [SGA 2] and Hartshorne [1] about the duality in algebraic geometry, Verdier [1], [2] and Iversen [1] about the duality in topology. The theory of *DG*-algebras (see Chap. V) can be considered as an attempt to introduce the multiplicative structure "from scratch". About deeper results see Boardman – Vogt [1], Shechtman [1], Hinich – Shechtman [1]. Classical theory of cohomological operations (Steenrod powers, Massey operations) also can be considered from such viewpoint.

e) Homological algebra and K-theory. The literature on *K*-theory is very ample; see the basic papers by Quillen [1], [2], [4], the review by Suslin [2], as well as [KT1], [KT2], where one can find further references.

f) Miscelleneous. Applications of Galois cohomology in number theory are based, first of all, on class field theory; see the classical exposition in Artin – Tate [1] and subsequent papers by Tate [1], Mazur [1], among others. There exists a large literature on homological methods in commutative algebra; see [ATT], Serre [6], André [2], [3], Avramov – Halperin [1], Quillen [3]. About some other applications of homological algebra see [AN], [ES], [SD].

3. To Chapter I. Sect. I.1–I.3: About further results of simplicial algebra, and, in particular, about its applications to homological algebra, see Gabriel – Zisman [1] and May [1]; see also the remarks to Chap. V below. Its application to the derivation of non-additive functors see in Dold – Puppe [1] and Illusie [1]. Deligne extensively used symplicial methods in the theory of mixed Hodge structures, see Deligne [1], Beilinson [2]. About Exercises 2, 3 to Sect. I.2 see Duskin [1], [2].

Sect. I.4: Algebraic topology is only slightly mentioned here, see Sect. 1 of this guide.

Sect. I.5: About the classical sheaf theory see Serre [3], [4], Godement [1], Bredon [1], Golovin [1], Iversen [1]. For sheaf theory in general topoi, as well as in ètale, cristalline, and other topoi of algebraic geometry see [SGA 4], Artin [1], Berthelot [1], Milne [1].

The most important development of the sheaf theory in the last ten years is related to the notion of a perverse sheaf and the corresponding cohomological formalism which is well suited to the study of singular varieties. Perverse sheaves are objects of the derived category of usual sheaves, so that a perverse sheaf is a complex of usual sheaves. See Goresky – MacPherson [1], Beilinson – Bernstein – Deligne [1], and [IH], [ES].

Sect. I.6: An exact sequence is the main tool of homological algebra. See further development in the framework of derived and triangulated categories in Sect. III.3 and IV.1.

Sect. I.7: There exists a large list of papers whose authors introduce and study important specific resolutions and complexes such as de Rham, Čech, Koszul, Hochshild, and bar-resolutions, cyclic complexes, complexes of continuous cochains, etc. See, in particular, Priddy [1], [2], Karoubi [1], [2], [3], Connes [1], Fuks [2], Hochshild [1].

4. To Chapter II. Sect. II.1: See MacLane [1], Goldblatt [1], Faith [1]; about 2-categories see Gabriel – Zisman [1].

Sect. II.2: About the theory of fundamental group in algebraic geometry see [SGA 2], about Gelfand duality see Gelfand – Shilov [1], about Morita equivalence see Morita [1], Faith [1]. A classical example of the non-trivial equivalence is the description of coherent sheaves on projective algebraic manifolds by corresponding modules over homogeneous coordinate ring, see Serre [3] and a generalization in Grothendieck – Dieudonné [1, EGA 3].

Further generalizations of these ideas lead to remarkable equivalences between some derived categories, see Sect. IV.3.

Sect. II.3: Several important theorems give an abstract characterization of representable functors. About Freyd's theorem in the general category theory framework see MacLane [2]. A lot of important spaces (like moduli spaces, i.e. bases of universal deformations) in algebraic and analytic geometry are introduced using the notion of a representable functor. In this context the characterization of representable functors by a short list of easily verified properties leads to some fundamental existence theorems, see Grothendieck [1], [2], [4], Artin [1], Knutson [1].

The fundamental notion of the adjoint functor was introduced by Kan [1]. Several important constructions in algebra, geometry, and topology can be described using this notion, see examples in André [1], Faith [1], MacLane [2].

Sect. II.4: For details about ringed spaces see Grothendieck – Dieudonné [1, Chap. 0], [2]. For the nerve of a category see Quillen [4], Suslin [1]. For quadratic algebras (Ex. 5) see Manin [1].

Sect. II.5–II.6: This is a classical part of the theory of abelian categories, see Cartan – Eilenberg [1], MacLane [1], [2], Grothendieck [1]. For the development of these ideas in the context of derived categories see Sect. III.6, IV.1. About ex. 1–7 in Sect. II.5 see MacLane [2]. About ex. 9 in Sect. II.5 see Serre [2].

5. To Chapter III. Sect. III.1–III.4: See Hartshorne [1], Verdier [3]. The fundamental diagram in Lemma 3 is taken from Bourbaki [1].

It seems that the main deficiency of the definition of a derived category is in the bad definition of distinguished triangles. The problem of what should be a good definition is discussed in unpublished notes of Deligne. See also the discussion about the functor det in Knudsen – Mumford [1], and in [SGA 6] and the definition of Tot in Exercises to IV.2.

Sect. III.5: The classical theory of the functors Ext in terms of complexes is due to Yoneda [1] (it generalizes the Baer's theory of Ext^1). Homological dimension was studied in algebraic geometry (Serre [4]), in depth theory ([SHA 2]), in group theory (Brown [1] and several papers in [HG]).

About theorem 21 see Hartshorne [1]. This theorem can be considered as one of the theorems establishing the equivalence between a derived category and a category of complexes modulo homotopic equivalence, see Beilinson [1], Bernstein – Gelfand – Gelfand [1], Kapranov [1], [2], [3]. For results related to ex. 4 see Happel [1].

Sect. III.6: The main references here are the same as in Sects. III.1–III.4. For ex. 1–5 see Deligne's paper in Grothendieck et al. [SGA 4, XVII], for ex. 6 see Roos [1], [2], about ex. 7–10 see Spaltenstein [1].

Sect. III.7: While an exact sequence can be considered as the main tool in the study of cohomology dependence on the abelian variable, a spectral sequence plays a similar role in the study of the dependence on the non-abelian variable. The first spectral sequence was introduced, presumably, by Leray [1]; the classical exposition of Serre [1] remains an excellent introduction into the subject. The standard construction of the spectral sequence associated to a filtered complex is given in Cartan – Eilenberg [1], and the one associated to an exact couple is given in Massey [1] (see also Eckmann – Hilton [1], [2]). Grothendieck [1] showed that some standard spectral sequences relate derived functors of the composition to the derived functors of factors. However, spectral sequences in homotopic topology are of different nature, see McCleary [1]. Fuks [1] gives a fascinating description of the Adams spectral sequence. See also exercises to IV.2.

Sect. III.8: This section, together with exercises to it and to IV.4, presents sheaf cohomology theory, as it is seen nowadays. The main difference from the status fixed in Godement [1] is the appearence of the functor $f^!$, which can be defined only using derived categories. This functor leads to the Verdier duality, which also can be formulated only in derived categories, and to the "six functors" formalism (see exercises to IV.4). References are Verdier [1], [2], [4], the volume [IH], Iversen [1]. The parallel theory in algebraic geometry is presented in Hartshorne [1] for coherent sheaves and in [SGA 4] (especially XVII) for étale topology.

6. To Chapter IV. Sect. IV.1–IV.2: The main sources for us were Verdier [3], Hartshorne [2] and Kapranov [3]. See also Happel [1], Iversen [1]. Exercises to Sect. IV.2 were composed by Kapranov.

Sect. IV.3: The description of derived categories of coherent sheaves on on projective spaces was initiated in Beilinson [1] and Bernstein – Gelfand – Gelfand [1]. A series of consecutive generalizations of this theory was found by Kapranov [1], [2], [3]; see also Meltzer [1]. The "$S - \Lambda$ duality" described in this section possesses far reaching generalizations, see Priddy 1], [2], Lofwall [1], Happel [1], Gorodentsev – Rudakov [1].

Sect. IV.4: We present here some ideas from Beilinson – Bernstein – Deligne [1]. The main application of this theory, that is, the construction of perverse sheaves, is not discussed in this book. About ex. 1–5 see Beilinson – Bernstein – Deligne [1], about ex. 6 see Bernstein – Gelfand – Ponomarev [1], Brenner – Butler [1], about ex. 7 see Happel [1].

7. To Chapter V. In this chapter we study homotopic algebra (algebraic foundations of homotopy theory), which is much less developed than homological algebra.

Sect. V.1–V.2: Here we introduce the main axiomatic notion that of a closed model category (Quillen [1]), which axiomatizes the main homotopic properties of topological spaces. Since we view simplicial sets as a bridge between topology and algebra, we give a (rather lengthy) proof that simplicial sets form a closed model category. We hope that these two sections will help an interested reader to study deeper parts of the book by Quillen [1], as well as further literature: Quillen [2], [4], May [1], Bousfield – Gugenheim [1], Tanré [1].

Sect. V.3–V.4: The second part of this chapter introduces to the reader some ideas of the famous paper by Sullivan [1] where he shows that the rational homotopic type of a manifold can be determined by its algebra of differential forms.

We prove that differential graded algebras form a closed model category and study minimal models in this category.

Exercises to these sections are based, mainly, on results from Tanré [1].

Sect. V.5: Here we present (without proofs) main results of the theory of rational homotopic type. The proofs, together with further details and references, can be found in Lemann [1], Bousfield – Gugenheim [1], Deligne – Griffiths – Morgan – Sullivan [1], Morgan [1], Halperin [1], Avramov – Halperin [1], Tanré [1].

Contents

I. **Simplicial Sets** 1
    I.1 Triangulated Spaces 1
    I.2 Simplicial Sets 6
    I.3 Simplicial Topological Spaces and the
         Eilenberg–Zilber Theorem 17
    I.4 Homology and Cohomology 23
    I.5 Sheaves 31
    I.6 The Exact Sequence 40
    I.7 Complexes 46

II. **Main Notions of the Category Theory** 57
    II.1 The Language of Categories and Functors 57
    II.2 Categories and Structures, Equivalence of Categories 69
    II.3 Structures and Categories. Representable Functors 78
    II.4 Category Approach to the Construction
         of Geometrical Objects 93
    II.5 Additive and Abelian Categories 109
    II.6 Functors in Abelian Categories 122

III. **Derived Categories and Derived Functors** 139
    III.1 Complexes as Generalized Objects 139
    III.2 Derived Categories and Localization 144
    III.3 Triangles as Generalized Exact Triples 153
    III.4 Derived Category as the Localization of Homotopic Category 159
    III.5 The Structure of the Derived Category 164
    III.6 Derived Functors 185
    III.7 Derived Functor of the Composition. Spectral Sequence ... 200
    III.8 Sheaf Cohomology 218

IV. **Triangulated Categories** 239
    IV.1 Triangulated Categories 239
    IV.2 Derived Categories Are Triangulated 251
    IV.3 An Example: The Triangulated Category of Λ-Modules ... 267
    IV.4 Cores 278

V. Introduction to Homotopic Algebra 291
   V.1 Closed Model Categories 291
   V.2 Homotopic Characterization of Weak Equivalences 299
   V.3 *DG*-Algebras as a Closed Model Category 333
   V.4 Minimal Algebras 342
   V.5 Equivalence of Homotopy Categories 352

References 357

Index 369

I. Simplicial Sets

I.1 Triangulated Spaces

1. Main Definitions

Figure I.1 depicts three triangulated spaces. Their main property is that they are glued from simplices: points, segments, triangles, tetrahedra and their higher-dimensional generalizations. Such spaces can be described combinatorially: one has to specify how many simplices of any dimension should be taken and how they should be glued together. Let us give precise definitions.

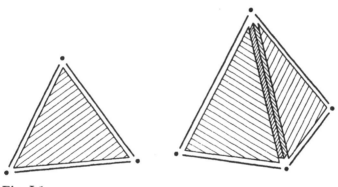

Fig. I.1.

a) The *n-dimensional simplex* is a topological space

$$\Delta_n = \left\{(x_0,\ldots,x_n) \in \mathbb{R}^{n+1} \,\middle|\, \sum_{i=0}^{n} x_i = 1, x_i \geq 0 \right\}.$$

The point e_i for which $x_i = 1$ is called the *i-th vertex* of Δ_n; the set of vertices is ordered: $e_0 < e_1 < \ldots < e_n$. More generally, with each subset $I \subset [n] \stackrel{\text{df}}{=} \{0, 1, \ldots, n\}$ one associates the *I-th face* of Δ_n defined as the set of all points $(x_0, \ldots, x_n) \in \Delta_n$ with $x_i = 0$ for $i \notin I$. Sometimes it is convenient to replace I by the monotone (increasing) mapping $f : [m] \to [n]$ with the image I, where card $I = m + 1$. It is clear that there exists a unique linear mapping $\Delta_f : \Delta_m \to \Delta_n$ that preserves the order of vertices and has the I-th face as its image.

b) By *gluing data* we mean the following set X of structures:

What to glue: $X_{(0)}$ points, $X_{(1)}$ segments, $X_{(2)}$ triangles, ..., $X_{(n)}$ n-dimensional simplices. (Elements of $X_{(n)}$ serve as indices to number simplices.)

How to glue: for any pair $(n, I \subset [n])$, card $I = m+1$, a map $X_{(n)} \to X_{(m)}$ is given that specifies which m-dimensional simplex should be identified with the I-th face of the corresponding n-dimensional simplex.

More precisely, let a face correspond to an increasing map $f : [m] \to [n]$ and let $X(f) : X_{(n)} \to X_{(m)}$ be the corresponding gluing map. The family $\{X(f)\}$ should satisfy the two following conditions:

$$X(\mathrm{id}) = \mathrm{id}, \qquad X(g \circ f) = X(f) \circ X(g)$$

(where id is the identity map). This means that to different elements of $X_{(n)}$ there correspond different simplices and that "a face of a face is a face".

What one gets after gluing is a topological space $|X|$ with the underlying set $\coprod_{n=0}^{\infty} (\Delta_n \times X_{(n)}) / R$, where R is the weakest equivalence relation that identifies $(s, x) \in \Delta_n \times X_{(n)}$ and $(t, y) \in \Delta_m \times X_{(m)}$ with

$$y = X(f)x, \qquad s = \Delta_f(t) \tag{I.1}$$

for some increasing mapping $f : [m] \to [n]$. We shall denote the situation as in (I.1) by $(t, y) \stackrel{f}{\mapsto} (s, x)$. The canonical topology on $|X|$ is the weakest topology for which the canonical mapping $\coprod_{n=0}^{\infty} (\Delta_n \times X_{(n)}) \to |X|$ is continuous.

The space $|X|$ together with the corresponding gluing data is called a *triangulated space*, and the gluing data themselves are called its *triangulation*.

2. Examples

a) *The n-dimensional simplex with the standard triangulation.*

Here

$$X_{(i)} = \{\text{cardinality } i+1 \text{ subsets in } [n]\}$$
$$= \{\text{increasing maps } [i] \to [n]\};$$

$$X\left([i] \stackrel{f}{\to} [j]\right) \text{ maps } g : [j] \to [n] \text{ to}$$
$$g \circ f : [i] \to [n].$$

Thus, the simplex is dismantled and reassembled.

b) *The sphere S^n with the standard triangulation.*

It is obtained from the standard triangulation of Δ_{n+1} by deleting the (unique) $(n+1)$-dimensional simplex.

The above examples suggest that any triangulated space is (set-theoretically) the disjoint union of interiors of its simplices. This is indeed the case. Namely, let

$$\overset{\circ}{\Delta}_n = \begin{cases} \text{the interior of } \Delta^n & \text{for } n \geq 1; \\ \Delta_0 & \text{for } n = 0. \end{cases}$$

Let $(X_{(i)}, X(f))$ be some gluing data and
$$\tau: \coprod_{n=0}^{\infty} (\Delta_n \times X_{(n)}) \to |X|$$
be the corresponding triangulation map. It induces the mapping
$$\overset{\circ}{\tau}: \coprod_{n=0}^{\infty} \left(\overset{\circ}{\Delta}_n \times X_{(n)}\right) \to |X|.$$

3. Proposition. $\overset{\circ}{\tau}$ *is a (set-theoretic) bijection.*

Proof. Let us associate with each point $(s, x) \in \Delta_n \times X_{(n)}$ its index $k(s, x)$ defined as the minimum dimension of a face of Δ_n containing s. It is clear that R-equivalent points in $\coprod (\Delta_n \times X_{(n)})$ have equal indices, so that $k(s, x)$ can be considered as a function $k(p)$ on $|X|$. Now, $k(s, x) = k$ if $s \in \overset{\circ}{\Delta}_k$ and, conversely, any point $p \in |X|$ with $k(p) = k$ has at least one representative in $\overset{\circ}{\Delta}_k \times X_{(k)}$. Therefore, $\overset{\circ}{\tau}$ is surjective.

Let us prove that $\overset{\circ}{\tau}$ is injective. It is clear that two points (s, x), (s', x') from $\coprod \left(\overset{\circ}{\Delta}_n \times X_{(n)}\right)$ will be identified in $|X|$ iff x and x' lie in the same $X_{(k)}$ and any chain of elementary equivalences (I.1) joining (s, x) and (s', x') contains only $x_i \in X_{(l_i)}$ with $l_i \geq k$. Any such chain can be transformed to the following form in which no two neighbouring arrows point in the same direction:
$$(s, x) \overset{f_1}{\longmapsto} (s_1, x_1) \overset{f_2}{\longleftarrow\!\shortmid} (s_2, x_2) \longmapsto \ldots \longleftarrow\!\shortmid (s', x'),$$
$x_i \in X_{(l_i)}$, $s_i \in \Delta_{(l_i)}$, $l_i \geq k$. Using this chain we construct another chain that joins (s, x) with (s', x') and has a smaller length. Let us recall that $f_1: [k] \to [l_1]$ and $f_2: [l_2] \to [l_1]$ are increasing mappings. Equalities $s = \Delta_{f_1}(s_1)$, $s_2 = \Delta_{f_2}(s_1)$ and $s \in \overset{\circ}{\Delta}_k$ imply that s belongs to the f_2-face of Δ_{l_1} so that there exists an increasing mapping $f: [k] \to [l_2]$ satisfying $f_1 = f_2 \circ f$ and $x = X(f)x_2$. Now we can replace our chain by a shorter chain
$$(s, x) \overset{g}{\longmapsto} (s_3, x_3) \overset{f_4}{\longleftarrow\!\shortmid} (s_4, x_4) \longmapsto \ldots \longleftarrow\!\shortmid (s', x'),$$
where $g = f_3 \circ f : [k] \to [l_3]$. Repeating this construction and recalling that $(s', x') \in \overset{\circ}{\Delta}_k \times X_{(k)}$ we get $(s, x) = (s', x')$. □

4. Skeleton

The *k-skeleton* of the triangulation $(X_{(i)}, X(f))$ is the gluing data $(X_{(i)}, i \leq k; X(f))$. The corresponding triangulated space $\text{sk}_k|X|$ is called the *k*-skeleton of the triangulated space $|X|$. Using the previous proposition we get the following properties of skeleta:

a) $|X| = \text{sk}_\infty|X| = \cup_{k \geq 0} \text{sk}_k|X|$.
b) Natural mappings $\text{sk}_k|X| \to \text{sk}_l|X|$, $k \leq l$, are closed embeddings.

c) $\mathrm{sk}_{k+1}|X|$ is obtained from $\mathrm{sk}_k|X|$ by gluing several $(k+1)$-dimensional open simplices along their common boundaries.

5. Triangulation of the Product of Two Simplices

The product of two segments $[0,1] \times [0,1]$ is not a triangle but a square; it can be naturally divided into two triangles by a diagonal, of two diagonals one is singled out by the fact that its vertices are naturally ordered: $[00, 11]$.

Generalizing this construction, we define the *canonical triangulation* $(X_{(n)}, X(f))$ of the product $\Delta_p \times \Delta_q$.

a) One element of $X_{(n)}$ ("multidimensional diagonal") is a set of $n+1$ different pairs of integers $\{(i_0, j_0), \ldots, (i_n, j_n)\}$, where

$$0 \leq i_0 \leq i_1 \leq \ldots \leq i_n \leq p,$$
$$0 \leq j_0 \leq j_1 \leq \ldots \leq j_n \leq q.$$

Such a set can be conveniently represented as a sequence of nodes in a two-dimensional square lattice with each node lying both to the right and above (nonstrictly) from the previous one.

b) For any increasing mapping $f : [m] \to [n]$ define $X(f)$ as follows:

$$X(f)\{(i_0, j_0), \ldots, (i_n, j_n)\} = \{(i'_0, j'_0), \ldots, (i'_m, j'_m)\}$$

where $i'_k = i_{f(k)}$, $j'_k = j_{f(k)}$.

Conditions from (1) are trivially satisfied.

c) Define a mapping

$$\theta_n : \Delta_n \times X_{(n)} \to \Delta_p \times \Delta_q$$

as follows: with the x-th simplex,

$$x = \{(i_0, j_0), \ldots, (i_n, j_n)\} \in X_n,$$

θ_n associates the simplex $\tilde{\Delta}_n$ in $\Delta_p \times \Delta_q \in \mathbb{R}^{p+q+2}$ spanned by the points (e_{i_a}, e'_{j_a}), $0 \leq a \leq n$, where e_i (resp. e'_i) is the i-th vertex of Δ_p (resp. Δ_q). Or, more formally, $\theta_n(\cdot, x) : \Delta_n \to \Delta_p \times \Delta_q$ is a (unique) linear order-preserving mapping with the image $\tilde{\Delta}_n$.

d) Let now $|X|$ be the triangulated space corresponding to the data $(X_{(n)}, X(f))$ from a), b).

We claim that *there exists a commutative diagram*

$$\begin{array}{ccc} & \coprod_n (\Delta_n \times X_{(n)}) & \\ & \swarrow \qquad \searrow & \theta = \coprod_n \theta_n \\ |X| & \xrightarrow[\varphi]{\sim} & \Delta_p \times \Delta_q \end{array}$$

where φ is a bijection.

First, from the definition of $X(f)$ in b) we see that θ maps R-equivalent points in $\coprod (\Delta_n \times X_{(n)})$ into one point in $\Delta_p \times \Delta_q$. Hence at least one such φ exists.

To prove that φ is unique and one-to-one, it suffices, by Proposition I.1.3, to check that any point $a \in \Delta_p \times \Delta_q$ has exactly one preimage $\theta^{-1}(a)$ in $\coprod (\overset{\circ}{\Delta}_n \times X_{(n)})$.

d) Let
$$\Delta_p = \left\{(x_0, \ldots, x_p) \,\Big|\, \sum x_i = 1, x_i \geq 0\right\},$$
$$\Delta_q = \left\{(y_0, \ldots, y_q) \,\Big|\, \sum y_j = 1, y_j \geq 0\right\}.$$

Let us introduce in Δ_p and Δ_q new coordinates as follows:

$\xi_1 = x_0,$ $\quad\quad\quad\quad\quad\quad\quad$ $\eta_1 = y_0,$
$\xi_2 = x_0 + x_1,$ $\quad\quad\quad\quad\quad$ $\eta_2 = y_0 + y_1,$
$\ldots$ $\quad\quad\quad\quad\quad\quad\quad\quad\quad$ $\ldots$
$\xi_p = x_0 + \ldots + x_{p-1}$ $\quad\quad\quad$ $\eta_q = y_0 + \ldots + y_{q-1}.$

Then
$$\Delta_p = \{(\xi_1, \ldots, \xi_p) \,|\, 0 \leq \xi_1 \leq \xi_2 \leq \ldots \leq \xi_p \leq 1\},$$
$$\Delta_q = \{(\eta_1, \ldots, \eta_q) \,|\, 0 \leq \eta_1 \leq \eta_2 \leq \ldots \leq \eta_q \leq 1\},$$

Now the vertices $\{e_i\}$ and $\{e'_j\}$ of simplices Δ_p and Δ_q have the following coordinates ξ and η respectively:
$$e_i = (\underbrace{0, \ldots, 0}_{i}, \underbrace{1, \ldots, 1}_{p-i}),$$
$$e'_j = (\underbrace{0, \ldots, 0}_{j}, \underbrace{1, \ldots, 1}_{q-j}).$$

Let
$$x = \{(i_0, j_0), \ldots, (i_{p+q}, j_{p+q})\} \in X_{(p+q)}$$
be a simplex of the maximum dimension, so that
$$(i_0, j_0) = (0, 0), \ldots, (i_{p+q}, j_{p+q}) = (p, q),$$
and for any $k, 0 \leq k \leq p + q - 1$, either $i_{k+1} = i_k + 1$, $j_{k+1} = j_k$ or $i_{k+1} = i_k$, $j_{k+1} = j_k + 1$. The image $\theta(\Delta_{p+q} \times x)$ consists of all pairs $(\xi_1, \ldots, \xi_p)$, $(\eta_1, \ldots, \eta_q)$ satisfying the following conditions:
$$0 \leq \xi_1 \leq \ldots \leq \eta_q \leq 1 \tag{I.2}$$
where ξ_i, η_j are ordered as follows:

(i) if $i < j$, then ξ_i precedes ξ_j and η_i precedes η_j.
(ii) if $j_{k+1} = j_k$, then the $(k+1)$-th place in (I.2) is occupied by ξ (with some index), otherwise by η (with some index). The index is uniquely determined by (i).

This description shows that θ is surjective so that φ is unique and also surjective.

e) Let $r = ((\xi_1, \ldots, \xi_p), (\eta_1, \ldots, \eta_q)) \in \Delta_p \times \Delta_q$. To define the unique element in $\coprod \overset{\circ}{\Delta}_n \times X_{(n)}$ mapped into r by θ, we break up $p+q+2$ numbers $0, \xi_i, \eta_j, 1$ into groups of equal numbers, and enumerate the group by indices $0, 1, \ldots, l+1, 0 \leq l \leq p+q+1$ in increasing order. Let numbers in the k-th group be equal γ_k, so that $0 = \gamma_0 < \gamma_1 < \ldots < \gamma_{l+1} = 1$. Define the element $x = \{(i_0, j_0), \ldots, (i_l, j_l)\} \in X_{(l)}$ corresponding to the point r as follows:

$$i_k = \max\{i : \xi_i \text{ belongs to the } k\text{-th group}\},$$
$$j_k = \max\{j : \eta_j \text{ belongs to the } k\text{-th group}\}.$$

If the k-th group contains no ξ's we let $i_k = i_{k-1}$ (and $i_k = 0$ for $k = 0$). Similarly, if the k-th group contains no η's we let $j_k = j_{k-1}$ (and $j_k = 0$ for $k = 0$).

Let now $s \in \Delta_l$ be a point with coordinates $z_i = \gamma_{i+1} - \gamma_i$, $0 \leq i \leq l$. All γ_i being distinct, we have $s \in \overset{\circ}{\Delta}_l$.

To complete the proof we must only check that $\theta(s, x) = r$. This is left to the reader.

Exercises

1. Draw triangulations of the Möbius band, of the real projective plane, and of the two-dimensional torus.

I.2 Simplicial Sets

1. Definition. *A simplicial set is a family of sets* $X_\bullet = (X_n)$, $n = 0, 1, \ldots$ *and of maps* $X(f) : X_n \to X_m$, *one for each nondecreasing map* $f : [m] \to [n]$ *such that the following conditions are satisfied:*

$$X(\mathrm{id}) = \mathrm{id}, \quad X(g \circ f) = X(f) \circ X(g).$$

The only difference with the gluing data introduced in I.1.1b) is that f need not be strictly increasing. Elements of X_n are called *n-simplices*. Instead of $X_\bullet$ we will often write simply X.

For any nondecreasing map $f : [m] \to [n]$ we define the "f-th face" Δ_f as the linear map $\Delta_m \to \Delta_n$ that maps any vertex $e_i \in \Delta_m$ into the vertex $e_{f(i)} \in \Delta_n$, $i = 0, \ldots, m$.

Contrary to the situation in I.1.1a), Δ_f is not necessarily an embedding: if f is not strictly increasing, then Δ_f decreases the dimension gluing together some vertices of Δ_m.

2. Definition. *The geometric realization* $|X|$ *of a simplicial set* (X_n) *is the topological space with the underlying set* $\coprod_{n=0}^\infty (\Delta_n \times X_n)/R$, *where* R *is the*

weakest equivalence relation identifying points $(s,x) \in \Delta_n \times X_n$ and $(t,y) \in \Delta_m \times X_m$ if
$$y = X(f)x, \qquad s = \Delta_f(t)$$
for some nondecreasing map $f : [m] \to [n]$. As in I.1 we shall write these conditions as
$$(t,y) \mapsto (s,x).$$

The topology on $|X|$ is the weakest one for which the factorization by R is continuous.

3. Nerve of a Covering

Let Y be a topological space, $U = (U_\alpha)$ be its covering (either by open or by closed sets), α running over some set A. Let
$$X_n = \{(\alpha_0, \ldots, \alpha_n) \,|\, U_{\alpha_0} \cap \ldots \cap U_{\alpha_n} \neq \emptyset\},$$
$$X(f)(\alpha_0, \ldots, \alpha_n) = (\alpha_{f(0)}, \ldots, \alpha_{f(m)}) \quad \text{for} \quad f : [m] \to [n].$$

This simplicial set reflects the combinatorial structure of a covering. One can show that if the covering U is locally finite and all nonempty finite intersections $U_{\alpha_0} \cap \ldots \cap U_{\alpha_n}$ are contractible, then the geometric realization $|X|$ of X is homotopically equivalent to Y, so that the topology can be efficiently encoded into combinatorical data.

The similar role is played by the following construction.

4. Singular Simplices

Let Y be a topological space. A *singular n-simplex* of Y is a continuous map $\varphi : \Delta_n \to Y$. Let

X_n = the set of all singular n-simplices of Y, $n = 0, 1, \ldots,$
$$X(f)(\varphi) = \varphi \circ \Delta_f \quad \text{where} \quad f : [m] \to [n], \quad \Delta_f : \Delta_m \to \Delta_n.$$

The sets X_n are, in general, very large; if Y possesses an additional structure it might be reasonable to consider only those singular simplices that are compatible with this structure (smooth if Y is a smooth manifold, or linear if Y is a polyhedron). Finally, if Y is a triangulated space with the gluing data $(X'_{(n)}, X'(f))$ then, by Proposition I.1.3, Y is a disjoint union of open simplices (including vertices) of its triangulation. A singular simplex $\varphi : \Delta_n \to Y$ is said to be *compatible with the triangulation* if $\varphi(\Delta_n)$ coincides with one of its simplices and $\varphi : \Delta_n \to \varphi(\Delta_n)$ is linear and preserves the order of vertices. Let

X'_n = the set of all singular simplices compatible
with the triangulation $(X'_{(n)}, X'(f));$
$$X'(f)(\varphi) = \varphi \circ \Delta_f.$$

This construction enables us to associate with each triangulation a simplicial set with the same geometric realization.

5. Simplicial Set $\Delta[p]$

Let
$$\Delta[p]_n = \text{the set of nondecreasing mapping } g : [n] \to [p]$$
$$\Delta[p](f)(g) = g \circ f \quad \text{for} \quad g : [n] \to [p], f : [m] \to [n].$$

The geometric realization of the simplicial set $\Delta[p]$ is the p-dimensional simplex Δ_p. The reader is encouraged to construct explicitly an isomorphism $\Delta_p \to |\Delta[p]|$.

Alternatively, one can describe $\Delta[p]$ as the simplicial set of all singular simplices of Δ_p that are compatible with the standard triangulation of Δ_p.

6. The Simplicial Set Associated with a Triangulated Space

Let $|X|$ be a triangulated space corresponding to the gluing data $X_{(n)}$, $n = 0, 1, \ldots$ and $X(f) : X_{(n)} \to X_{(m)}$ for increasing maps $f : [m] \to [n]$. Using these data, we construct a simplicial set $\tilde{X} = (\tilde{X}_n, \tilde{X}(f))$ as follows. As $\tilde{X}_m$ we take the set of all pairs (x, g) where $x \in X_{(k)}$ and $g : [m] \to [k]$ is a nondecreasing surjection. Let now $(x, g) \in \tilde{X}_m$ and $f : [n] \to [m]$ be a nondecreasing map. Let us represent $g \circ f : [n] \to [k]$ as a composition $g \circ f = f_1 \circ f_2$, where $f_1 : [l] \to [k]$ is an injection and $f_2 : [n] \to [l]$ is a surjection, and set $\tilde{X}(f)(x, g) = (X(f_1)x, f_2) \in \tilde{X}_n$. We leave it to the reader to verify that $\tilde{X}(\text{id}) = \text{id}$ and $\tilde{X}(f' \circ f) = \tilde{X}(f) \circ \tilde{X}(f')$.

In Corollary I.2.13 we shall show that the geometric realization $|\tilde{X}|$ of this simplicial set $\tilde{X}$ is homeomorphic to the triangulated space $|X|$.

Let us remark also that not every simplicial set can be obtained from some gluing data. There are "fewer" gluing data than simplicial sets. We leave to the reader the proof of the following statement:

7. Proposition. *A simplicial set $\tilde{X}$ can be obtained from some gluing data X iff for any nondegenerate (see I.2.9 below) simplex $x \in \tilde{X}_n$ and for any increasing map $f : [m] \to [n]$ the simplex $\tilde{X}(f)(x) \in \tilde{X}_m$ is also nondegenerate. In this case $\tilde{X}$ determines X uniquely.*

8. Classifying Space of a Group

Let G be a group. Let
$$(BG)_n = G^n$$
and for $f : [m] \to [n]$ let
$$BG(f)(g_1, \ldots, g_n) = (h_1, \ldots, h_m),$$
where

I.2 Simplicial Sets

$$h_i = \prod_{j=f(i-1)+1}^{f(i)} g_j, \quad h_i = e \quad \text{if} \quad f(i-1) = f(i).$$

The following diagram shows what this formula looks like for the mapping $f : [3] \to [4]$ with $f(0) = 0$, $f(1) = f(2) = 2$, $f(3) = 4$:

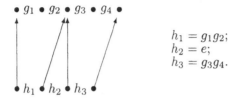

$h_1 = g_1 g_2;$
$h_2 = e;$
$h_3 = g_3 g_4.$

The geometric realization $|BG|$ of BG is called *the classifying space* of G.

The structure of the geometric realization of a simplicial set X can be clarified using an analogue of Proposition I.1.3. To formulate such an analogue we need the notion of a nondegenerate simplex.

9. Nondegenerate Simplices

Let X be a simplicial set. An n-simplex $x \in X_n$ is said to be *degenerate* iff there exists a surjective nondecreasing map $f : [n] \to [m], m < n$, and an element $y \in X_m$ such that $x = X(f)y$. One can easily check that if x is nondegenerate and $x = X(f)y$ for some f and y, then f is an injection.

Let now

$$X_{(n)} = \text{the set of nondegenerate } n\text{-simplices of } X.$$

Consider the canonical map

$$\overset{\circ}{\tau} : \coprod_n \overset{\circ}{\Delta}_n \times X_{(n)} \to |X|.$$

10. Proposition. $\overset{\circ}{\tau}$ *is one-to-one.*

We need the following lemma.

11. Lemma. *For any $x \in X_n$ there exists a unique pair (f, y) consisting of a nondegenerate simplex $y \in X_m$ and a surjective nondecreasing map $f : [n] \to [m]$ such that $x = X(f)y$.*

Proof. It is clear that at least one such pair exists. Assume that there exist two such pairs $(f, y), f : [n] \to [m]$ and $(f', y'), f' : [n] \to [m']$. Let $g : [m] \to [n]$ be some nondecreasing section of the surjective map f, so that $f \circ g : [m] \to [m]$ is the identity map. It is clear that $y = X(g)x$, so that

$$y = X(g)X(f')y' = X(f' \circ g)y.$$

As Y is nondegenerate, the map $f' \circ g : [m] \to [m']$ is injective, so that $m \leq m'$. Similarly $m' \leq m$ and, therefore, $m = m'$. Thus $f' \circ g$ is a nondecreasing one-to-one map of the set $[m]$ into itself, so that $f' \circ g = \mathrm{id}$, $y = X(f' \circ g)y' = y'$. Moreover, $f' \circ g = \mathrm{id}$ for any section g of the map f, so that $f = f'$. □

12. Corollary. *Assume we are given a simplex $x \in X_n$, a nondegenerate simplex $y \in X_m$ and a nondecreasing surjection $f : [n] \to [m]$ such that $x = X(f)y$. Assume also a simplex $z \in X_l$ and a nondecreasing surjection $y : [n] \to [l]$ satisfy $x = X(g)z$. Then f can be decomposed as $f = h \circ g$ for some $h : [l] \to [m]$ such that $z = X(h)y$.*

Proof. Let (h', y') be a pair satisfying the conditions of Lemma I.2.11 for $z \in X_l$. Then $h' \circ g$ is a nondecreasing surjection, y' is a nondegenerate simplex and $x = X(g)z = X(g)X(h')y' = X(h' \circ g)y'$. By Lemma I.2.11, we have $y' = y$ and $h' \circ g = f$ so that we can take $h = h'$. □

Proof (of Proposition I.2.10). a) $\overset{\circ}{\tau}$ *is surjective*. Indeed, let $p \in |X|$. Let k be the smallest dimension for which there exists $(s, x) \in \Delta_k \times X_k$ such that $\tau(s, x) = p$. We will show that in such a pair, x is nondegenerate and $s \in \overset{\circ}{\Delta}_k$. If $k = 0$, there is nothing to prove.

Next, if x is degenerate and $x = X(f)y$ for $f : [k] \to [l]$, $l < k$, then $\tau(s, x) = \tau(\Delta_f(s), y)$, contradicting the choice of k. Similarly, if s is not in $\overset{\circ}{\Delta}_k$ then s lies in one of the faces of Δ_k of dimension $l < k$, again in contradiction with the choice of k.

b) To prove that $\overset{\circ}{\tau}$ *is injective* we have to show that if the images of two points $(s, x) \in \overset{\circ}{\Delta}_k \times X_{(k)}$, $(s', x') \in \overset{\circ}{\Delta}_l \times X_{(l)}$ in $|X|$ coincide, then these two points also coincide.

By the definition of the equivalence relation R, these points can be joined by a chain of equivalences from Definition I.2.2,

$$(s, x) = (s_0, x_0) \sim (s_1, x_1) \sim \ldots \sim (s_N, x_N) = (s', x'),$$

where each equivalence in the chain, say $(s_i, x_i) \sim (s_{i+1}, x_{i+1})$ is either $(s_i, x_i) \overset{f_i^+}{\longmapsto} (s_{i+1}, x_{i+1})$ or $(s_i, x_i) \overset{f_i^-}{\longleftarrow} (s_{i+1}, x_{i+1})$. We may assume that any two consecutive arrows have opposite directions (otherwise we can replace two maps by their composition). Let us remark that $x_1, \ldots, x_{N-1}$ may be degenerate and $s_1, \ldots, s_{N-1}$ may lie at the boundaries of the corresponding simplices.

We will prove that for $N = 1$ we necessarily have $f_1 = \mathrm{id}$ and for $N \geq 2$ there exists a chain of smaller length joining (s, x) and (s', x'). These assertions imply injectivity of $\overset{\circ}{\tau}$.

c) The following remark will be useful in the realization of the above plan: if $(s, x) \in \overset{\circ}{\Delta}_k \times X_{(k)}$, then f_0^+ is an injection and f_0^- is a surjection. Indeed, for f_0^+ we have $x = X(f_0^+)x_1$, so that f_0^+ is injective in view of the nondegeneracy of x. For f_0^- we have $s = \Delta_{f_0^-}(s_1)$, and, as $s \in \overset{\circ}{\Delta}_k$, we conclude that $\Delta_{f_0^-}$ is a surjection implying that f_0^- is also a surjection.

This remark immediately implies that for $N = 1$ our map is both injective and surjective, i.e. it is the identity map.

d) The key step in decreasing the length of the chain for $N \geq 2$ is based on the following observation. Let a segment of the chain be of the following form:

$$(s_i, x_i) \xrightarrow{f_i^+} (s_{i+1}, x_{i+1}) \xleftarrow{f_{i+1}^-} (s_{i+2}, x_{i+2}),$$

$$[m_i] \xrightarrow{f_i^+} [m_{i+1}] \xleftarrow{f_{i+1}^-} [m_{i+2}],$$

where f_i^+ is an injection. Then (for some l) there exists a chain

$$(s_i, x_i) \xleftarrow{g} (t, y) \xrightarrow{h} (s_{i+2}, x_{i+2}),$$

$$[m_i] \xleftarrow{g} [l] \xrightarrow{h} [m_{i+2}].$$

To prove this assertion denote by $I \subset [m_{i+2}]$ the preimage (under f_{i+1}^-) of the set $f_i^+([m_i]) \subset [m_{i+1}]$. Let $l+1$ be the cardinality of I and $h: [l] \to [m_{i+2}]$ be the injection with the image I. It is clear that there exists a unique map $g: [l] \to [m_i]$ such that $f_i^+ \circ g = f_{i+1}^- \circ h$. Since the image of h coincides with $(f_{i+1}^-)^{-1}(f_i^+([m_i]))$, the image of Δ_h coincides with $(\Delta_{f_{i+1}^-})^{-1}(\Delta_{f_i^+}(\Delta_{m_i}))$. But $\Delta_{f_{i+1}^-}(s_{i+2}) = s_{i+1} = \Delta_{f_i^+}(s_i)$. Therefore there exists $t \in \Delta_l$ such that $\Delta_h(t) = s_{i+2}$. Moreover, $\Delta_{f_i^+}(\Delta_g(t)) = \Delta_{f_{i+1}^-}(\Delta_h(t)) = s_{i+1}$ and $\Delta_g(t) = s_i$ because $\Delta_{f_i^-}$ is an injection.

Letting $y = X(h)x_{i+2} = X(g)x_i$, we obtain the required chain.

e) Let now $N \geq 2$. If the chain joining (s, x) and (s', x') starts as

$$(s, x) = (s_0, x_0) \xrightarrow{f_0^+} (s_1, x_1),$$

then, according to c), f_0^+ is an injection and we can replace the first two maps in the chain, obtaining a new chain of length 2 for $N = 2$ and of length $N - 1$ for $N \geq 3$, which starts as follows:

$$(s_0, x_0) \xleftarrow{f_0^-} (s_1, x_1) \xrightarrow{f_1^+} (s_2, x_2),$$

$$[k] \xleftarrow{f_0^-} [m_1] \xrightarrow{f_1^+} [m_2].$$

According to c), f_0^- is a surjection.

f) If $N = 2$, so that $(s_2, x_2) \in \overset{\circ}{\Delta}_l \times X_{(l)}$, then f_1^+ is also a surjection. Since x_0 and x_2 are nondegenerate, Lemma I.2.11 implies that $f_0^- = f_1^+$, $x_0 = x_2$, and hence $s_0 = s_2$.

g) Now let $N \geq 3$. We can write $f_1^+ = i \circ p$ where $p : [m_1] \to [l]$ is a nondecreasing surjection and $i : [l] \to [m_2]$ is an increasing injection. Since s_0 is nondegenerate and f_0^- is a surjection, by Corollary 12, we have $f_0^- = g \circ p$ for some $g : [l] \to [k]$. Replacing the maps

$$(s_0, x_0) \xleftarrow{f_0^-} (s_1, x_1) \xmapsto{f_1^+} (s_2, x_2),$$

$$[k] \xleftarrow{f_0^-} [m_1] \xrightarrow{f_1^+} [m_2],$$

with

$$(s_0, x_0) \xleftarrow{g} (p(s_1), g(x_0)) \xmapsto{i} (s_2, x_2),$$

$$[k] \xleftarrow{g} [l] \xrightarrow{i} [m_2],$$

we see that f_1^+ can be assumed to be an injection.

Now, by d), we can replace

$$(s_1, x_1) \xmapsto{f_1^+} (s_2, x_2) \xleftarrow{f_2^-} (s_3, x_3)$$

with

$$(s_1, x_1) \xleftarrow{k} (t, y) \xmapsto{h} (s_3, x_3).$$

Taking $f_0^- \circ k$ and $f_3^+ \circ h$ as new f_0^- and f_1^+ (and performing the corresponding renumeration of other f_i's) we obtain a chain of length $N - 2$ joining (s, x) and (s', x'). □

13. Corollary. *Let $|X|$ be a triangulated space with gluing data $(X_{(n)}, X(f))$ and $\tilde{X}$ the corresponding simplicial set (see I.2.6). Then $|\tilde{X}|$ is homeomorphic to $|X|$.*

Proof. The set $\tilde{X}_n$ of n-simplices of $\tilde{X}$ consists of all pairs (x, g), where $x \in X_{(m)}$, $g : [n] \to [m]$ is a nondecreasing surjection. Define

$$\varphi : \coprod_{n=0}^{\infty} \Delta_n \times \tilde{X}_n \to \coprod_{n=0}^{\infty} \Delta_n \times X_{(n)}$$

by the formula

$$\varphi(s, \tilde{x}) = (\Delta_g(s), x),$$

where $(s, \tilde{x}) \in \Delta_n \times \tilde{X}_n$, $(\Delta_g(s), x) \in \Delta_m \times X_{(m)}$.

It is clear that φ maps equivalent points in $\coprod \Delta_n \times \tilde{X}_n$ into equivalent points in $\coprod \Delta_n \times X_{(n)}$, so that it induces a continuous map $\tilde{\varphi} : |\tilde{X}| \to |X|$. By Propositions I.1.3 and I.2.10, to prove that $\tilde{\varphi}$ is a homeomorphism it suffices to verify that nondegenerate n-simplices in $\tilde{X}$ are exactly the pairs $(x, \mathrm{id}_{[n]})$, $x \in X_{(n)}$. This is left to the reader. □

14. Skeleton and Dimension

Let X be a simplicial set. The n-skeleton of X is a simplicial set $\operatorname{sk}_n X$ with

$$(\operatorname{sk}_n X)_p = \{x \in X_p \mid \exists q \leq n, \exists f : [p] \to [q], \exists y \in X_q \text{ such that } x = X(f)y\}$$

$(\operatorname{sk}_n, X)(g)$ is the restriction of $X(g)$ on $\operatorname{sk}_n X$.

Hence, p-simplices of the n-skeleton of X are precisely degenerations of q-simplices of X with $q \leq n$. We leave to the reader to verify that $X(g), g : [p] \to [q]$, maps $(\operatorname{sk}_n X)_q$ into $(\operatorname{sk}_n X)_p$.

From the definition and from the proof of Proposition I.2.10 we obtain that the mapping

$$\overset{\circ}{\tau} : \coprod_{m \leq n} \overset{\circ}{\Delta}_m \times X_{(m)} \to |\operatorname{sk}_n X|$$

is one-to-one. Therefore, $|\operatorname{sk}_n X|$ is a closed subspace in $|X|$.

A simplicial set X is said to have the dimension n if $X = \operatorname{sk}_n X \neq \operatorname{sk}_{n-1} X$. This means that any simplex of dimension $> n$ is degenerate, while nondegenerate n-simplices do exist.

15. Mappings of Simplicial Sets

Let X, X' be two simplicial sets. A *simplicial map* $F : X \to X'$ is a family of maps $F_n : X_n \to X'_n, n = 0, 1, \ldots$, satisfying the following condition: for any nondecreasing $f : [m] \to [n]$ the diagram

$$\begin{array}{ccc} X_n & \xrightarrow{F_n} & X'_n \\ {\scriptstyle X(f)}\downarrow & & \downarrow{\scriptstyle X'(f)} \\ X_m & \xrightarrow{F_m} & X'_m \end{array}$$

is commutative.

Let us show that such an F induces a continuous map of geometric realizations $|F| : |X| \to |X'|$. Let

$$\tilde{F} : \coprod_n \Delta_n \times X_n \to \coprod_n \Delta_n \times X'_n, \quad \tilde{F}(s, x) = (s, F_n(x)).$$

It is clear that $\tilde{F}$ maps equivalent points into equivalent ones; therefore, it induces a required map $|F|$. The reader can easily see that $|\operatorname{id}| = \operatorname{id}$, $|F \circ G| = |F| \circ |G|$.

Now we give some examples of natural maps for simplicial sets from I.2.3, I.2.4, and I.2.8.

16. Refinements of Coverings

Let Y be a topological space, $(U_\alpha | \alpha \in A)$, $(V_\beta | \beta \in B)$ two coverings of Y. We say that U is a refinement of V if there exists a map $\psi : A \to B$ such that $U_\alpha \subset V_{\psi(\alpha)}$ for all $\alpha \in A$. Let X be the nerve of U, X' the nerve of V. Then ψ induces a simplicial map $F : X \to X'$ as follows (in the notation of I.2.3):

$$F_n(\alpha_0, \ldots, \alpha_n) = (\psi(\alpha_0), \ldots, \psi(\alpha_n)).$$

The definition makes sense since $U_{\alpha_0} \cap \ldots \cap U_{\alpha_n} \neq \emptyset$ implies $V_{\psi(\alpha_0)} \cap \ldots \cap V_{\psi(\alpha_n)} \neq \emptyset$.

17. Continuous Maps

Let Y, Y' be two topological spaces, $\psi : Y \to Y'$ a continuous map and X, X' the sets of singular simplices of Y, Y' respectively. Define a simplicial map $F : X \to X'$ as follows (in the notation of I.2.4):

$$F_n(\varphi) = \psi \circ \varphi, \quad \Delta_n \xrightarrow{\varphi} Y \xrightarrow{\psi} Y'.$$

18. Group Homomorphisms

Let $\psi : G \to H$ be a homomorphism. Define a simplicial map $F : BG \to BH$ as follows (in the notation of I.2.8):

$$F(g_1, \ldots, g_n) = (\psi(g_1), \ldots, \psi(g_n)).$$

Exercises

1. Faces and Degeneration. Let us consider the following maps:

- "i-th face": $\partial_n^i : [n-1] \to [n]$ the only strictly increasing map not taking the value i.
- "i-th degeneration": $\sigma_n^i : [n+1] \to [n]$ the nondecreasing surjection taking the value i twice.

Prove the following assertions:

a) Any strictly increasing map is a composition of face maps. Any nondecreasing surjection is a composition of degeneration maps. Any nondecreasing map is a composition of face and degeneration maps.

b) Face and degeneration maps satisfy following relations:

$$\partial_{n+1}^j \partial_n^i = \partial_{n+1}^i \partial_n^{j-1} \quad \text{for} \quad i < j;$$

$$\sigma_n^j \sigma_{n+1}^i = \sigma_n^i \sigma_{n+1}^{j+1} \quad \text{for} \quad i \leq j;$$

$$\sigma_{n-1}^j \partial_n^i = \begin{cases} \partial_{n-1}^i \sigma_{n-2}^{j-1} & \text{for } i < j; \\ \mathrm{id}_{[n-1]} & \text{for } i = j \text{ or } i = j+1; \\ \partial_{n-1}^{i-1} \sigma_{n-2}^j & \text{for } i > j+1. \end{cases}$$

c) Formulate and prove the exact statement expressing the fact that the above relations generate all relations between faces and degenerations. (*Hint*: first verify that any nondecreasing map $f : [m] \to [n]$ can be uniquely written in the form
$$f = \partial_n^{i_1} \partial_{n-1}^{i_2} \cdots \partial_{n-s+1}^{i_s} \sigma_{m-t}^{j_t} \cdots \sigma_{m-2}^{j_2} \sigma_{m-1}^{j_1}$$
where $n \geq i_1 > \ldots > i_s \geq 0$, $m > j_1 > \ldots > j_t \geq 0$, $n = m - t + s$.)

d) Prove that a strictly increasing map $f : [m] \to [n]$ has exactly $(n-m)!$ different representations in the form
$$f = \partial_n^{i_1} \cdots \partial_{m+1}^{i_{n-m}}.$$

2. Inductive Construction of the Skeleton. Let us consider the simplicial set $\dot{\Delta}[n]$ with
$$\dot{\Delta}[n]_m = \left\{ \begin{array}{c} f : [m] \to [n] \mid f \text{ is nondecreasing map,} \\ \operatorname{Im} f \neq [n] \end{array} \right\}.$$

This set is the simplicial $(n-1)$-sphere; its geometric realization is the boundary of Δ_n. There exists a canonical (simplicial) embedding $\dot{\Delta}[n] \subset \Delta[n]$. For any simplicial set X and for any nondegenerate $x \in X_n$ there exists an embedding $\tilde{x} : \Delta[n] \to \operatorname{sk}_n X$, $\tilde{x}(f) = X(f)x$ for $f : [m] \to [n]$, $f \in \Delta[n]_m$. (It is clear that for $m > n$ the simplex $\tilde{x}(f)$ is a degeneration of the n-simplex x.) The image of $\dot{\Delta}[n] \subset \Delta[n]$ under this embedding lies in $\operatorname{sk}_{n-1} X \subset \operatorname{sk}_n X$ (verify this). Now let $X_{(n)}$ be the set of all nondegenerate n-simplices of X. Let us consider two commutative diagrams of simplicial sets and simplicial maps,

$$\begin{array}{ccc} \coprod_{x \in X_{(n)}} \dot{\Delta}[n] & \longrightarrow & \operatorname{sk}_{n-1} X \\ \downarrow & & \downarrow \\ \coprod_{x \in X_{(n)}} \Delta[n] & \longrightarrow & \operatorname{sk}_n X \end{array}$$

$$\begin{array}{ccc} \coprod_{x \in X_{(n)}} \dot{\Delta}[n] & \longrightarrow & \operatorname{sk}_{n-1} X \\ \downarrow & & \downarrow \\ \coprod_{x \in X_{(n)}} \Delta[n] & \longrightarrow & Y \end{array}$$

(for some simplicial set Y).

a) Prove that there exists a unique map $\varphi : \operatorname{sk}_n X \to Y$ such that the second diagram is obtained from the first one by combining maps going to the right lower corner with the map φ.

b) Let also $Z \subset X$ be a simplicial subset (i.e., $Z_n \subset X_n$ and $Z(f)$ are compatible with $X(f)$). Consider similar diagrams

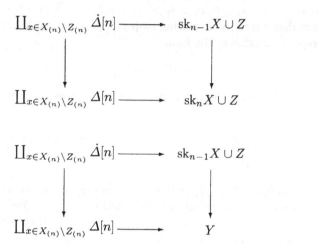

and prove for them a similar statement (we will need it in Sect. V.1).

3. Truncated Simplicial Sets and Coskeleton. An *N-truncated simplicial set* $(N \geq 0)$ is a family of data $(X_n, X(f))$ as in Definition I.2.1, but defined only for $n \leq N$ and for $f : [m] \to [n]$ with $m, n \leq N$. The N-truncation $\operatorname{Tr}^N X$ of a simplicial set X is defined in an obvious way, as well as the N-truncation of an M-truncated X for $M \geq N$.

a) Let $Y^{(N)}$ be an N-truncated simplicial set. Define an $(N+1)$-truncated simplicial set $Y^{(N+1)}$ as follows:

$$\operatorname{Tr}^N Y^{(N+1)} = Y^{(N)},$$

$$Y_{N+1}^{(N+1)} = \left\{ (y_0, \ldots, y_{N+1}) \in \left(Y_N^{(N)} \right)^{N+2} \,\Big|\, d_N^i y_j = d_N^{j-1} y_i, i < j \right\},$$

where $d_n^i = Y^{(N)}(\partial_n^i)$, $n \leq N$, and

$$Y^{(N+1)}\left(\partial_{N+1}^i\right) = d_{N+1}^i : (y_0, \ldots, y_{N+1}) \mapsto y_i$$

for $0 \leq i \leq N+1$. Intuitively, $Y_{N+1}^{(N+1)}$ consists of $(N+1)$-simplices filling all "simplicial holes" in $Y_N^{(N)}$.

Show that $Y^{(N+1)}(\sigma_{N+1}^i)$ can be defined in a unique way.

b) Prove that $Y^{(N+1)}$ is the universal $(N+1)$-extension of $Y^{(N)}$ in the following sense. Let Z be an $(N+1)$-truncated set and let $F^N : \operatorname{Tr}^N Z \to Y^{(N)}$

be any (N-truncated) simplicial map. Then it extends uniquely to an $(N+1)$-truncated simplicial map $Z \to Y^{(N+1)}$.

c) Let $X^{(N)}$ be an arbitrary N-truncated simplicial set. Its coskeleton is defined as a simplicial set $X = \operatorname{cosk} X^{(N)}$ such that

$$\operatorname{Tr}^N X = X^{(N)},$$

$\operatorname{Tr}^{M+1} X$ is the universal $(M+1)$-extension of $\operatorname{Tr}^M X$ for any $M \geq N$.

Establish, for any simplicial set X, a natural bijection between the set of simplicial maps of $\operatorname{Tr}^N Z$ into $X^{(N)}$ and of Z into X. (Using the language of §3 in Chap. II, we can say that the functor cosk^N is right adjoint to the functor Tr^N.)

d) Let $X^{(0)}$ be an 0-truncated simplicial set consisting of $p+1$ point. Prove that $\operatorname{cosk} X^{(0)}$ is isomorphic to $\Delta[p]$ (see I.2.5).

I.3 Simplicial Topological Spaces and the Eilenberg–Zilber Theorem

1. Three Descriptions of $\Delta_p \times \Delta_q$

In Sect. I.1.5 we have already remarked that the product of two geometric simplices is not a simplex, and we have constructed the canonical triangulation of this product by "diagonal" simplices. Thus we take the following view at $\Delta_p \times \Delta_q$.

a) $\Delta_p \times \Delta_q$ is the geometric realization of the simplicial set X whose *nondegenerate* N-simplices are "shuffles" $\{(i_0, j_0), \ldots, (i_n, j_n)\}$, $0 \leq i_0 \leq \ldots \leq i \leq p$, $0 \leq j_0 \leq \ldots \leq j_n \leq q$, all pair (i_k, j_k) are distinct.

On the other hand we have set-theoretically

b) $\Delta_p \times \Delta_q = \cup_{x \in \Delta_q} (\Delta_p \times x)$.

c) $\Delta_p \times \Delta_q = \cup_{y \in \Delta_p} (y \times \Delta_q)$.

Two last descriptions are rather similar to the definition of $\Delta_p \times \Delta_q$ by some gluing data. There are two differences: a nonessential one, and an essential one. The nonessential one is that Δ_p (resp. Δ_q) on the right-hand side of b) (resp. of c)) should be replaced by its standard triangulation. The essential difference is that we must take into account the topology on the set of indices $x \in \Delta_q$ (resp. $y \in \Delta_p$); otherwise, the corresponding "geometric realization" of $\Delta_p \times \Delta_q$ splits into a continuum of connected components. Therefore, we must change the definition of geometric realization in such a way that it would take into account the topology on the set of indices.

2. Definition.
a) *A simplicial topological space is a family of topological spaces $X = (X_n)$, $n = 0, 1, 2, \ldots$ and of continuous maps $X(f) : X_n \to X_m$, one for each nondecreasing map $f : [m] \to [n]$, such that*

$$X(\operatorname{id}) = \operatorname{id}, \quad X(g \circ f) = X(f) \circ X(g).$$

18   I. Simplicial Sets

b) The geometric realization $|X|$ of a simplicial topological space is a topological space with the set of points $\coprod_{n=0}^{\infty} (\Delta_n \times X_n)/R$ where R is the equivalence relation from I.2.2. The topology of $|X|$ is the weakest topology for which the factorization map

$$\coprod_{n=0}^{\infty} (\Delta_n \times X_n) \to |X|$$

is continuous; on the left-hand side we have the disjoint union of products of topological spaces.

This definition is a reasonable axiomatization of the viewpoints I.3.1b) and I.3.1c).

On the other hand, a reasonable generalization of the combinatorial data for the decomposition of a topological space into bisimplices $\Delta_p \times \Delta_q$ is the notion of a bisimplicial set.

3. Definition. *A bisimplicial set is a family of sets (X_{mn}), $m, n = 0, 1, 2, \ldots$, and of maps $X(f, g) : X_{pq} \to X_{mn}$, one for each pair of nondecreasing maps $f : [m] \to [p]$, $g : [n] \to [q]$, such that*

$$X(\mathrm{id}, \mathrm{id}) = \mathrm{id}, \quad X(f \circ f', g \circ g') = X(f', g') \circ X(f, g).$$

4. Example. Let X, Y be two simplicial sets, and let

$$Z_{mn} = X_m \times Y_m, \quad Z(f, g) = (X(f), Y(g)).$$

This bisimplicial set is called the *direct product* of X and Y.

5. Definition. *The diagonal of a bisimplicial set $X = (X_{mn}, X(f, g))$ is the simplicial set DX with*

$$(DX)_n = X_{nn}, \quad DX(f) = X(f, f).$$

6. Geometric Realization of a Bisimplicial Set

This can be defined in three different ways according to the three descriptions of $\Delta_p \times \Delta_q$.

a) For a simplicial set X we define $|X|^D = |DX|$ to be the geometric realization of the diagonal of X.

b) First we construct a simplicial topological space X^I that can be considered as a "geometric realization of X with respect to the first index":

$$X^I_n = |X_{\bullet n}|, \quad X^I(g) = |X(\mathrm{id}, g)|.$$

More explicitly, for a fixed second index n the family of sets and of mappings

$$X_{\bullet n} = (X_{mn}, X(f, \mathrm{id}_{[n]}))$$

3. Simplicial Topological Spaces and the Eilenberg–Zilber Theorem

is a simplicial set and the n-th component of $X_\bullet^I$ is the geometric realization of this simplicial set.

Next, for any nondecreasing map $g : [n] \to [n']$ the family

$$X\left(\mathrm{id}_{[m]}, g\right) : X_{mn'} \to X_{mn}, \quad m = 0, 1, \ldots,$$

is a simplicial map of simplicial sets $X_{\bullet n'} \to X_{\bullet n}$; we denote it by $X(\mathrm{id}, g)$. Its geometric realization (see I.2.12) $|X(\mathrm{id}, g)|$ is a continuous map of the corresponding topological spaces. Clearly,

$$X(\mathrm{id}, \mathrm{id}) = \mathrm{id}, \quad X(\mathrm{id}, g \circ g') = X(\mathrm{id}, g') \circ X(\mathrm{id}, g).$$

Hence we have constructed a simplicial topological space X^I and we can consider its geometric realization. Define now

$$|X|^{I,II} = |X^I|.$$

c) We can make similar construction starting from the second index:

$$X_n^{II} = |X_{n\bullet}|, \quad X^{II}(f) = |X(f, \mathrm{id})|,$$

and

$$|X|^{II,I} = |X^{II}|.$$

7. Theorem (Eilenberg–Zilber). *The three geometric realizations of a bisimplicial set X are canonically isomorphic.*

8. The Plan of the Proof

Denote by $D[m, n]$ the following simplicial set:

$$D[m, n] = D(\Delta[m] \times \Delta[n]).$$

Let us consider the following disjoint union of simplices:

$$Z = \coprod_{m,n,k} (X_{mn} \times D[m, n]_k \times \Delta_k).$$

Introduce on Z three equivalence relations R^I, R^{II}, R^D generated by the following identifications:

$$\begin{aligned} R^I \; : \; & (x \in X_{mn}, (f, g) \in D[m, n]_k, s \in \Delta_k) \\ & \sim (x' \in X_{m'n}, (f', g) \in D[m', n]_k, s \in \Delta_k), \end{aligned}$$

if there exists $h : [m] \to [m']$ such that

$$X\left(h, \mathrm{id}_{[n]}\right)(x') = x, \quad h \circ f = f'.$$

(Recall that $D[m, n]_k$ consists of pairs of nondecreasing maps $f : [k] \to [m]$, $g : [k] \to [n]$.)

Similarly,

$$R^{II} : (x \in X_{mn}, (f,g) \in D[m,n]_k, s \in \Delta_k)$$
$$\sim (x' \in X_{mn'}, (f,g') \in D[m,n']_k, s \in \Delta_k),$$

if there exists $h : [n] \to [n']$ such that

$$X(\mathrm{id}_{[m]}, h)(x') = x, \quad h \circ g' = g.$$

Finally, R^D is generated by identifications

$$R^D : (x \in X_{mn}, (f,g) \in D[m,n]_k, s \in \Delta_k)$$
$$\sim (x \in X_{mn}, (f',g') \in D[m,n]_{k'}, s' \in \Delta_{k'}),$$

if there exists a nondecreasing map $h : [k] \to [k']$ such that

$$f' \circ h = f, \quad g' \circ h = g, \quad \Delta_h(s) = s'.$$

We show that factorizing Z with respect to R^I, R^{II}, R^D in different orders we get respectively $|X|^{I,II}$, $|X|^{II,I}$, $|X|^D$. On the other hand, we can factorize Z by the equivalence relation generated by R^I, R^{II}, R^D obtaining the same space.

9. Consecutive Factorizations

We describe the last assertion in somewhat different form and in a more general situation. Let Z be an arbitrary set, R_1 and R_2 two equivalence relations on Z, R the equivalence relation generated by R_1 and R_2. We want to show that the following three factorizations coincide:

a) Let R_2/R_1 be the equivalence relation on Z/R_1 generated by the following identifications: $(x \bmod R_1) \sim (y \bmod R_1) \bmod R_2/R_1$ if in R_1-classes of x,y there exist R_2-equivalent representatives x',y'.
Now we can take $(Z/R_1)/(R_2/R_1)$.
b) Similarly we can define $(Z/R_2)/(R_1/R_2)$.
c) We can take also Z/R.

We leave it to the reader to prove that these three quotient spaces are naturally isomorphic, and any of them can be obtained as follows: $x, y \in Z$ lie in the same class iff there exists a chain $x = x_0, x_1, \ldots, x_n = y$ such that for any i either $x_i \sim x_{i+1} \bmod R_1$ or $x_i \sim x_{i+1} \bmod R_2$.

One can prove a similar statement for any finite family of equivalences.

Finally, if Z is a topological space and any quotient inherits the weakest topology for which the factorization is continuous, then all three quotient spaces as above are naturally homeomorphic. In fact, a subset in any of these quotients is open iff its preimage in Z is open.

Now we can describe the results of all intermediate factorizations of the space Z from I.3.8.

10. Lemma. *Let R be the equivalence relation generated by R^I and R^{II}. Then*
$$Z/R = \coprod_k (X_{kk} \times \Delta_k),$$
$$(Z/R) / (R^D/R) = |X|^D.$$

Proof. Let $\tilde{p} : Z \to \coprod(X_{kk} \times \Delta_k)$ be the map defined by $\tilde{p}(x, (f,g), s) = (X(f,g)(x), s)$. More precisely, if $x \in X_{mn}$, $f : [k] \to [m]$, $g : [k] \to [n]$, then $X(f,g)$ maps x into X_{kk}. One can easily check that R^I- or R^{II}-equivalent points have the same image under $\tilde{p}$. So $\tilde{p}$ determines a map $p : Z/R \to \coprod_k (X_{kk} \times \Delta_k)$. On the other hand, we have the embedding
$$i : \coprod_k (X_{kk} \times \Delta_k) \to Z,$$
$$(x \in X_{kk}, s \in \Delta_k) \mapsto (x, (\mathrm{id}_{[k]}, \mathrm{id}_{[k]}), s).$$

The following facts can be easily verified:
a) $p \circ i$ is the identity map of $\coprod_k (X_{kk} \times \Delta_k)$. Therefore, p is surjective.
b) Any point $(x \in X_{mn}, (f,g), s \in \Delta_k)$ is R-equivalent to the point $(X(f,g)(x) \in X_{kk}, (\mathrm{id}_{[k]}, \mathrm{id}_{[k]}), s \in \Delta_k)$ (one must take the composition of an R^I-equivalence and of an R^{II}-equivalence).

a) and b) imply that p is a homeomorphism.

To complete the proof we must compute the equivalence relation R^D/R on Z/R. The easiest way to do this is to compute the images of elementary R^D-equivalences. For two R^D-equivalent points we get
$$p : (x, (f,g), s) \mapsto (X(f,g)(x), s),$$
$$p : (x, (f',g'), s') \mapsto (X(f',g')(x), s'),$$
$$f' \circ h = f, \quad g' \circ h = g, \quad \Delta_h(s) = s'.$$
Therefore,
$$X(f,g)(x) = X(h,h)\,[X(f',g')(x)],$$
and the equivalence relation R^D/R on $\coprod_k (X_{kk} \times \Delta_k)$ coincides with the one used to define the geometric realization $|X|^D$. □

11. Lemma. *The geometric realization $|D[m,n]|$ of the simplicial set $D[m,n]$ is canonically homeomorphic to $\Delta_m \times \Delta_n$.*

Proof. With a k-simplex $(f,g) \in D[m,n]_k$ we associate the $(k+1)$-tuple $\{(i_0, j_0), \ldots, (i_k, j_k)\}$, where $i_l = f(l)$, $j_l = g(l)$, $0 \le l \le k$. It is clear that the simplex (f,g) is nondegenerate if and only if all pairs (i_k, j_k) are distinct. Moreover, it is clear that if $h : [k'] \to [k]$ is injective and (f,g) is nondegenerate, then $D[m,n](h)(f,g) = (f \circ h, g \circ h)$ is also nondegenerate. Therefore, by Corollary I.2.13, the geometric realization of $D[m,n]$ is homeomorphic to the triangulated space described in I.1.5a,b, i.e. by I.1.5d, $|D[m,n]|$ is homeomorphic to $\Delta_m \times \Delta_n$. □

12. Corollary. We have
$$Z/R^D = \coprod_{m,n} (X_{mn} \times \Delta_m \times \Delta_n)$$
and R^I/R^D is generated by the following identifications:
$$(x \in X_{mn}, s \in \Delta_m, t \in \Delta_n) \sim (x' \in X_{m'n}, s' \in \Delta_{m'}, t \in \Delta_n)$$
if there exists $h : [m] \to [m']$ such that $s' = \Delta_h(s)$, $X\left(h, \mathrm{id}_{[n]}\right)(x') = x$. Similarly, R^{II}/R^D is generated by the identifications
$$(x \in X_{mn}, s \in \Delta_m, t \in \Delta_n) \sim (x' \in X_{mn'}, s \in \Delta_m, t' \in \Delta'_n)$$
if there exists $h : [n] \to [n']$ such that $t' = \Delta_h(t)$, $X\left(\mathrm{id}_{[m]}, h\right)(x') = x$.

Proof. Let $h : [m] \to [m']$ be a nondecreasing map. Denote by $F_h : F[m, n] \to D[m', n]$ the following simplicial map:
$$F_h(f, g) = (h \circ f, g),$$
$$f : [k] \to [m], \quad g : [k] \to [n].$$

The corollary follows from the fact that under the identification $|D[m, n]| \cong \Delta_m \times \Delta_n$ the geometric realization of F_h (see I.2.15) becomes
$$|F_h| = (\Delta_h, \mathrm{id}) : \Delta_m \times \Delta_n \to \Delta_{m'} \times \Delta_n. \qquad \square$$

To simplify the notations denote the space Z/R^D by $\tilde{Z}$ and the equivalence relations R^I/R^D and R^{II}/R^D on it by $\tilde{R}^I$ and $\tilde{R}^{II}$. The description of $\tilde{Z}$, $\tilde{R}^I$, $\tilde{R}^{II}$ given in Corollary I.3.12 implies:

13. Corollary. We have
$$\left(\tilde{Z}/\tilde{R}^I\right) / \left(\tilde{R}^{II}/\tilde{R}^I\right) = |X|^{I,II}$$
and, similarly,
$$\left(\tilde{Z}/\tilde{R}^{II}\right) / \left(\tilde{R}^I/\tilde{R}^{II}\right) = |X|^{II,I}.$$

The Eilenberg–Zilber theorem follows from Lemma I.3.10, Corollary I.3.13 and the isomorphism of consecutive factorization (see I.3.9).

Exercises

1. Triangulations of Prisms. Enumerate all nondegenerate simplices of $D[1, 2]$, $D[1, n]$, $D[2, 2]$ (see I.3.8).

2. Homotopies of Simplicial Maps. Let X, Y be two simplicial sets. Denote by $X \times Y$ the diagonal of the direct product of these sets (see I.3.4, I.3.5).

a) Prove that simplicial maps $X \to Y \times Z$ are in one-to-one correspondence with pairs of simplicial maps $X \to Y$, $X \to Z$.

b) Define "projections to vertices 0 and 1" as simplicial maps $p_0, p_1 : \Delta[1] \times X \to X$.

Maps $f, g : X \to Y$ are said to be *simply homotopic* if there exists $h : \Delta[1] \times X \to Y$ such that either $f = h \circ p_0, g = h \circ p_1$ or $f = h \circ p_1, g = h \circ p_0$. Maps f, g are said to be *homotopic* if there exists a chain $f := f_0, f_1, \ldots, f_{n+1} = g$ such that f_i, f_{i+1} are simply homotopic for any $i, 0 \leq i \leq n$.

c) For any i, $0 \leq i \leq n$, define the map "projection onto the i-th vertex" $pr_i : \Delta[n] \to \Delta[n]$ as follows: for any $f : \Delta[n]_k$, $f : [k] \to [n]$, $(pr_i)_k(f)$ is the constant map $[k] \to [n]$ taking the value i. Prove that id $: \Delta[n] \to \Delta[n]$ and $pr_i : \Delta[n] \to \Delta[n]$ are homotopic if and only if $i = n$.

d) Consider the diagram of simplicial sets

$$X' \xrightarrow{h} X \underset{g}{\overset{f}{\rightrightarrows}} Y \xrightarrow{k} Y'.$$

Prove that if f and g are homotopic, then $k \circ f \circ h$ are $k \circ g \circ h$ are also homotopic.

I.4 Homology and Cohomology

1. Chains and Cochains

The boundary of the geometric simplex Δ_1 is the difference of its vertices $(1) - (0)$. In such a form the boundary appears in the Leibniz formula $\int_0^1 f'(x)dx = f(1) - f(0)$. Similarly, the boundary of Δ_n is the alternating sum of its faces.

To make these definitions precise we need the following notions.

An *n-dimensional chain* (or simply *n-chain*) of a simplicial set X is an element of the free abelian group $C_n(X)$ generated by all n-simplices of X. So an n-dimensional chain is a formal linear combination $\sum_{x \in X_n} a(x)x$, where $a(x) \in \mathbb{Z}$ and $a(x) \neq 0$ for a finite number of simplices x.

Let $\partial_n^i : [n-1] \to [n]$ be the unique strictly increasing mapping whose image does not contain $i \in [n]$.

The *boundary* of an n-chain $c \in C_n(x)$ is the $(n-1)$-chain $d_n c$ defined by the following formula:

$$d_n \left(\sum_{x \in X_n} a(x)x \right) = \sum_{x \in X_n} a(x) \sum_{i=0}^n (-1)^i X\left(\partial_n^i\right)(x). \tag{I.3}$$

The so defined *boundary operator* $d_n : C_n(X) \to C_{n-1}(X)$ is clearly a group homomorphism. For $n = 0$ we set $d_0 = 0$.

There exists an obvious generalization of this construction, namely, *chains with coefficients in an abelian group* A. Such a chain is a formal linear combination $\sum_{x \in X_n} a(x)x, a(x) \in A$. In other words,

$$C_n(X, A) = C_n(X) \underset{\mathbb{Z}}{\otimes} A,$$

so that $C_n(X) = C_n(X, \mathbb{Z})$.

The *boundary operator* $d_n : C_n(X, A) \to C_{n+1}(X, A)$ is again defined by the formula (I.3).

Dually, one can define *cochains with coefficients* in A: $C^n(X, A)$ is the group of functions on X_n with values in A.

The *coboundary* $d^n : C^n(X, A) \to C^{n+1}(X, A)$ is given by the formula

$$(d^n f)(x) = \sum_{i=0}^{n+1} (-1)^i f\left(X\left(\partial^i_{n+1}\right)(x)\right). \tag{I.4}$$

Formally, chains can be considered as special cases of cochains: there exists an inclusion $C_n(X, A) \subset C^n(X, A)$ that maps a chain $\sum_{x \in X_n} a(x)x$ into the function $a : X_n \to A$. However, this inclusion is incompatible with the action of d_n and d^n (they act in opposite directions) and, even more important, it is incompatible with the behaviour of C_n and C^n under simplicial maps $X \to Y$ (we will consider this later).

Let us note also that in (I.3) we cannot allow the infinite number of nonzero coefficients $a(x)$ because the right-hand side can become undefined. On the contrary, in (I.4) we can consider a subgroup $C^n_f(X, A)$ of cochains taking nonzero values only for a finite number of simplices x; the coboundary operator maps this group into $C^{n+1}_f(X, A)$.

2. Lemma. a) $d_{n-1} \circ d_n = 0$ for $n \geq 1$.
b) $d^{n+1} \circ d^n = 0$ for $n \geq 0$.

Proof. Note first that for any $0 \leq j < i \leq n - 1$ we have

$$\partial^i_n \circ \partial^j_{n-1} = \partial^j_n \circ \partial^{i-1}_{n-1};$$

indeed, both sides of the equality give unique increasing mapping of $[n-2]$ into $[n]$ not taking values i and j.

To prove part a) of the lemma it suffices to check that $d_{n-1} \circ d_n(x) = 0$ for any $x \in X_n$. But

$$d_{n-1} \circ d_n(x) = \sum_{j=0}^{n-1} \sum_{i=0}^{n} (-1)^{i+j} X\left(\partial^j_{n-1}\right) X\left(\partial^i_n\right)(x)$$

$$= \sum_{j=0}^{n-1} \sum_{i=0}^{n} (-1)^{i+j} X\left(\partial^i_n \circ \partial^j_{n-1}\right)(x).$$

Compositions $\partial_n^i \circ \partial_{n-1}^j$ for different i, j all yield increasing maps of $[n-2]$ into $[n]$, and the map whose image does not contain i and j appears exactly twice: the first time as $\partial_n^i \circ \partial_{n-1}^j$ with the sign $(-1)^{i+j}$ and the second time as $\partial_n^j \circ \partial_{n-1}^{i-1}$ with the opposite sign $(-1)^{i+j-1}$. Hence $d_{n-1} \circ d_n(x) = 0$.

Similarly one proves part b). □

3. Complexes

Let us define several algebraic notions. A *chain complex* is a sequence of abelian groups and homomorphisms

$$C_\bullet : \ldots \xrightarrow{d_{n+1}} C_n \xrightarrow{d_n} C_{n-1} \xrightarrow{d_{n-1}} \ldots$$

with the property $d_n \circ d_{n+1} = 0$ for all n. Homomorphisms d_n are called *boundary maps* or *boundary operators*.

A *cochain complex* is a similar sequence

$$C^\bullet : \ldots \xrightarrow{d^{n-1}} C^n \xrightarrow{d^n} C^{n+1} \xrightarrow{d^{n+1}} \ldots,$$

$d^n \circ d^{n-1} = 0$. Any chain complex can be transformed into a cochain complex by setting $D^n = C_{-n}$, $d^n = d_{-n-1}$. So we will usually consider only cochain complexes.

The following definition is the central one in homological algebra.

4. Definition.
a) *Homology groups of a chain complex* $C_\bullet$ are

$$H_n(C_\bullet) = \operatorname{Ker} d_n / \operatorname{Im} d_{n+1}.$$

b) *Cohomology groups of a cochain complex* $C^\bullet$ are

$$H^n(C^\bullet) = \operatorname{Ker} d^n / \operatorname{Im} d^{n-1}.$$

A substantial part of homological algebra can be considered as a collection of methods for computing (co)homology of various complexes. In this section we will give the simplest examples. For a simplicial set X we will use the following notations:

$$H_n(X, A) = H_n(C_\bullet(X, A)), \quad H^n(X, A) = H^n(C^\bullet(X, A)).$$

Elements of the group $H_n(X, A)$ are called *homology classes*, and those of $H^n(X, A)$ are called *cohomology classes* (of the simplicial set X with coefficients in A). Each homology (resp. cohomology) class is represented by an n-chain c (resp. cochain f) such that $d_n c = 0$ (resp. $d^n f = 0$). Such chains (resp. cochains) are called *cycles* (resp. *cocycles*). A cycle c in a given homology class is defined up to a summand of the form $b = d_{n+1}c'$; such chains are called *boundaries*. Similarly, cochains of the form $d^{n-1}c'$ are called *coboundaries*. Two chains whose difference is a boundary are said to be *homological*.

The same terminology applies to general complexes.

Let us clarify the meaning of the above definitions in simple examples. We will consider for a moment only triangulated spaces and chains constructed from nondegenerate simplices (this will be justified by Ex. I.7.1d).

5. Geometry of Chains

a) Why is the boundary of a boundary zero? Look at the picture of a tetrahedron Δ_3 (Figure I.2).

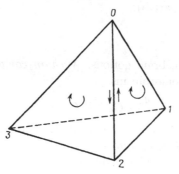

Fig. I.2.

On any edge two adjacent faces induce opposite orientation. Therefore the corresponding terms in $d_2 d_3(\Delta_3)$ have opposite signs.

b) Where do nontrivial (non homological to zero) cycles come from? Consider the triangulated space $S^2 = \text{sk}_2 \Delta[3]$. In the group of 2-chains of S^2 there exists a cycle, namely, the boundary of the deleted 3-simplex. This cycle is definitely non homological to 0, because $C_3(S^2) = 0$ (Recall that we consider nondegenerate chains!).

One can show that the class of this cycle generates the group $H_2(S^2, \mathbb{Z})$.

In geometrical terms one can say that cycles are boundaries of holes (possibly of rather complicated shape).

c) Zero-dimensional homology yields somewhat different information. Since, $d_0 = 0$, any 0-chain is a cycle. Let us show that there exists a natural isomorphism

$$H_0(X, \mathbb{Z}) = \left\{ \begin{array}{l} \text{free abelian group generated by piecewise} \\ \text{connected components of } |X| \end{array} \right\}.$$

Denote for a moment the group on the right-hand side by $\Pi_0(X)$. Define a map $\Pi_0(X) \to H_0(X, \mathbb{Z})$ by associating with a component of $|X|$ the class of a chain consisting of one (arbitrary) point in this component.

The reader can easily check that this map is well-defined and gives an isomorphism if he/she uses the following description of zero-dimensional boundaries: a 0-chain $\sum a(x) x$ is a boundary if and only if

$$\sum_{x \in L} a(x) = 0$$

for any connected component $L \subset |X|$.

d) An important role in geometry is played by various modifications of topological spaces that eliminate some (co)homology classes or generate new

I.4 Homology and Cohomology

ones. Below we describe a universal construction that eliminates all homology groups except H_0, which becomes equal to $\mathbb{Z}$.

Let X be a triangulated space. The *cone* CX over X is a triangulated space obtained from X in the following manner:

$$\{\text{Vertices of } CX\} = \{\text{vertices of } X\} \cup \{*\}$$

and for $n \geq 1$

$$\{n\text{-simplices of } CX\} = \{n\text{-simplices of } X\} \cup$$
$$\{\text{cones of } (n-1)\text{-simplices of } X$$
$$\text{with the vertex } *\}.$$

(so that $*$ is the vertex of the cone).

More formally, let

$$(CX)_{(0)} = X_{(0)} \cup \{*\},$$
$$(CX)_{(n)} = X_{(n)} \cup (X_{(n-1)} \times \{*\}) \quad \text{for} \quad n \geq 1,$$

and for any increasing map $f : [m] \to [n]$ let

$$(CX)(f)(x) = X(f)x \quad \text{for} \quad x \in X_{(n)},$$

$$(CX)(f)(x, *) = \begin{cases} X(g)(x) & \text{where } g : [m] \to [n-1], \\ & g(i) = f(i) - 1, \quad \text{if} \quad f(0) > 0, \\ (X(h)(x), *) & \text{where } h : [m-1] \to [n-1], \\ & h(i) = f(i+1) - 1, \quad \text{if} \quad f(0) = 0. \end{cases}$$
(I.5)

We claim that any hole in X is filled in CX (by the cone over the boundary of the hole) and no new holes appear.

Indeed, let us define the complex of chains $\tilde{C}_\bullet(X)$ of a triangulated space X as follows:

$$\tilde{C}_n(X) = \text{the free abelian group generated by } X_{(n)},$$

the boundary operator on $\tilde{C}$ is given by (I.3), and let $\tilde{H}_n(X) = H_n\left(\tilde{C}_\bullet(X)\right)$.
We claim that

$$H_n\left(\tilde{C}_\bullet(CX)\right) = \begin{cases} 0 & \text{for} \quad n > 0, \\ \mathbb{Z} & \text{for} \quad n = 0. \end{cases}$$

We have $\tilde{C}_n(CX) \cong C_n(X) \oplus C_{n-1}(X)$ and the formula (I.5) shows that with respect to this decomposition the boundary operator takes the form

$$\tilde{d}_n \begin{pmatrix} c_n \\ c_{n-1} \end{pmatrix} = \begin{pmatrix} d_n & 1 \\ 0 & -d_{n-1} \end{pmatrix} \begin{pmatrix} c_n \\ c_{n-1} \end{pmatrix} \quad \text{for} \quad n \geq 1.$$

28   I. Simplicial Sets

This formula implies that, first, any chain of the form $\binom{c_n}{0}$ is homological to some chain of the form $\binom{0}{c_{n-1}}$ (because $\binom{c_n}{0} = \binom{0}{dc_n} + \tilde{d}_{n+1}\binom{0}{c_n}$) and, second, any cycle of the form $\binom{0}{c_{n-1}}$ equals zero:

$$\tilde{d}_n \binom{0}{c_{n-1}} = \binom{c_{n-1}}{-d_{n-1}c_{n-1}} = 0 \Longrightarrow c_{n-1} = 0.$$

The case $n = 1$ follows from c) because CX is obviously connected.

Of course, later we define the cone of any simplicial set, as well as the cone of any complex, and prove a similar result about its (co)homology.

6. Geometry of Cochains

Regretfully, we have to announce that cochains are nongeometric beasts. Their main role is to translate geometry into algebra. We will try to substantiate this claim in Chap. III.

7. Coefficient Systems

We can construct chains and cochains of a simplicial set using as coefficients something more involved than just abelian groups. There are two types of coefficient systems: for homology and for cohomology.

8. Definition. *a) A homological coefficient system $\mathcal{A}$ on a simplicial set X is a family of abelian groups $\{\mathcal{A}_x\}$, one for each simplex $x \in X_n$, and a family of homomorphisms $\mathcal{A}(f, x) : \mathcal{A}_x \to \mathcal{A}_{X(f)x}$, one for each pair $x \in X_n$, $f : [m] \to [n]$, such that the following conditions are satisfied:*

$$\mathcal{A}(\mathrm{id}, x) = \mathrm{id};$$
$$\mathcal{A}(f \circ g, x) = \mathcal{A}(g, X(f)x)\mathcal{A}(f, x). \tag{I.6}$$

The second equality means that the following diagram is commutative:

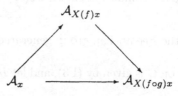

b) A cohomological coefficient system $\mathcal{B}$ on a simplicial set X is a family of abelian groups $\{\mathcal{B}_x\}$, one for each simplex $x \in X_n$, and a family of homomorphisms $\mathcal{B}(f, x) : \mathcal{B}_{X(f)x} \to \mathcal{B}_x$, one for each pair (x, f) as in a), such that the following conditions are satisfied:

$$\mathcal{B}(\mathrm{id}, x) = \mathrm{id}; \quad \mathcal{B}(f \circ g, x) = \mathcal{B}(f, x)\mathcal{B}(g, X(f)x). \tag{I.7}$$

The second equality is equivalent to the commutativity of a diagram similar to the one above.

9. Remarks and Examples

a) Let $\mathcal{A}_x = A$ for each x, $\mathcal{A}(f,x) = \text{id}$ for each f, x. Such a coefficient system is called *constant*. It is both homological and cohomological.

b) Let Y be a topological space, $U = (U_\alpha)$ an open covering, and X the nerve of U defined in I.2.3. The following data form a cohomological coefficient system:

$\mathcal{F}_{\alpha_0,\ldots,\alpha_n}$ is the group of continuous function (under addition) on $U_{\alpha_0} \cap \ldots \cap U_{\alpha_n}$,

$\mathcal{F}(f,(\alpha_0,\ldots,\alpha_n))$ maps a function φ on $U_{\alpha_{f(0)}} \cap \ldots \cap U_{\alpha_{f(m)}}$ into its restriction to $U_{\alpha_0} \cap \ldots \cap U_{\alpha_n}$.

To verify the axioms we need only trivial properties of the restriction of a function to a subset. So, instead of all functions we can take a subset stable under addition and restriction, e.g., smooth functions for a differentiable manifold, analytic functions for a complex manifold, etc. We can also take the group of invertible functions under multiplication. This idea will be further pursued in Sects. I.5 and I.7.

c) Let G be a group (not necessarily abelian) and A a left G-module, i.e., an abelian group on which G acts by automorphisms. We can construct the following cohomological coefficient system $\mathcal{B}$ on the simplicial set BG described in I.2.8:

$$\mathcal{B}_x = A \quad \text{for all} \quad x,$$
$$\mathcal{B}(f,x)(a) = ha, \quad \text{where} \quad h = \prod_{j=1}^{f(0)} g_j \qquad (\text{I.8})$$
$$\text{for} \quad f:[m] \to [n], \quad x = (g_1,\ldots,g_n) \in (BG)_n, \quad a \in A.$$

One can construct also a homological coefficient system $\mathcal{A}$ on BG:

$$\mathcal{A}_x = A \quad \text{for all} \quad x,$$
$$\mathcal{A}(f,x)(a) = h^{-1}a, \quad h \text{ as in (I.8)}.$$

10. Homology and Cohomology with a Coefficient System

Let $\mathcal{A}$ be a homological coefficient system on a simplicial set X.

An *n-dimensional chain* of X with coefficients in $\mathcal{A}$ is a formal linear combination

$$\sum_{x \in X_n} a(x)x, \quad a(x) \in \mathcal{A}_x.$$

Such chains form an abelian group (under addition) which is denoted by $C_n(X,\mathcal{A})$. The boundary of an n-dimensional chain $c = \sum a(x)x \in C_n(X,\mathcal{A})$ is an $(n-1)$-dimensional chain $d_n c \in C_{n-1}(X,\mathcal{A})$ defined by

$$d_n c = \sum_{x \in X_n} \sum_{i=0}^{n} \mathcal{A}\left(\partial_n^i, x\right) (a(x)) (-1)^i X\left(\partial_n^i\right)(x),$$

$$n \geq 1 \tag{I.9}$$

$$d_0 c = 0.$$

As before (see I.4.2) one can easily verify that

$$C_\bullet(X, \mathcal{A}) : \ldots \xrightarrow{d_{n+1}} C_n(X, \mathcal{A}) \xrightarrow{d_n} C_{n-1}(X, \mathcal{A}) \longrightarrow \ldots$$

is a chain complex, i.e. $d_{n-1} \circ d_n = 0$.

Homology groups of the complex $C_\bullet(X, \mathcal{A})$ are called the *homology groups of the simplicial set X with coefficients in $\mathcal{A}$*; they are denoted by $H_n(X, \mathcal{A})$.

Similarly, let $\mathcal{B}$ be a cohomology coefficient system on X. Let

$$C^n(X, \mathcal{B}) = \{\text{functions } f \text{ on } X_n \text{ with } f(x) \in \mathcal{B}_x\}.$$

The coboundary operator

$$d^n : C^n(X, \mathcal{B}) \longrightarrow C^{n+1}(X, \mathcal{B})$$

is given by

$$(d^n f)(x) = \sum_{i=0}^{n+1} (-1)^i \mathcal{B}\left(\partial_{n+1}^i, x\right) \left(f\left(X\left(\partial_{n+1}^i\right) x\right)\right), x \in X_{n+1}. \tag{I.10}$$

Cohomology groups of the cochain complex

$$C^\bullet(X, \mathcal{B}) : \ldots \xrightarrow{d^{n-1}} C^n(X, \mathcal{B}) \xrightarrow{d^n} C^{n+1}(X, \mathcal{B}) \longrightarrow \ldots$$

are called the *cohomology groups of X with coefficients in $\mathcal{B}$* and denoted by $H^n(X, \mathcal{B})$.

11. Examples

a) Homology and cohomology of a simplicial set X with coefficients in the constant system $\mathcal{A} = (A, \text{id}_A)$ (see 9a) coincide with $H_n(X, A)$ and $H^n(X, A)$ respectively.

b) Let Y be a topological space, X the nerve of an open covering $U = (U_\alpha)$, $\mathcal{F}$ a cohomological coefficient system from 9b). Cohomology groups $H^n(X, \mathcal{F})$ are called the *Čech cohomology groups of the sheaf of continuous functions on Y with respect to the covering (U_α)*. Taking other coefficient systems we get similar groups for sheaves of smooth, analytic, etc. functions (see also I.7.4).

c) Let G be a group, A a left G-module, and $\mathcal{A}$, $\mathcal{B}$ the homological and the cohomological coefficient system on $X = BG$ constructed in 9c). Groups $H_n(X, \mathcal{A})$ and $H^n(X, \mathcal{B})$ are called homology and cohomology groups respectively of the group G with coefficients in A.

Exercise

1. Homology of Triangulated Spaces. Compute homology (with coefficients in $\mathbb{Z}$) of simplicial sets corresponding to the following triangulated sets:

a) the n-dimensional simplex $\Delta[n]$,
b) the $(n-1)$-sphere S^{n-1} (the boundary of $\Delta[n]$);
c) the two-dimensional torus;
d) the real projective plane.

Hint: In each case, use a triangulation (as simple as possible) and consider (see Ex. I.7.9,1d)) only linear combination of nondegenerate simplices.

I.5 Sheaves

1. Examples of Sheaves

a) Holomorphic Functions on the Riemann Sphere. The Riemann sphere is the topological space $\mathbb{C} \cup \{\infty\}$ with the base of neighbourhoods of infinity formed by exteriors of circles in $\mathbb{C}$. The structure of the Riemann sphere includes also a family of complex-valued functions $f : U \to \mathbb{C}$ on open subsets U. Such functions (called holomorphic) can be characterized by the following property: any point $z_0 \in U$ has a neighbourhood V, $z_0 \in V \subset U$, in which f is represented by a convergent power series $\sum_{i=0}^{\infty} a_i(z - z_0)^i$ for $z_0 \in \mathbb{C}$ or $\sum_{i=0}^{\infty} a_i z^{-i}$ for $z_0 = \infty$. The family of all such functions (more precisely, of pairs (f, V) consisting of a function f and its domain V – not necessary the maximal one!) is called the sheaf $\mathcal{O}$ of holomorphic functions on $\mathbb{C} \cup \{\infty\}$. Due to the locality of the definition of holomorphicity, the following two problems can be far from trivial.

Problem 1. Can a given holomorphic function $f : U \to \mathbb{C}$ be extended to a larger open set $V \supset U$?

(Example: the holomorphic continuation of the Riemann zeta-function $\zeta(z) = \sum_{n=1}^{\infty} n^{-z}$ from the region $\{z \in \mathbb{C} | Re\, z > 1\}$ to $\mathbb{C} \setminus \{1\}$. For some other interesting Dirichlet series $\sum_{n=1}^{\infty} a_n n^{-z}$ with integral coefficients, this problem is still unsolved.)

Problem 2. Describe the set $\Gamma(U, \mathcal{O})$ of all functions holomorphic in a given domain U.

For example, the Liouville theorem asserts that $\Gamma(\mathbb{C} \cup \{\infty\}, \mathcal{O})$ consists of constant functions only.

The theorem about the convergence radius of a Taylor series says that

$$\Gamma(C, \mathcal{O}) = \left\{ \sum_{i=0}^{\infty} a_i z^i \,\middle|\, \forall r > 0, a_i = o(r^i) \right\}.$$

The main point of the sheaf theory is that we should not pay attention to such particular problems, considering instead the sheaf $\mathcal{O}$ as a single object and compare it with other similar objects.

b) The Sheaf of Solutions of a Linear Differential Equation. Let $U \subset \mathbb{C} \cup \{\infty\}$ be an open set, $a_i(z) \in \Gamma(U, \mathcal{O})$, $i = 0, 1, \ldots, n-1$. Denote by $\mathcal{S}$ the set of pairs (V, f) consisting of an open subset $V \subset U$ and a holomorphic function f in V satisfying in V the equation

$$Lf \stackrel{df}{=} \frac{d^n f}{dz^n} + \sum_{i=0}^{n-1} a_i(z) \frac{d^i f}{dz^i} = 0.$$

This gives the sheaf of holomorphic solutions of the differential equation $Lf = 0$.

In the case where V is connected and simply connected, the theorem on the existence and uniqueness of a solution implies that the space $\Gamma(V, \mathcal{S})$ of solutions holomorphic in V is an n-dimensional linear space over $\mathbb{C}$. For more complicated regions V the answer ceases to be that simple, and to formulate it one has to introduce of the notion of monodromy. Let, for example, $Lf = d^2f/dz^2 + z^{-1}df/dz$, $U = \mathbb{C}\setminus 0$. Solutions of the equation $Lf = 0$ are $c_1 \log z + c_2$, where $\log z$ is "any branch" of the logarithm function. For example, in an annulus $V: 0 \leq r_1 < |z| < r_2$ there is no single-valued branch of the logarithm, hence $\Gamma(V, \mathcal{S}) = \{\text{constants}\}$. The computation of $\Gamma(V, \mathcal{S})$ for an operator L of the second order with three singular points of simplest type on the Riemann sphere constitutes a substantial part of the theory of hypergeometric functions.

The sheaf $\mathcal{S}$ is, in an obvious sense, a subsheaf of $\mathcal{O}$ and L, and in the same obvious sense, acts from $\mathcal{O}$ to $\mathcal{O}$ in such a way that $\mathcal{S}$ is the kernel of this action. The formalism we present later in this section axiomatizes the structures in the examples above.

2. Definition. *a) A presheaf of sets $\mathcal{F}$ on a topological space Y consists of the following data:*

- *a set $\mathcal{F}(U)$ (of sections of the presheaf $\mathcal{F}$) for any open subset $U \subset Y$;*
- *a (restriction) map $r_{UV} : \mathcal{F}(U) \to \mathcal{F}(V)$ for any pair $V \subset U$.*

These data should satisfy the following conditions:

$$r_{UU} = \mathrm{id}, \quad r_{VW} \circ r_{UV} = r_{UW} \quad \text{for} \quad W \subset V \subset U.$$

b) A presheaf $\mathcal{F}$ is said to be a sheaf if the following additional condition is satisfied:

- *For any open covering $U = \cup_{i \in I} U_i$ and for any family of sections $s_i \in \mathcal{F}(U_i)$ such that for all $i, j \in I$*

$$r_{U_i, U_i \cap U_j}(s_i) = r_{U_j, U_i \cap U_j}(s_j),$$

there exists a unique section $s \in \mathcal{F}(U)$ such that

$$s_i = r_{U, U_i}(s)$$

for all $i \in I$.

c) A *morphism* $f : \mathcal{F} \to \mathcal{G}$ *of presheaves on* Y is a family of maps $f(U) : \mathcal{F}(U) \to \mathcal{G}(U)$, one for any open set $U \subset Y$, commuting with the restrictions:
$$r_{UV} \circ f(U) = f(V) \circ r_{UV}.$$

A morphism of sheaves is a morphism of corresponding presheaves.

Instead of $\mathcal{F}(U)$ one uses also the notation $\Gamma(U, \mathcal{F})$, as was done in I.5.1.

For any presheaf $\mathcal{F}$ one can define the presheaf $\mathcal{F}|U$, the restriction of $\mathcal{F}$ on an open set U, by setting $(\mathcal{F}|U)(V) = \mathcal{F}(U \cap V)$. If $\mathcal{F}$ is a sheaf, $\mathcal{F}|U$ is also a sheaf.

3. Presheaves and Sheaves of Structured Sets

A presheaf $\mathcal{F}$ can be a presheaf of groups, rings, topological spaces, etc.; by definition, this means that each set $\mathcal{F}(U)$ is endowed with the corresponding structure and each restriction map is a morphism of corresponding structures.

Similarly one can define external composition laws: a presheaf of modulus $\mathcal{M}$ over a presheaf of rings $\mathcal{O}$ consists of a family of $\mathcal{O}(U)$-modules $\mathcal{M}(U)$ such that the composition laws commute with restrictions in an obvious sense.

Here is an example that will play an important role later: let $\mathcal{F}, \mathcal{G}$ be two presheaves of abelian groups on a topological space Y. A morphism $f : \mathcal{F} \to \mathcal{G}$ consists of a family of group homomorphisms $f(U) : \mathcal{F}(U) \to \mathcal{G}(U)$ commuting with restrictions. Let
$$\mathcal{K}(U) = \operatorname{Ker}(U), \quad \mathcal{C}(U) = \mathcal{G}(U)/f(\mathcal{F}(U)).$$

Defining $\mathcal{K}(U) \to \mathcal{K}(V)$, $\mathcal{C}(U) \to \mathcal{C}(V)$ for $V \subset U$ in an obvious way we get the presheaves $\mathcal{K}$ and $\mathcal{C}$ that are called respectively the *kernel* and the *cokernel* of the morphism f.

A sequence of presheaves $\mathcal{F} \xrightarrow{f} \mathcal{G} \xrightarrow{g} \mathcal{H}$ is said to be *exact* at $\mathcal{G}$ if for any open $U \subset Y$ the sequence of abelian groups $\mathcal{F}(U) \xrightarrow{f(U)} \mathcal{G}(U) \xrightarrow{g(U)} \mathcal{H}(U)$ is exact at $\mathcal{G}(U)$ (i.e., $\operatorname{Ker} g(U) = \operatorname{Im} f(U)$, see Sect. I.6).

Let us look through all the constructions of this subsection under the assumption that all involved presheaves are in fact sheaves.

Definitions of sheaves of rings, groups, etc. and of their morphisms do not change.

We have to be careful only in those cases where we construct new presheaves from old ones: even if old presheaves were sheaves, the new one might fail to be a sheaf. Let us consider a typical situation.

4. Proposition. *a) The kernel* $\mathcal{K}$ *of a morphism of sheaves of abelian groups* $f : \mathcal{F} \to \mathcal{G}$ *is a sheaf of abelian groups.*

b) The cokernel of a morphism of sheaves of abelian groups is always a presheaf but might not be a sheaf.

Proof. a) Let $U = \cup U_i$ and $s_i \in \mathcal{K}(U_i)$ be a family of sections of $\mathcal{K}$ that agree with each other on pairwise intersections. As $\mathcal{K}(U_i) \subset \mathcal{F}(U_i)$ and $\mathcal{F}$ is a sheaf, there exists a unique section s such that $s_i = r_{U,U_i}(s)$. Let us verify that $s \in \mathcal{K}(U)$. We have $r_{U,U_i}(f(s)) = f(r_{U,U_i}(s)) = 0$. As $\mathcal{G}$ is a sheaf, there exists a unique section of $\mathcal{G}$ over U whose restriction on each U_i is zero, and this is clearly the zero section. Therefore, $f(s) = 0$ and $s \in \mathcal{K}(U)$.

b) Let us give an example. Let $Y = \mathbb{C}\backslash\{0\}$, and let $\mathcal{O}_Y$ be the sheaf of holomorphic functions on Y (see I.5.1a)). Define $f : \mathcal{O}_Y \to \mathcal{O}_Y$ by $f(\varphi) = d\varphi/dz$ for a holomorphic function φ on an open set $U \subset X$. One can easily see that any $y \in Y$ has a neighbourhood V_y such that Coker $f(V_y) = 0$ (more precisely, this property holds for any $V \subset Y$ that does not encircle the point 0). On the other hand, dim(Coker $f(Y)) = 1$: the equation $d\varphi/dz = \Psi$ for a holomorphic function Ψ on Y with the Laurent series $\Psi = \sum_{i=-\infty}^{\infty} a_i z^i$ has a solution iff $a_{-1} = 0$. Therefore, the presheaf Coker f on Y is not a sheaf. □

In Sect. II.5 we will show that the definition of the cokernel of a morphism of sheaves can and should be changed in such a way that the cokernel will always be a sheaf.

In Sect. I.7 of this chapter we introduce the notion of Čech cohomology with coefficients in presheaves and in sheaves; in particular, 0-cohomology measures the degree of violation of the axiom 2b) of a sheaf.

Returning to general definitions, we introduce the following useful notions.

5. Germs and Fibres

Let $y \in Y$, $\mathcal{F}$ be a presheaf on Y. A *germ* s_y of a section of $\mathcal{F}$ at the point y is an equivalence class of pairs (s, V) consisting of an open neighbourhood V of y and of $s \in \Gamma(V, \mathcal{F})$, by the relation:

$$(s, V) \sim (s', V') \Leftrightarrow \exists W \subset V \cap V',$$
$$r_{V,W}(s) = r_{V',W}(s').$$

A *fibre* of $\mathcal{F}$ at a point y is the set $\mathcal{F}_y$ of all germs of $\mathcal{F}$ at this point. (On the language of inductive limits, $\mathcal{F}_y = \varinjlim \Gamma(V, \mathcal{F})$ where the limit is taken over the system of all neighbourhoods $V \ni y$.)

For any $V \ni y$ there exists an obvious map

$$r_{V,y} : \mathcal{F}(V) \to \mathcal{F}_y.$$

Similarly one can define a germ of sections of $\mathcal{F}$ over an arbitrary subset $Z \subset Y$; this is an element of the set $\Gamma(Z, \mathcal{F}) = \varinjlim \Gamma(V, \mathcal{F})$, the limit being taken over all open $V \supset Z$. Usually elements of $\Gamma(Z, \mathcal{F})$ are called simply "sections over Z".

The total space of a presheaf $\mathcal{F}$ is the set

$$F = \bigsqcup_{y \in Y} \mathcal{F}_y.$$

For any $s \in \mathcal{F}(U)$, $U \subset Y$, let $s_y = r_{U,y}(s)$ and then
$$F(s) = \{s_y, y \in U\} \subset F.$$
Introduce on F the weakest topology for which $F(s)$ (for all open $U \subset Y$ and all $s \in \mathcal{F}(U)$) are open subsets in F. In this topology the natural projection $\pi : F \to Y$ is continuous (check!). Moreover, π is a local homeomorphism in the sense that any point of F has an open neighbourhood that is homeomorphic to its image under π.

Let us remark now that for any map $f : Z \to Y$ of topological spaces one can define the sheaf Γ of local continuous sections of f as follows: $\Gamma(U) = \{\sigma : U \to Z \mid f \circ \sigma = \mathrm{id}_U\}$. (Check that Γ is indeed a sheaf.) The following construction now looks quite natural.

6. Definition (– Lemma). *a) The sheaf $\mathcal{F}^+$ associated with a presheaf $\mathcal{F}$ is the sheaf of local continuous sections of $\pi : F \to Y$ (where F is the total space of $\mathcal{F}$). There exists a canonical morphism of presheaves $\mathcal{F} \to \mathcal{F}^+$:*
$$\mathcal{F}(U) \ni s \mapsto \{s_y \mid y \in U\} \in \mathcal{F}^+(U).$$

b) If $\mathcal{F}$ is a sheaf then this canonical morphism is an isomorphism.

Proof. It follows from the definition of the topology on F that for any $s \in \mathcal{F}(U)$ the map $\sigma : U \to F$, $\sigma(y) = s_y \in \mathcal{F}_y$, is continuous. Therefore we get a map $\mathcal{F}(U) \to \mathcal{F}^+(U)$. Since these maps commute with restrictions, we obtain a morphism of presheaves $\iota : \mathcal{F} \to \mathcal{F}^+$.

Let us prove b). Let $\mathcal{F}$ be a sheaf. We have to prove that for any $U \subset Y$ the map $\iota(U) : \mathcal{F}(U) \to \mathcal{F}^+(U)$ is one-to-one. Let $s, s' \in \mathcal{F}(U)$ be such that $\iota(U)s = \iota(U)s'$. This means that $r_{U,y}(s) = r_{U,y}(s')$ for any y, so that any point $y \in U$ has a neighbourhood $V_y \subset U$ such that
$$r_{U,V_y}(s) = r_{U,V_y}(s').$$
As $\{V_y, y \in U\}$ is a covering of U and $\mathcal{F}$ is a sheaf, we get $s = s'$.

Let now $\sigma \in \mathcal{F}^+(U)$, i.e., we have a continuous map $\sigma : U \to F$ with $\sigma(y) \in \mathcal{F}_y$. Let $\sigma(y)$ be represented by a section $s_y \in \mathcal{F}(V_y)$ over some open set $V_y \subset U$ with $y \in V_y$. For any $y \in U$ let us choose a neighbourhood G_y of $\sigma(y) \in \mathcal{F}_y \subset F$ in F such that $\pi|G_y$ is a homeomorphism. We can assume that $W_y = \pi(G_y) \subset V_y$ and it is clear that $\sigma(z) = r_{V_y, W_y}(s_y)$ for any $z \in W_y$.

The last equality implies that the sections $r_{V_y, W_y}(s_y) \in \mathcal{F}(W_y)$ are compatible with each other on intersections $W_y \cap W_{y'}$. Hence there exists a section $s \in \mathcal{F}(U)$ with $s_y = r_{U,V_y}(s)$. Clearly, $\sigma = \iota(U)s$. □

7. Main Classes of Sheaves

Roughly speaking, we can divide sheaves arising in various mathematical problems into two large classes: a) sheaves similar to sheaves of functions (holomorphic, as in example I.5.1a), or smooth, continuous, algebraic, etc.);

b) sheaves similar to constant sheaves (as the solution sheaf in the example I.5.1b)).

We give some examples and definitions starting with class b).

8. Definition. *Let A be a set, Y a topological space.*

a) The constant presheaf $\mathbf{A}$ on Y with fibre A is defined by $\mathbf{A}(U) = A$ for all $U \subset Y$, $r_{U,V} = \mathrm{id}$ for all $V \subset U \subset Y$.

b) The constant sheaf $\mathcal{A}$ on Y with fibre A is $\mathcal{A} = \mathbf{A}^+$.

This example clearly illustrates the influence of the topology of the space Y on the structure of a sheaf. The space of $\mathbf{A}$ is, obviously, $Y \times A$ (with the discrete topology on A). Therefore sections of $\mathcal{A} = \mathbf{A}^+$ over an open set U are locally constant functions on U with values in A. In particular, for a connected U we have $\mathcal{A}(U) = A$.

The following, more interesting, class of sheaves contains, in particular, sheaves of holomorphic solutions of linear differential equations from I.5.1b).

9. Definition. *A sheaf $\mathcal{F}$ on Y is said to be locally constant if any point Y has an open neighbourhood U such that $\mathcal{F} \mid U$ is a constant sheaf.*

Let us remark here that Definitions I.5.8 and I.5.9 can be extended to sheaves of structured sets, provided that the structure is stable under the limits required to pass from $\mathbf{A}$ to $\mathbf{A}^+$.

However, it might happen that a sheaf that is constant as a sheaf of sets ceases to be constant when considered as a sheaf of structures. A simple example is a sheaf of Lie algebras depending on parameters.

The classification of locally constant sheaves over a topological space is closely related to the structure of the fundamental group of this space (see Definition II.2.9).

Further generalization of the notion of locally constant sheaves gives constructive sheaves. We shall not enter into details here; let us remark only that these sheaves enable us to describe some types of "jumps" and of "singularities" like, for example, those arising as singularities of solutions of linear differential equation.

10. Sheaves of Functions and Extension of Sections

Sheaves of functions can be considered, first of all, as a glue that enables one to construct global objects (smooth manifolds, analytic spaces, schemes, etc.) from local models. This viewpoint is studied in detail in II.1.4, and we advise the reader to look through the beginning of II.1.4 before going further.

Here we consider only one property of sheaves of functions, namely the rigidity of a sheaf. The simplest question about the rigidity is the following one. Let $U \subset V$ be two nonempty open sets; can a section of a sheaf on U be extended to V and what can be said about the uniqueness of such an extension? Let us consider the full spectrum of possibilities.

a) A function on U might be nonextendable even to the closure (or on the part of the boundary) of U, to say nothing about V. Example: $1/(x^2-1)$ for $U = (-1, 1) \subset \mathbb{R}$. This remark can be applied equally to continuous, smooth, or analytic functions. Therefore the natural extension question should be asked about closed, rather than open, sets. We remind the reader, that, according to I.5.5, a section of $\mathcal{O}$ over Z is a class of sections of $\mathcal{O}$ over open neighbourhoods $U' \supset Z$.

b) So, let $Z \subset X$ be a closed subset. If Y is a Hausdorff space and $\mathcal{O}$ is the sheaf of continuous real-valued functions on Y, then any section of $\mathcal{O}$ over Z can be extended to a section of $\mathcal{O}$ over X. In the other words, the canonical map (restriction) $\Gamma(X, \mathcal{O}) \to \Gamma(Z, \mathcal{O})$ is surjective.

The same is true if X is a smooth (C^∞) manifold and $\mathcal{O}$ is the sheaf of C^∞-functions on X.

The degree of nonuniqueness is illustrated already by the fact that for any closed $Z' \subset X \setminus Z$ and for any continuous (resp. smooth) function on Z' the extension can be chosen in such a way that its restriction on Z' would coincide with this function. (Indeed, the restriction $\Gamma(X, \mathcal{O}) \to \Gamma(Z \cup Z', \mathcal{O})$ is again surjective.)

If $\pi : E \to X$ is a vector bundle, then the same extendability properties hold for the sheaf of germs of sections of π.

c) Now let X be a connected complex analytic manifold, $Z \subset X$ a closed subset, and $\mathcal{O}$ the sheaf of holomorphic functions. Then $\Gamma(X, \mathcal{O}) \to \Gamma(Z, \mathcal{O})$ is injective because if a function on Z can be extended to X at all, it can be extended uniquely: the difference of two extensions is zero in some neighbourhood of Z, and hence vanishes on X (recall that a section of $\mathcal{O}$ over Z is not an analytic function on Z, but an analytic function on some neighbourhood of Z).

A lot of functions are nonextendable; for example, if Z is a point and X is compact, then only the germs of constant functions can be extended to X.

Rigidity properties of sheaves of holomorphic functions (and of sections of holomorphic bundles) are close to those of locally constant sheaves.

Exercises

1. Prove that if $0 \to \mathcal{F} \to \mathcal{G} \to \mathcal{H} \to 0$ is an exact sequence of presheaves and $\mathcal{F}, \mathcal{H}$ are sheaves, then $\mathcal{G}$ is also a sheaf.

2. Flabby, Soft, Fine Sheaves. a) A sheaf of sets $\mathcal{F}$ on a topological space X is said to be *flabby* if for any open $U \subset X$ the restriction map $\Gamma(X, \mathcal{F}) \to \Gamma(U, \mathcal{F})$ is a surjection. Let $f : X \to Y$ be a map of topological spaces. Define the sheaf $\mathcal{C}_f$ of all (not necessarily continuous) local sections of f by
$$\Gamma(U, \mathcal{C}_f) = \{\sigma : U \to X, f \circ \sigma = \mathrm{id}_U\}$$
(let us emphasize once more that σ need not be continuous). Prove that for a surjective f the sheaf $\mathcal{C}_f$ is flabby. Conclude that any sheaf of sets $\mathcal{F}$ is a subsheaf of a flabby sheaf (namely, of the sheaf $\mathcal{C}_\pi$ for $\pi : F \to X$; see I.5.5).

b) Prove that if
$$0 \to \mathcal{F} \to \mathcal{G} \to \mathcal{H} \to 0 \tag{I.11}$$
is an exact sequence of abelian groups and $\mathcal{F}$ is a flabby sheaf, then for any open U the sequence of abelian groups
$$0 \to \Gamma(U, \mathcal{F}) \to \Gamma(U, \mathcal{G}) \to \Gamma(U, \mathcal{H}) \to 0$$
is exact. Conclude that if in (I.11) $\mathcal{F}$ and $\mathcal{G}$ are flabby sheaves, so is $\mathcal{H}$.

c) Let X be a paracompact topological space (i.e., X is separable and any open covering of X has a locally finite subcovering). A sheaf $\mathcal{F}$ on X is said to be *soft* if for any closed $Y \subset X$ the restriction map $\Gamma(X, \mathcal{F}) \to \Gamma(Y, \mathcal{F})$ (see I.5.5) is a surjection. Prove that a flabby sheaf on a paracompact space X is soft. Prove the following analogues of statements from a): if in the exact sequence (I.11) $\mathcal{F}$ is a soft sheaf, then for any closed $Y \subset X$ the sequence
$$0 \to \Gamma(Y, \mathcal{F}) \to \Gamma(Y, \mathcal{G}) \to \Gamma(Y, \mathcal{H}) \to 0$$
is exact, and if $\mathcal{F}$ and $\mathcal{G}$ in (I.11) are soft sheaves, then $\mathcal{H}$ is also soft.

d) *Decomposition of Sections.* Let $\mathcal{F}$ be a sheaf of abelian groups on a topological space X, s a section of $\mathcal{F}$ over the whole X, and $(U_i)_{i \in I}$ an open covering of X. A decomposition of the section s subordinated to the covering (U_i) is a family of sections s_i of the sheaf $\mathcal{F}$ over open sets $V_i \subset U_i$ which is locally finite (that is, for any $x \in X$ only a finite number of stalks $(s_i)_x$ are nonzero) and satisfies the equality $s_x = \sum_{i \in I}(s_i)_x$ for all $x \in X$. Prove that if $\mathcal{F}$ is a soft sheaf on a paracompact space X, then for any $s \in \Gamma(X, \mathcal{F})$ and for any covering $(U_i)_{i \in I}$ of X there exists a decomposition of s subordinated to (U_i).

e) Prove that if $\mathcal{F}$ is a soft sheaf of rings then any sheaf of $\mathcal{F}$-modules is also soft.

f) A sheaf $\mathcal{F}$ of abelian groups on a paracompact topological space is said to be *fine* if for any two disjoint closed subsets Y_1, Y_2 of X there exists an automorphism $\varphi : \mathcal{F} \to \mathcal{F}$ which induces the zero map on some neighbourhood of Y_1 and the identity map on some neighbourhood of Y_2. Prove that any fine sheaf is soft. Prove that the sheaves of germs of continuous functions on X (either real- or complex-valued) are fine (this is, essentially, the classical Uryson theorem: for any two disjoint closed sets Y_1, Y_2 there exists a continuous function f which is 1 on some neighbourhood of Y_1 and 0 on some neighbourhood of Y_2). Prove that the sheaf of germs of smooth functions on a smooth manifold X is fine.

g) Let $\mathcal{F}, \mathcal{G}$ be two presheaves of sets on a space X. The *presheaf of morphisms* $\mathcal{H} = Hom(\mathcal{F}, \mathcal{G})$ is defined as follows: $\mathcal{H}(U) = \mathrm{Hom}(\mathcal{F}|U, \mathcal{G}|U)$ where Hom is the set of morphisms of the presheaf $\mathcal{F}|U$ into the presheaf $\mathcal{G}|U$, with natural restriction maps $\mathcal{H}(U) \to \mathcal{H}(V)$. Prove that if $\mathcal{F}$ and $\mathcal{G}$ are sheaves, then $\mathcal{H}$ is also a sheaf. If $\mathcal{F}, \mathcal{G}$ are sheaves with a structure (e.g., sheaves of abelian groups), then $Hom(\mathcal{F}, \mathcal{G})$ shall denote the sheaf of morphisms that preserve this structure. Prove the following characterization of fine sheaves: a sheaf of abelian groups is fine if and only if the sheaf of rings $Hom(\mathcal{F}, \mathcal{F})$ is flabby.

h) Prove that the properties of flabbiness, of softness, and of fineness are local: a sheaf $\mathcal{F}$ on a space X (resp. on a paracompact space X) is flabby (resp. soft, resp. fine) iff any point $x \in X$ has a neighbourhood U such that $\mathcal{F}|U$ is flabby (resp. soft, resp. fine).

3. Affine Schemes. a) Let A be a commutative unitary ring, $\operatorname{Spec} A$ be the set of all simple ideals in A (recall that an ideal $p \subset A$ is said to be simple if $ab \in p$ implies that either $a \in p$ or $b \in p$). For any $f \in A$ let $D(f) = \{p \in \operatorname{Spec} A, f \notin p\}$. Prove that $D(gf) = D(f) \cap D(g)$. Define the topology on $\operatorname{Spec} A$ for which $D(f), f \in A$, form a base of open sets. Prove the following properties:

(i) $D(g) \subseteq D(f)$ iff $g^n = uf$ for some $n > 0$, $u \in A$.
(ii) $D(f) = \cup_{i=1}^{k} D(f_i)$ iff there exists $b_1, \ldots, b_k \in A$ and $n > 0$ such that $f^n = b_1 f_1 + \ldots + b_k f_k$.
(iii) The closure of a point $p \in \operatorname{Spec} A$ consists of all $p' \in \operatorname{Spec} A$ containing p; in particular, closed points correspond to maximal ideals.

b) Let $S \subseteq A$ be a multiplicatively closed subset (i.e. $a, b \in S$ implies $ab \in S$; standard examples are sets $\{f^n, n \geq 0\}$ for $f \in A$ and $A \setminus p$ for $p \in \operatorname{Spec} A$). The localization $S^{-1}A$ of the ring A by S is the set of equivalence classes of fractions a/s, $a \in A$, $s \in S$, under the equivalence relation $a_1/s_1 \sim a_2/s_2$ iff there exists $b \in S$ with $b(a_1 s_2 - a_2 s_1) = 0$. Prove that natural operations for the addition and for the multiplication of fractions make $S^{-1}A$ a ring. Standard notations for $S^{-1}A$ in cases $S = \{f^n\}$ and $S = A \setminus p$ are A_f and A_p.

c) Prove that A_f depends only on $D(f)$ and not on f itself.

d) Let $V = D(g) \subset U = D(f)$, so that $g^n = uf$ (see a),(i)). Define $r_{UV} : A_f \to A_g$ by $r_{UV}(a/f^m) = $ (the class of au^m/g^{mn} in A_g). Prove that r_{UV} depends only on U and V (and not on f, u, and g), that r_{UV} is a homomorphism of rings, and that $r_{UU} = \operatorname{id}$,

$$r_{VW} \circ r_{UV} = r_{UW} \quad \text{for} \quad W = D(h) \subset V = D(g) \subset U = D(f).$$

e) Prove that on $\operatorname{Spec} A$ there exists a unique sheaf of rings $\mathcal{O}$ with $\Gamma(D(f), \mathcal{O}) = A_f$, $r_{D(f)D(g)}$ as in d).

Hint. First prove that the axioms of a sheaf are satisfied for open sets of the form $D(f)$. Then define $\Gamma(U, \mathcal{O})$ for an arbitrary $U = \cup D(f_i)$ as the set of equivalence classes of compatible families

$$\{s_i \in A_{f_i}, \ r_{D(f_i)D(f_i f_j)}(s_i) = r_{D(f_j)D(f_i f_j)}(s_j)\}$$

under the natural equivalence relation for families $\{s_i\}$ and $\{t_i\}$ corresponding to two different coverings $U = \cup D(f_i) = \cup D(g_j)$.

f) Prove that the stalk of $\mathcal{O}$ at the point $p \in \operatorname{Spec} A$ is A_p (notations of b)).

g) Let M be an A-module. For a multiplicatively closed subset $S \subseteq A$ the localization $S^{-1}M$ is defined (similarly to b)) as the set of equivalence classes of ratios $\{m/s, m \in M, s \in S\}$ by the equivalence relations: $m_1/s_1 \sim m_2/s_2$ iff $b(s_1 m_2 - s_2 m_1) = 0$ for some $b \in S$. Prove that $S^{-1}M$ is a module over $S^{-1}A$.

h) Prove that on Spec A there exists a unique sheaf of modules $\tilde{M}$ over the sheaf of rings $\mathcal{O}$ for which $\Gamma\left(D(f), \tilde{M}\right) = M_f$ (localization of M by $\{f^n, n \geq 0\}$) and restriction maps are defined similarly to d). Find the stalk of $\tilde{M}$ at a point $p \in \operatorname{Spec} A$.

The sheaves of the form $\tilde{M}$ are called *quasicoherent*.

i) Let $\tilde{M}_1, \tilde{M}_2$ be two quasicoherent sheaves on Spec A. Prove that morphisms $\tilde{M}_1 \to \tilde{M}_2$ (as sheaves of $\mathcal{O}$-modules) are in one-to-one correspondence with homomorphisms $M_1 \to M_2$ of corresponding A-modules.

I.6 The Exact Sequence

1. Homology as a Function in Two Variables

In Sect. I.4 we defined groups $H_n(X, \mathcal{A})$ and $H^n(X, \mathcal{B})$, where X is a simplicial set, and $\mathcal{A}$ and $\mathcal{B}$ are coefficient systems. In some simple cases these groups can be computed directly. But the main technique consists in the study of the behaviour of these groups under the change of X or the change of $\mathcal{A}$.

We shall call $\mathcal{A}$ (or $\mathcal{B}$) an *abelian* variable and X a *nonabelian* variable. In the next chapter we shall try to show that any (co)homology theory contains (explicitly or implicitly) these two arguments.

In this section we fix X and study the dependence of homology and cohomology on coefficients. The main tool here is the theorem about the exact sequence.

2. Exact Sequences

An exact sequence of abelian groups is a complex $C^\bullet$ with all cohomology groups vanishing (for chain complexes the definition is the same). This means that $\operatorname{Ker} d^n = \operatorname{Im} d^{n-1}$ for all n.

An exact triple (or a short exact sequence) of abelian groups is an exact sequence of the form

$$\cdots \to 0 \to 0 \to A \xrightarrow{i} B \xrightarrow{p} C \to 0 \to 0 \to \cdots$$

Usually such a sequence is written as

$$0 \to A \xrightarrow{i} B \xrightarrow{p} C \to 0. \tag{I.12}$$

To give such a triple is the same as to give an abelian group B and its subgroup A. The homomorphism theorem says that

$$\operatorname{Im} i = \operatorname{Ker} p \quad \text{implies} \quad C = B/A.$$

3. Theorem. *Let X be a simplicial set. Any exact triple of abelian groups (I.12) canonically determines a cohomology exact sequence*

$$0 \longrightarrow H^0(X,A) \longrightarrow H^0(X,B) \longrightarrow H^0(X,C) \longrightarrow H^1(X,A) \longrightarrow$$
$$\longrightarrow H^1(X,B) \longrightarrow \ldots \longrightarrow H^n(X,A) \longrightarrow H^n(X,B) \longrightarrow \quad \text{(I.13)}$$
$$\longrightarrow H^n(X,C) \longrightarrow H^{n+1}(X,A) \longrightarrow \ldots$$

and a similar homology sequence

$$\ldots \longrightarrow H_n(X,A) \longrightarrow H_n(X,B) \longrightarrow H_n(X,C) \longrightarrow H_{n-1}(X,A) \longrightarrow$$
$$\quad \text{(I.14)}$$
$$\ldots \longrightarrow H_1(X,C) \longrightarrow H_0(X,A) \longrightarrow H_0(X,B) \longrightarrow H_0(X,C) \longrightarrow 0.$$

4. Remarks

a) This theorem has an almost obvious special case. If the exact triple (I.12) splits, i.e., has the form

$$0 \longrightarrow A \xrightarrow{i} B = A \oplus C \xrightarrow{p} C \longrightarrow 0,$$

$$i(a) = (a,0), \quad p(a,c) = c,$$

then the sequence (I.13) can be decomposed into split exact triples of the form

$$0 \longrightarrow H^n(X,A) \longrightarrow H^n(X,B) =$$
$$H^n(X,A) \oplus H^n(X,C) \longrightarrow H^n(X,C) \longrightarrow 0.$$

The main difference of the general case from this special one the is nontrivial phenomenon of the "connecting" (or boundary) homomorphisms

$$H^n(X,C) \longrightarrow H^{n+1}(X,A).$$

b) Theorem I.6.3 has an analogue for coefficient systems; it will be stated later.

c) The proof of the theorem comprises two steps: the construction of all homomorphisms in (I.13) and the proof of the exactness. It is more convenient to perform both steps in a somewhat more general setting. Namely, we show that the exact triple (I.12) gives rise to a similar exact triple of chains (and of cochains) and then we prove an analogue of Theorem I.6.3 for exact triples of complexes, independent of their origin.

5. Morphisms of Complexes

Let $B^\bullet, C^\bullet$ be two complexes. A *morphism* $f^\bullet : B^\bullet \to C^\bullet$ is a family of homomorphisms $f^n : B^n \to C^n$ commuting with differentials:

$$d^n \circ f^n = f^{n+1} \circ d^n. \tag{I.15}$$

Given $f^\bullet : B^\bullet \to C^\bullet$, let us construct a family of homomorphisms

$$H^n(f) : H^n(B^\bullet) \longrightarrow H^n(C^\bullet)$$

as follows. Let $b \in H^n(B^\bullet)$ be represented by a cocycle $\tilde{b} \in \operatorname{Ker} d^n \subset B^n$. Then, by (I.15), $f^n(\tilde{b}) \in \operatorname{Ker} d^n \subset C^n$, and we define $H^n(f)(b)$ to be the class of $f^n(\tilde{b})$ in $H^n(C^\bullet)$. Again by (I.15), this class does not depend on the choice of a representative of b modulo $\operatorname{Im} d^{n-1}$.

It is clear also that if $g^\bullet : A^\bullet \to B^\bullet$ is another morphism of complexes then

$$H^n(f \circ g) = H^n(f) \circ H^n(g). \tag{I.16}$$

Let again $f^\bullet : B^\bullet \to C^\bullet$ be a morphism of complexes. Let

$$\operatorname{Ker} f^\bullet = (\operatorname{Ker} f^n),$$
$$\operatorname{Coker} f^\bullet = (\operatorname{Coker} f^n).$$

By (I.15), $\operatorname{Ker} f^\bullet$ and $\operatorname{Coker} f^\bullet$ are complexes (with differentials induced by differentials in $B^\bullet$ and in $C^\bullet$).

A sequence of complexes and morphisms of complexes

$$0 \longrightarrow A^\bullet \xrightarrow{i} B^\bullet \xrightarrow{p} C^\bullet \longrightarrow 0$$

is said to be an *exact triple of complexes* if all $0 \longrightarrow A^n \xrightarrow{i^n} B^n \xrightarrow{p^n} C^n \longrightarrow 0$ are exact triples of abelian groups.

Now let $f : B \to C$ be a homomorphism of groups and X a simplicial set. Then we can define in an obvious way natural homomorphisms

$$f_n : C_n(X, B) \longrightarrow C_n(X, C), \quad f^n : C^n(X, B) \longrightarrow C^n(X, C).$$

It is clear that these homomorphisms commute with differentials, and, therefore, form morphisms of corresponding complexes.

6. Lemma. *Let* $0 \longrightarrow A \xrightarrow{i} B \xrightarrow{p} C \longrightarrow 0$ *be an exact triple of abelian groups. Then the sequences of groups of chains and of cochains*

$$0 \longrightarrow C_\bullet(X, A) \xrightarrow{i_\bullet} C_\bullet(X, B) \xrightarrow{p_\bullet} C_\bullet(X, C) \longrightarrow 0,$$
$$0 \longrightarrow C^\bullet(X, A) \xrightarrow{i^\bullet} C^\bullet(X, B) \xrightarrow{p^\bullet} C^\bullet(X, C) \longrightarrow 0$$

are exact.

Proof. An element of $C_n(X, A)$ is a formal linear combination $\sum_{x \in X_n} a(x)x, a(x) \in A$. The image of this element under the mapping $i_n : C_n(X, A) \longrightarrow C_n(X, B)$ is $\sum_{x \in X_n} i(a(x))x$ and since i is an injection, i_n is also an injection. Similarly one proves that p_n is a surjection. Further, $p_n i_n (\sum a(x)x) = \sum (p \circ i)(a(x))x = 0$ for $\sum a(x)x \in C_n(X, A)$. Let now $\beta = \sum b(x)x \in C_n(X, B)$ and $p_n(\beta) = 0$. Then $p(b(x)) = 0$ for all $x \in X_n$, i.e., $b(x) = i(a(x))$, $a(x) \in A$ and $\beta = i_n(\alpha)$ for $\alpha = \sum a(x)x \in C_n(X, A)$. The second sequence is treated similarly. □

Using this lemma one can deduce Theorem I.6.3 from the more general Theorem I.6.8.

7. The Construction of the Boundary Homomorphism

Let
$$0 \longrightarrow A^\bullet \longrightarrow B^\bullet \longrightarrow C^\bullet \longrightarrow 0 \tag{I.17}$$

be an exact sequence of cochain complexes. For any n we define a homomorphism
$$\delta^n = \delta^n(i^\bullet, p^\bullet) : H^n(C^\bullet) \longrightarrow H^{n+1}(A^\bullet)$$

as follows. Let $c \in H^n(C^\bullet)$ be represented by a cocycle $\tilde{c} \in C^n$. The homomorphism p^n being surjective, we have $\tilde{c} = p^n(\tilde{b})$ for some $\tilde{b} \in B^n$. Now, $p^n(d\tilde{b}) = dp^n(\tilde{b}) = d\tilde{c} = 0$ so that $d\tilde{b} = i^{n+1}(\tilde{a})$ for some (unique) $\tilde{a} \in A^{n+1}$. This $\tilde{a}$ is a cocycle because
$$i^{n+1}(d\tilde{a}) = di^{n+1}(\tilde{a}) = d(d\tilde{b}) = 0$$

and $d\tilde{a} = 0$ because i^{n+1} is an injection.

Different choices of $\tilde{c}$ and $\tilde{b}$ change $\tilde{a}$ by a coboundary:
$$\tilde{c}' = \tilde{c} + d\tilde{c}_1 = p^n \left(\tilde{b} + d\tilde{b}_1 + i^n \tilde{a}_1 \right)$$

where $\tilde{c}_1 \in C^{n-1}$, $p^{n-1}(\tilde{b}_1) = \tilde{c}_1, \tilde{a}_1 \in A^n$. Hence,
$$\tilde{a}' = \tilde{a} + d\tilde{a}_1$$

and the cohomology class of $\tilde{a}$ does not depend on choices. Now we define
$$\delta^n(i^\bullet, p^\bullet)(c) = \tilde{a} \bmod \operatorname{Im} d^{n+1} \in H^{n+1}(A^\bullet).$$

One can easily verify (similarly to the above) that
$$\delta^n(i^\bullet, p^\bullet) : H^n(C^\bullet) \to H^{n+1}(A^\bullet)$$

is a homomorphism of abelian groups.

8. Theorem. *In the setup of I.6.7 the following sequence is exact:*
$$\cdots \longrightarrow H^n(A^\bullet) \xrightarrow{H^n(i^\bullet)} H^n(B^\bullet) \xrightarrow{H^n(p^\bullet)} H^n(C^\bullet) \xrightarrow{\delta^n(i^\bullet, p^\bullet)} H^{n+1}(A^\bullet) \longrightarrow \cdots$$

Proof. a) *Exactness at* $H^n(B^\bullet)$. First of all, $H^n(p^\bullet) \circ H^n(i^\bullet) = H^n(p^\bullet \circ i^\bullet) = 0$ because $p^\bullet \circ i^\bullet = 0$. Next, let $b \in H^n(B^\bullet)$ and $H^n(p^\bullet)(b) = 0$. We construct $a \in H^n(A^\bullet)$ with $b = H^n(i^\bullet)(a)$ as follows. Let $\tilde{b} \in B^n$ be a representative of b so that $d\tilde{b} = 0$. Since $H^n(p^\bullet)(b) = 0$ we have $p^n(\tilde{b}) = d\tilde{c}$ for some $\tilde{c} \in C^{n-1}$ and, $p^{n-1}: B^{n-1} \to C^{n-1}$ being a surjection, $\tilde{c} = p^{n-1}(\tilde{b}_1)$ for some $\tilde{b}_1 \in B^{n-1}$. It is clear that $p^n(\tilde{b} - d\tilde{b}_1) = 0$ so that by the exactness of (I.17), $\tilde{b} - d\tilde{b}_1 = i^n(\tilde{a})$ for some $\tilde{a} \in A^n$; moreover, $i^{n+1}(d\tilde{a}) = di^n(\tilde{a}) = d\tilde{b} = 0$. As i^{n+1} is injective, $d\tilde{a} = 0$. Now one can easily check that $a = \tilde{a} \mod \operatorname{Im} d^{n-1} \in H^n(B^\bullet)$ satisfies the required property.

b) $\delta^n(i^\bullet, p^\bullet) \circ H^n(p^\bullet) = 0$. Let $c = H^n(p^\bullet)(b)$ for some $b \in H^n(B^\bullet)$ and let $\tilde{b} \in B^n$, $\tilde{c} \in C^n$ be representatives of b, c respectively, so that $d\tilde{b} = 0$. Then the definition of $\delta^n(i^\bullet, p^\bullet)$ in 7 shows that $\delta^n(i^\bullet, p^\bullet)(c) = 0$.

c) $\operatorname{Ker} \delta^n(i^\bullet, p^\bullet) \subset \operatorname{Im} H^n(p^\bullet)$. Let $\delta^n(i^\bullet, p^\bullet)(c) = 0$ and let $\tilde{c} \in C^n$ be a cocycle representing the cohomology class c. We follow the construction of $\delta^n(i^\bullet, p^\bullet)(c)$ as in 7. Let $\tilde{c} = p^n(\tilde{b})$ and $d\tilde{b} = i^{n+1}(\tilde{a})$ for $\tilde{b} \in B^n$, $\tilde{a} \in A^{n+1}$. Then $\delta^n(i^\bullet, p^\bullet)(c) = 0$ implies $\tilde{a} = d\tilde{a}_1$ for some $\tilde{a}_1 \in A^n$. Let $\tilde{b}_1 = \tilde{b} - i^n(\tilde{a}_1)$. Then $d\tilde{b}_1 = 0$ and $p^n(\tilde{b}_1) = p^n(\tilde{b}) - p^n \circ i^n(\tilde{a}_1) = \tilde{c}$.

Hence $c = H^n(p^\bullet)(b_1)$, where $b_1 = \tilde{b}_1 \mod \operatorname{Im} d^n \in H^n(B^\bullet)$.

d) $H^{n+1}(i^\bullet) \circ \delta^n(i^\bullet, p^\bullet) = 0$. By 7, $a = \delta^n(i^\bullet, p^\bullet)(c) \in H^{n+1}(A^\bullet)$ is represented by a cocycle $\tilde{a} \in A^{n+1}$ such that $i^{n+1}(\tilde{a}) = d\tilde{b}$ for some $\tilde{b} \in B^n$. Hence $H^{n+1}(i^\bullet)(a) = 0$.

e) $\operatorname{Ker} H^{n+1}(i^\bullet) \subset \operatorname{Im} \delta^n(i^\bullet, p^\bullet)$. Let $a \in H^{n+1}(A^\bullet)$ be such that $H^{n+1}(i^\bullet)(a) = 0$ and let $\tilde{a} \in A^{n+1}$ be a cocycle representing a. Then $i^{n+1}(\tilde{a}) = d\tilde{b}$ for some $\tilde{b} \in B^n$. Let $\tilde{c} = p^n(\tilde{b}) \in C^n$. Then $d\tilde{c} = dp^n(\tilde{b}) = p^{n+1}(d\tilde{b}) = p^{n+1}i^{n+1}(\tilde{a}) = 0$, so that $\tilde{c}$ is a cocycle. Then $a = \delta^n(i^\bullet, p^\bullet)(c)$, where $c \in H^n(C^\bullet)$ is the class of the cocycle $\tilde{c}$. □

9. Generalization to a Coefficient System

Let $\mathcal{B}$, $\mathcal{B}'$ be two cohomological coefficient systems on a simplicial set X. A *morphism* $\varphi : \mathcal{B} \to \mathcal{B}'$ is a family of homomorphisms

$$\varphi_x : \mathcal{B}_x \longrightarrow \mathcal{B}'_x,$$

one for each simplex $x \in X$, commuting with maps $\mathcal{B}(f, x)$ and $\mathcal{B}'(f, x)$, i.e., such that for any $f : [m] \to [n]$, $x \in X_n$ the diagram

$$\begin{array}{ccc} \mathcal{B}_{X(f)x} & \xrightarrow{\mathcal{B}(f,x)} & \mathcal{B}_x \\ \varphi_{X(f)x} \downarrow & & \downarrow \varphi_x \\ \mathcal{B}'_{X(f)x} & \xrightarrow{\mathcal{B}'(f,x)} & \mathcal{B}'_x \end{array}$$

is commutative.

Similarly one defines a morphism of homological coefficient systems.

A morphism $\varphi : \mathcal{B} \to \mathcal{B}'$ yields a morphism of cochains $\varphi^n : C^n(X, \mathcal{B}) \to C^n(X, \mathcal{B}')$:
$$\varphi^n(f)(x) = \varphi_x(f(x)), \quad x \in X_n.$$

Formula (I.10) from Sect. I.4 shows that $\varphi^\bullet = (\varphi^n)$ is a morphism of complexes. Using similarly (I.9) from Sect. I.4 one can define a morphism of complexes of chains.

A sequence of coefficient systems and their morphisms
$$0 \longrightarrow \mathcal{B}' \xrightarrow{\iota} \mathcal{B} \xrightarrow{\pi} \mathcal{B}'' \longrightarrow 0$$
is said to be *exact* if for any $x \in X_n$ the sequence
$$0 \longrightarrow \mathcal{B}'_x \xrightarrow{\iota_x} \mathcal{B}_x \xrightarrow{\pi_x} \mathcal{B}''_x \longrightarrow 0$$
is exact. Similar definitions apply to homology coefficient system.

In this setup the following analogue of Lemma I.6.6 (and its homological version) is true: the sequence of complexes
$$0 \longrightarrow C^\bullet(X, \mathcal{B}') \longrightarrow C^\bullet(X, \mathcal{B}) \longrightarrow C^\bullet(X, \mathcal{B}'') \longrightarrow 0$$
is exact.

Applying Theorem I.6.8 to this sequence we get an analogue of Theorem I.6.3.

Exercise

1. Functoriality of the Exact Sequence. Assume we are given the following diagram of complexes and their morphisms:

$$\begin{array}{ccccccccc} 0 & \longrightarrow & A^\bullet & \xrightarrow{i} & B^\bullet & \xrightarrow{p} & C^\bullet & \longrightarrow & 0 \\ & & \downarrow f & & \downarrow g & & \downarrow h & & \\ 0 & \longrightarrow & \tilde{A}^\bullet & \xrightarrow{\tilde{i}} & \tilde{B}^\bullet & \xrightarrow{\tilde{p}} & \tilde{C}^\bullet & \longrightarrow & 0 \end{array} \quad (\text{I}.18)$$

which is commutative (i.e., $gi = \tilde{i}f$, $hp = \tilde{p}g$) and has exact rows. Prove that the homomorphisms $H^n(f)$, $H^n(g)$, $H^n(h)$ yield the following commutative diagram:

$$\begin{array}{ccccccccc} \cdots \longrightarrow & H^n(A^\bullet) & \longrightarrow & H^n(B^\bullet) & \longrightarrow & H^n(C^\bullet) & \longrightarrow & H^{n+1}(A^\bullet) & \longrightarrow \cdots \\ & \downarrow H^n(f) & & \downarrow H^n(g) & & \downarrow H^n(h) & & \downarrow H^{n+1}(f) & \\ \cdots \longrightarrow & H^n(\tilde{A}^\bullet) & \longrightarrow & H^n(\tilde{B}^\bullet) & \longrightarrow & H^n(\tilde{C}^\bullet) & \longrightarrow & H^{n+1}(\tilde{A}^\bullet) & \longrightarrow \cdots \end{array}$$

whose rows are long exact sequences corresponding to rows of (I.18).

For other properties of exact sequences see Exs. II.5.6 and II.5.7.

I.7 Complexes

1. Where Do Complexes Come from?

In this section we describe several classes of complexes of either algebraic or geometric origin. All these complexes can be roughly divided into two groups: combinatorial complexes and differential geometry complexes. An example of a complex from the first group is the chain complex of a simplicial set, that from the second group is the de Rham complex. We describe also some algebraic operations with complexes.

2. Definition. *A simplicial abelian group is a simplicial set* $A = (A_n)$, $n = 0, 1, \ldots$ *(see Definition I.2.1) such that each A_n is an abelian group and all $A(f) : A_n \to A_m$ are group homomorphisms.*

In Sect. I.2 we constructed groups of chains $C_n(X, \mathcal{A})$ for any simplicial set X and any homology coefficient system $\mathcal{A}$ on it. For any nondecreasing $f : [m] \to [n]$ we define a homomorphism

$$C_n(f) : C_n(X, \mathcal{A}) \longrightarrow C_m(X, \mathcal{A})$$

by the formula

$$C_n(f)\left(\sum_{x \in X_n} a(x)x\right) = \sum_{x \in X_n} \mathcal{A}(f, x)\left(a(x)\right) X(f)(x).$$

Formulas (I.6), (I.7) from Sect. I.4 show that $C_\bullet(X, \mathcal{A}) = (C_n(X, \mathcal{A}))$ becomes a simplicial abelian group.

Similarly to formula (I.9) in Sect. I.4, one can construct a complex for any simplicial abelian group as follows.

3. Definition (– Lemma). *Let A be a simplicial abelian group. Define $d_n : A_n \to A_{n-1}$ by*

$$d_n(a) = \sum_{i=0}^{n} (-1)^i A(\partial_n^i)(a).$$

Then $d_{n-1} \circ d_n = 0$ so that (A_n, d_n) is a chain complex.

Proof. The proof is similar to that of Lemma I.4.2. □

The cochain complex of a simplicial complex is a special case of a similar construction that uses cosimplicial abelian groups. A *cosimplicial abelian group* is a family of abelian groups, (B^n), $n = 0, 1, \ldots$, and of homomorphisms $B(f) : B^m \to B^n$, one for each nondecreasing map $f : [m] \to [n]$ satisfying the conditions $B(\mathrm{id}) = \mathrm{id}$, $B(f \circ g) = B(f) \circ B(g)$.

We leave to the reader to formulate and prove an analogue of Definition – Lemma I.7.3.

4. The Čech Complex

Let Y be a topological space, $U = (U_\alpha)$ a (not necessarily open) covering of X, and $\mathcal{F}$ a sheaf of abelian groups on Y. Define the cosimplicial abelian group of Čech cochains $\check{C}(U, \mathcal{F})$ as follows.

An element from $\check{C}^m(U, \mathcal{F})$ is a family of sections $\varphi = \{\varphi_{\alpha_0 \ldots \alpha_m} \in \mathcal{F}(U_{\alpha_0} \cap \ldots \cap U_{\alpha_m})\}$, one for any $(m+1)$-tuple of indices $(\alpha_0, \ldots, \alpha_m)$. (Note that for an empty subset $\emptyset \subset X$ we set $\mathcal{F}(\emptyset) = \{0\}$.)

The homomorphism $\check{C}(g) : \check{C}^m \to \check{C}^n$ for a nondecreasing map $f : [m] \to [n]$ takes a cochain φ into the cochain with components

$$(\check{C}(g)\varphi)_{\alpha_0 \ldots \alpha_n} = \mathrm{res}\left(\varphi_{\alpha_{g(0)} \ldots \alpha_{g(m)}}\right)$$

where res is the restriction map corresponding to the inclusion

$$U_{\alpha_{g(0)}} \cap \ldots \cap U_{\alpha_{g(m)}} \hookrightarrow U_{\alpha_0} \cap \ldots \cap U_{\alpha_m}.$$

The differential in the complex associated with this cosimplicial abelian group is given by the formula

$$(d\varphi)_{\alpha_0 \ldots \alpha_{m+1}} = \sum_{i=0}^{m+1} (-1)^i \mathrm{res}\left(\varphi_{\alpha_0 \ldots \hat{\alpha}_i \ldots \alpha_{m+1}}\right).$$

Several initial terms of the Čech complex have the following form:

$$\check{C}^0(U, \mathcal{F}) \ni \{\varphi_\alpha \in \mathcal{F}(U_\alpha)\} \xrightarrow{d} \{(d\varphi)_{\alpha_0\alpha_1} = \mathrm{res}(\varphi_{\alpha_1}) - \mathrm{res}(\varphi_{\alpha_0})\},$$

$$\check{C}^1(U, \mathcal{F}) \ni \{\varphi_{\alpha_0\alpha_1} \in \mathcal{F}(U_{\alpha_0} \cap U_{\alpha_1})\} \xrightarrow{d}$$
$$\{(d\varphi)_{\alpha_0\alpha_1\alpha_2} = \mathrm{res}(\varphi_{\alpha_1\alpha_2}) - \mathrm{res}(\varphi_{\alpha_0\alpha_2}) + \mathrm{res}(\varphi_{\alpha_0\alpha_1})\}.$$

These formulas show that the Čech complex can be completed to the following complex:

$$0 \longrightarrow \mathcal{F}(X) \xrightarrow{\varepsilon} \check{C}^0(U, \mathcal{F}) \xrightarrow{d^0} \check{C}^1(U, \mathcal{F}) \longrightarrow \ldots$$

where $\varepsilon(\varphi)_\alpha = \mathrm{res}_{XU_\alpha}(\varphi)$. Definition I.5.2 shows that if U is an open covering then this complex is exact at terms $\mathcal{F}(X)$ and $\check{C}^0(U, \mathcal{F})$.

Cohomology groups $\check{H}^i(U, \mathcal{F})$ of the complex $\check{C}^\bullet(U, \mathcal{F})$ are called Čech cohomology groups of $\mathcal{F}$ with respect to the covering U. In particular, $\check{H}^0(U, \mathcal{F}) = \Gamma(X, \mathcal{F})$ does not depend on an open covering U.

5. Complex of Singular Chains

Let Y be a topological space, X_n the set of its singular n-simplices (see I.2.4), and $C_n(X, A)$ a simplicial abelian group of its chains with coefficients in an abelian group A. Construct the complex as in I.7.3. Its homology is called the *singular homology* of Y with coefficient in A and is denoted by $H_n^{\mathrm{sing}}(Y, A)$. Similarly one defines singular cohomology.

6. Homology and Cohomology of Groups

Let G be a group, A a left G-module. In I.4.9c) we used A to construct a homological and a cohomological coefficient system on the simplicial set BG.

The homology and cohomology of BG with these coefficient systems are called respectively the *homology* and *cohomology of G with coefficients in A*. They are denoted by $H_n(G, A)$ and $H^n(G, A)$ respectively.

7. The de Rham Complex

Let X be a C^∞-manifold, $C(X)$ the ring of C^∞-functions on X, and $\Omega^i(X)$ the $C(X)$-module of i-forms on X, so that $\Omega^0(X) = C(X)$. The exterior differential $d: \Omega^k(X) \to \Omega^{k+1}(X)$ is defined in local coordinates $(x^1, \ldots, x^n)$ by the formula

$$d\left(\sum_{|I|=k} f_I dx^I\right) = \sum_{|I|=k} \sum_i \frac{\partial f_I}{\partial x^i} dx^i \wedge dx^I,$$

$$I = (i_1, \ldots, i_k), \quad |I| = k, \quad dx^I = dx^{i_1} \wedge \ldots \wedge dx^{i_k}.$$

It is completely determined by the following properties:

a) $d(\omega^k \wedge \omega^l) = d\omega^k \wedge \omega^l + (-1)^k \omega^k \wedge d\omega^l$, $\omega^k \in \Omega^k(X)$, $\omega^l \in \Omega^l(X)$;
b) $d^2 = 0$;
c) $d: C(X) \to \Omega^1(X)$ associates with each function its differential.

Cohomology groups of the complex $\Omega^\bullet(X)$ are called the de Rham cohomology groups of X and are sometimes denoted by $H^\bullet_{DR}(X)$.

This construction allows several important modifications. Associating with an open set $U \subset X$ the linear space $\Omega^k(U)$ and defining the restrictions $\Omega^k(U) \to \Omega^k(V)$ for $V \subset U$ in a natural way we get a sheaf that is denoted by Ω^k. Exterior differentials commute with restrictions, so that we get a complex of sheaves $\Omega^\bullet$.

To get another modification we can consider a complex analytic manifold and holomorphic differential forms on it.

8. Homology and Cohomology of a Lie Algebra

Apply the above construction to the case when X is a connected Lie group G. The group G acts by right translations on $C(G)$ and on $\Omega^k(G)$, the action on $\Omega^k(G)$ being uniquely determined by commutativity with d. Denote by $\Omega^\bullet_{\mathrm{inv}}(G)$ the subcomplex of $\Omega^\bullet(G)$ consisting of G-invariant forms. There exists a purely algebraic description of this complex in terms of the Lie algebra $\mathfrak{g}$ of the group G. Namely, let us represent by $\mathfrak{g}$ the space of right invariant vector fields on G. Then

$$\Omega^k_{\mathrm{inv}}(G) = L\left(\wedge^k \mathfrak{g}, \mathbb{R}\right)$$

where on the right-hand side we have the space of anti-symmetric k-linear forms on $\mathfrak{g}$. The exterior differential of $\omega^k \in \Omega^k(G)$ considered as a polylinear function on vector fields is given by the following formula of E. Cartan:

$$d\omega^k(\xi_1,\ldots,\xi_{k+1}) =$$
$$\sum_{1\leq j<l\leq k+1}(-1)^{j+l-1}\omega^k\left([\xi_j,\xi_l],\xi_1,\ldots,\hat{\xi}_j,\ldots,\hat{\xi}_l,\ldots,\xi_{k+1}\right)$$
$$+\sum_{j=1}^{k+1}(-1)^j\xi_j\left(\omega^k\left(\xi_1,\ldots,\hat{\xi}_j,\ldots,\xi_{k+1}\right)\right).$$

Applying this formula to elements from the subcomplex $\Omega^k_{\text{inv}}(G)$, we get the exterior differential in $C^\bullet(\mathfrak{g}) = \mathcal{L}(\wedge^\bullet\mathfrak{g},\mathbb{R})$,

$$dc(g_1,\ldots,g_{k+1}) =$$
$$\sum_{1\leq j<l\leq k+1}(-1)^{j+l-1}c([g_j,g_l],g_1,\ldots,\hat{g}_j,\ldots,\hat{g}_l,\ldots,g_{k+1}).$$

The cohomology of this complex is denoted by $H^\bullet(\mathfrak{g},\mathbb{R})$. An important point is that this construction does not assume the existence of the group G associated to the algebra $\mathfrak{g}$. In particular, it can be applied to infinite-dimensional Lie algebras.

More generally, let A be a $\mathfrak{g}$-module. Set $C^k(\mathfrak{g},A) = L(\wedge^k\mathfrak{g},A)$ and define the differential by a general Cartan formula,

$$dc(g_1,\ldots,g_{k+1}) =$$
$$\sum_{1\leq j<l\leq k+1}(-1)^{j+l-1}c([g_j,g_l],g_1,\ldots,\hat{g}_j,\ldots,\hat{g}_l,\ldots,g_{k+1})$$
$$+\sum_{j=1}^{k+1}(-1)^j g_j c((g_1,\ldots,\hat{g}_j,\ldots,g_{k+1}).$$

The cohomology groups of this complex are denoted, as you might guess, by $H^\bullet(\mathfrak{g},A)$.

9. Homotopic Mappings of Complexes

In addition to the exact sequence discussed in Sect. I.6, there exists another important technical tool to compute homology: changes of complexes that preserve homology. More generally, let $f_\bullet, g_\bullet : B_\bullet \to C_\bullet$ be two morphisms of chain complexes, and $H_\bullet(f), H_\bullet(g)$ be corresponding morphisms of homologies as in I.6.5. Morphisms $f_\bullet$ and $G_\bullet$ are said to be *homotopic* if there exists a sequence of group homomorphisms $k = (k_n)$, $k_n : B_n \to C_{n+1}$, such that

$$f_n - g_n = k_{n-1}d_n^B + d_{n+1}^C k_n \qquad (I.19)$$

for all n (we do not require that k_n commute with d's). Similarly, two morphisms $f^\bullet, g^\bullet : B^\bullet \to C^\bullet$ of cochain complexes are said to be homotopic if $f^n - g^n = k^{n+1}d_B^n + d_C^{n-1}k^n$ for some group homomorphisms $k^n : B^n \to C^{n-1}$.

10. Lemma. *If $f_\bullet, g_\bullet$ are homotopic, then $H_n(f) = H_n(g)$ for all n.*

Proof. It suffices to verify that if $f_\bullet - g_\bullet$ is homotopic to the zero map of complexes, then $H_n(f - g) = 0$, i.e., $f - g$ takes any cycle $b \in B_n$ into a boundary. But
$$(f - g)b = (kd + dk)b = d(k(b)).$$
□

A similar statement is of course, true for homotopic morphisms of cochain complexes.

11. Theorem. *a) Let $\varphi, \psi : X \to Y$ be two topologically homotopic continuous maps of topological spaces. They induce equal maps of singular homology groups with any coefficients.*

b) Let $\varphi, \psi : X \to Y$ be two smoothly homotopic mappings of C^∞-manifolds. They induce equal mappings of the de Rham cohomology.

12. Corollary. *If X is contractible, that is, if the identity map $\mathrm{id} : X \to X$ is homotopic to the map $X \to \{\text{point}\}$, then $H_0^{\mathrm{sing}}(X, A) = A$, $H_i^{\mathrm{sing}}(X, A) = 0$ for $i > 0$. Similar equalities hold for the de Rham cohomology.*

Proof (of Theorem I.7.11). The proof consists in constructing algebraic homotopy between morphisms $C(\varphi)$ and $C(\psi)$ of corresponding (co)chain complexes.

a) *Singular Homology*. Denote by $C_n^{\mathrm{sing}}(X, A)$, $C_n^{\mathrm{sing}}(Y, A)$ groups of singular chains of spaces X, Y respectively with coefficients in an abelian group A. Let a homotopy between φ and ψ be given by a continuous map $F : X \times I \to Y$ (where I is the segment $[0, 1]$), so that
$$F(x, 0) = \varphi(x), \quad F(x, 1) = \psi(x). \tag{I.20}$$

Denote by $f_\bullet, g_\bullet : C_\bullet^{\mathrm{sing}}(X, A) \to C_\bullet^{\mathrm{sing}}(Y, A)$ morphisms of complexes induced by φ and ψ respectively.

Roughly speaking, an algebraic homotopy $k_n : C_n^{\mathrm{sing}}(X, A) \to C_{n+1}^{\mathrm{sing}}(Y, A)$ associates with any singular n-simplex $s : \Delta_n \to X$ "the prism over s", i.e., a singular $(n+1)$-chain in Y which is an appropriate triangulation of the image of the prism $\Delta_n \times I$ in Y under the mapping $(\lambda, t) \mapsto F(s(\lambda), t), \lambda \in \Delta_n, t \in I$. The formula $dk + kd = f - g$ means now that the boundary of the prism (term dk) consists of side faces (term kd), and of upper (g) and lower (f) bases taken with appropriate signs.

Let us give now a rigorous proof. First, for every p we construct a fixed singular $(p + 1)$-chain δ_p (with integer coefficients) of the space $\Delta_p \times I$. Introducing in Δ_p coordinates $(x_0, \ldots, x_p)$, $x_i \geq 0$, $\sum x_i = 1$, let us consider for every l, $0 \leq l \leq p$ the following singular $(p + 1)$-simplex:
$$\delta_{p,l} : \Delta_{p+1} \longrightarrow \Delta_p \times I : (x_0, \ldots, x_{p+1}) \mapsto$$
$$\left\{ x_0, \ldots, x_{l-1}, x_l + x_{l+1}, x_{l+2}, \ldots, x_{p+1}, \sum_{i=l+1}^{p+1} x_i \right\} \in \Delta_p \times I.$$

I.7 Complexes 51

We remark that the family of simplices $\delta_{p,l}(\Delta_{p+1}) \subset \Delta_p \times I$ defines the triangulation of $\Delta_p \times I$ considered in I.1.5.

We set
$$\delta_p = \sum_{l=0}^{p+1}(-1)^l \delta_{p,l}.$$

To compute the boundary of δ_p, consider in $\Delta_p \times I$ the following p-chains:

$\varepsilon_0 : \Delta_p \longrightarrow \Delta_p \times I, \quad \lambda \mapsto (\lambda, 0) \quad$ (lower base),
$\varepsilon_1 : \Delta_p \longrightarrow \Delta_p \times I, \quad \lambda \mapsto (\lambda, 1) \quad$ (upper base),

$$\delta_{p-1}^{(i)} = \sum_{l=0}^{p}(-1)^l \left(\Delta_{\partial_p^i} \times \mathrm{id}_I\right) \circ \delta_{p-1,l} \quad \text{(triangulation of the i-th side face)}$$

where $\Delta_{\partial_p^i} : \Delta_{p-1} \to \Delta_p$ is the embedding of the i-th face (see 1) and $\delta_{p-1,l} : \Delta_p \to \Delta_{p-1} \times I$ is defined similarly to $\delta_{p,l}$.

We leave to the reader the proof of the following formula:
$$d\delta_p = -\sum_{i=0}^{p}(-1)^i \delta_{p-1}^{(i)} + \varepsilon_1 - \varepsilon_0. \tag{I.21}$$

It can be easily proved by computing coefficients on the left- and right-hand sides for all singular p-simplices entering $d\delta_{p,l}$ for some l. The corresponding picture for $p = 1$ looks like this:

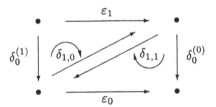

To construct $k_n : C_n^{\mathrm{sing}}(X, A) \to C_{n+1}^{\mathrm{sing}}(Y, A)$ we define it first on each singular n-simplex in X and then extend by linearity. Let $h : \Delta_n \to X$ be a singular simplex in X. Then the singular $(n+1)$-chain $k_n(h)$ in Y is defined as follows:
$$k_n(h) = \sum_{l=0}^{n+1}(-1)^l F \circ (h \times \mathrm{id}_I) \circ \delta_{n,l}.$$

It suffices to verify (I.19) for each singular n-simplex in X; in this case it easily follows from (I.20), (I.21).

b) *Case of the de Rham Cohomology.* Let $F : X \times I \to Y$ be a smooth homotopy, $F(x, 0) = \varphi(x)$, $F(x, 1) = \psi(x)$.

We must construct maps $k^n : \Omega^n(Y) \to \Omega^{n-1}(X)$ such that
$$dk^n \omega + k^{n+1} d\omega = \psi^*(\omega) - \varphi^*(\omega)$$
for any $\omega \in \Omega^n(Y)$.

To construct k^n, we use local coordinates $x^1, \ldots, x^n$ on X. The reader can verify that all constructions are invariant under changes of coordinates, so that they are defined globally on I.

Let $\omega \in \Omega^n(Y)$. Then $F^*(\omega) \in \Omega^n(X \times I)$. Write $F^*(\omega)$ in the form
$$F^*(\omega) = \omega_1 + \omega_2 \wedge dt,$$
where an n-form ω_1 and an $(n-1)$-form ω_2 are locally of the form
$$\omega_1 = \sum_{|I|=n} \alpha_I(x,t) dx^I,$$
$$\omega_2 = \sum_{|J|=n-1} \beta_J(x,t) dx^J.$$

Let
$$\gamma_J(x) = \int_0^1 \beta_J(x,t) dt,$$
$$k^n \omega = (-1)^{n+1} \sum_{|J|=n-1} \gamma_J(x) dx^J \in \Omega^{n-1}(x).$$

Let us verify that k^n satisfies the required conditions. First, the equalities $F(x,0) = \varphi(x)$, $F(x,1) = \psi(x)$ imply
$$\sum_{|I|=n} \alpha_I(x,t) dx^I = \begin{cases} \varphi^*(\omega) & \text{for } t = 0, \\ \psi^*(\omega) & \text{for } t = 1. \end{cases}$$

Using the expression for d in local coordinates, we get
$$dk^n \omega = (-1)^{n+1} \sum_{|J|=n-1} \sum_i \left(\int_0^1 \frac{\partial \beta_J(x,t)}{\partial x^i} \right) dx^i \wedge dx^J. \tag{I.22}$$

Furthermore,
$$F^*(d\omega) = dF^*(\omega) = d\omega_1 + d\omega_2 \wedge dt = \tilde{\omega}_1 + \tilde{\omega}_2 \wedge dt,$$
where
$$\tilde{\omega}_2 = (-1)^n \sum_{|I|=n} \frac{\partial \alpha_I(x,t)}{\partial t} dx^I + \sum_{|J|=n-1} \sum_i \frac{\partial \beta_J(x,t)}{\partial x^i} dx^i \wedge dx^J. \tag{I.23}$$

Computing now $k^{n+1} d\omega$ we see that the expression obtained from the second sum in (I.23) differs from (I.22) by the sign only. Therefore,
$$k^{n+1} d\omega + dk^n \omega = \sum_{|I|=n} \left(\int_0^1 \frac{\partial \alpha_I(x,t)}{\partial t} dt \right) dx^I$$
$$= \sum_{|I|=n} \alpha_I(x,t) dx^I \big|_{t=0}^{t=1} = \psi^*(\omega) - \varphi^*(\omega). \qquad \square$$

Proof (of Corollary I.7.12). It follows from the theorem because
$$H_0^{\text{sing}}(\{\text{point}\}, A) = A, \quad H_i^{\text{sing}}(\{\text{point}\}, A) = 0 \quad \text{for} \quad i > 0$$
(see Ex. 1e) below). □

Exercises

1. Degenerate Chains. Let C be a simplicial abelian group (see I.7.2) and $\tilde{C} = (C_n, d_n)$ the complex defined in I.7.3. Let
$$D_n = \sum_{i=0}^{n-1} \text{Im}\left(C\left(\sigma_{n-1}^i\right) : C_{n-1} \longrightarrow C_n\right).$$

a) Prove that $d_n(D_n) \subset D_{n-1}$.
b) Prove that the complex $\tilde{D} = (D_n, d_n)$ is homotopically trivial, so that
$$H_i(\tilde{C}) \simeq H_i(\tilde{C}/\tilde{D}).$$

c) Let X be a simplicial set and $\mathcal{A}$ be a homological coefficient system on X. Prove that $C_n(X, \mathcal{A})$ with natural definition of simplicial operations (extended by linearity from operations on I and on $\mathcal{A}$) form a simplicial abelian group.

d) Let X be a simplicial set for which the faces of nondegenerate simplices $x \in X_n$ (i.e., $\partial_n^i x \in X_{n-1}$, $0 \leq i \leq n$) are also nondegenerate (such simplicial sets correspond to triangulated spaces, see I.2.7). Then $C_i(X, \mathcal{A})$ contains a subcomplex $N_i(X, \mathcal{A})$ consisting of linear combinations of nondegenerate simplices. Prove that the inclusions $N_i(X, \mathcal{A}) \to C_i(X, \mathcal{A})$ induce an isomorphism
$$H_i(N_\bullet(X, \mathcal{A})) \simeq H_i(X, \mathcal{A}).$$

e) The dimension $\dim X$ of a simplicial set X is the largest n such that X has nondegenerate n-simplices. Prove that for $i > \dim X$ we have $H_i(X, \mathcal{A}) = 0$ for any homological coefficient systems on X.

2. Relations Among Various Notions of Homotopy. At the present point we have three different notions of homotopy: a homotopy of continuous maps of topological spaces $\varphi, \psi : X \to Y$, a homotopy of simplicial maps of simplicial sets $\varphi_\bullet, \psi_\bullet : X_\bullet \to Y_\bullet$ (see problem 1), and a homotopy of morphisms of complexes (see I.7.9). Prove that if $\varphi, \psi : X \to Y$ are homotopic, then the corresponding maps of simplicial sets $\text{sing}(\varphi), \text{sing}(\psi) : \text{sing}(X) \to \text{sing}(Y)$ (see I.2.4) are also homotopic. Prove that if maps of simplicial sets $\varphi_\bullet, \psi_\bullet : X_\bullet \to Y_\bullet$ are homotopic, then the morphism of chain complexes $C(\varphi_\bullet), C(\psi_\bullet) : C_\bullet(X_\bullet, A) \to C_\bullet(Y_\bullet, A)$ are homotopic for any coefficient group A.

Hint: Adjust the proof of Theorem I.7.11a).

3. Hochschild Cohomology. Let k be a commutative unitary ring, A a k-algebra, and M an A-bimodule. Prove that the formulas

$$C_n(A, M) = M \otimes A^{\otimes n} \quad \text{(tensor product over } k\text{),}$$

$$d(m \otimes a_1 \otimes \ldots \otimes a_n) = ma_1 \otimes a_2 \otimes \ldots \otimes a_n$$
$$+ \sum_{i=1}^{n-1} (-1)^i m \otimes a_1 \otimes \ldots \otimes a_i a_{i+1} \otimes \ldots \otimes a_n$$
$$+ (-1)^n a_n m \otimes a_1 \otimes \ldots \otimes a_{n-1},$$

define a chain complex. Its homology is called the *Hochschild homology* of the algebra A with coefficients in the bimodule M, and is denoted by $H_n(A, M)$.

Similarly, the *Hochschild cohomology* $H^n(A, M)$ of the algebra A with coefficients in M is defined as the cohomology of the following cochain complex:

$$C^n(A, M) = \mathrm{Hom}_k(A^{\otimes n}, M),$$

$$df(a_1, \ldots, a_{n+1}) = a_1 f(a_2, \ldots, a_{n+1}) +$$
$$+ \sum_{i=1}^{n} (-1)^i f(a_1, \ldots, a_i a_{i+1}, \ldots, a_{n+1}) +$$
$$+ (-1)^{n+1} f(a_1, \ldots, a_n) a_{n+1}.$$

Compute $H_n(A, A)$ and $H^n(A, A^*)$ (where $A^* = \mathrm{Hom}_k(A, k)$).

4. Cyclic Homology of Algebras. Under the assumptions of the previous problem let $k \supset \mathbb{Q}$ and $M = A$. Define the cyclic shift on $C_\bullet(A, A)$ as follows:

$$t(a_0 \otimes \ldots \otimes a_n) = (-1)^n a_n \otimes a_0 \otimes \ldots \otimes a_{n-1}.$$

Prove that the image of $1 - t$ is a subcomplex in $C_\bullet(A, A)$. (More explicitly, prove that $d(1-t) = (1-t)d'$, where

$$d'(a_0 \otimes \ldots \otimes a_n) = \sum_{i=0}^{n-1} (-1)^i a_0 \otimes \ldots \otimes a_i a_{i+1} \otimes \ldots \otimes a_n.)$$

Cyclic homology $H_n^\lambda(A)$ of the algebra A is the homology of the complex

$$C_n^\lambda(A) = C_n(A, A)/\mathrm{Im}(1-t)$$
$$d^\lambda = d \bmod \mathrm{Im}(1-t).$$

Similarly, *cyclic cohomology* $H_\lambda^n(A)$ is the cohomology of the subcomplex $C_\lambda^n(A) \subset C^\bullet(A, A)$ consisting of t-invariant chains.

5. Koszul complex. Let A be a ring (not necessarily commutative), M a (say) left A-module and $\varphi_1, \ldots, \varphi_p$ be a family of pairwise commuting A-module morphisms $M \to M$. The *Koszul complex*

$$\mathcal{M} = \left\{ 0 \longrightarrow M^{(0)} \xrightarrow{d} M^{(1)} \xrightarrow{d} \cdots \xrightarrow{d} M^{(p)} \longrightarrow 0 \right\}$$

associated with the data $(M, \varphi_1, \ldots, \varphi_p)$ is defined as follows:

$$M^{(k)} = M \underset{\mathbb{Z}}{\otimes} \wedge^k(\mathbb{Z}^p)$$

$$d\left(x \otimes (e_{j_1} \wedge \ldots \wedge e_{j_k})\right) = \sum_{j=1}^{p} \varphi_j(x) \otimes (e_j \wedge e_{j_1} \wedge \ldots \wedge e_{j_k})$$

(where $(e_1, \ldots, e_p)$ is the canonical basis in $\mathbb{Z}^p$).

Prove (by induction in p) the following facts:

a) Let for any $j, 1 \leq j \leq p$, φ_j be surjective as an automorphism of the submodule $\operatorname{Ker} \varphi_1 \cap \ldots \cap \operatorname{Ker} \varphi_{j-1} \subset M$. Then $H^j(\mathcal{M}) = 0$ for $j \neq 0$ and $H^0(\mathcal{M}) = \operatorname{Ker} \varphi_1 \cap \ldots \cap \operatorname{Ker} \varphi_p$.

b) Let for any $j, 1 \leq j \leq p$, the morphism φ_j be injective as a morphism of the quotient module $M/(\operatorname{Im} \varphi_1 + \ldots + \operatorname{Im} \varphi_{j-1})$. Then $H^j(\mathcal{M}) = 0$ for $j \neq p$, and $H^p(\mathcal{M}) = M/(\operatorname{Im} \varphi_1 + \ldots + \operatorname{Im} \varphi_p)$.

If one of the conditions a) or b) is satisfied, then the sequence $(\varphi_1, \ldots, \varphi_p)$ is said to be *regular*. An important special case: A is a commutative ring, and $\varphi_j, 1 \leq j \leq p$, is the multiplication by an element $a_j \in A$.

II. Main Notions of the Category Theory

II.1 The Language of Categories and Functors

1. Definition. *A category $\mathcal{C}$ consists of the following data:*

  a) A class $\mathrm{Ob}\,\mathcal{C}$ whose elements are called objects *of $\mathcal{C}$.*

  b) A collection of sets $\mathrm{Hom}(X, Y)$, one for each ordered pair of objects $X, Y \in \mathrm{Ob}\,\mathcal{C}$, whose elements are called morphisms *(from X to Y); they are denoted by $\varphi : X \to Y$.*

  c) A collection of mappings

$$\mathrm{Hom}(X, Y) \times \mathrm{Hom}(Y, Z) \longrightarrow \mathrm{Hom}(X, Z),$$

one for each ordered triple of objects $X, Y, Z \in \mathrm{Ob}\,\mathcal{C}$. Any mapping in this collection associates with a pair $\varphi : X \to Y$, $\psi : Y \to Z$ a morphism from X to Z, denoted by $\psi \circ \varphi$ or $\psi\varphi : X \to Z$, and called the composition *or* product *of φ and ψ.*

  These data should satisfy the following conditions:

  A) Any morphism φ uniquely determines $X, Y \in \mathrm{Ob}\,\mathcal{C}$ such that $\varphi \in \mathrm{Hom}(X, Y)$. In other words, the sets $\mathrm{Hom}(X, Y)$ are pairwise disjoint.

  B) For any $X \in \mathrm{Ob}\,\mathcal{C}$ there exists the identity morphism $\mathrm{id}_X : X \to X$ of X; it is determined uniquely by the conditions $\mathrm{id}_X \circ \varphi = \varphi$, $\psi \circ \mathrm{id}_X = \psi$ whenever these compositions are defined.

  C) The composition of morphisms is associative:

$$(\xi\psi)\varphi = \xi(\psi\varphi)$$

for any $\varphi : X \to Y$, $\psi : Y \to Z$, $\xi : Z \to U$.

2. About Notations

We shall sometimes write $X \in \mathcal{C}$ instead of $X \in \mathrm{Ob}\,\mathcal{C}$, and $\mathrm{Hom}_{\mathcal{C}}(X, Y)$ or $\mathcal{C}(X, Y)$ instead of $\mathrm{Hom}(X, Y)$. A morphism $\varphi \in \mathrm{Hom}(X, Y)$ may sometimes be called an *arrow* starting at X and ending at Y. The set $\cup_{X, Y \in \mathcal{C}} \mathrm{Hom}(X, Y)$ is denoted by $\mathrm{Mor}(\mathcal{C})$.

Defining properties of a category include manipulations only with morphisms, not with objects. The simplest concrete statement in a category is that some composition of morphisms is equal to some other composition, say

$\psi\varphi = \varphi'\psi'$. Instead of speaking about this identity, it is often more convenient to say that the diagram

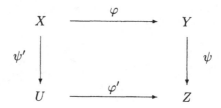

is commutative.

3. The Category of Sets and Categories of Sets with a Structure

A mathematician spends most of his working time in the category *Set* of "all" sets and "all" maps of sets. The composition of morphisms is the composition of maps; the identity morphisms are identity maps. In standard axiomatic set theory (say, Zermelo–Frankel) the collection of all sets forms a class, but not a set, so that some operations with sets are forbidden. Thus the above naive definition of *Set* forbids some categorical constructions we will consider later. The standard way of dealing with the situation is to introduce the Universe, i.e., a large set of sets which is closed under all necessary operations, and to consider only the sets belonging to the Universe. Later in this book we will always assume, whenever necessary, that all the required hygiene regulations are obeyed.

An important class of categories is formed by categories whose objects are sets with some additional structure and whose morphisms are maps respecting this structure. Here are some examples:

- *Top*, the category of topological spaces with continuous maps between them;
- *Diff*, the category of C^∞-manifolds and C^∞-maps;
- *Ab*, the category of abelian groups and homomorphisms;
- *A*-mod, the category of left modules over some fixed ring A;
- *Gr*, the category of groups and homomorphisms.

4. Examples of Categories from Chapter I

a) The category Δ:

$$\mathrm{Ob}\,\Delta = \{[n] \mid n = 0, 1, 2, \ldots\},$$
$$\mathrm{Hom}_\Delta([m], [n]) = \{\text{the set of nondecreasing maps from } \{0, 1, \ldots, m\} \text{ to } \{0, 1, \ldots, n\}\}.$$

b) The category $\Delta°Set$ of simplicial sets:

$$\text{Ob } \Delta°Set = \{\text{simplicial sets}\} \quad (\text{see I.2.1}),$$
$$\text{Hom}(X_\bullet, Y_\bullet) = \{\text{simplicial maps from } X_\bullet \text{ to } Y_\bullet\} \quad (\text{see I.2.11}).$$

c) The category $\Delta°Top$ of simplicial topological spaces:

$$\text{Ob } \Delta°Top = \{\text{simplicial topological spaces}\} \quad (\text{see I.3.2}),$$
$$\text{Hom}(X_\bullet, Y_\bullet) = \{\text{families of continuous maps } \varphi_\bullet = \{\varphi_n\}, \varphi_n : X_n \to Y_n, \text{ such that } Y(f)\varphi_n = \varphi_m X(f) \text{ for any nondecreasing } f : [m] \to [n]\,\}.$$

d) The category $\Delta°\mathcal{C}$ of simplicial objects in a fixed category $\mathcal{C}$. The reader is advised to give the definition by analogy with b) and c).

e) The category *Kom Ab* of complexes of abelian groups:

$$\text{Ob } Kom\ Ab = \{\text{cochain complexes } C^\bullet \text{ of abelian groups}\} \quad (\text{see I.4.3}),$$
$$\text{Hom}(B^\bullet, C^\bullet) = \{\text{morphisms of complexes } f^\bullet : B^\bullet \to C^\bullet\} \quad (\text{see I.6.5}).$$

5. More Examples

In this second group of examples, the objects of the categories are sets with some additional structure (as in II.1.3), but morphisms are defined differently.

a) The category *Toph*:

$$\text{Ob } Toph = \text{Ob } Top = \{\text{topological spaces}\},$$
$$\text{Hom}_{Toph}(X, Y) = \{\text{the set of homotopy classes of continuous mapping from } X \text{ to } Y\}.$$

The category *Toph* is the category one uses most in homotopy theory.

b) The category of relations *Rel*:

$$\text{Ob } Rel = \text{Ob } Set = \{\text{sets in the given Universe}\},$$
$$\text{Hom}_{Rel}(X, Y) = \{\text{subsets of the direct product } X \times Y\}.$$

The composition of $\varphi : X \to Y$ and $\psi : Y \to Z$ is defined as follows:

$$\psi \circ \varphi = \{(x, z) \in X \times Z \mid \text{there exists } y \in Y \text{ such that } (x, y) \in \varphi, (y, z) \in \psi\} \subset X \times Z.$$

The identity morphism is the diagonal

$$\text{id}_X = \{(x, x), x \in X\} \subset X \times X.$$

c) The category *Rel Ab* of additive relations:

$$\text{Ob } Rel\,Ab = \text{Ob } Ab = \{\text{abelian groups}\},$$
$$\text{Hom}_{Rel\,Ab}(X,Y) = \{\text{the set of subgroups in } X \times Y\}.$$

The composition of morphisms and the identity morphism are defined as in b).

The examples of the third group are some classical structures which can be conveniently considered as categories.

d) The category $\mathcal{C}(I)$ of a partially ordered set I:

$$\text{Ob}\,\mathcal{C}(I) = I,$$
$$\text{Hom}_{\mathcal{C}(I)}(i,j) \quad \text{consists of one element if } i \leq j \text{ and is empty otherwise}.$$

The composition of morphisms and the identity morphism are defined in the only possible way.

An important special case of $\mathcal{C}(I)$ is the following one.

e) The category Top_X. Let X be a topological space. Define

$$\text{Ob } Top_X = \{\text{open subsets of } X\},$$
$$\text{Hom}(U,V) \quad \text{consists of the inclusion } U \to V \text{ if } U \subset V,$$
$$\text{Hom}(U,V) \quad \text{is empty if } U \not\subset V.$$

Now we introduce the second most important notion of category theory.

6. Definition. *A functor F from a category $\mathcal{C}$ to a category $\mathcal{D}$ (notation: $F: \mathcal{C} \to \mathcal{D}$) consists of the following data:*

a) A mapping $\text{Ob}\,\mathcal{C} \to \text{Ob}\,\mathcal{D} : X \to F(X)$.
b) A mapping $\text{Mor}\,\mathcal{C} \to \text{Mor}\,\mathcal{D} : \varphi \to F(\varphi)$ such that for $\varphi \in \text{Hom}_{\mathcal{C}}(X,Y)$ we have $F(\varphi) \in \text{Hom}_{\mathcal{D}}(F(X), F(Y))$.

These data should satisfy the following conditions: $F(\varphi\psi) = F(\varphi)F(\psi)$ for any $\varphi, \psi \in \text{Mor}\,\mathcal{C}$ for which $\varphi\psi$ is defined. In particular, $F(id_X) = id_{F(X)}$.

7. Examples of Functors from Chapter I

a) Geometric realization:

$$|\bullet| : \Delta°Set \longrightarrow Top.$$

The values of this functor on objects in $\Delta°Set$ were defined in I.2.2, and on morphisms in $\Delta°Set$ are defined in I.2.15. The equality $|gf| = |f| \cdot |g|$ is clear.

b) Singular simplicial set:

$$\text{Sing} : Top \longrightarrow \Delta°Set.$$

The values of this functor on objects were defined in I.2.4:

$$(\text{Sing } Y)_n = \{\text{the set of singular } n\text{-simplices of } Y\},$$
$$(\text{Sing } Y)(f)(\varphi) = \varphi \circ \Delta_f \text{ where } \Delta_f : \Delta_m \to \Delta_n \text{ is the mapping associated with } f : [m] \to [n] \text{ (see 1), and } \varphi : \Delta_n \to Y.$$

The value of Sing on a morphism (continuous mapping) $a : Y \to Y'$ is defined as the composition: $\text{Sing}(a)$ maps a singular simplex $\varphi : \Delta_n \to Y$ to the singular simplex $a \circ \varphi : \Delta_n \to Y'$. The details are left to the reader.

This example can be used to illustrate the pros and cons of two methods for dealing with morphisms: the one using equalities and the one using commutative diagrams.

c) n-th cohomology group:

$$H^n : \text{Kom Ab} \longrightarrow \text{Ab}.$$

Values of H^n on objects are given in I.4.4, and on morphisms in I.6.5.

d) Classifying space:

$$B : \text{Gr} \longrightarrow \Delta°\text{Set}.$$

The definition of BG is given in 8. The other details are left to the reader.

e) n-skeleton. The mapping $X \to \text{sk}_n X_\bullet$, which associates with each simplicial set $X_\bullet$ its n-skeleton (see I.2.14), determines the functor $\text{sk}_n : \Delta°\text{Set} \to \Delta°\text{Set}$. The jobs of constructing the action of sk_n on morphisms and of checking all the necessary properties are left to the reader.

8. Remarks

The notion of a functor formalizes what was once called the "natural construction", and the examples in II.1.7 present several such natural constructions. In II.1.9 we give more examples of functors. A very special role is played by the functor $h_X : \mathcal{C} \to \text{Set}$ defined for an arbitrary category $\mathcal{C}$ and for an arbitrary $X \in \text{Ob}\,\mathcal{C}$ as follows:

$$h_X(Y) = \text{Hom}_\mathcal{C}(X, Y),$$
$$h_X(f)(\varphi) = f \circ \varphi, \text{ where } f : Y \to Y', \varphi \in \text{Hom}_\mathcal{C}(X, Y).$$

We will return to this functor later in Sect. II.3.

b) The functors $F : \mathcal{C} \to \mathcal{D}$ defined in II.1.7 are sometimes called *covariant functors*. There exist also *contravariant functors*; such a functor G is defined as a pair of mappings $G : \text{Ob}\,\mathcal{C} \to \text{Ob}\,\mathcal{D}, \text{Mor}\,\mathcal{C} \to \text{Mor}\,\mathcal{D}$ such that $G(\varphi) \in \text{Hom}_\mathcal{D}(G(Y), G(X))$ for $\varphi \in \text{Hom}_\mathcal{C}(X, Y)$, and $G(\varphi\psi) = G(\psi)G(\varphi)$. However, it is common nowadays to attribute this "inversion of arrows" to the initial category $\mathcal{C}$.

Formally, define the *dual category* $\mathcal{C}°$ as follows: $\text{Ob}\,\mathcal{C}° = \text{Ob}\,\mathcal{C}$ (but $X \in \text{Ob}\,\mathcal{C}$ considered as an object of $\mathcal{C}°$ will be denoted $X°$); $\text{Hom}_{\mathcal{C}°}(X°, Y°) =$

$\mathrm{Hom}_{\mathcal{C}}(Y, X)$ (with $\varphi : Y \to X$ we associate $\varphi^\circ : X^\circ \to Y^\circ$); finally $\psi^\circ \circ \varphi^\circ = (\varphi\psi)^\circ$, $\mathrm{id}_{X^\circ} = (\mathrm{id}_X)^\circ$.

Now a "contravariant" functor $G : \mathcal{C} \to \mathcal{D}$ is defined as a (covariant) functor $G : \mathcal{C}^\circ \to \mathcal{D}$. As an example, let us view $\mathrm{Hom}_\mathcal{C}$ as a functor of its first argument, $h'_X : \mathcal{C}^\circ \to \mathrm{Set}$:

$$h'_X(Y) = \mathrm{Hom}_\mathcal{C}(Y, X),$$
$$h'_X(f)(\varphi) = \varphi \circ f \quad \text{where} \quad f : Y' \to Y, \; \varphi \in \mathrm{Hom}_\mathcal{C}(Y, X).$$

The notation $\Delta^\circ \mathrm{Set}$, introduced in II.1.4b), reminds us that any simplicial set $X_\bullet$ can be considered as a functor $X_\bullet : \Delta^\circ \to \mathrm{Set}$:

$$X_\bullet([n]) = X_n, \quad X_\bullet(f) = X(f).$$

c) It would be natural to consider the functor $\mathrm{Hom}_\mathcal{C}$ as a functor of two variables. Instead of formally introducing the corresponding definition, we nowadays use the notion of the direct product of categories.

Let, for example, $\mathcal{C}$ and $\mathcal{C}'$ be two categories. Define

$$\mathrm{Ob}(\mathcal{C} \times \mathcal{C}') = \mathrm{Ob}\,\mathcal{C} \times \mathrm{Ob}\,\mathcal{C}',$$
$$\mathrm{Hom}_{\mathcal{C} \times \mathcal{C}'}((X, X'), (Y, Y')) = \mathrm{Hom}_\mathcal{C}(X, Y) \times \mathrm{Hom}_{\mathcal{C}'}(X', Y'),$$
$$(\varphi, \varphi') \circ (\psi, \psi') = (\varphi \circ \varphi', \psi \circ \psi'),$$
$$\mathrm{id}_{(X, X')} = (\mathrm{id}_X, \mathrm{id}_{X'}).$$

It is easy to check that $\mathcal{C} \times \mathcal{C}'$ is a category. In a similar way one can define the product of any indexed family of categories. A functor in several variables is, by definition, a functor on the corresponding product of categories. We have already seen two examples:

$$\mathrm{Hom}_\mathcal{C} : \mathcal{C}^\circ \times \mathcal{C} \longrightarrow \mathrm{Set},$$

and bisimplicial sets $X_{\bullet\bullet}$ (see I.3.3)

$$X_{\bullet\bullet} : \Delta^\circ \times \Delta^\circ \longrightarrow \mathrm{Set}.$$

d) By composing (set-theoretically) two functors $\mathcal{C} \xrightarrow{F} \mathcal{D} \xrightarrow{G} \mathcal{E}$ we get the functor $\mathcal{C} \xrightarrow{G \circ F} \mathcal{E}$. The identity map $\mathrm{id}_\mathcal{C} : \mathrm{Ob}\,\mathcal{C} \to \mathrm{Ob}\,\mathcal{C}$, $\mathrm{Mor}\,\mathcal{C} \to \mathrm{Mor}\,\mathcal{C}$ is a functor. So we can consider any set of categories as a new category with functors as new morphisms.

9. More Examples of Functors

a) Any presheaf of abelian groups on a topological space X is a functor

$$F : \mathrm{Top}_X \longrightarrow \mathrm{Ab}$$

(the category Top_X was defined in II.1.5).

b) Let I be a partially ordered set, $\mathcal{C}(I)$ the corresponding category (see II.1.5). A functor $G : \mathcal{C}(I) \to \mathrm{Ab}$ is a family of abelian groups

$\{G(i) | i \in I\}$ and of maps $g_{ij} : G_i \to G_j$, one for each ordered pair $i \leq j$; these maps should satisfy the conditions $g_{jk}g_{ij} = g_{ik}$ for $i \leq j \leq k$, $g_{ii} = \mathrm{id}_{G_i}$. Such families usually appear as raw material for inductive and/or projective limits.

c) *Forgetful Functors.* One can obtain a lot of functors in the following way: just forget one (or several) structures on objects of the initial category. Among the examples of these functors are the "set of elements" functors

$$\mathrm{Top},\ \mathrm{Diff},\ \mathrm{Ab},\ \mathrm{Gr} \longrightarrow \mathrm{Set},$$

as well as the functors

$$\mathrm{Diff} \longrightarrow \mathrm{Top},\ A\text{-mod} \longrightarrow \mathrm{Ab}.$$

The last definition of this section is that of a morphism of functors (it is sometimes called a "natural transformation of natural constructions").

10. Definition. *Let F, G be two functors from $\mathcal{C}$ to $\mathcal{D}$. A morphism of functors from F to G (notation $f : F \to G$) is a family of morphisms in $\mathcal{D}$:*

$$f(X) : F(G) \longrightarrow G(X),$$

one for each $X \in \mathrm{Ob}\,\mathcal{C}$, satisfying the following condition: for any morphism $\varphi : X \to Y$ in $\mathcal{C}$ the diagram

$$\begin{array}{ccc} F(X) & \xrightarrow{f(X)} & G(X) \\ F(\varphi) \downarrow & & \downarrow G(\varphi) \\ G(X) & \xrightarrow{f(Y)} & G(Y) \end{array}$$

is commutative.

The composition of morphisms of functors, as well as the identity morphism of functors, are defined in the obvious way. As a result, functors from $\mathcal{C}$ to $\mathcal{D}$ form a category, which is usually denoted by $\mathrm{Funct}(\mathcal{C}, \mathcal{D})$.

11. Examples from Chapter I

a) If simplicial sets $X_\bullet, Y_\bullet \in \Delta^\circ\mathrm{Set}$ are considered as functors $X, Y : \Delta^\circ \to \mathrm{Set}$ (see II.1.7b)), then any simplicial map $f : X_\bullet \to Y_\bullet$ is the same as a morphism of these functors.

b) Consider the category *Exc* (exact sequences of complexes) whose objects are exact triples S of complexes of abelian groups:

$$S : 0 \longrightarrow A^\bullet \xrightarrow{i^\bullet} B^\bullet \xrightarrow{p^\bullet} C^\bullet \longrightarrow 0$$

(see I.5.5) and morphisms are commutative diagrams of the form

$$\begin{array}{ccccccccc} 0 & \to & A^\bullet & \to & B^\bullet & \to & C^\bullet & \to & 0 \\ & & \downarrow f^\bullet & & \downarrow g^\bullet & & \downarrow h^\bullet & & \\ 0 & \to & \tilde{A}^\bullet & \to & \tilde{B}^\bullet & \to & \tilde{C}^\bullet & \to & 0 \end{array} \qquad \text{(II.1)}$$

where $f^\bullet, g^\bullet, h^\bullet$ are morphisms of complexes. Fix an integer n and consider two functors

$$F(S) = H^n(C^\bullet), \quad G(S) = H^{n+1}(A^\bullet).$$

The connecting homomorphism $\delta^n(S)$ defined in Theorem I.6.7 (there denoted $\delta^n(i^\bullet, p^\bullet)$) yields a morphism of functors

$$\delta^n : F \longrightarrow G.$$

To check this one must prove that for any morphism $(f^\bullet, g^\bullet, h^\bullet) : S \to \tilde{S}$ in the category Exc (see (II.1)) the diagram

$$\begin{array}{ccc} H^n(C^\bullet) & \xrightarrow{\delta^n(S)} & H^{n+1}(A^\bullet) \\ H^n(h^\bullet) \downarrow & & \downarrow H^{n+1}(f^\bullet) \\ H^n(\tilde{C}^\bullet) & \xrightarrow{\delta^n(\tilde{S})} & H^{n+1}(\tilde{A}^\bullet) \end{array}$$

commutes. The proof uses the explicit construction of $\delta^n(S)$ (see I.6.7) and is left to the reader.

12. Several Definitions

Finally, we give some more definitions that will be constantly used throughout this book.

A category $\mathcal{C}$ is called a *subcategory* of a category $\mathcal{D}$ if

a) $\text{Ob}\,\mathcal{C} \subset \text{Ob}\,\mathcal{D}$;
b) $\text{Hom}_\mathcal{C}(X, Y) \subset \text{Hom}_\mathcal{D}(X, Y)$ for any $X, Y \in \text{Ob}\,\mathcal{C}$;
c) The composition of morphisms in $\mathcal{C}$ coincides with their composition in $\mathcal{D}$; for $X \in \text{Ob}\,\mathcal{C}$ the identity morphism id_X in $\mathcal{C}$ coincides with the identity morphism id_X in $\mathcal{D}$.

A subcategory $\mathcal{C}$ is said to be *full* if $\text{Hom}_\mathcal{C}(X, Y) = \text{Hom}_\mathcal{D}(X, Y)$ for any $X, Y \in \text{Ob}\,\mathcal{C}$.

A functor $f : \mathcal{C} \to \mathcal{D}$ is said to be *faithful* if for any $X, Y \in \text{Ob}\,\mathcal{C}$ the map

$$F : \text{Hom}_\mathcal{C}(X, Y) \longrightarrow \text{Hom}_\mathcal{D}(FX, FY)$$

is injective, and *full* if this map is surjective.

In particular, an embedding of a full subcategory is a fully faithful functor. One can prove that, conversely, any fully faithful functor can be obtained in such a way (see Ex. II.2.1).

An object α of a category $\mathcal{C}$ is said to be an *initial object* if $\mathrm{Hom}_{\mathcal{C}}(\alpha, X)$ is an one-element set for any $X \in \mathrm{Ob}\,\mathcal{C}$. Similarly, an object ω is said to be a *final object* in $\mathcal{C}$ if $\mathrm{Hom}_{\mathcal{C}}(X, \omega)$ is an one-element set for any $X \in \mathrm{Ob}\,\mathcal{C}$. Clearly, both an initial and a final object (if they exist) are determined uniquely up to an isomorphism.

13. Example. In the category *Set* the empty set is the initial object and any one-element set is a final object.

Exercises

1. Another Definition of a Category. Let us consider a class Mor with the following structures:

a) Maps $\alpha, \omega : \mathrm{Mor} \to \mathrm{Mor}$;
b) A composition law $\mathrm{Mor} \times \mathrm{Mor} \to \mathrm{Mor}$, $(g, f) \to g \circ f$, defined for pairs (g, f) with $\omega(f) = \alpha(g)$.

Let the following conditions be satisfied:

c) $\omega^2 = \omega$, $\alpha^2 = \alpha$, $\omega\alpha = \alpha$, $\alpha\omega = \omega$;
d) $\alpha(g \circ f) = \alpha(f)$, $\omega(g \circ f) = \omega(g)$;
e) $h \circ (g \circ f) = (h \circ g) \circ f$;
f) if $f = \alpha(f) = \omega(f)$, then $f \circ g = g$, $g \circ f = g$, provided these compositions are defined.

Show that this definition of a category is equivalent to the usual one in the following sense. For a category $\mathcal{C}$ let $\mathrm{Mor} = \mathrm{Mor}\,\mathcal{C}$, $\alpha(f : X \to Y) = \mathrm{id}_X$, $\omega(f : X \to Y) = \mathrm{id}_Y$. Show also that any Mor as above is obtained from a unique category $\mathcal{C}$.

2. Polycategory. Let J be a nonempty set; one can imagine J as the set of coordinate directions in the J-dimensional vector space. By a J-category we understand a class Mor with the following structures:

a) *Mappings* $\alpha_j, \omega_j : \mathrm{Mor} \to \mathrm{Mor}$, $j \in J$;
b) *Composition laws* $\underset{j}{\circ} : \mathrm{Mor} \times \mathrm{Mor} \to \mathrm{Mor}$; $G \underset{j}{\circ} F$ is defined for all pairs (F, G) such that $\omega_j(F) = \alpha_j(G)$.

These data should satisfy the following conditions:

c) *Relations between α and ω:*

$$\begin{array}{ll} \text{for } j \neq k & \text{for } j = k \\ \omega_j \omega_k = \omega_k \omega_j, & \omega_j \omega_j = \alpha_j \omega_j = \omega_j, \\ \alpha_j \alpha_k = \alpha_k \alpha_j, & \alpha_j \alpha_j = \omega_j \alpha_j = \alpha_j. \\ \alpha_j \omega_k = \omega_k \alpha_j, & \end{array}$$

d) *Relations between* α, ω *and the composition*:

for $j \neq k$

$$\alpha_j(G \underset{k}{\circ} F) = \alpha_j(G) \underset{k}{\circ} \alpha_j(F),$$

$$\omega_j(G \underset{k}{\circ} F) = \omega_j(G) \underset{k}{\circ} \omega_j(F),$$

for $j = k$

$$\alpha_j(G \underset{j}{\circ} F) = \alpha_j(F),$$

$$\omega_j(G \underset{j}{\circ} F) = \omega_j(F).$$

e) *Associativity*:
three morphisms in one direction:

$$H \underset{k}{\circ} \left(G \underset{k}{\circ} F\right) = \left(H \underset{k}{\circ} G\right) \underset{k}{\circ} F, \qquad F \longrightarrow G \longrightarrow H \atop k$$

four morphisms in two directions:

$$\left(G \underset{k}{\circ} F\right) \underset{j}{\circ} \left(K \underset{k}{\circ} H\right) = \left(G \underset{j}{\circ} K\right) \underset{k}{\circ} \left(F \underset{j}{\circ} H\right),$$

$$\begin{array}{c} j\uparrow \quad F \quad \uparrow \quad G \\ \longrightarrow \quad \longrightarrow \\ \uparrow \quad H \quad \uparrow \quad K \\ \longrightarrow \quad \longrightarrow_k \end{array}$$

f) *Identities*: if $F = \alpha_j(F) = \omega_j(F)$ then

$$G \underset{j}{\circ} F = G, \quad F \underset{j}{\circ} G = F,$$

provided these compositions are defined. (Let us note that, as follows from c), for any morphism H the morphism $F = \alpha_j(H)$ satisfies $F = \alpha_j(F) = \omega_j(F)$ and similarly for $F' = \omega_j(H)$.)

Elements of Mor are called, naturally, (poly)morphisms.

The following property can be called the multi-dimensional associativity: *the product of any set of morphisms does not depend on the order*. More precisely, let $j_1, \ldots, j_n \in J$ be a finite subset, $j_a \neq j_b$ for $a \neq b$, and $m_1, \ldots, m_n$ be some positive integers. Let for any set of integers $x = (x_1, \ldots, x_n)$, $1 \leq x_r \leq m_r$, an element $F(x_1, \ldots, x_n) \in$ Mor be given such that

$$\omega_{j_r}(F(x_1, \ldots, x_r, \ldots, x_n)) = \alpha_{j_r}(F(x_1, \ldots, x_r + 1, \ldots, x_n))$$

for $r = 1, \ldots, n$, $1 \leq x_r \leq m_r - 1$. Such a family of morphisms is called *composable*. Prove that:

a) One can define the composition $\underset{x}{\circ} F(x)$ over all x using only pairwise products.

b) The result does not depend on the order of these pairwise products.

Each particular way to compute $\underset{x}{\circ} F(x)$ is something like a "multi-dimensional bracket distribution in the parallelotope of the size $m_1 \times \ldots \times m_n$". What is the total number of ways? (For $n = 1$ this number is called the m_1-th Catalana number.)

II.1 The Language of Categories and Functors

3. Category of Categories as a $\{1,2\}$-Category. Let $\mathcal{C}$ be a category whose objects are such categories $\mathcal{D}$ that $\operatorname{Ob}\mathcal{D}$ is a set, and whose morphisms are functors. Let

$$\operatorname{Mor} = \{(F \circ f) | F : X \longrightarrow Y, f \in \operatorname{Mor}(X); X, Y \in \operatorname{Ob}\mathcal{C}\}$$

and then

$$a_1(F, f) = (\operatorname{Id}_{\text{beginning of } F}, f), \quad \omega_1(F, f) = (\operatorname{Id}_{\text{end of } F}, f),$$
$$a_2(F, f) = (F, \operatorname{id}_{\text{beginning of } f}), \quad \omega_2(F, f) = (F, \operatorname{id}_{\text{end of } f}).$$
$$(G, g = F(f)) \underset{1}{\circ} (F, f) = (G \circ F, f),$$
$$(F, g) \underset{2}{\circ} (F, f) = (F, g \circ f).$$

Verify that the conditions of Ex. II.1.2 are satisfied.

4. Ordered Sets as Categories. Let P, P' be two partially ordered sets, $\mathcal{C}, \mathcal{C}'$ be the corresponding categories (see 5d)).

a) Prove that functors $F : \mathcal{C} \to \mathcal{C}'$ correspond to nondecreasing mappings $f : P \to P'$ (i.e. $p_1 \le p_1 \Rightarrow f(p_1) \le f(p_2)$).

b) Prove that if $F_1, F_2 : \mathcal{C} \to \mathcal{C}'$ are two functors then there exists at most one morphism $F_1 \to F_2$; such a morphism exists if and only if the mappings $f_1, f_2 : P \to P'$ corresponding to F_1, F_2 satisfy the condition $f_1(p) \le f_2(p)$ for all $p \in P$.

In the following exercises we describe some categories whose objects are finite sets. First of all we mention the category $\mathcal{F}\text{Set}$ of all finite sets and all their maps. A more interesting example is given by the simplicial category Δ (see 4a)).

5. The Category Σ. Objects of Σ are finite sets $[n] = \{0, \dots, n\}$ for all $n \ge 0$. Next, for $[m], [n] \in \operatorname{Ob} \Sigma$ let

$$\operatorname{Hom}_\Sigma([m], [n]) = \left\{ \begin{array}{l} \text{the set of pairs } \varphi = \langle f, \sigma \rangle, \text{ where } f : [m] \to [n], \sigma \\ \text{is a collection of complete orderings of all the sets} \\ f^{-1}(i), i \in [n] \end{array} \right\}.$$

Composition is defined as follows. Let $\varphi = \langle f, \sigma \rangle \in \operatorname{Hom}_\Sigma([m], [n])$, $\psi = \langle g, \tau \rangle \in \operatorname{Hom}_\Sigma([n], [p])$. Define $\psi \circ \varphi = \langle g \circ f, \rho \rangle$ where $i <_\rho j$ if either $f(i) \ne f(j)$ and $f(i) <_\tau f(j)$ or $f(i) = f(j)$ and $i <_\sigma j$.

Prove that Σ is a category (i.e., prove the existence of the identity morphisms and the associativity of the composition).

6. The Cyclic Category Λ. a) Define on the set $[n]$ the standard cyclic ordering $0 < 1 < 2 < \dots < n < 0$ (it might be convenient to represent $[n]$ as a subset of the circle $S^1 = [z \in \mathbf{C}, |z| = 1\}$ identifying $k \in [n]$ with the root of unity $e^{2\pi i k/(n+1)}$). The category Λ is defined as a subcategory of Σ:

$$\operatorname{Ob} \Lambda = \operatorname{Ob} \Sigma;$$

68   II. Main Notions of the Category Theory

$\mathrm{Hom}_\Lambda([m],[n])$ consists of those $\varphi = \langle f, \sigma \rangle \in \mathrm{Hom}_\Sigma([m],[n])$ which have the cyclic ordering on $[m]$ induced by the standard cyclic ordering on $[n]$ and by the ordering σ coinciding with the standard cyclic ordering on $[m]$.

Prove that Λ is a category. Prove that for a given $f : [m] \to [n]$ there exists of at least one pair $\langle f, \sigma \rangle \in \mathrm{Hom}_\Lambda([m], [n])$ if and only if f is cyclically nondecreasing, i.e., the condition $i \leq j \leq k \leq i$ in $[m]$ implies $f(i) \leq f(j) \leq f(k) \leq f(i)$ in $[n]$. Prove that if f is nonconstant then σ is uniquely determined by f, and if f is a constant mapping then there exist $n+1$ choices for σ.

b) *Other Definitions of Λ.* Prove that Λ can be defined in either one of the following two ways:

(i) $\mathrm{Hom}_\Lambda([m],[n])$ is the set of homotopy classes of continuous cyclically nondecreasing mappings $F : S^1 \to S^1$ such that $F([m]) = [n]$. (This condition should not be violated during a homotopy.)

(ii) Morphisms in Λ are given by generators

$$\partial_n^i : [n-1] \longrightarrow [n], \quad \sigma_n^i : [n+1] \longrightarrow [n], \quad 0 \leq i \leq n, \quad \tau : [n] \longrightarrow [n]$$

and relations

$$\partial_{n+1}^j \partial_n^i = \partial_{n+1}^i \partial_n^{j-1} \quad \text{for} \quad i < j;$$

$$\sigma_n^j \sigma_{n+1}^i = \sigma_n^i \sigma_{n+1}^{j+1} \quad \text{for} \quad i \leq j;$$

$$\sigma_{n-1}^j \partial_n^i = \begin{cases} \partial_{n-1}^i \sigma_{n-2}^{j-1} & \text{for} \quad i < j, \\ \mathrm{id}_{[n-1]} & \text{for} \quad i = j, j+1, \\ \partial_{n-1}^{i-1} \sigma_{n-2}^j & \text{for} \quad i > j+1, \end{cases}$$

$$\tau_n \partial_n^i = \partial_n^{i-1} \tau_{n-1} \quad \text{for} \quad i = 1, 2, \ldots, n,$$

$$\tau_n \sigma_n^i = \sigma_n^{i-1} \tau_{n+1} \quad \text{for} \quad i = 1, 2, \ldots, n,$$

$$\tau_n \partial_n^0 = \partial_n^n, \quad \tau_n \sigma_n^0 = \sigma_n^n, \quad (\tau_n)^{n+1} = \mathrm{id}_{[n]}.$$

Here ∂_n^i misses the value i, σ_n^i takes the value i twice, $\tau_n(j) = j+1$, and $\tau_n(n) = 0$. The ordering on the preimages should be prescribed only for σ_0^0: it is $0 < 1$).

c) *Self-duality of Λ.* Let us assign to a morphism $\varphi = \langle f, \sigma \rangle \in \mathrm{Hom}_\Lambda([m], [n])$ the morphism $\varphi^* = \langle g, \tau \rangle \in \mathrm{Hom}_\Lambda([n], [m])$ as follows:

$g(i)$ is the σ-minimal element in $f^{-1}(j)$, where j is the maximum (in the standard cyclic ordering in $[n]$) element from $f([m])$ satisfying the condition $j \leq i$.

The ordering τ should be defined only in the case when g is a constant mapping: this happens exactly when f is constant and in this case τ is uniquely determined by the requirement that $f([m])$ is the τ-minimum element in $[n]$. Prove that the mappings $[n] \mapsto [n]$, $\phi \mapsto \phi^*$ determine a functor from Λ to the dual category Λ° which gives an equivalence $* : \Lambda \to \Lambda^\circ$.

d) Prove that the simplicial category Δ (see 4a) is a subcategory of Λ.

e) Similarly to the construction of simplicial sets, and, more generally, of simplicial objects of any category $\mathcal{C}$ starting from the simplicial category Δ (as functors $\Delta^\circ \to Set$ or $\Delta^\circ \to \mathcal{C}$), one can define cyclic objects as functors $\Lambda^\circ \to \mathcal{C}$. The definition of cyclic homology and cohomology (see Ex. I.7.4) is based on cyclic objects.

II.2 Categories and Structures, Equivalence of Categories

1. Isomorphisms

Many mathematical problems are what are called classification problems: classification of simple groups, of singularities, etc. In the next two sections we will discuss categorical aspects of these problems. Usually classification means classification up to an (or the) isomorphism.

2. Definition. *a) A morphism $\varphi : X \to Y$ in a category $\mathcal{C}$ is said to be* an isomorphism *if there exists a morphism $\psi : Y \to X$ such that $\varphi \circ \psi = \mathrm{id}_Y$, $\psi \circ \varphi = \mathrm{id}_X$.*

b) Objects X, Y in a category $\mathcal{C}$ are said to be isomorphic *if there exists at least one isomorphism between them.*

The reader can easily check that the relation "to be isomorphic" is an equivalence relation in $\mathrm{Ob}\,\mathcal{C}$. Two morphisms $\psi : X \to Y$ and $\psi : Y \to X$ with the properties indicated in the definition above are called mutually inverse. The inverse of a given isomorphism is determined uniquely.

Applying the above notion to the category *Funct* $(\mathcal{C}, \mathcal{D})$ of functors (see II.1.10) we obtain the important notion of an isomorphism of two functors $F : \mathcal{C} \to \mathcal{D}$ and $G : \mathcal{C} \to \mathcal{D}$: it is a morphism of functors $\varphi : F \to G$ admitting the inverse morphism $\psi : G \to F$, $\psi\varphi = \mathrm{id}_F$, $\varphi\psi = \mathrm{id}_G$.

The reader can easily check that the existence of the inverse morphism $\psi : G \to F$ in this definition can be replaced by a more natural condition: for any $X \in \mathrm{Ob}\,\mathcal{C}$ the morphism $\varphi(X) : F(X) \to G(X)$ is an isomorphism.

A meaningful example of an isomorphism of functors is "double dualization" as described below.

3. Example. Let k be a field and *Vect*$_k$ the category of linear spaces over k with linear mappings as morphisms. Let *Vect*$_k^f$ be the full subcategory of *Vect*$_k$ formed by the finite-dimensional vector spaces. Let $*$ be the dualization functor

$$* : \mathit{Vect}_k \longrightarrow (\mathit{Vect}_k)^\circ;$$

it is defined on objects by

$$*(L) = \mathrm{Hom}_{Vect_k}(L, k) = \{\text{the space of linear functionals on } L\},$$

and transforms a morphism $\varphi : L \to M$ into

$$*(\varphi) : \mathrm{Hom}(M, k) \longrightarrow \mathrm{Hom}(L, k)$$
$$(f : M \longrightarrow k) \longmapsto (f \circ \varphi : L \longrightarrow k).$$

Instead of writing $*(L)$ and $*(\varphi)$ we will write, as usual, L^* and φ^*. The fact that $*$ is a contravariant functor is expressed by the important formula

$$(\varphi \circ \psi)^* = \psi^* \circ \varphi^*.$$

We also have $*(Vect_k^f) \subset (Vect_k^f)^\circ$.

Now we note that $*$ can be considered also as a functor from $(Vect)^\circ$ to $Vect_k$,

$$* : (Vect_k)^\circ \longrightarrow Vect_k.$$

(A very cautious reader may feel that the distinction between these two functors should be reflected in the notation.) The composition of these two dualizations gives the functor

$$** : Vect_k \longrightarrow Vect_k.$$

This functor is isomorphic to the identity functor when restricted to the subcategory $Vect_k^f$. More explicitly, we have a morphism of functors

$$\mathrm{Id}_{Vect_k} = \mathrm{Id} \longrightarrow **$$

given by

$$L \longrightarrow L^{**}, \; l \; \mapsto \; (l \text{ considered as a functional on the space of}$$
$$\text{functionals on } L).$$

It is well known that $L \to L^{**}$ is an isomorphism for a finite-dimensional L, so that the morphism $\mathrm{Id} \to **$ is an isomorphism on $Vect_k^f$. However, it definitely ceases to be an isomorphism on the whole of $Vect_k$. Indeed, if L is a space with a countably infinite basis, then L^*, and hence L^{**}, has uncountable bases.

4. A Useless Notion: Isomorphism of Categories

In an attempt to apply the definition from II.2.2 to the category of categories (see II.1.7d)), one would obtain the following notion. *An isomorphism between categories* $\mathcal{C}$ *and* $\mathcal{D}$ *is given by a pair of functors* $F : \mathcal{C} \to \mathcal{D}$, $G : \mathcal{D} \to \mathcal{C}$, *such that* $FG = \mathrm{Id}_\mathcal{D}$, $GF = \mathrm{Id}_\mathcal{C}$. Contrary to expectations, this

more or less useless, the main reason being that neither of the requirements $FG = \mathrm{Id}_\mathcal{D}$, $GF = \mathrm{Id}_\mathcal{C}$ is realistic. Whenever we apply two natural constructions to an object, the most we can ask for is to get a new object which is canonically isomorphic to the old one; it would be too much to hope for the new object to be identical to the old one. An illuminating example is the double dualization.

A much more useful notion is given by the following definition.

5. Definition. *a) A functor $F : \mathcal{C} \to \mathcal{D}$ is said to be an* equivalence of categories *if there exists a functor $G : \mathcal{D} \to \mathcal{C}$ such that the functor GF is isomorphic to $\mathrm{Id}_\mathcal{C}$, and the functor FG is isomorphic to $\mathrm{Id}_\mathcal{D}$.*

b) Categories $\mathcal{C}$ and $\mathcal{D}$ are said to be equivalent *if there exists an equivalence $F : \mathcal{C} \to \mathcal{D}$.*

In the situation of a), one sometimes says that the functor G is quasi-inverse *to F.*

6. Example. Let Vect^n_k be the full subcategory of Vect_k consisting of all n-dimensional vector spaces over k, and let $\mathcal{V}^n_k$ be the category with one object k^n and all linear mappings of k^n to itself as morphisms. There exists an obvious inclusion functor $\mathcal{V}^n_k \to \mathit{Vect}^n_k$, and it is an equivalence of categories.

This example displays the following typical features:

a) Equivalent categories have "the same" isomorphism classes of objects and "the same" morphisms between these classes.
b) A quasi-inverse functor is usually not unique, and to define such a functor one often must use the axiom of choice (in our example one must choose a basis in every n-dimensional vector space).

In proving that a given functor is an equivalence of categories the following result is sometimes useful.

7. Theorem. $F : \mathcal{C} \to \mathcal{D}$ *is an equivalence of categories if and only if:*

a) F is a fully faithful functor.
b) Any object $Y \in \mathrm{Ob}\,\mathcal{D}$ is isomorphic to an object of the form $F(X)$ for some $X \in \mathrm{Ob}\,\mathcal{C}$.

Proof. ($\Rightarrow$). Let F be an equivalence of categories and $G : \mathcal{D} \to \mathcal{C}$ be a quasi-inverse functor. Let

$$f(X) : GFX \longrightarrow X, \quad X \in \mathrm{Ob}\,\mathcal{C},$$
$$g(Y) : FGY \longrightarrow Y, \quad Y \in \mathrm{Ob}\,\mathcal{D}.$$

be isomorphisms of functors $f : GF \to \mathrm{Id}_\mathcal{C}$, $g : FG \to \mathrm{Id}_\mathcal{D}$. First note that an object $Y \in \mathrm{Ob}\,\mathcal{D}$ is isomorphic to the object FX, where $X = GY \in \mathrm{Ob}\,\mathcal{C}$; this proves b).

Next, for any $\varphi \in \mathrm{Hom}_{\mathcal{C}}(X, X')$ the diagram

$$\begin{array}{ccc} GFX & \xrightarrow{f(X)} & X \\ {\scriptstyle GF(\varphi)}\downarrow & & \downarrow{\scriptstyle \varphi} \\ GFX' & \xrightarrow{f(X')} & X' \end{array}$$

is commutative. Therefore, φ can be reconstructed from $F(\varphi)$ by the formula

$$\varphi = f(X') \circ GF(\varphi) \circ (f(X))^{-1} \tag{II.2}$$

so that F is a faithful functor. Similarly, G is a faithful functor. To prove that F is fully faithful, consider an arbitrary morphism $\psi \in \mathrm{Hom}_{\mathcal{D}}(FX, FX')$, and set

$$\varphi = f(X') \circ G(\psi) \circ (f(X))^{-1} \in \mathrm{Hom}_{\mathcal{C}}(X, X').$$

Then (see (II.2)) $\varphi = f(X') \circ GF(\varphi) \circ (f(X))^{-1}$, and $G(\psi) = GF(\varphi)$ because $f(X), f(X')$ are isomorphisms. Since G is a faithful functor, $\psi = G(\varphi)$, so that F is fully faithful.

($\Leftarrow$). Assume a) and b) are satisfied. For any $Y \in \mathrm{Ob}\,\mathcal{D}$ we fix $X_Y \in \mathrm{Ob}\,\mathcal{C}$ and an isomorphism $g(Y): FX_Y \to Y$. To construct a functor $G: \mathcal{D} \to \mathcal{C}$ which is quasi-inverse to F, we set $GY = X_Y$ for $Y \in \mathrm{Ob}\,\mathcal{D}$. Next define G on morphisms $\psi \in \mathrm{Hom}_{\mathcal{D}}(Y, Y')$ by

$$G(\psi) = g(Y')^{-1} \circ \psi \circ g(Y) \in \mathrm{Hom}_{\mathcal{D}}(FGY, FGY') = \mathrm{Hom}_{\mathcal{C}}(GY, GY')$$

(the last equality follows from a)). It is easy to check that G is a functor and that $g = \{g(Y)\}: FG \to \mathrm{Id}_{\mathcal{D}}$ is an isomorphism of functors. Finally, $g(FX): FGFX \to FX$ is an isomorphism for any $X \in \mathrm{Ob}\,\mathcal{C}$. Therefore, a) implies that $g(FX) = F(f(X))$ for a unique isomorphism $f(X): GFX \to X$. An easy calculation shows that $f: \{f(X)\}: GF \to \mathrm{Id}\,\mathcal{C}$ is an isomorphism of functors. Therefore G is quasi-inverse to F. $\square$

A meaningful theorem about an equivalence of two categories can often be interpreted as giving two complementary descriptions of mathematical object. We will try to illustrate this using as examples several small mathematical theories that can be thought of as equivalence theorems.

8. Galois Theory

Let k be a field; for simplicity assume that $\mathrm{char}\,k = 0$. Denote by G the Galois group of the algebraic closure $\bar{k}/k$ with the Krull topology. A substantial part of Galois theory can be summarized as follows:

II.2 Categories and Structures, Equivalence of Categories

The dual category of commutative semisimple k-algebras $(k\text{-Alg})^\circ$ is equivalent to the category of finite topological G-sets G-Set.

Let us give some explanations.

a) A finite-dimensional commutative k-algebra L is semisimple if it contains no nonzero nilpotent elements. Any such algebra is isomorphic to a direct sum of finite algebraic extensions of the field k, $L \simeq \oplus J_i$.

b) The Krull topology on G is defined as follows: a neighbourhood basis of the identity element in G consists of subgroups $G_K = \{g \in G \mid \forall x \in K, gx = x\}$, where K runs over all finite extensions of k inside $\bar{k}$.

c) A functor $F : G\text{-Set} \to (k\text{-Alg})^\circ$ that establishes the equivalence is constructed as follows:

$$F(S) = Map_G(S, \bar{k}) = \text{the algebra of } G\text{-equivariant } \bar{k}\text{-valued functions on } S \text{ such that } f(gs) = g(f(s)) \text{ for all } s \in S, g \in G.$$

The action of the functor F on morphisms is the natural one: for $\varphi : T \to S$ the morphism $F(\varphi) : F(S) \to F(T)$ transforms a function f on S into the function $f \circ \varphi$ on T. The algebra $F(S)$ consists of $\bar{k}$-valued functions, and, therefore, does not contain nilpotent elements. Its dimension over k is finite because the intersection of stabilizers of all the points in S, $H = \{g \in G \mid gs = s, \forall s \in S\}$, is open in G, so that the field $\bar{k}^H$ has finite degree over k, and the values of equivariant functions on S lie in this field.

It is clear that F transforms disjoint unions into direct sums. If S is an irreducible G-Set (i.e., with one G-orbit), then the k-algebra $F(S)$ is isomorphic to the field $\bar{k}^H$, where H is the stabilizer of some point $s \in S$. This implies that any object of k-Alg is isomorphic to an object of the form $F(S)$. We won't explain the full faithfulness of the functor F; this would mean reproducing a substantial part of Galois theory. We will only say that the quasi-inverse functor $F' : (k\text{-Alg})^\circ \to G\text{-Set}$ is given on objects by

$$F'(L) = \text{Hom}_{k\text{-Alg}}(L, \bar{k})$$

with the action of an element $g : \bar{k} \to \bar{k}$ being the composition $L \to \bar{k} \xrightarrow{g} \bar{k}$. The definition of F' on morphisms also uses this composition.

9. Fundamental Group à la Poincaré

Let X be a path-connected Hausdorff topological space with a distinguished point $x_0 \in X$. A covering is a morphism $p : Y \to X$ of topological spaces satisfying two conditions:

a) p is a local homeomorphism; that is, any point $y \in Y$ has a neighbourhood U such that the restriction of p to U is a homeomorphism.

b) p satisfies the path extension condition: for any map $g : [0,1] \to X$ with $g(0) = x_0$ and for any $y_0 \in Y$ over x_0 ($p(y_0) = x_0$) there exists a lifting $\tilde{g} : [0,1] \to Y$ such that $\tilde{g}(0) = y_0, p \circ \tilde{g} = g$.

A morphism φ of one covering $p_1 : Y_1 \to X$ into another covering $p_2 : Y_2 \to X$ is a commutative diagram,

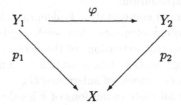

So we have defined the category Cov_X of coverings of X.

On the other hand, denote by $\pi_1(X, x_0)$ the group of homotopy classes of closed loops $g : [0, 1] \to X$, $g(0) = g(1) = x_0$, and let π_1-Set be the category of all left $\pi_1(X, x_0)$-sets. The theory of covering spaces can be summarized as follows:

The category Cov_X is equivalent to the category π_1-Set.

Quasi-inverse functors establishing this equivalence can be defined as follows:

$$Cov_X \longrightarrow \pi_1\text{-}Set \; : \; \begin{cases} \text{(a covering } p : Y \to X) \to \text{(the fibre } p^{-1}(x_0) \text{ with the obvious action of} \\ \pi_1(X, x_0). \\ \text{(morphism of coverings)} \to \text{(induced} \\ \text{mapping of fibres } p_1^{-1}(x_0) \to p_2^{-1}(x_0)). \end{cases}$$

$$\pi_1\text{-}Set \longrightarrow Cov_X \; : \; \begin{cases} (\pi_1\text{-set } S) \to \tilde{X} \underset{\pi_1}{\times} S \\ \text{(morphism of } \pi_1\text{-sets)} \to \text{(morphism of} \\ \text{fibre products)}. \end{cases}$$

Here $\tilde{X}$ is the universal covering, i.e. the space of homotopy classes of paths starting at x_0, and $\tilde{X} \underset{\pi_1}{\times} S$ is the fibre product, i.e. the quotient space of the direct product $\tilde{X} \times S$ (with the discrete topology in S) by the equivalence

$$(\tilde{x}, s) \sim (g\tilde{x}, gs), \quad g \in \pi_1(X, x_0).$$

10. Commutative Banach Algebras

A commutative Banach algebra is a commutative $\mathbb{C}$-algebra A with the unit 1 and with the norm $\|x\|$ making A into a complete Banach space; the product in A is required to be continuous with respect to this norm. An involution in A is an anti-linear homomorphism $* : A \to A$ with $(X^*)^* = x$, $\|x\| = \|x^*\|$, $\|xx^*\| = \|x\|^2$ for any $x \in A$. Define the category Ban as follows:

II.2 Categories and Structures, Equivalence of Categories

$$\text{Ob } Ban = \{\text{commutative Banach algebras with involution}\},$$

$$\text{Hom}_{Ban}(A, A') = \{\text{algebra homomorphisms } \varphi : A \to A' \text{ respecting the norm and the involution}\}.$$

On the other hand, denote by *Haus* the full subcategory of the category *Top* consisting of all compact Hausdorff spaces. One of main results in the theory of commutative Banach algebras can be stated as follows:

Categories Ban° and Haus are equivalent.

The corresponding quasi-inverse functors $Ban° \underset{G}{\overset{F}{\rightleftarrows}} Haus$ can be constructed as follows:

$$GX = \{\text{the ring of continuous functions on } X \text{ with the norm } \|f\| = \max_{x \in X} |f(x)| \text{ and with the involution } f^*(x) = \overline{f(x)}\}.$$

To define F, we introduce the space of maximal ideals in a commutative Banach algebra A. The ideal $m \subset A$ is said to be *maximal* if it is not contained in any other proper ideal in A. Such an m is necessarily closed, and $A/m = \mathbb{C}$. Let $\mathcal{M}(A)$ be the set of maximal ideals invariant under the involution in A. Any element $a \in A$ determines a complex-valued function f_a on $\mathcal{M}(A) = f_a(m) = \{\text{the class of } a \text{ in } A/m = \mathbb{C}\}$. Endow $\mathcal{M}(A)$ with the weakest topology such that all functions $f_a, a \in A$ are continuous. With this topology $\mathcal{M}(A)$ becomes a locally compact Hausdorff space. Define $F : Ban° \to Haus$ by $F(A) = \mathcal{M}(A)$.

The theory of commutative Banach algebras says that the functors F and G are quasi-inverse. In particular, the equivalence of functors $\text{Id}_{Ban} \simeq GF$ means that each commutative Banach algebra with involution is isomorphic to the algebra of continuous functions on the space of its maximal ideals.

11. Pontryagin Duality

Let $\mathcal{C}$ be the following category:

$$\text{Ob } \mathcal{C} = \{\text{commutative locally compact groups}\}$$
$$\text{Hom}_{\mathcal{C}}(A_1, A_2) = \{\text{continuous group homomorphisms } A_1 \to A_2\}.$$

The Pontryagin duality theorem can be formulated as follows:

The category $\mathcal{C}$ is equivalent to the dual category $\mathcal{C}°$.

The equivalence is established via the theory of characters. Let S be the group of complex numbers with modulus 1 (under multiplication). A character of A is an element $\chi \in \text{Hom}_{\mathcal{C}}(A, S)$, i.e., a continuous homomorphism

$A \to S$. Pointwise multiplication makes the set of all characters of A into a group denoted by $\hat{A}$. Supply $\hat{A}$ with the topology of uniform convergence on compact subsets. Any continuous homomorphism $\varphi : A_1 \to A_2$ determines a continuous homomorphism $\hat{\varphi} : \hat{A}_2 \to \hat{A}_1$ by the formula

$$\hat{\varphi}(\chi)(x_1) = \chi(\varphi(x_1)), \quad x_1 \in A_1, \quad \chi \in \hat{A}_2.$$

So we obtain the functor $F : \mathcal{C} \to \mathcal{C}^\circ$:

$$FA = (\hat{A})^\circ, \quad F(\varphi) = (\hat{\varphi})^\circ.$$

Let $F^\circ : \mathcal{C}^\circ \to \mathcal{C}$ be the dual functor, i.e.,

$$F^\circ(A^\circ) = \hat{A}, \quad A^\circ \in \mathrm{Ob}\,\mathcal{C}^\circ, \quad \hat{A} \in \mathrm{Ob}\,\mathcal{C}.$$

A somewhat more precise form of the duality theorem says that *the functors $F : \mathcal{C} \to \mathcal{C}^\circ$, $F^\circ : \mathcal{C}^\circ \to \mathcal{C}$ are quasi-inverse.*

12. Final Remarks

The examples in II.2.8–II.2.11 show several typical patterns of the category theory. The usual way to learn patterns is to consider a lot of examples; however, we will try to formulate several rules.

   a) A good classification theorem should describe not only objects, but also morphisms of these objects. From this point of view the complete theory of semisimple Lie algebra should contain not only Cartan classification (in terms of root systems), but also the theory of finite-dimensional representations of these algebras (H. Weyl's highest weight theory), as well as some other problems of Lie algebra theory.

   b) The interrelations between algebra and geometry are given by a system of functors of the type

$$\begin{aligned} \text{spaces} &\longleftrightarrow \text{rings of functions}, \\ \text{spaces} &\longleftrightarrow \text{cohomology rings}, \\ \text{sheaves} &\longleftrightarrow \text{modules over rings of functions}. \end{aligned}$$

Good categorical properties of such functors (e.g., equivalence) are so important that to save them one is often forced to change old structures or to introduce new ones. This is how affine schemes, nuclear vector spaces, rigid analytic spaces, and objects of derived categories appeared in mathematics.

   c) The formal inversion of arrows furnished in the definition of dual category in concrete examples often produces *duality theorems* (when categories coincide or are close to each other), or some relations of the type *geometry vs algebra* (when categories are not alike, e.g., rings and their spectra).

Exercises

1. Full Subcategories. Prove that any fully faithful functor $F : \mathcal{C} \to \mathcal{D}$ determines an equivalence of the category $\mathcal{C}$ with a full subcategory in $\mathcal{D}$.

2. Quotient Categories.
Let $\mathcal{C}$ be a category and let for any pair of objects X, Y of $\mathcal{C}$ an equivalence relation $\sim = \sim_{X,Y}$ in $\mathrm{Hom}_\mathcal{C}(X, Y)$ be given. Then there exists a category $\mathcal{D} = \mathcal{C}/\sim$ and a functor $Q : \mathcal{C} \to \mathcal{D}$ such that:

(i) if $f \sim f'$ in $\mathcal{C}$ then $Qf = Qf'$ in $\mathcal{D}$;
(ii) if $F : \mathcal{C} \to \mathcal{D}'$ is a functor such that $Ff = Ff'$ for any $f \sim f'$ then there exists a unique functor $G : \mathcal{D} \to \mathcal{D}'$ such that $F = G \circ Q$.

Sketch of the Proof. Let us assume first that the family of equivalence relations $\sim_{X,Y}$ satisfies the following condition: $f \sim_{X,Y} f'$ implies $afb \sim_{U,V} af'b$ for any $a : Y \to V$, $b : U \to X$. Then define $\mathcal{D} = \mathcal{C}/\sim$ as follows: $\mathrm{Ob}\,\mathcal{D} = \mathrm{Ob}\,\mathcal{C}$, $\mathrm{Hom}_\mathcal{D}(X, Y) = \mathrm{Hom}_\mathcal{C}(X, Y)/\sim$ with the natural composition, $Q : \mathcal{C} \to \mathcal{D}$ is the natural functor (identity on objects). In the general case $\sim_{X,Y}$ generates the weakest equivalence relation $\approx_{X,Y}$ satisfying the above condition and we set $\mathcal{C}/\sim = \mathcal{C}/\approx$. Note that in any case $\mathrm{Ob}\,\mathcal{D} = \mathrm{Ob}\,\mathcal{C}$.

3. Morita Equivalence.
a) An object X of a category $\mathcal{C}$ is said to be a *generator* if $h'_X : Y \mapsto \mathrm{Hom}_\mathcal{C}(X, Y)$ is a faithful functor from $\mathcal{C}$ to Set (i.e., for any $Y_1, Y_2 \in \mathrm{Ob}\,\mathcal{C}$ the map $h'_X : \mathrm{Hom}_\mathcal{C}(Y_1, Y_2) \to \mathrm{Hom}_\mathrm{Set}(h'_X Y_1, h'_X Y_2)$ is an embedding).

b) Morita's theorem says that for two rings A and B the following conditions are equivalent:

(i) Categories A-mod and B-mod are equivalent.
(ii) Categories mod-A and mod-B are equivalent.
(iii) There exist a finitely generated projective generator P of mod-A and a ring isomorphism $B \cong \mathrm{End}_A P$.

Proof of (iii) $\Rightarrow$ (ii): an equivalence of mod-A and mod-B is given by the functor $h_P : X \mapsto \mathrm{Hom}_{\mathrm{mod}-A}(P, X)$, and the inverse equivalence is given by the functor $h_{P^*} : Y \mapsto \mathrm{Hom}_{\mathrm{mod}-B}(P^*, Y)$, where $P^* = \mathrm{Hom}_{\mathrm{mod}-A}(P, A)$.

Rings A and B are said to be *Morita equivalent* if they satisfy the equivalent conditions (i)–(iii). Example: matrix rings $M_n(k)$ (k is a fixed field) for different n are Morita equivalent.

c) A property T of a ring A is said to be *Morita invariant* if for any two Morita equivalent rings it is always either satisfied or not satisfied for both of them. The following properties of a ring A are Morita invariant: A is simple; A is semisimple; A is finite; each right ideal in A is projective; each right ideal in A is injective. The following properties of a ring A are not Morita invariant: A does not have zero divisors; A is a field; A is commutative; each projective A-module is free.

d) The *centre* $Z_\mathcal{C}$ of a category $\mathcal{C}$ is the ring of all morphisms of the identity functor $\mathrm{Id}_\mathcal{C} : \mathcal{C} \to \mathcal{C}$. Prove that the centre of the category mod-A is isomorphic to the centre of the ring A. Therefore two commutative rings are Morita equivalent if and only if they are isomorphic.

II.3 Structures and Categories. Representable Functors

1. What Should We Do?

We have to learn how to treat an object of a category as if this object were a set endowed with some structure. We have to be able to define the direct product or the limit of a projective system of objects, to define what one would call a group object, and so on. In classical constructions we use the fact that objects are composed of elements (points), and that these points can be processed in various manners: one can form pairs or sequences, choose elements with a given property, etc.

To work with an abstract category $\mathcal{C}$ we must either describe set-theoretic constructions in the diagram language and transfer the result to $\mathcal{C}$, or else we must find substitutes in the category language for the notions of points or elements.

We describe the second approach first. It is based on the simple but useful remark that any set X in the category Set can be identified with the set $\mathrm{Hom}_{Set}(e, X)$, where e is a one-point set. In an arbitrary category $\mathcal{C}$ an analogue of e does not necessarily exist. However, by considering instead $\mathrm{Hom}_{\mathcal{C}}(X, Y)$ for *all Y simultaneously*, we can recover complete information about the object X (up to an isomorphism). In this language morphisms $\varphi : Y \to X$ are sometimes called *Y-points* of an object X, and $\mathrm{Hom}_{\mathcal{C}}(Y, X)$ is denoted by $X(Y)$.

To be more precise we introduce the category of functors (see II.1.10)

$$\hat{\mathcal{C}} = Funct(\mathcal{C}^\circ, Set),$$

and consider the functor $h_X : \mathcal{C}^\circ \to Set$ defined by $h_X(Y^\circ) = \mathrm{Hom}_{\mathcal{C}}(Y, X)$ (see II.1.7b)) as an object of $\hat{\mathcal{C}}$.

2. Definition. *A functor $F \in \mathrm{Ob}\,\hat{\mathcal{C}}$ is said to be representable if it is isomorphic to a functor of the form h_X for some $X \in \mathrm{Ob}\,\mathcal{C}$; one says also that the object X represents the functor F.*

Let $\varphi : X_1 \to X_2$ be a morphism in $\mathcal{C}$. We associate with φ the morphism of functors $h_\varphi : h_{X_1} \to h_{X_2}$ defined by the property that, for any $Y \in \mathrm{Ob}\,\mathcal{C}$, the associated map

$$h_\varphi(Y) : h_{X_1}(Y) \longrightarrow h_{X_2}(Y)$$

sends a morphism $\theta \in \mathrm{Hom}_{\mathcal{C}}(Y, X_1) = h_{X_1}(Y)$ into the composition $\varphi \circ \theta \in \mathrm{Hom}_{\mathcal{C}}(Y, X_2) = h_{X_2}(Y)$. It is clear that $h_{\varphi\psi} = h_\varphi h_\psi$.

3. Theorem. *In the above notations, the map $\varphi \mapsto h_\varphi$ yields an isomorphism of sets*

$$\mathrm{Hom}_{\mathcal{C}}(X, Y) \xrightarrow{\sim} \mathrm{Hom}_{\hat{\mathcal{C}}}(h_X, h_Y).$$

Moreover, this map is an isomorphism of functors, $\mathcal{C}° \times \mathcal{C} \to \text{Set}$, in two variables X and Y. Therefore, the functor $h : \mathcal{C} \to \hat{\mathcal{C}}$ defined by $h(X) = h_X$, $h(\varphi) = h_\varphi$, yields an equivalence of $\mathcal{C}$ with the full subcategory of $\hat{\mathcal{C}}$ formed by representable functors.

4. Corollary. *If a functor $F \in \text{Ob}\,\hat{\mathcal{C}}$ is representable, then the representing object in $\mathcal{C}$ is defined uniquely up to a unique isomorphism.*

Proof (of Theorem 3). Define a map
$$i : \text{Hom}_{\hat{\mathcal{C}}}(h_X, h_Y) \longrightarrow \text{Hom}_{\mathcal{C}}(X, Y)$$
which associates with a morphism of functors $h_X \to h_Y$ the image of $\text{id}_X \in h_X(X)$ in $h_Y(X) = \text{Hom}_{\mathcal{C}}(X, Y)$ under the map $h_X(X) \to h_Y(X)$ determined by this morphism of functors. Let us check that the maps $\varphi \mapsto h_\varphi$ and i are mutually inverse.

a) $i(h_\varphi) = h_\varphi(\text{id}_X) = \varphi$ by the definition of h_φ.

b) On the other hand, let $g : h_X \to h_Y$ be a morphism of functors. It is determined by the family of morphisms $g(Z) : h_X(Z) \to h_Y(Z)$ for all $Z \in \text{Ob}\,\mathcal{C}$. By definition, $i(g) = g(X)(\text{id}_X)$, and we must show that
$$h_{i(g)}(Z) = g(Z). \tag{II.3}$$
Now, $h_{i(g)}(Z) : h_X(Z) \to h_Y(Z)$ maps a morphism $\varphi : Z \to X$ to the composition $i(g) \circ \varphi : Z \to Y$. Therefore, it suffices to show that
$$g(Z)(\varphi) = i(g) \circ \varphi.$$
Using the commutativity of the diagram (see II.1.10)

$$\begin{array}{ccc} h_X(X) & \xrightarrow{g(X)} & h_Y(X) \\ h_X(\varphi) \downarrow & & \downarrow h_Y(\varphi) \\ h_X(Z) & \xrightarrow{g(Z)} & h_Y(Z) \end{array}$$

we can send the element $\text{id}_X \in h_X(X)$ into $h_Y(Z)$ in two different ways. Following the upper path (via $h_Y(X)$) we map this element first to $i(g)$ and then to $i(g) \circ \varphi$. Following the lower path (via $h_X(Z)$) we map it first to $h_X(\varphi)(\text{id}_X) = \varphi$ and then to $g(Z)\varphi$. Formula (II.3) is proved.

So we have proved that the functor h is fully faithful. This implies immediately that h determines an equivalence of $\mathcal{C}$ with the image of $\hat{\mathcal{C}}$. The other assertions in this theorem are trivial. □

5. Example (Direct and Fibre Products). Let us recall that the direct product $X \times Y$ of two sets X, Y is defined as the set of all ordered pairs $(x, y), x \in X, y \in Y$. We give two definitions of the direct product of two objects $X, Y \in \text{Ob}\,\mathcal{C}$ for an arbitrary category $\mathcal{C}$ and show that they are equivalent.

a) The direct product $X \times Y$ "is" the object Z representing the functor $U \mapsto \{$direct product of sets $X(U) \times Y(U)\}$ (if this functor is representable).

b) The direct product $X \times Y$ "is" the object Z together with two projection morphisms $X \xleftarrow{p_X} Z \xrightarrow{p_Y} Y$ such that for any pair of morphism $X \xleftarrow{p'_X} Z' \xrightarrow{p'_Y} Y$ there exists a unique morphism $q : Z' \to Z$ with $p'_X = p_X \circ q, p'_Y = p_Y \circ q$ (again, if (Z, p_X, p_Y) with such properties exists). This second definition is the result of translating the set-theoretic construction into the language of category Set.

a) $\Rightarrow$ b) Assume that the functor $U \mapsto X(U) \times Y(U)$ is represented by an object Z, and consider the collection of set isomorphisms,

$$\mathrm{Hom}_{\mathcal{C}}(U, Z) \xrightarrow{\sim} \mathrm{Hom}_{\mathcal{C}}(U, X) \times \mathrm{Hom}_{\mathcal{C}}(U, Y),$$

defining an isomorphism of this functor with $U \mapsto Z(U)$. Take $U = Z$. In the left-hand set we have the distinguished element id_Z; let its image in the right-hand set be $(p_X, p_Y) \in X(U) \times Y(U)$. We claim that the diagram $X \xleftarrow{p_X} Z \xrightarrow{p_Y} Y$ satisfies the property in definition b), and leave the job of checking this to the reader.

b) $\Rightarrow$ a) Using the diagram $X \xleftarrow{p_X} Z \xrightarrow{p_Y} Y$ we construct the following morphism of functors (when applied to U):

$$\begin{array}{ccc} \mathrm{Hom}_{\mathcal{C}}(U, Z) & \longrightarrow & \mathrm{Hom}_{\mathcal{C}}(U, X) \times \mathrm{Hom}_{\mathcal{C}}(U, Y) \\ \cup\!\!\!\cup & & \cup\!\!\!\cup \\ q & \longmapsto & (p_X \circ q, p_X \circ q) \end{array}$$

The universality of the diagram $X \xleftarrow{p_X} Z \xrightarrow{p_Y} Y$ means that this mapping is a bijection for any U. Therefore the morphism of functors is an isomorphism.

Every proof of the equivalence of two definitions (the "structural" one and the "diagrammatical" one) is similar to the one given above. Let us note also that setting $Z = X = Y$, $p'_X = p'_Y = \mathrm{id}$ in b) we get the so-called *diagonal morphism* $\delta : X \to X \times X$.

An easy generalization of this construction allows us to give a categorical version of the fibre product. To do this we recall that if $\varphi : X \to S$, $\psi : Y \to S$ are two maps of sets, then the *fibre product of X and Y over S* is the following set of pairs:

$$X \underset{S}{\times} Y = \{(x, y) \in X \times Y \mid \varphi(x) = \psi(y)\} \subset X \times Y.$$

This notion generalizes several set-theoretical constructions:

1) the usual direct product (if S is a point);
2) the intersection of subsets (if φ, ψ are embeddings);
3) the preimage of a subset under the mapping φ (ψ is the embedding of a subset).

II.3 Structures and Categories. Representable Functors

As before, one can give two definitions of $X \underset{S}{\times} Y$ in an arbitrary category $\mathcal{C}$:

a') $X \underset{S}{\times} Y$ represents the functor $U \mapsto X(U) \underset{S(U)}{\times} Y(U)$.

b') $X \underset{S}{\times} Y$ "is" the usual direct product in the new category $\mathcal{C}_S$ whose objects are morphisms $\varphi : X \to S$ in $\mathcal{C}$ and whose morphisms are commutative diagrams

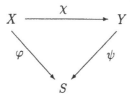

(χ is a morphism in $\mathcal{C}_S$ of $X \xrightarrow{\varphi} S$ to $Y \xrightarrow{\psi} S$). The diagram b) in the category $\mathcal{C}_S$ is represented in the category $\mathcal{C}$ by the diagram

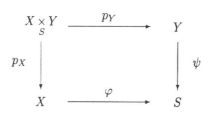

which satisfies the obvious universality condition, and is called a *cartesian square*.

6. Arrow Inversion

Given any categorical construction, we can create the dual construction by making the initial construction in the dual category $\mathcal{C}^\circ$ and interpreting the result in $\mathcal{C}$; in other words, we invert all arrows in the original construction. In such a way one can obtain the *amalgamated sum* (or *coproduct, cocartesian square*) $X \underset{S}{\coprod} Y$ corresponding to a diagram of the form $X \xleftarrow{\varphi} S \xrightarrow{\psi} Y$:

The universality property of this square is expressed as follows: for any diagram $X \xrightarrow{j_X} Z \xleftarrow{j_Y} Y$ such that $j_X \varphi = j_Y \psi$ there exists a unique morphism $X \coprod_S Y \xrightarrow{g} Z$ such that $j_X = q \circ i_X, j_Y = q \circ i_Y$. Later in the book we will often skip such explanations.

7. Example (Tensor Product of A-Algebras). Any morphism $\varphi : A \to B$ in the category of commutative rings with unit can be considered as an A-algebra structure on B. Let B, C be two A-algebras. Consider the diagram

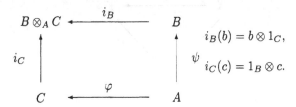

$$i_B(b) = b \otimes 1_C,$$
$$i_C(c) = 1_B \otimes c.$$

This diagram is a cocartesian square. It becomes a cartesian square in the dual category, called *the category of affine schemes*. Despite its simplicity, this is one of the most fundamental constructions in algebraic geometry.

8. Example (Groups in a Category). Let $\mathcal{C}$ be a category and $X \in \text{Ob}\mathcal{C}$.

9. Definition. *A group structure on an object X is a family of group structures on all sets $h_X(Y) = \text{Hom}_{\mathcal{C}}(Y, X)$; these structures should be compatible in the following sense:*

For any morphism $\varphi : Y_1 \to Y_2$ in $\mathcal{C}$ the corresponding mapping $h_X(\varphi) : h_X(Y_2) \to h_X(Y_1)$ is a group homomorphism.

An object X with a group structure is often called a group in the category $\mathcal{C}$. A morphism $X_1 \to X_2$ in $\mathcal{C}$ is said to be a morphism of the corresponding groups *if all mappings $h_{X_1}(Y) \to h_{X_2}(Y), Y \in \text{Ob}\mathcal{C}$, are group homomorphisms.*

10. Groups in a Category: The Diagram Definition

In the second definition of a group structure on an object one deals only with X, and not with all objects of $\mathcal{C}$.

We assume that $\mathcal{C}$ contains:

a) the final object E;
b) all the products $X \times X, X \times X \times X$ (see II.3.18b) and Ex. II.3.7 below).

A group structure on the object X is determined by three morphisms in $\mathcal{C}$:

$$\begin{aligned} m &: X \times X \longrightarrow X & \text{(multiplication)}, \\ i &: X \longrightarrow X & \text{(inversion)}, \\ e &: E \longrightarrow X & \text{(identity)}, \end{aligned}$$

satisfying the following conditions:

II.3 Structures and Categories. Representable Functors

Associativity: the diagram

$$\begin{array}{ccc} X \times X \times X & \xrightarrow{(m, \mathrm{id}_X)} & X \times X \\ {\scriptstyle (\mathrm{id}_X, m)}\downarrow & & \downarrow {\scriptstyle m} \\ X \times X & \xrightarrow{m} & X \end{array}$$

is commutative.

Left inversion: the diagram

$$\begin{array}{ccc} X \times X & \xrightarrow{(i, \mathrm{id}_X)} & X \times X \\ {\scriptstyle \delta}\uparrow & & \downarrow {\scriptstyle m} \\ X & \xrightarrow{} E \xrightarrow{e} & X \end{array}$$

is commutative (here δ is the diagonal morphism, as in II.3.5).

Left identity: the diagram

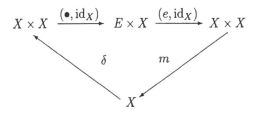

is commutative.

The reader might find it very useful to check the equivalence of the two definitions of group in a category.

11. Example (Affine Group Schemes). Let Alg be the category of commutative unitary algebras. The dual category $\mathit{Aff} = (Alg)^\circ$ is called (by algebraic geometers) the *category of affine schemes*. The object of Aff corresponding to an algebra $A \in \mathrm{Ob}\, Alg$ is denoted by $\mathit{Spec}\, A$. We study how to construct a group structure on an affine scheme $\mathit{Spec}\, A$.

12. Definition. *A structure of* bialgebra *on an algebra A is given by three algebra homomorphisms:*

$$\begin{array}{rll} \mu & : A \longrightarrow A \otimes A & \text{(comultiplication)}, \\ \iota & : A \longrightarrow A & \text{(coinversion)}, \\ \varepsilon & : A \longrightarrow \mathbb{Z} & \text{(coidentity)} \end{array}$$

(the tensor product is taken over $\mathbb{Z}$) satisfying the following conditions:

Coassociativity: *the diagram*

$$
\begin{array}{ccc}
A \otimes A \otimes A & \xleftarrow{\mu \otimes id_A} & A \times A \\
{\scriptstyle id_A \otimes \mu} \uparrow & & \uparrow {\scriptstyle \mu} \\
A \otimes A & \xleftarrow{\mu} & A
\end{array}
$$

is commutative.

Left coinversion: *the diagram*

$$
\begin{array}{ccc}
A \otimes A & \xleftarrow{\iota \otimes id_A} & A \otimes A \\
\downarrow & & \uparrow {\scriptstyle \mu} \\
A & \longleftarrow \mathbb{Z} \xleftarrow{\varepsilon} & A
\end{array}
$$

is commutative (the left vertical arrow is the multiplication: $a \otimes b \mapsto ab$, the left lower horizontal arrow is induced by the embedding of the identity: $a \mapsto a \cdot 1$).

Left coidentity: *the diagram*

$$
\begin{array}{ccc}
A \otimes A & \longleftarrow & A \\
\downarrow & & \uparrow {\scriptstyle \varepsilon \otimes id_A} \\
A & \xrightarrow{\mu} & A \otimes A
\end{array}
$$

is commutative (the left arrow is the multiplication; the upper arrow is given by $a \mapsto 1 \otimes a$).

A bialgebra is said to be commutative if $\mu = s \circ \mu$, where $s : A \otimes A \to A \otimes A$ is the permutation of two factors.

Using the description of cartesian squares in the category **Aff** (see II.3.7) we see that group structures on an affine scheme $X = Spec\, A$ are in one-to-one correspondence with the bialgebra structures on the algebra A.

13. Cartier Duality

Let us remark that the structure of commutative algebra on an abelian group A can be defined via the following homomorphisms of abelian groups:

$$\bar{\mu} : A \otimes A \longrightarrow A, \quad \bar{\mu}(a \otimes b) = ab,$$

$$\bar{\varepsilon} : \mathbb{Z} \longrightarrow A, \quad \bar{\varepsilon}(m) = m \cdot 1_A.$$

with the commutativity and associativity axioms for $\bar{\mu}$ and the identity axiom for $\bar{\varepsilon}$. Therefore, the corresponding structure of commutative bialgebra is described by homomorphisms of abelian groups

$$\begin{array}{ccccc} A \otimes A & \xrightarrow{\bar{\mu}} & A & \xrightarrow{\mu} & A \otimes A, \\ \mathbb{Z} & \xrightarrow{\bar{\varepsilon}} & A & \xrightarrow{\varepsilon} & \mathbb{Z}, \\ & & A & \xrightarrow{\iota} & A, \end{array} \qquad (\text{II.4})$$

together with various axioms. One part of the axioms is contained in II.3.13; another part can be obtained from it by inverting the arrows and substituting $\bar{\mu}$ and $\bar{\varepsilon}$ for μ and ε; an additional axiom requires $\mu : A \to A \otimes A$ to be a homomorphism of algebras with the multiplications $\bar{\mu}$ and $\bar{\mu} \otimes \bar{\mu}$ respectively.

Now let A be a free abelian group of finite type and $A^* = \mathrm{Hom}_{\mathbb{Z}}(A, \mathbb{Z})$. Identifying $(A \otimes A)^*$ with $A^* \otimes A^*$ and $\mathbb{Z}^*$ with $\mathbb{Z}$, we obtain a bialgebra structure on A^* by dualizing the diagrams in (II.4):

$$\begin{array}{ccccc} A^* \otimes A^* & \xrightarrow{\mu^*} & A^* & \xrightarrow{\bar{\mu}^*} & A^* \otimes A^*, \\ \mathbb{Z} & \xrightarrow{\varepsilon^*} & A^* & \xrightarrow{\bar{\varepsilon}^*} & \mathbb{Z}, \\ & & A^* & \xrightarrow{\iota^*} & A^*. \end{array}$$

The bialgebra axioms for A^* trivially follow from those for A.

The group scheme $X^* = \mathrm{Spec}\, A^*$ is called the *Cartier dual* to the commutative group scheme $X = \mathrm{Spec}\, A$.

14. Relative Group Schemes

One can relativize all the discussion in II.3.11–II.3.13, considering algebras A over a fixed commutative ring K. All we need to do is to replace $\mathbb{Z}$ by K, and to assume that all mappings are homomorphisms of K-modules and all tensor products are taken over K.

15. Limits

Taking limits (projective, injective, etc.) can be interpreted as a universal functor.

We fix a category $\mathcal{J}$ (it is often called an index category). We shall always assume $\mathcal{J}$ to be small (Ob $\mathcal{J}$ is a set); sometimes $\mathcal{J}$ is finite (has a finite number of objects and of morphisms). We recall that $\text{Funct}(\mathcal{J}, \mathcal{C})$ denotes the category of all functors $F : \mathcal{J} \to \mathcal{C}$ (see II.1.10).

The *diagonal functor* $\Delta : \mathcal{C} \to \text{Funct}(\mathcal{J}, \mathcal{C})$ is defined as follows:

on objects: $\Delta X =$ {the constant functor $\mathcal{J} \to \mathcal{C}$ taking value X}, i.e., $\Delta X(j) = X$ for any $j \in \text{Ob}\,\mathcal{J}$, $\Delta X(\varphi) = \text{id}_X$ for any $\varphi \in \text{Mor}\,\mathcal{J}$;

on morphisms: $\Delta \psi : \Delta X \to \Delta X'$ for $\psi : X \to X'$ in $\mathcal{C}$ is defined as follows: $\Delta \psi(j) = \psi : X = \Delta X(j) \to X' = \Delta X'(j)$, $j \in \text{Ob}\,\mathcal{J}$.

It is clear that $\Delta \psi$ is a morphism of functors and $\Delta(\psi \circ \psi') = \Delta \psi \circ \Delta \psi'$ so that Δ is indeed a functor from $\mathcal{C}$ to $\text{Funct}(\mathcal{J}, \mathcal{C})$.

16. Definition. *Let $F : \mathcal{J} \to \mathcal{C}$ be a functor. The projective limit of F in the category $\mathcal{C}$ is an object $X \in \text{Ob}\,\mathcal{C}$ representing the functor*

$$Y \mapsto \text{Hom}_{\text{Funct}(\mathcal{J},\mathcal{C})}(\Delta Y, F) : \mathcal{C}^\circ \longrightarrow \text{Set}.$$

The projective limit of F is denoted by $X = \varprojlim F$; sometimes it is called the inverse limit, or simply the limit.

According to Definition II.3.2, $X = \varprojlim F$ is characterized by the property

$$\text{Hom}_\mathcal{C}(Y, X) = \text{Hom}_{\text{Funct}(\mathcal{J},\mathcal{C})}(\Delta Y, F). \tag{II.5}$$

Theorem II.3.3 shows that if $\varprojlim F$ exists it is defined uniquely up to a unique isomorphism.

17. The Universal Property of the Limit

Any functor $F : \mathcal{J} \to \mathcal{C}$ is determined by the set of objects $F(j) = X_j \in \text{Ob}\,\mathcal{C}$ (one for every $j \in \text{Ob}\,\mathcal{J}$) and the set of morphisms $f(\varphi) : X_j \to X_{j'}$ in $\mathcal{C}$ (one for every morphism $\varphi : j \to j'$ in $\mathcal{J}$).

Assume that the limit $X = \varprojlim F$ exists. Take $Y = X$ in (II.5). To the identity morphism $\text{id}_X : X \to X$ there corresponds the morphism of functors $f : \Delta X \to F$, that is, the set of morphisms $f(j) : X \to X_j$ in $\mathcal{C}$ (one for every $j \in \text{Ob}\,\mathcal{J}$) satisfying the conditions

$$F(\varphi)f(j) = f(j') \quad \text{for each} \quad \varphi : j \longrightarrow j' \quad \text{in} \quad \mathcal{J}. \tag{II.6}$$

Next, an arbitrary morphism of functors $g : \Delta Y \to F$ is a set of morphisms $g(j) : Y \to X_j$ in $\mathcal{C}$, $j \in \text{Ob}\,\mathcal{J}$, satisfying similar conditions:

$$F(\varphi)g(j) = g(j') \quad \text{for any} \quad \varphi : j \longrightarrow j'. \tag{II.7}$$

Equation (II.5) shows that the definition of $\varprojlim F$ can be given in the form of the following universality condition:

An object $X \in \mathrm{Ob}\,\mathcal{C}$ is the projective limit of the functor $F : \mathcal{J} \to \mathcal{C}$ in the category $\mathcal{C}$ if there is a family of morphisms $f(j) : X \to X_j = F(j)$ in $\mathcal{C}$ (one for every $j \in \mathrm{Ob}\,\mathcal{J}$) satisfying (II.6) and such that for any family $g(j) : Y \to X_j$, $j \in \mathrm{Ob}\,\mathcal{J}$, satisfying (II.7) there exists a unique morphism $\psi : Y \to X$ in $\mathcal{C}$ with $g(j) = f(j) \circ \psi$.

18. Examples

a) Let $\mathcal{J} = \emptyset$ be the empty category (no objects, no morphisms). There exists only one functor $F : \emptyset \to \mathcal{C}$ and for any $Y \in \mathrm{Ob}\,\mathcal{C}$ the set $\mathrm{Hom}_{\mathrm{Funct}(\emptyset,\mathcal{C})}(\Delta Y, F)$ contains exactly one element. Therefore, $\varprojlim F = \omega$ (if the limit exists) has the property that $\mathrm{Hom}_\mathcal{C}(Y,\omega)$ for any $Y \in \mathrm{Ob}\,\mathcal{C}$ consists of exactly one element. Such an object ω is called the *final element* of $\mathcal{C}$.

b) Let $\mathcal{J} = \{0,1\}$ be the category with two objects 0, 1 and two morphisms $\mathrm{id}_0 : 0 \to 0$, $\mathrm{id}_1 : 1 \to 1$. A functor $F : \mathcal{J} \to \mathcal{C}$ is determined by two objects $X_0 = F(0)$, $X_1 = F(1)$ in $\mathcal{C}$. The limit $X = \varprojlim F$ is an object of $\mathcal{C}$ given together with a pair of morphisms $X \to X_0$, $X \to X_1$. It is easy to check that X is the direct product $X = X_0 \times X_1$ in $\mathcal{C}$ (see II.3.5).

One can give a similar definition for the direct product of any set of objects in $\mathcal{C}$.

c) Let $\mathcal{J} = \{1 \to 0 \leftarrow 2\}$ be the category with three objects and two nonidentity morphisms $1 \to 0$, $2 \to 0$. A functor $F : \mathcal{J} \to \mathcal{C}$ is a diagram in $\mathcal{C}$ of the form

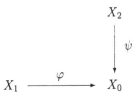

One can easily check that $X = \varprojlim F$ is the fibre product $X = X_1 \underset{X_0}{\times} X_2$ in $\mathcal{C}$ (see II.3.5).

d) *Equalizers.* Let $\mathcal{J} = (0 \rightrightarrows 1)$ be the category with two objects and two nonidentity morphisms, both from 0 to 1. A functor $F : \mathcal{J} \to \mathcal{C}$ is a diagram in $\mathcal{C}$:

$$X_0 \underset{\varphi'}{\overset{\varphi}{\rightrightarrows}} X_1.$$

The limit $X = \varprojlim F$ is an object $X \in \mathrm{Ob}\,\mathcal{C}$ and a morphism $\theta : X \to X_0$ such that $\varphi \circ \theta = \varphi' \circ \theta$ and such that the following universality condition is satisfied.

For any $\psi : Y \to Y_0$ with $\varphi \circ \psi = \varphi' \circ \psi$ there exists a unique morphism $\rho : Y \to X$ with $\psi : \theta \circ \rho$.

The morphism $\theta : X \to X_0$ is called the *equalizer* of the morphisms φ and φ'. Similarly one can define the equalizer of an arbitrary set of morphisms between two objects. For other examples of limits, as well as their construction in concrete categories (Set, Grp, Top, Ab) see the exercises for this section.

19. The Dual Notion: Colimits

As before, let $\mathcal{J}$ be some index category and let $F : \mathcal{J} \to \mathcal{C}$ be a functor. The *inductive limit* (or direct limit, or colimit) of the functor F in the category $\mathcal{C}$ is an object $X = \varinjlim F$ in $\mathcal{C}$ representing the functor

$$Y \mapsto \mathrm{Hom}_{\mathrm{Funct}(\mathcal{J},\mathcal{C})}(F, \Delta Y) : \mathcal{C} \longrightarrow \mathrm{Set}.$$

In other words, $X = \varinjlim F$ if the equality

$$\mathrm{Hom}_{\mathcal{C}}(X, Y) = \mathrm{Hom}_{\mathrm{Funct}(\mathcal{J},\mathcal{C})}(F, \Delta Y)$$

holds functorially in Y.

The inductive limit can also be defined via its universal property (dual to the definition of the limit in II.3.17). Here we list some special cases:

a) The initial object of a category $\mathcal{C}$ (for $\mathcal{J} = \emptyset$, see II.3.18a)).
b) The direct sum $X_0 \coprod X_1$ of two objects in $\mathcal{C}$ (for $\mathcal{J} = \{0, 1\}$, see II.3.18b)).
c) The amalgamated sum (6) $X_1 \coprod_{X_0} X_2$ in $\mathcal{C}$ (for $\mathcal{J} = (0 \to 1 \leftarrow 2)$, see II.3.18c)).
d) The coequalizer of two morphisms $\varphi, \varphi' : X_0 \to X_1$ in $\mathcal{C}$ (for $\mathcal{J} = (0 \rightrightarrows 1)$, see II.3.18d)).

In some applications it is essential to know that the limit $\varprojlim F$ exists for some class of functors $F : \mathcal{J} \to \mathcal{C}$. The following theorem gives a useful tool for proving its existence.

20. Theorem. *Assume that a category $\mathcal{C}$ contains a final object ω, the equalizer of any pair of morphisms, and the direct product of any pair of objects. Then all finite limits (i.e., all limits $\varprojlim F$ for any $F : \mathcal{J} \to \mathcal{C}$ with a finite index category $\mathcal{J}$) exist in $\mathcal{C}$.*

One has, of course, a similar theorem for colimits.

Proof. Note first that the existence of pairwise limits $X_0 \times X_1$ in $\mathcal{C}$ implies the existence of the product $\prod_{\alpha \in A} X_\alpha$ for any finite set of objects (see Ex. II.3.7).
(For empty A this is the final object ω of $\mathcal{C}$.)

For any morphism $\varphi : j \to k$ in $\mathcal{J}$ write $j = a(\varphi)$ (the beginning of φ), $k = z(\varphi)$ (the end of φ). Let $\mathcal{J}$ be a category with a finite number of objects and morphisms, and $F : \mathcal{J} \to \mathcal{C}$ a functor. For each $j \in \operatorname{Ob} \mathcal{J}$ let $X_j = F(j)$. Consider the following two products in $\mathcal{C}$:

$$Y = \prod_{j \in \operatorname{Ob} \mathcal{J}} X_j, \quad Z = \prod_{\varphi \in \operatorname{Mor} \mathcal{C}} X_{z(\varphi)}.$$

For $j \in \operatorname{Ob} \mathcal{J}$ denote the corresponding projection by $p_j : Y \to X_j$; similarly, for $\varphi \in \operatorname{Mor} \mathcal{J}$ denote the corresponding projection by $q_\varphi : Z \to X_{z(\varphi)}$.

According to the universal property of Z, any morphism $g : Y \to Z$ is determined by a family of morphisms $g_\varphi : Y \to X_{z(\varphi)}$, one for each $\varphi \in \operatorname{Mor} \mathcal{J}$, such that $g_\varphi = q_\varphi \circ g$.

Define $\{g_\varphi\}, \{g'_\varphi\}$ by

$$\begin{aligned} g_\varphi &= p_{z(\varphi)} : Y \longrightarrow X_{z(\varphi)} \\ g'_\varphi &= F(\varphi) \circ p_{a(\varphi)} : Y \longrightarrow X_{z(\varphi)}. \end{aligned} \quad (\text{II.8})$$

Let $g, g' : Y \to Z$ be the corresponding morphisms, and let $\theta : X \to Y$ be their equalizer. For $j \in \operatorname{Ob} \mathcal{J}$ let $f(j) = p_j \circ \theta : X \to X_j$.

We claim that the object X together with the family $f(j) : X \to X_j$, $j \in \operatorname{Ob} \mathcal{J}$, gives the limit $X = \varprojlim F$ in the category $\mathcal{C}$. To prove this claim, we have to check the equalities (II.6) and to prove the universality property, both given in II.3.17. We check (II.6), leaving the universality property to the reader.

Let $\varphi : j \to j'$ be a morphism in $\mathcal{J}$. Since θ is the equalizer of g and g', we have $g \circ \theta = g' \circ \theta$ and hence $q_\varphi \circ g \circ \theta = q_\varphi \circ g' \circ \theta : X \to X_{j'}$. But according to (II.8), $q_\varphi \circ g = p_{j'}$, $q_\varphi \circ g' = F(\varphi) \circ p_j$. Therefore, $p_{j'} \circ \theta = F(\varphi) \circ p_j \circ \theta$, i.e., $f(j') = F(\varphi) f(j)$. □

21. Corollary. *In the category Set all finite limits and colimits exist.*

Proof. The final object in Set is a set consisting of one element, the direct product in Set is the usual direct product of sets, and the equalizer of two mappings $\psi, \psi' : X_0 \to X_1$ is the set $X = \{x \in X_0, \psi(x) = \psi'(x)\}$ with the natural inclusion $X \hookrightarrow X_0$ (check it!). A similar proof holds for colimits. □

22. Adjoint Functor

Let $\mathcal{C}, \mathcal{D}$ be two categories and $F : \mathcal{C} \to \mathcal{D}$ a functor between them. For any $Y \in \operatorname{Ob} \mathcal{D}$ the mapping $T \mapsto \operatorname{Hom}_\mathcal{D}(F(T), Y)$ determines a functor $\mathcal{C}^\circ \to \text{Set}$. Let us assume that this functor is representable by $X \in \operatorname{Ob} \mathcal{C}$. This means that there exists a family of set isomorphisms

$$a_T : \operatorname{Hom}_\mathcal{C}(T, X) \xrightarrow{\sim} \operatorname{Hom}_\mathcal{D}(F(T), Y), \quad T \in \operatorname{Ob} \mathcal{C}, \quad (\text{II.9})$$

which is functorial in T, that is, for any morphism $\varphi : S \to T$ in $\mathcal{C}$ we have

$$a_S(u \circ \varphi) = a_T(u) \circ F(\varphi), \quad u \in \operatorname{Hom}_\mathcal{C}(T, X). \quad (\text{II.10})$$

In particular, substituting in (II.9) $T = X$ we get a canonical morphism $\sigma_Y = a_X(\mathrm{id}) : F(X) \to Y$ and (II.10) yields the canonical decomposition

$$a_T(u) = \sigma_Y \circ F(u) \qquad (\text{II}.11)$$

for any $u \in \mathrm{Hom}_\mathcal{C}(T, X)$.

23. Definition (– Lemma). *Assume that the functor $\mathrm{Hom}_\mathcal{D}(F(T), Y)$ is representable for any $Y \in \mathrm{Ob}\,\mathcal{D}$ and let $X_Y = X \in \mathrm{Ob}\,\mathcal{C}$ be the representing object. The mapping $Y \mapsto X_Y$ can be extended to a unique functor $G : \mathcal{D} \to \mathcal{C}$ which determines an isomorphism of bifunctors $\mathrm{Hom}_\mathcal{C}(T, G(Y)) \xrightarrow{\sim} \mathrm{Hom}_\mathcal{D}(F(T), Y)$ from $\mathcal{C}^\circ \times \mathcal{D}$ to Set. The functor G is said to be the right adjoint to F (and F is said to be the left adjoint to G).*

Proof. Let $v : Y \to Y'$ be an arbitrary morphism in $\mathcal{D}$. Define $G(v)$ by the formula

$$G(v) = a_{G(Y)}^{-1}(v \circ \sigma_Y) : G(Y) \longrightarrow G(Y'). \qquad (\text{II}.12)$$

Using the decomposition (II.11) one can easily see that the diagram

$$\begin{array}{ccc} FG(Y) & \xrightarrow{FG(v)} & FG(Y') \\ {\scriptstyle \sigma_Y} \downarrow & & \downarrow {\scriptstyle \sigma_{Y'}} \\ Y & \xrightarrow{v} & Y' \end{array} \qquad (\text{II}.13)$$

commutes. Moreover, the commutativity of this diagram distinguishes $G(v)$ among all morphisms $G(Y) \to G(Y')$ in $\mathcal{C}$. Indeed, let $\tilde{u} : G(Y) \to G(Y')$ be a morphism such that the substitution of $F(\tilde{u})$ for $FG(v)$ leaves the diagram commutative, i.e.,

$$v \circ \sigma_Y = \sigma_{Y'} \circ F(\tilde{u}).$$

By (II.11) and (II.12) we have

$$G(v) = a_{G(Y)}^{-1}(v \circ \sigma_Y) = a_{G(Y)}^{-1}\left(a_{G(Y)}(\tilde{u})\right) = \tilde{u}.$$

The just-proved property of $G(v)$ immediately shows that G is a functor, i.e., that $G(v_1 \circ v_2) = G(v_1) \circ G(v_2)$, as well as the uniqueness of G. □

24. Adjoint Morphisms

Let F and G be a pair of adjoint functors, so that we have functorial in X and Y isomorphisms of sets,

$$a : \mathrm{Hom}_\mathcal{C}(X, G(Y)) \longrightarrow \mathrm{Hom}_\mathcal{D}(F(X), Y). \qquad (\text{II}.14)$$

Substituting, as in II.3.22, $X = G(Y)$ we get a morphism
$$\sigma_Y : FG(Y) \longrightarrow Y.$$
On the other hand, substituting $Y = F(X)$ we get a morphism
$$\tau_X = a^{-1}\left(\mathrm{id}_{F(X)}\right) : X \longrightarrow GF(X).$$
The commutative diagram (II.13), together with a similar commutative diagram for τ, show that $\{\sigma_Y\}$ and $\{\tau_X\}$ determine functor morphisms
$$\sigma : FG \longrightarrow \mathrm{Id}_{\mathcal{D}}, \quad \tau : \mathrm{Id}_{\mathcal{C}} \longrightarrow GF. \tag{II.15}$$
These morphisms are called the *adjunction morphisms* corresponding to the pair of adjoint functors F, G. One can easily verify that they satisfy the following condition: the compositions
$$F \xrightarrow{F \circ \tau} FGF \xrightarrow{\sigma \circ F} F, \quad G \xrightarrow{\tau \circ G} GFG \xrightarrow{G \circ \sigma} G \tag{II.16}$$
are identity morphisms of functors F and G respectively.

One can prove that the existence of adjunction morphisms is equivalent to the adjointness of F and G. More precisely, if $F : \mathcal{C} \to \mathcal{D}$ and $G : \mathcal{D} \to \mathcal{C}$ are two functors such that there exist functor morphisms (II.15) satisfying (II.16), then F and G are adjoint to each other: the isomorphism (II.12) is $a(u) = \sigma_Y \circ F(u)$ and the inverse isomorphism is $a^{-1}(v) = G(v) \circ \tau_X$.

Exercises

In the following exercises we give examples of pairs of adjoint functors.

1. Forgetful Functors. Find the left adjoint functors to the following forgetful functors: a) Gr $\to$ Set; b) Top $\to$ Set; c) R-mod $\to$ Ab (R is a fixed ring); d) Ab $\to$ Set; e) k-Alg (associative algebras over a fixed field k) $\to$ Vect$_k$ (vector spaces of k) (Answer: the tensor algebra of a space V); f) Commet (complete metric spaces) $\to$ Met (metric spaces) (Answer: the completion of a metric space X); g) k-Alg $\to$ Lie$_k$ (Lie algebras over k), the functor makes a k-algebra A into the Lie algebra with the bracket $[a,b] = ab - ba$; (Answer: the enveloping algebra $U(\mathfrak{G})$ of a Lie algebra $\mathfrak{G}$).

2. Let R, S be two rings, M an (R,S)-bimodule (left over R, right over S). Prove that the functor $X \mapsto X \otimes_S M$ from S-mod to R-mod is left adjoint to the functor $Y \mapsto \mathrm{Hom}_R(M,Y)$.

3. Verify that $\varinjlim$ is left adjoint and $\varprojlim$ is right adjoint to the diagonal functor $\Delta : C \to \mathrm{Funct}(J,C)$.

4. Let F be a functor from the category $SCat$ of small categories (i.e., such that $\mathrm{Ob}\,\mathcal{C}$ is a set) to Set that maps $\mathcal{C}$ to the set $\mathrm{Ob}\,\mathcal{C}$. Prove that F has the left adjoint G_l which associates with a set X the discrete category $\mathcal{C}_X$ ($\mathrm{Ob}\,\mathcal{C}_X = X$, $\mathrm{Hom}_{\mathcal{C}_X}(x,y) = \{\mathrm{id}_X\}$ for $x = y$ and $= \emptyset$ for $x \neq y$), and the right adjoint G_r that associates with X the category $\tilde{\mathcal{C}}_X$ with $\mathrm{Ob}\,\tilde{\mathcal{C}}_X = X$, $\mathrm{Hom}_{\tilde{\mathcal{C}}_X}(x,y)$ is a one-element set for all $x, y \in X$. Prove that in turn, G_l has the left adjoint functor that associates with a small category $\mathcal{C}$ the set of its connected components (i.e., the set of equivalence classes of $\mathrm{Ob}\,\mathcal{C}$ by the relation $x \sim y$ iff $\mathrm{Hom}_{\mathcal{C}}(x,y)$ is nonempty).

5. Skeleton and Coskeleton. Let $(\Delta^\circ Set)_N$ be the category of N-truncated simplicial sets (see Sect. I.2, Ex. 3), $\mathrm{Tr}^N : \Delta^\circ Set \to (\Delta^\circ Set)_N$ be the truncation functor. Prove that Tr^N has the left adjoint $G : (\Delta^\circ Set)_N \to \Delta^\circ Set$; G has the property that $G \circ \mathrm{Tr}^N$ is the skeleton functor sk_N (see I.2.14 and II.1.7d)) and the corresponding adjunction morphism τ is the natural map $X \mapsto \mathrm{sk}_N X$. Prove that the coskeleton functor $\mathrm{Cosk}^N = (\Delta^\circ Set)_N \to \Delta^\circ Set$ (see Sect. I.2, Ex. 3) is the right adjoint to Tr^N.

6. Prove that the duality functor $* : Vect_k \to (Vect_k)^\circ$ (see II.2.3) is the right adjoint to itself (or, more precisely, to the dual functor $*^\circ : (Vect_k)^\circ \to Vect_k$).

7. Associativity of the Product. Prove that if at least one of the products $(X_1 \times X_2) \times X_3$, $X_1 \times (X_2 \times X_3)$, $X_1 \times X_2 \times X_3$ exists, then the other two also exist and all the three are naturally isomorphic.

8. Inductive and Projective Limits. a) Prove that for any categories $\mathcal{C}, \mathcal{D}$ we have
$$\mathrm{Funct}(\mathcal{C}^\circ, \mathcal{D}^\circ) = [\mathrm{Funct}(\mathcal{C}, \mathcal{D})]^\circ.$$

b) Prove that for any functor $F : \mathcal{J} \to \mathcal{C}$ (which, according to a)) can be considered as a functor $F^\circ : \mathcal{J}^\circ \to \mathcal{C}^\circ$) we have $\varinjlim F$ (in $\mathcal{C}$) $= (\varprojlim F^\circ)^\circ$.

9. Limit Over a Partially Ordered Set. An important special case (historically the first one) of limits is the case where the index category $\mathcal{J}$ is the category $\mathcal{C}(I)$ for some partially ordered set I (see I.1.5d)).

a) Let $\mathcal{C} = Set$ be the category of sets. To give a functor $F : \mathcal{C}(I) \to Set$ is the same as to give a collection of sets $X_\alpha, \alpha \in I$, and of mappings $f_{\alpha\beta} : X_\alpha \to X_\beta$, $\alpha \leq \beta$, such that $f_{\alpha\alpha} = \mathrm{id}$, $f_{\beta\gamma}f_{\alpha\beta} = f_{\alpha\gamma}$.

Prove that $\varinjlim F$ in Set can be constructed as follows. A subset $L \subset I$ is called *complete* if $\alpha \in L, \beta > \alpha \Rightarrow \beta \in L$, i.e., together with any element L contains all larger elements. A collection $\{x_\alpha \in X_\alpha, \alpha \in L\}$ for some complete L is called a *chain* if $f_{\alpha\beta}x_\alpha = x_\beta$ for $\alpha, \beta \in L, \alpha \leq \beta$. Then $\varinjlim F$ is the set of equivalence classes of chains with respect to the following equivalence relation: $\{x_\alpha \in X_\alpha, \alpha \in L\} \sim \{x'_\beta \in X_\beta, \beta \in L'\}$ if and only if for any $\alpha \in L, \beta \in L'$ there exists $\gamma, \gamma \geq \alpha, \gamma \geq \beta$, such that $f_{\alpha\gamma}x_\alpha = f_{\beta\gamma}x'_\beta$.

b) Prove that if $I' \subset I$ is a *filtered* subset (i.e., for any $\alpha \in I$ there exists $\beta \in I'$ with $\beta \geq \alpha$) then $\varinjlim F$ is equal to $\varinjlim F'$, where F' is the restriction of F to $\mathcal{C}(I') \subset \mathcal{C}(I)$.

c) Prove the statements similar to those in a) for the categories Gr, Ab, Top.

In the case where a partially ordered set I is directed (i.e., for any $\alpha, \beta \in I$ there exists $\gamma \in I$ which is larger than both of them; a classical example is the set $\mathbb{Z}^+$ of positive integers) $\varinjlim F$ for $F : \mathcal{C}(I) \to \mathcal{C}$ was called by the classics *the limit of direct spectrum* and $\varprojlim F$ for $F : \mathcal{C}(I) \to \mathcal{C}^\circ$ (see Ex. 8b)) was called the *limit of inverse spectrum*.

10. Limits of Cyclic Groups. a) Let $F : \mathcal{C}(\mathbb{Z}^+) \to \text{Ab}$ be given as the family of groups $A_n = \mathbb{Z}/p^n\mathbb{Z}$ and of monomorphisms $A_n \to A_m$, $X \mapsto p^{m-n}x$ for $m \geq n$. Prove that $\varinjlim F$ is the group of p-rational numbers in $\mathbb{Q}/\mathbb{Z}$ (i.e., the group of ratios of the form a/p^n modulo the integers).

b) Let $F : \mathcal{C}(\mathbb{Z}^+) \to \text{Ab}^\circ$ be given by the same groups A_n and by the reduction morphisms $A_m \to A_n$, $m \geq n$. Prove that $\varprojlim F$ is the group $\mathbb{Z}_p$ of p-adic integers.

c) Let I be a partially ordered set given by the divisibility condition in $\mathbb{Z}^+$: $m \geq n$ iff n divides m. Define $F : \mathcal{C}(I) \to \text{Ab}$ by the collection $A_n = \mathbb{Z}/n\mathbb{Z}$, $p_{nm} : A_n \to A_m$, $x \to (m/n)x$. Prove that $\varinjlim F = \mathbb{Q}/\mathbb{Z}$.

d) Let $F : \mathcal{C}(I) \to \text{Ab}^\circ$ (I as before) be given by the family $A_n = \mathbb{Z}/n\mathbb{Z}$, $q_{nm} : A_m \to A_n$, where q_{nm} is the reduction morphism for $n|m$. Prove that $\varprojlim F = \prod_p \mathbb{Z}_p$ (this group is called the completion of $\mathbb{Z}$).

11. Localization as a Limit. Let M be a module over a commutative ring A, and $f \in A$. Define $F : \mathcal{C}(\mathbb{Z}^+) \to \text{Ab}$ by setting $F(n) = M$ for all n, $p_{nm} : x \mapsto f^{m-n}x$ for $x \in M = F(n)$, $m \geq n$. Prove that $\varinjlim F = M_f$ (see Ex. I.5.3h)).

II.4 Category Approach to the Construction of Geometrical Objects

1. Three Geometrical Categories

We recall the definition of three classes of manifolds: topological (C^0), smooth (C^∞) and complex analytic (An).

A manifold of each class is a pair $(M, \mathcal{O}_M)$ consisting of a topological space M and a sheaf $\mathcal{O}_M$ of (partially defined) functions on M. A manifold of each class above is described using the following scheme.

a) First, we describe explicitly some manifolds from this class (the so-called local models).

b) Next, we say that a pair $(M, \mathcal{O}_M)$ belongs to the given class if it is locally isomorphic to a local model, that is, if any point $x \in M$ has a neighbourhood U such that $(U, \mathcal{O}_M/U)$ is isomorphic to some local model from the given class.

Models of C^0 are pairs (an open set $U \subset \mathbb{R}^n$, continuous partially defined real-valued functions on U).

Models of C^∞ are pairs (an open set $U \subset \mathbb{R}^n$, partially defined C^∞ real-valued functions on U).

Models of An are pairs (an open set $U \subset \mathbb{C}^n$, complex-valued partially defined functions on U which can be represented in a neighbourhood of any point of its the definition domain by a convergent power series).

Here a partially defined function is a pair $(D(f), f)$ where $D(f) \subset U$ is an open subset, $f : D(f) \to \mathbb{R}$ (or $\mathbb{C}$) is a function on $D(f)$. The sheaf $\mathcal{O}_M$ is called the structure sheaf of a manifold M. Usually M is assumed to be a Hausdorff space.

2. Atlases

Several decades ago, manifolds were usually defined in a slightly different way, using atlases. An atlas on M is an open covering $M = \cup_{i \in I} U_i$ together with a coordinate system $\left(z_1^{(i)}, \ldots, z_n^{(i)}\right)$ on each U_i (n may depend on a connected component of M) such that the following conditions are satisfied:

a) The map $\varphi_i : U_i \to \mathbb{R}^n$ for C^0 and C^∞ (resp. $\varphi_i : U_i \to \mathbb{C}^n$ for An) given by $\varphi_i(x) = \left(z_1^{(i)}(x), \ldots, z_n^{(i)}(x)\right)$ is an open embedding.

b) For any pair i, j with nonempty intersections $U_i \cap U_j$ coordinates $z^{(i)}$ are continuous (resp. smooth, resp. complex analytic) functions in coordinates $z^{(j)}$.

This definition is equivalent to the previous one. To construct a sheaf from the atlas we set $f \in \mathcal{O}_M(U)$ if for any nonempty intersection $U \cap U_j$ the restriction of f to $U \cap U_j$ is a continuous (resp. smooth, resp. complex analytic) function in local coordinates $z^{(i)}$. To construct an atlas from the sheaf one must choose local models covering M.

3. Morphisms

To make C^0, C^∞, and An categories one must define morphisms. A morphism $\Phi : (M, \mathcal{O}_M) \to (N, \mathcal{O}_N)$ is defined uniquely by a continuous map $\varphi : M \to N$. However, the existence of structure sheaves imposes some restrictions on admissible φ. Namely, let $f \in \mathcal{O}_M(U)$. Define a function $\varphi^\bullet(f)$ on $\varphi^{-1}(U) \subset M$ by
$$\varphi^\bullet(f)(x) = f(\varphi(x)).$$
Then Φ is a morphism if and only if $\varphi^\bullet(f) \in \mathcal{O}_M\left(\varphi^{-1}(U)\right)$ for all U, f. The reader can easily check that if M and N are given by atlases $M = \cup V_i$, $N = \cup U_j$, and $\varphi(V_i) \subset U_{j(i)}$, then the above condition means that the functions expressing local coordinates on $U_{j(i)}$ in terms local coordinates on

II.4 Category Approach to the Construction of Geometrical Objects

V_i belong to the given class. We recommend to verify that this definition makes C°, C^∞, and An into categories.

4. Structure Sheaves Versus Sheaves of Functions

The language of category theory enables one to introduce geometrical objects of nonclassical nature: the so-called locally ringed spaces. We describe two classes of local models that give a useful illustration of special features of these objects.

a) *Super-regions in* $\mathbb{R}^{m|n}$. Let m, n be two nonnegative integers. A super-region in $\mathbb{R}^{m|n}$ is a pair $(U, \mathcal{O}_U^s)$, where $U \subset \mathbb{R}^m$ is an open set and for $V \subset U$

$\mathcal{O}_U^s(V)$ is the ring of formal sums of the form
$$\sum_{1 \leq i_1 < \ldots < i_k \leq m} f_{i_1 \ldots i_k} \xi_{i_1} \ldots \xi_{i_k},$$ where $f_{i_1 \ldots i_k}$ are smooth functions on V, ξ_j are anticommuting formal variables.

In other words, $\mathcal{O}_U^s = \Lambda_{\mathcal{O}_U}^\bullet (\mathcal{O}_U \xi_1 \oplus \ldots \oplus \mathcal{O}_U \xi_n)$ where $\Lambda^\bullet$ is the exterior, or Grassmann, algebra, and $\mathcal{O}_U$ is the sheaf of smooth function on U. The definition of the structure sheaf also includes a $\mathbb{Z}_2$-grading: $\deg \mathcal{O}_U = 0$, $\deg \xi_j = 1$.

b) *Affine Schemes.* Let A be a commutative ring. Define Spec A to be the set of all prime ideals in A (see Ex. I.5.3). Recall that the Zarisky topology on Spec A is defined as follows: any closed subset is of the form $V(I)$ where $I \subset A$ and $V(I) = \{\mathfrak{p} | I \subset \mathfrak{p}\}$. In particular, for any element $f \in A$ an open set $D(f)$ is defined by $D(f) = \text{Spec } A \setminus V(f) = \{\mathfrak{p} | f \notin \mathfrak{p}\}$. These open sets form a base; obviously, $D(fg) = D(f) \cap D(g)$.

The ring of fractions A_f can be defined by $A_f = A[T]/(fT-1)$. By Ex. I.5.3, there exists on the topological space $X = \text{Spec } A$ a unique sheaf $\mathcal{O}_X$ such that
$$\Gamma(D(f), \mathcal{O}_X) = A_f,$$
and the restriction map $\Gamma(D(f), \mathcal{O}_X) \to \Gamma(D(fg), \mathcal{O}_X)$ is induced by the map

$$A[T_f]/(f \cdot T_f - 1) \longrightarrow A[T_{fg}]/(fg\, T_{fg} - 1),$$
$$\text{the class of } T_f \longmapsto g \cdot \text{the class of } T_{fg}$$

(the class of T_f in A_f should be thought of as $1/f$, therefore the class of T_{fg} is $1/fg$).

In both examples, sections of the structure sheaf are not functions anymore. So we cannot directly repeat the definition of a morphism from II.4.3, because a continuous map of spaces does not automatically define a transfer of sections of structure sheaves. In particular, we cannot glue together global spaces from local ones, because we do not know what an isomorphism is.

A natural remedy is to define separately morphisms of spaces and of sheaves, and to impose minimal compatibility conditions. This approach leads to the category of ringed spaces.

5. Definition. *a) A ringed space is a pair $(M, \mathcal{O}_M)$, where M is a topological space and $\mathcal{O}_M$ is a sheaf of rings on M (the structure sheaf).*

b) A morphism of ringed spaces $\Phi : (M, \mathcal{O}_M) \to (N, \mathcal{O}_N)$ is a pair (φ, θ), where $\varphi : M \to N$ is a continuous map and θ is a set of ring homomorphisms $\theta_U : \mathcal{O}_N(U) \to \mathcal{O}_M(\varphi^{-1}(U))$, one for every open set $U \subset N$, such that for any pair $U_1 \subset U_2 \subset N$ we have

$$\mathrm{res}_{\varphi^{-1}(U_2), \varphi^{-1}(U_1)} \circ \theta_{U_2} = \theta_{U_1} \circ \mathrm{res}_{U_2, U_1}.$$

One can substitute a set $\theta = (\theta_U)$ by a more standard object. Namely, there exist two ways to make θ a morphism of sheaves. We describe one of these ways. For $U \subset N$ and $U_1 \subset U_2 \subset N$ let

$$\varphi_\bullet(\mathcal{O}_M)(U) = \mathcal{O}_M\left(\varphi^{-1}(U)\right),$$

$$\mathrm{res}_{U_2, U_1} = \mathrm{res}_{\varphi^{-1}(U_2), \varphi^{-1}(U_1)} : \varphi_\bullet(\mathcal{O}_M)(U_2) \longrightarrow \varphi_\bullet(\mathcal{O}_M)(U_1).$$

It is easy to check that $\varphi_\bullet(\mathcal{O}_M)$ is a sheaf of rings on N, and θ defines a morphism of sheaves (denoted by the same letter) $\theta : \mathcal{O}_N \to \varphi_\bullet(\mathcal{O}_M)$.

The class of ringed spaces with the above notion of a morphism becomes a category. We recommend to the reader to define the composition and to verify the axioms.

There exist natural constructions of ringed spaces whose structure sheaves have properties very different from those of sheaves of functions. One of the most important constructions yields the spaces $(M, \mathcal{D}_M)$, where M is a manifold of the class C^∞ or An, and $\mathcal{D}_M$ is the sheaf of linear differential operators, i.e., of maps $P : \mathcal{O}_M \to \mathcal{O}_M$ which have in local coordinates $(z_1, \ldots, z_n)$ the form

$$P = \sum f_I(z) \frac{\partial^I}{\partial z^I}, \quad z^I = z_1^{i_1} \ldots z_n^{i_n}, \quad I = (i_1, \ldots, i_n).$$

On the other hand, ringed spaces in II.4.4 are so close to spaces with functions that there exists an axiomatic definition of a class of ringed spaces which includes classical manifolds as well as super-regions and affine spaces.

6. Definition. *a) A ringed space $(M, \mathcal{O}_M)$ is called a locally ringed space if $\mathcal{O}_M$ is a sheaf of local rings, i.e., for any $x \in M$ the fibre $\mathcal{O}_{M,x}$ is a local ring.*

b) A morphism $\Phi = (\varphi, \theta) : (M, \mathcal{O}_M) \to (N, \mathcal{O}_N)$ is called a morphism of locally ringed spaces if for any pair $x \in M$, $y \in N$ with $y = \varphi(x)$ the corresponding homomorphism $\theta_{y,x} : \mathcal{O}_{N,y} \to \mathcal{O}_{M,x}$ is a local homomorphism of rings.

II.4 Category Approach to the Construction of Geometrical Objects 97

7. Commentary

a) Recall first of all that a ring A is said to be a local ring if it contains the unique maximal ideal $\mathfrak{m}_A$. In this case $\mathfrak{m}_A$ is a prime ideal, and $A/\mathfrak{m}_A$ is a field. A homomorphism $f : A \to B$ of local rings is called a local homomorphism if $f(\mathfrak{m}_A) \subset \mathfrak{m}_B$.

Next, the fibre $\mathcal{O}_{M,x}$ at a point $x \in M$ is the direct limit of rings $\mathcal{O}_M(U)$, where U runs over the set of neighbourhoods of the point x, ordered by inclusion. Therefore, an element of $\mathcal{O}_{M,x}$ is a germ of a section of $\mathcal{O}_M$ over x: it is represented by a section over some neighbourhood of x and two such representatives determine the same germ if they coincide on some smaller neighbourhood of x contained in the regions of definitions of both sections.

b) If M is a manifold of the class C°, C^∞, or An, then the unique maximal ideal $\mathfrak{m}_x$ in the local ring $\mathcal{O}_{M,x}$ consists of all germs of functions vanishing at x (the proof follows from the fact that any germ, nonvanishing at x, is invertible in $\mathcal{O}_{M,x}$). Therefore, such manifolds are locally ringed spaces. Next, morphisms of the manifold, defined in II.4.3, preserve values of functions in the sense that if $\varphi = \varphi(x)$, then $\theta(f)(x) = f(y)$ for any germ $f \in \mathcal{O}_{N,y}$. Therefore, these morphisms are morphisms of locally ringed spaces.

Conversely, consider a morphism $(\varphi, \theta) : (M : \mathcal{O}_M) \to (N, \mathcal{O}_N)$ of, say, C°-manifolds in the category of locally ringed spaces. The sheaves $\mathcal{O}_M$ and $\mathcal{O}_N$ are sheaves of $\mathbb{R}$-algebras. We show that if θ respects the structure of an $\mathbb{R}$-algebra (i.e., θ is identical on constant functions), then (φ, θ) is a classical morphism, i.e., θ can be recovered from φ as in II.4.5. Indeed, the function $\theta(f)$ is determined by its values at all points of M; next, $\theta(f)(x) = f(y)$ if $f(y) = 0$ (and, as before, $y = \varphi(x)$); finally, if $f(y) = c$, then $\theta(f - c)(x) = (f - c)(y)$ by locality, and, as $\theta(c) = c$, $\theta(f)(x) = f(y)$ in the general case.

Let us note that one can construct nontrivial morphisms (nonidentical on constants) even between "points", i.e., between manifolds $(x, \mathbb{C})$, where x is a single point. Such morphisms are in one-to-one correspondence with the algebraic automorphisms of the field $\mathbb{C}$.

c) *Affine schemes* $(X = \operatorname{Spec} A, \mathcal{O}_X)$ are also locally ringed spaces. The fibre of $\mathcal{O}_X$ at a point x corresponding to a prime ideal $\mathfrak{p} \subset A$ is the ring of fractions A_S, where $S = A\backslash\mathfrak{p}$; the image of $\mathfrak{p}$ in A_S is the maximal ideal in A_S. Every morphism $f : A \to B$ induces a morphism (in the opposite direction) of locally ringed spaces $\varphi(\mathfrak{p}) = f^{-1}(\mathfrak{p}), \mathfrak{p} \subset B$. Any morphism between two spectra is obtained in such a way.

d) Now let $(M, \mathcal{O}_M)$ be a generic locally ringed space. Any section of the structure sheaf $f \in \mathcal{O}_M(U)$ determines a function with variable range:

$$f(x) = f \bmod \mathfrak{m}_x \in \mathcal{O}_{M,x}/\mathfrak{m}_x = k(x), \quad x \in U.$$

These values are also preserved by morphisms of locally ringed spaces. More precisely, a local morphism $\theta_{y,x} : \mathcal{O}_{N,y} \to \mathcal{O}_{M,x}$ yields the morphism of fields $\tilde{\theta}_{y,x} : k(y) \to k(x)$, and the latter transforms the value of f at y into the value of $\theta(f)$ at x. However, a function is not determined by its values at all points: for example, nilpotent elements of the structure sheaf take only zero values.

e) *Super-regions* in $\mathbb{R}^{m|n}$ for $n \neq 0$ do not satisfy the conditions of Definition II.4.6, because Grassmann algebras are noncommutative. However, one can easily give a necessary extension of the notion of commutativity.

8. Supercommutativity

Let $A = A_0 + A_1$ be a $\mathbb{Z}_2$-graded ring. This means that the additive group of A is the direct sum of two subgroups consisting of even (A_0) and odd (A_1) elements, and if $f \in A_i, g \in A_j$, then $fg \in A_k$, where $k = i + j \pmod 2$. If $f \in A_i$, we write $\tilde{f} = i$. We define the *supercommutator* $[f, g] = fg - (-1)^{\tilde{f}\tilde{g}} gf$ for any pair of homogeneous elements $f, g \in A$, and then extend it to A by additivity.

A ring A is said to be *supercommutative* if $[f, g] = 0$ for all f, g. In other words, odd elements anticommute among themselves, and even elements commute with both even and odd elements. Assume that 2 is invertible in A. Then for any odd f we have $0 = [f, f] = 2f^2$, so that $f^2 = 0$.

By a *morphism* of $\mathbb{Z}_2$-graded (in particular, supercommutative) rings we mean any ring homomorphism preserving $\mathbb{Z}_2$-grading. Definitions of a local ring and of a local morphism remain the same as in the commutative case. Now we can give the definition of a superspace that is completely parallel to Definition II.4.6, and in fact contains it as a special case.

9. Definition. *a) A* locally ringed superspace *is a ringed space* $(M, \mathcal{O}_M)$ *such that* $\mathcal{O}_M$ *is a sheaf of supercommutative rings (all restriction maps preserve $\mathbb{Z}_2$-degree) and all fibres* $\mathcal{O}_{M,x}$ *are local rings.*

b) A morphism of locally ringed superspaces *is a morphism of corresponding ringed spaces such that all $\theta_{y,x}$'s are local morphisms of $\mathbb{Z}_2$-graded rings.*

Using Definitions II.4.6 and II.4.9 we can add to the list of geometrical categories those defined by the following local models: affine schemes, super-regions in $\mathbb{R}^{m|n}$ or analytic super-regions in $\mathbb{C}^{m|n}$ (the reader is requested to give the corresponding definitions).

10. Definition. *A* scheme *is a locally ringed space* $(M, \mathcal{O}_M)$ *such that any point $x \in M$ has a neighbourhood U with the property that $(U, \mathcal{O}_M \mid U)$ is isomorphic to an affine scheme in the category of locally ringed superspaces.*

11. Definition. *A* differentiable supermanifold *is a locally ringed superspace* $(M, \mathcal{O}_M)$ *such that any point $x \in M$ has a neighbourhood U with the property that $(U, \mathcal{O}_M \mid U)$ is isomorphic to a superdomain in $\mathbb{R}^{m|n}$ in the category of superspaces locally ringed by $\mathbb{R}$-algebras.*

(For the significance of the $\mathbb{R}$-algebra condition see II.4.7b)).

12. How Can We Describe Topology in the Category Theory Language?

Until now we have studied some extensions of classical geometrical categories. Now we change our viewpoint and describe additional data one must introduce on a category so that it might play the role of an analogue of a topological space.

Recall first of all, that open sets in an arbitrary topological space X are objects of the category Top_X whose morphisms are inclusions (see II.1.5e)). This category has a final object X: for any open set U there exists a unique morphism $U \to X$. The combinatorial structure of Top_X is described by intersections; in the category theory language intersections are fibre products over X (see II.3.5)

$$U_1 \cap U_2 = U_1 \underset{X}{\times} U_2.$$

Finally, an important notion of covering is described in terms of sums in Top_X (more precisely, in terms of the amalgamated sum over the empty open set, see II.3.6):

a) for any family (U_i) the sum $\coprod_i U_i$ exists and coincides with the set-theoretic union $\bigcup_i U_i$.

b) a family of inclusions $U_i \to U$, $i \in I$ is a covering if and only if the canonical morphism $\coprod_i U_i \to U$ is an isomorphism.

Historically, the first categorical generalization of this categorical description of a topological space was the notion of the Grothendieck topology on a scheme. Consider an affine scheme Spec A, where A is an integral domain, say $A = \mathbb{F}_q[x_1, \ldots, x_n]$. In the Zariski topology the intersection of any two nonempty open subsets of this scheme is again nonempty. Therefore, the nerve of any finite open covering has the combinatorial type of a simplex. The same is true for any irreducible algebraic variety. Therefore, purely topological invariants cannot distinguish these varieties. The consideration of cohomology with coefficients in coherent sheaves improves the situation, but this improvement appears to be insufficient. For example, one still lacks a good Lefschetz-type formula for the number of fixed points of a mapping.

The Grothendieck idea to overcome this insufficiency was to extend the notion of topology: he suggested to consider as "open sets" not just open embeddings but certain more general mappings $f : U \to X$ such as, for example, nonramified coverings, flat morphisms (in the category of schemes), etc. In such a generalization open sets become objects of some category. Intersections and preimages are defined using fibre products. The essential point is that the notion of the covering is not deduced from some structures in the category, but instead forms a part of the definition. A convenient intermediate notion is the notion of sieve.

13. Definition. *Let S be a category, $U \in \text{Ob}\,S$, $\Phi = \{\varphi_i : U_i \to U \mid i \in I\}$ be some family of morphisms in S.*

a) Φ is said to be a sieve over U if any composition $V \xrightarrow{\psi} U_i \xrightarrow{\varphi} U$ with $\varphi \in \Phi$ belongs to Φ.

b) The minimal sieve containing all morphisms $\psi_j : U_j \to U, j \in J$, is called the sieve generated by this family of morphisms.

c) Let Φ be a sieve over U and $\varphi : V \to U$ be a morphism. The restriction Φ_V of the sieve Φ on V is the family of all $\psi_i : V_i \to V$ such that $\varphi \circ \psi_i \in \Phi$ for all i.

14. Definition. *A Grothendieck topology on the category S consists of families $C(U)$ of sieves, one family for each $U \in \mathrm{Ob}\,S$. The sieves from $C(U)$ are called covering sieves over U. They should satisfy the following axioms.*

a) The set of all morphisms $\varphi : U' \to U$ in S is a covering sieve over U.

b) The restriction of a covering sieve is a covering sieve.

c) The notion of the covering sieve is local in the following sense. Let Φ be a covering sieve over U and let Ψ be another sieve over U such that the restriction of Ψ to any element of Φ is a covering sieve. Then Ψ itself is a covering sieve.

A covering sieve over U in the category $\mathrm{Top}\,X$ is, roughly speaking, a covering containing all its subdivisions.

15. Sheaves

A category together with a Grothendieck topology on it is called a *site*.

Axioms of Grothendieck topologies (contrary to, say, topologies in function spaces) are devised to tackle the processes of gluing global objects from local ones (rather than taking limits) and for studying the associated cohomology invariants. Sheaves are the main tools in such constructions. Now we show how to define a sheaf on a site.

A *presheaf* of sets (of abelian groups, of rings, etc.) on a site S is a (contravariant) functor $F : S^\circ \to \mathrm{Set}$ (resp. $S^\circ \to \mathrm{Ab}$, $S^\circ \to \mathrm{Ring}$). The standard language of the sheaf theory can be easily extended to the case of presheaves on a site: elements $s \in F(U)$ are called sections of F over U; the image of s in $F(V)$ under the mapping $F(f)$ associated with $f : V \to U$ is called the restriction of s to V. (It can be denoted by $s|V$, keeping in mind that V can be mapped to U by different morphisms f, and $s|V$ depends on f).

Let F be a presheaf, Φ a sieve over U. A family of sections $s_\varphi \in F(V)$, one for each element $\varphi : V \to U$ of the sieve Φ, is said to be compatible if for any $V' \xrightarrow{\psi} V \xrightarrow{\varphi} U$ we have $s_{\varphi \circ \psi} = s_\varphi | V'$.

A presheaf F is said to be a *sheaf* if for any object U, any covering sieve Φ over U, and any compatible family of sections (s_φ) over Φ there exists a unique section $s \in F(U)$ whose restrictions to elements of Φ are s_φ.

16. Example. Let G be a group, S_G a site of G-sets. An object of the corresponding category G-Set is a set with the left action of G, a morphism

II.4 Category Approach to the Construction of Geometrical Objects

in G-Set is a mapping of sets commuting with the G-action. A sieve Φ over U in $\mathcal{S}_G$ is said to be a covering sieve if $U = \bigcup_{\varphi \in \Phi} \varphi(V)$.

To understand better the structure of the site $\mathcal{S}_G$ it is useful to take into account the following facts:

a) any G-set S is a disjoint union of orbits, that is, of irreducible G-sets; b) any orbit is isomorphic to the coset G/H, where H is the stabilizer of a point in the orbit; c) morphisms $G/H_1 \to G/H_2$ are in one-to-one correspondence with elements $gH_2 \in G/H_2$ such that $H_1 \subset gH_2g^{-1}$.

In particular, $\mathcal{S}_G$ has the maximal irreducible object: namely $G_l = G$ with the left action of G. Any morphism $\varphi : G_l \to G_l$ is of the form $\varphi = \varphi_g : h \mapsto hg$ for some $g \in G$ and $\varphi_{g_1 g_2} = \varphi_{g_2} \varphi_{g_1}$. Let Sh_G be the category of sheaves on $\mathcal{S}_G$ and $F \in \mathrm{Ob}\, Sh_G$. Then $F(G_l)$ is a G-set: the action of G on $s \in F(G_l)$ is given by $gs = F(\varphi_g)s$. The sheaf F can be recovered from this G-set. More precisely, we have the following proposition.

17. Proposition. *The mapping $F \mapsto F(G_l)$ can be extended to the functor*

$$\alpha : Sh_G \to \mathcal{S}_G,$$

and this functor is an equivalence of categories.

To prove the proposition we need a lemma that describes some properties of sheaves on $\mathcal{S}_G$.

18. Lemma. *Let F be a sheaf on the site $\mathcal{S}_G$. Then:*

a) F transforms disjoint unions of G-sets into disjoint unions of sets.

b) Let H be a subgroup of G, $U = G/H \in \mathrm{Ob}\, \mathcal{S}_G$, $\varphi : G_l \to U$ the natural projection. Then $F(\varphi) : F(U) \to F(G_l)$ identifies $F(U)$ with the subset of $F(G_l)$ of H-invariant elements:

$$F(U) \xrightarrow{F(\phi)} \{s \in F(G_l) \mid F(\varphi_h)s = s,\ \forall h \in H\}.$$

Proof. We leave to the reader an easy proof of the first statement. Let us prove the second statement. Let Φ be a sieve over U generated by $\varphi : G_l \to U$, so that Φ consists of all morphisms $\psi : V \to U$ of the form $\psi = \varphi \circ \theta$ for some $\theta : V \to G_l$. It is clear that Φ is a covering sieve. It is easy to check that $\varphi \circ \theta_1 = \varphi \circ \theta_2$ for $\theta_1, \theta_2 : V \to G_l$ if and only if $\theta_1 = \varphi_h \theta_2$ for some $h \in H$. Hence any compatible set of sections $\{s_\psi\}, \psi \in \Phi$, is determined by a section $\bar{s} \in F(G_l)$ such that $F(\varphi_h)\bar{s} = \bar{s}$, so that $s_\psi = F(\psi)\bar{s}$. Since F is a sheaf, sections $s \in F(U)$ are in one-to-one correspondence with compatible families $\{s_\psi\}$ and we get the required statement. $\square$

Proof (of Proposition II.4.17). The functor $\alpha : Sh_G \to \mathcal{S}_G$ is given by

$\alpha(F) = F(G_l)$ with the natural action of G as described above;

$\alpha(\xi) = \xi(G_l)$ for a morphism of functors $\xi : F_1 \to F_2$.

To construct a quasi-inverse functor $\beta : \mathcal{S}_G \to Sh_G$ we define, for any $X \in \mathrm{Ob}\,\mathcal{S}_G$, the functor $\beta(X) : (\mathcal{S}_G)^\circ \to Set$ as follows:

$$\beta(X)(Y) = \mathrm{Hom}_{\mathcal{S}_G}(Y, X) \quad \text{for} \quad Y \in \mathrm{Ob}(\mathcal{S}_G)^\circ = \mathrm{Ob}\,\mathcal{S}_G,$$
$$\beta(X)(f) : \varphi \in \beta(X)(Y_1) \mapsto \varphi \circ f \in \beta(X)(Y_2)$$
$$\text{for} \quad f : Y_2 \to Y_1$$
$$(\text{so that} \quad f \in \mathrm{Hom}_{(\mathcal{S}_G)^\circ}(Y_1, Y_2)).$$

It is easy to check that $\beta(X)$ is a sheaf (and not only a presheaf) of sets on the site $\mathcal{S}_G$.

Next, the action of β on morphisms in $\mathcal{S}_G$ is given as follows: for $\psi \in \mathrm{Hom}_{\mathcal{S}_G}(X_1, X_2)$, the morphism of sheaves $\beta(\psi) : \beta(X_1) \to \beta(X_2)$ is defined by

$$\beta(\psi)(\varphi) = \psi \circ \varphi \quad \text{for} \quad \varphi \in \beta(X_1)(Y), \quad Y \in \mathrm{Ob}(\mathcal{S}_G)^\circ.$$

To prove that α and β are quasi-inverse functors we construct isomorphism of functors

$$\varepsilon : \alpha \circ \beta \to \mathrm{id}_{\mathcal{S}_G}, \qquad \delta : \mathrm{Id}_{Sh_G} \to \beta \circ \alpha.$$

To construct ε we must define compatible isomorphisms of G-sets $\varepsilon(Y) : \alpha \circ \beta(Y) \to Y$, $Y \in \mathrm{Ob}\,\mathcal{S}_G$. We have $\alpha \circ \beta(Y) = \mathrm{Hom}_{\mathcal{S}_G}(G_l, Y)$, and $\varepsilon(Y)$ is given by $\varepsilon(Y)\varphi = \varphi(e) \in Y$, where e is the unit element in G. The trivial verification shows that ε is an isomorphism of functors.

Let us next construct δ. To do this we must define, for any sheaf of sets F on $\mathcal{S}_G$, a morphism $\delta(F) : F \to \beta \circ \alpha(F)$ in Sh_G, that is, to construct a morphism $\delta(F)$ of the functor $F : (\mathcal{S}_G)^\circ \to Set$ to the functor $\beta \circ \alpha(F) : (\mathcal{S}_G)^\circ \to Set$. This means that for any $X \in \mathrm{Ob}(\mathcal{S}_G)^\circ$ we must define a map of sets

$$\delta(F, X) : F(X) \to \beta \circ \alpha(F)(X).$$

These maps $\delta(F, X)$ should satisfy the following conditions:
 a) $\delta(F, X)$ is a bijection for any F, X.
 b) If $\xi : F_1 \to F_2$ is a morphism of functors, then for any X the diagram

$$\begin{array}{ccc} F_1(X) & \xrightarrow{\delta(F_1, X)} & \beta \circ \alpha(F_1)(X) \\ {\scriptstyle \xi(X)}\downarrow & & \downarrow {\scriptstyle \beta \circ \alpha(\xi)(X)} \\ F_2(X) & \xrightarrow{\delta(F_2, X)} & \beta \circ \alpha(F_2)(X) \end{array}$$

is commutative.

c) If $\varphi : X_1 \to X_2$ is a morphism of G-sets, then for any F the diagram

$$\begin{array}{ccc} F(X_2) & \xrightarrow{\delta(F, X_2)} & \beta \circ \alpha(F)(X_2) \\ F(\varphi) \downarrow & & \downarrow \beta \circ \alpha(F)(\varphi) \\ F(X_1) & \xrightarrow{\delta(F, X_1)} & \beta \circ \alpha(F)(X_1) \end{array}$$

is commutative.

To construct $\delta(F, X)$ we note that

$$\beta \circ \alpha(F)(X) = \mathrm{Hom}_{\mathcal{S}_G}(X, F(G_l)).$$

Let $x \in X$ and $\theta_x : G_l \to X$ be the unique morphism of G-sets such that $\theta_x(e) = x$. Then $F(\theta_x)$ maps the set $F(X)$ to the set $F(G_l)$, and $\delta(F, X)$ is given by

$$(\delta(F, X)(a))(x) = F(\theta_x)(a) \in F(G_l) \quad \text{for} \quad a \in F(X), \quad x \in X.$$

To verify that $\delta(F, X)$ is an isomorphism one must note that any G-set X is a disjoint union, $X = \cup G/H_i$, where H_i are some subgroups of G. Part a) of Lemma II.4.18 enables us to reduce the verification to the case $X = G/H$, where one can use part b) of Lemma II.4.18. All the details of this verification, as well as the proof of properties b), c) of maps $\delta(F, X)$ are left to the reader. □

19. Remarks

a) Similarly one can prove that the category of sheaves of abelian groups on $\mathcal{S}_G$ is equivalent to the category of G-modules. In applications one needs some variants of these constructions where G is a topological group with the continuous action on elements of a site. Let, for example, k be a field, $G = \mathrm{Gal}(\bar{k}/k)$ with the Krull topology, and $\tilde{\mathcal{S}}_G$ the category of G-sets with continuous action of G. This is the étale Grothendieck topology on the point affine scheme $\mathrm{Spec}\, k$ (cf. II.3.11). A sheaf of abelian groups on $\mathrm{Spec}\, k$ is the same as a continuous G-module.

b) The proof of Proposition II.4.17 shows that any sheaf of sets on $\mathcal{S}_G$ is a representable functor.

For a generic site there is no direct relation between sheaves of sets and representable functors. However, the existence of such a relation is quite important. In particular, there exists a theorem that gives a characterization of those sites where sheaves of sets coincide with representable functors. Roughly speaking, these sites are themselves categories of sheaves with some canonical topology.

104   II. Main Notions of the Category Theory

Now we can describe the last geometrical construction of this section. It associates with any category a big (in general, infinite-dimensional) topological space whose homotopy properties form, for example, the basis for the algebraic K-theory. This space is the geometrical realization of some simplicial set (see Sect. I.2) called the nerve of the category, and we will use the language of simplicial sets. The nerve can be constructed only for small categories, i.e., those ones whose objects form a set.

20. Definition. *a) The nerve of a small category $\mathcal{C}$ is the simplicial set $N\mathcal{C}$ such that*

$$N\mathcal{C}_n = \quad \text{the set of diagrams of the form}$$

$$X_0 \xrightarrow{\varphi_0} X_1 \xrightarrow{\varphi_1} \ldots \xrightarrow{\varphi_{n-1}} X_n, \quad X_i \in \mathrm{Ob}\,\mathcal{C}, \quad \varphi_i \in \mathrm{Mor}\,\mathcal{C}. \qquad (\mathrm{II.17})$$

The map $N\mathcal{C}_n \to N\mathcal{C}_m$ corresponding to a nondecreasing map $f : [m] \to [n]$ transforms the diagram (II.17) into the diagram

$$Y_0 \xrightarrow{\psi_0} Y_1 \xrightarrow{\psi_1} \ldots \xrightarrow{\psi_{m-1}} Y_m,$$

where $Y_i = X_{f(i)}$, $\psi_i = \mathrm{id}$ if $f(i) = f(i+1)$, $\psi_i = \varphi_{f(i+1)-1} \circ \ldots \circ \varphi_{f(i)}$ otherwise.

b) Let $F : \mathcal{C} \to \mathcal{C}'$ be a functor from one small category to another. The nerve of this functor is the morphism of simplicial sets (see I.2.15), $NF : N\mathcal{C} \to N\mathcal{C}'$, that maps a simplex (X_i, φ_i) into the simplex $(F(X_i), F(\varphi_i))$.

21. Example. Let Σ_n be the category with $n+1$ objects, $\mathrm{Ob}\,\Sigma_n = \{0, 1, \ldots, n\}$ and with $\mathrm{Hom}_{\Sigma_n}(k, l)$ consisting of one element for $k \leq l$ and being empty for $k > l$. Then $N\Sigma_n = \Delta[n]$ is the simplicial set described in I.2.5 (check this).

It is easy to see that N is a functor from the category $S\mathrm{Cat}$ of small categories to the category $\Delta^\circ\mathrm{Set}$ of simplicial sets. Here are some of its properties.

22. Properties of the Nerve

a) The category $\mathcal{C}$ can be recovered from its nerve uniquely up to an isomorphism (not equivalence!). In fact, let $X = N\mathcal{C}$. Then $\mathrm{Ob}\,\mathcal{C} = X_0$, $\mathrm{Mor}\,\mathcal{C} = X_1$.

More explicitly, in the notations of I.4.1, an 1-simplex $x \in X_1$ represents a morphism from $X\left(\partial_1^1\right)(x)$ to $X\left(\partial_1^0\right)(x)$. Next, let $x_1, x_2 \in X_1$ be two 1-simplices with the common vertex $X\left(\partial_1^0\right)x_1 = X\left(\partial_1^1\right)x_2$. Then in X exists a unique 2-simplex $x_2 \circ x_1 \in X_2$ with $X\left(\partial_2^2\right)(x_2 \circ x_1) = x_1$, $X\left(\partial_2^0\right)(x_2 \circ x_1) = x_2$ (because X is the nerve of a category; an arbitrary X does not have this property). The composition of morphisms represented by x_1 and x_2 is represented by the simplex

$$X\left(\partial_2^1\right)(x_2 \circ x_1).$$

II.4 Category Approach to the Construction of Geometrical Objects

Finally, the identity morphism $\mathrm{id}_x : x \to x, x \in X_0$, is represented by the element $X(s_0^0) x$, where s_0^0 is the only mapping $[1] \to [0]$.

b) $N(\mathcal{C} \times \mathcal{C}') = D(N\mathcal{C} \times N\mathcal{C}')$ (see Definitions II.1.7c) and I.3.4, I.3.5). The proof is obvious.

c) The functor N is fully faithful: the mapping $F \mapsto NF$ induces a bijection

$$\mathrm{Hom}_{S\mathit{Cat}}(\mathcal{C}, \mathcal{C}') \xrightarrow{\sim} \mathrm{Hom}_{\Delta^\circ \mathit{Set}}(N\mathcal{C}, N\mathcal{C}').$$

To recover from $f : N\mathcal{C} \to N\mathcal{C}'$ a functor $F : \mathcal{C} \to \mathcal{C}'$ such that $f = NF$ is easy: the action of F on objects (resp. on morphisms) is the action of f on 0-simplices (resp. on 1-simplices). The fact that F is a functor follows, by a), from the commutativity of f with simplicial maps.

Note now that $\mathrm{Hom}_{S\mathit{Cat}}(\mathcal{C}, \mathcal{C}')$ (functors from $\mathcal{C}$ to $\mathcal{C}'$) is again a category with natural transformations as morphisms (see II.1.10 where we wrote *Funct* instead of $\mathrm{Hom}_{S\mathit{Cat}}$). Hence we may try to compute the nerve of $\mathrm{Hom}_{S\mathit{Cat}}(\mathcal{C}, \mathcal{C}')$ in terms of $N\mathcal{C}$ and $N\mathcal{C}'$. To state the answer it is useful to discuss one general notion.

23. The Object "Inner Hom"

Morphisms of one set into another form a set; morphisms of one abelian group into another form an abelian group. To axiomatize the situation when for two objects $Y, Z \in \mathrm{Ob}\,\mathcal{C}$ one can construct the third object "inner Hom" $\mathbf{Hom}(Y, Z) \in \mathrm{Ob}\,\mathcal{C}$, it is natural to construct the corresponding representing functor. In most cases it is given by a formula of the form

$$\mathrm{Hom}_\mathcal{C}(X, \mathbf{Hom}(Y, Z)) = \mathrm{Hom}_\mathcal{C}(X * Y, Z)$$

where $*$ is some product operation (so that $Z \mapsto \mathbf{Hom}(Y, Z)$ is right adjoint to $X \mapsto X * Y$). The reader can check that this formula is valid in the following categories: *Set* ($*$ is the direct product of sets), *Ab* ($*$ is the tensor product over $\mathbb{Z}$), and *Cat* ($*$ is the product of categories, cf. II.1.7c)). The next theorem shows the existence of "inner Hom" in the category $\Delta^\circ \mathit{Set}$ and its relation to the corresponding object in *Cat*.

For two simplicial sets X, Y define $X * Y = D(X \times Y)$ (see I.3.4, I.3.5) so that $(X * Y)_i = X_i \times Y_i$ and $(X * Y)(f) = (X(f), Y(f))$ for $f : [m] \to [n]$.

24. Theorem. *a) For any two objects Y, Z in $\Delta^\circ \mathit{Set}$ there exists a third object $\mathbf{Hom}(Y, Z)$ representing the functor*

$$X \mapsto \mathrm{Hom}_{\Delta^\circ \mathit{Set}}(X * Y, Z). \tag{II.18}$$

b) For any two small categories $\mathcal{C}, \mathcal{C}'$ there exists a natural isomorphism of simplicial sets

$$N\left(\mathrm{Hom}_{S\mathit{Cat}}(\mathcal{C}, \mathcal{C}')\right) = \mathbf{Hom}(N\mathcal{C}, N\mathcal{C}').$$

Proof. a) We give an explicit construction of the simplicial set $T = \mathbf{Hom}(Y, Z)$. Recall that for any n we have the simplex $\Delta[n] \in \mathrm{Ob}\,\Delta°\mathrm{Set}$ with $\Delta[n]_m = \{g : [m] \to [n]\}$ and $\Delta[n](f)(g) = g \circ f$ for $f : [k] \to [m]$.

For any $h : [k] \to [n]$ define the map $\varepsilon_h \in \mathrm{Hom}_{\Delta°\mathrm{Set}}(\Delta[k], \Delta[n])$ by $\varepsilon_h(\tilde{g}) = h \circ \tilde{g}$, $\tilde{g} : [m] \to [k]$.

Now let:

i) $T_n = \mathrm{Hom}_{\Delta°\mathrm{Set}}(\Delta[n] * Y, Z)$.
ii) For $f : [m] \to [n], t \in T_n$,

$$T(f)t = t \circ (\varepsilon_h * \mathrm{id}_Y)$$

where $\varepsilon_h * \mathrm{id}_Y \in \mathrm{Hom}_{\Delta°\mathrm{Set}}(\Delta[m] * Y, \Delta[n] * Y)$ is identical on the second factor.

To prove that

$$\mathrm{Hom}_{\Delta°\mathrm{Set}}(X, T) = \mathrm{Hom}_{\Delta°\mathrm{Set}}(X * Y, Z)$$

we construct two mutually inverse maps between these sets.

Let first $\varphi = \{\varphi_n\} \in \mathrm{Hom}_{\Delta°\mathrm{Set}}(X, T)$ where

$$\varphi_n : X_n \longrightarrow \mathrm{Hom}_{\Delta°\mathrm{Set}}(\Delta[n] * Y, Z).$$

Define $\psi = \{\psi_n\} \in \mathrm{Hom}_{\Delta°\mathrm{Set}}(X * Y, Z)$ by

$$\psi_n(x_n, y_n) = \varphi_n(x_n)\left(\mathrm{id}_{[n]}, y_n\right),$$

where $x_n \in X_n$, $\mathrm{id}_{[n]} \in \Delta[n]_n$, $y_n \in Y_n$.

Next, let $\psi = \{\psi_m\} \in \mathrm{Hom}_{\Delta°\mathrm{Set}}(X * Y, Z)$. Define $\varphi = \{\varphi_n\} \in \mathrm{Hom}_{\Delta°\mathrm{Set}}(X, T)$ by

$$(\varphi_n(x_n))_m (f, y_m) = \psi_m (X(f)x_n, y_m) \in Z_m$$

where $x_n \in X_n$, $f : [m] \to [n] \in \Delta[n]_m$, $y_m \in Y_m$. Here $(\varphi_n(x_n))_m$ is the m-th component of the map $\varphi_n(x_n) : \Delta[n] * Y \to Z$ of simplicial sets.

We leave to the reader the easy job to check that the maps $\varphi \mapsto \psi$ and $\psi \mapsto \varphi$ define an isomorphism of functors $h_T : X \mapsto \mathrm{Hom}_{\Delta°\mathrm{Set}}(X, T)$ and $X \mapsto \mathrm{Hom}_{\Delta°\mathrm{Set}}(X * Y, Z)$ from $(\Delta°\mathrm{Set})°$ to Set.

b) Computing $\mathbf{Hom}(N\mathcal{C}, N\mathcal{C}')$ we have

$$\begin{aligned}
\mathbf{Hom}(N\mathcal{C}, N\mathcal{C}')_n &= \mathrm{Hom}_{\Delta°\mathrm{Set}}(\Delta[n] * N\mathcal{C}, N\mathcal{C}') \quad \text{(by the definition of } \mathbf{Hom}) \\
&= \mathrm{Hom}_{\Delta°\mathrm{Set}}(N\Sigma_n * N\mathcal{C}, N\mathcal{C}') \quad \text{(see the example in II.4.19)} \\
&= \mathrm{Hom}_{\Delta°\mathrm{Set}}(N(\Sigma_n \times \mathcal{C}), N\mathcal{C}') \quad \text{(by II.4.20b))} \\
&= \mathrm{Hom}_{\mathrm{Cat}}(\Sigma_n \times \mathcal{C}, \mathcal{C}') \quad \text{(by II.4.20c)).}
\end{aligned}$$

Now, by the definitions of Σ_n and of the product of categories (II.1.7c)) we see that $\mathrm{Hom}_{\mathrm{Cat}}(\Sigma_n \times \mathcal{C}, \mathcal{C}')$ is the set of diagrams of the form

II.4 Category Approach to the Construction of Geometrical Objects

$$F_0 \xrightarrow{\varphi_i} F_1 \xrightarrow{\varphi_1} \ldots \xrightarrow{\varphi_{n-1}} F_n$$

where all F_i are functors $\mathcal{C} \to \mathcal{C}'$ (i.e., objects of $\mathrm{Hom}_{\mathrm{Cat}}(\mathcal{C}, \mathcal{C}')$) and all φ_i are morphisms of functors (i.e. morphisms in $\mathrm{Hom}_{\mathrm{Cat}}(\mathcal{C}, \mathcal{C}')$). Therefore,

$$\begin{aligned}\mathbf{Hom}(N\mathcal{C}, N\mathcal{C}')_n &= \mathrm{Hom}_{\mathrm{Cat}}(\Sigma_n \times \mathcal{C}, \mathcal{C}') \\ &= N\left(\mathrm{Hom}_{\mathrm{Cat}}(\mathcal{C}, \mathcal{C}')\right)_n.\end{aligned}$$

The reader can easily verify that this identification defines an isomorphism of simplicial sets $N(\mathrm{Hom}_{\mathrm{Cat}}(\mathcal{C}, \mathcal{C}'))$ and $\mathbf{Hom}(N\mathcal{C}, N\mathcal{C}')$. □

Exercises

1. Nerve of a Category and Homotopy. a) Let $\mathcal{C}, \mathcal{C}'$ be two categories, $F_0, F_1 : \mathcal{C} \to \mathcal{C}'$ two functors, and $\varphi : F_0 \to F_1$ a functor morphism. Prove that $NF_0, NF_1 : N\mathcal{C} \to N\mathcal{C}'$ are homotopic as mappings of simplicial sets. (Use φ to construct a homotopy explicitly).

b) Prove that adjoint functors $F : \mathcal{C} \to \mathcal{C}'$, $G : \mathcal{C}' \to \mathcal{C}$ lead to homotopically inverse mappings $NF : N\mathcal{C} \to N\mathcal{C}'$, $NG : N\mathcal{C}' \to N\mathcal{C}$.

c) Prove that if $\mathcal{C}$ has either an initial or a final object, then $N\mathcal{C}$ is contractible (i.e. the identity mapping is homotopic to the mapping to a point).

2. Barycentric Decomposition. Let X be a simplicial set. Introduce a category $\mathcal{B}_X$ as follows:

$$\mathrm{Ob}\,\mathcal{B}_X = \{\text{nondegenerate simplices of } X \text{ (of all dimensions)}\},$$

and for $x \in X_n$, $x' \in X_m$

$$\mathrm{Hom}_{\mathcal{B}_X}(x, x') = \{f : [m] \longrightarrow [n],\ f \text{ is strictly increasing and } X(f)(x) = x'\}.$$

Composition and identity morphisms are defined in a natural way.

Prove that a nondegenerate simplex of $N\mathcal{B}_X$ is a family $(x_0, \ldots, x_n; i_0, \ldots, i_{n-1})$, where x_j is a nondegenerate j-simplex of X and $X\left(\partial_j^{i_j}\right)(x_{j+1}) = x_j$. $N\mathcal{B}_X$ can be thought of as the barycentric decomposition of the simplicial set X (see Fig. II.1).

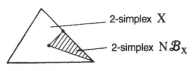

Fig. II.1.

3. Morphisms of Standard Simplices. Prove that

$$\mathbf{Hom}(\Delta[m], \Delta[n]) = \mathbf{Hom}(\Delta[n-1], \Delta[m+1]).$$

Hint. First of all, $\Delta[m] = N\Sigma_m$, where Σ_m is the category with $m+1$ objects $0, 1, \ldots, m$ corresponding to the ordered set $0 < 1 < \ldots < m$. Therefore, $\mathbf{Hom}(\Delta[m], \Delta[n]) = NFunct(\Sigma_m, \Sigma_n)$. Now use Ex. II.1.4.

4. Prove that for any $X, Y, Z \in \text{Ob}\,\Delta^\circ Set$ we have

$$\mathbf{Hom}(X * Y, Z) = \mathbf{Hom}(X, \mathbf{Hom}(Y, Z)).$$

5. **Quadratic Algebras.** a) *The category $\mathcal{QA}$.* Let k be a field. By a *quadratic algebra* we mean an associative $\mathbb{Z}$-graded algebra $A = \bigoplus_{i=0}^{\infty} A_i$ satisfying the following conditions:

(i) $A_0 = k$, $\dim A_1 < \infty$.
(ii) A is generated by A_0 and A_1, and the ideal of relations among elements of A_1 is generated by a subspace $R(A) \subset A_1 \otimes A_1$. We will denote such data by

$$A \longleftrightarrow \{A_1, R(A) \subset A_1 \otimes A_1\}.$$

A morphism $f : A \to B$ in $\mathcal{QA}$ is a homomorphism of graded k-algebras. Therefore, $\text{Hom}_{\mathcal{QA}}(A, B)$ is in one-to-one correspondence with the set of linear transformations $f_1 : A_1 \to B_1$ such that $(f \otimes f)(R(A)) \subset R(B)$.

b) *Duality.* For $A \in \text{Ob}\,\mathcal{QA}$ let

$$A^! \longleftrightarrow \{A_1^*, R(A)^\perp \subset A_1^* \otimes A_1^*\}$$

where $R(A)^\perp$ is the annihilator of $R(A)$ in $(A_1 \otimes A_1)^* = A_1^* \otimes A_1^*$. Prove that the mapping $A \mapsto A^*$ can be extended to an equivalence of categories $\mathcal{QA} \to \mathcal{QA}^\circ$. Prove that $S(V)^! = \Lambda(V^*)$, $\Lambda(V)^! = S(V^*)$ (here S and Λ are respectively the symmetric and the exterior algebra of a vector space; they are obviously quadratic).

c) *Products.* For $A, B \in \text{Ob}\,\mathcal{QA}$ let

$$A \circ B \longleftrightarrow \{A_1 \otimes B_1, S_{(23)}(R(A) \otimes B_1 \otimes B_1 + A_1 \otimes A_1 \otimes R(B))\},$$
$$A \bullet B \longleftrightarrow \{A_1 \otimes B_1, S_{(23)}(R(A) \otimes R(B))\}.$$

where $S_{(23)}$ denote the permutation of the second and the third factors in $A_1 \otimes A_1 \otimes B_1 \otimes B_1$. Prove that $(A \circ B)^! = A^! \bullet B^!$, $(A \bullet B)^! = A^! \circ B^!$ (an isomorphism of bifunctors).

d) *Inner* Hom. Construct an isomorphism of trifunctors (in A, B, C)

$$\mathbf{Hom}(A \bullet B, C) = \mathbf{Hom}(A, B^! \circ C)$$

that associate with a mapping $f : A_1 \otimes B_1 \to C_1$ the mapping $g : A_1 \to B_1^* \otimes C_1$ with the property $f(a \otimes b) = \langle g(a), (b) \rangle$ (convolution in b).
Let $\mathbf{Hom}(B, C) = B^! \circ C$ (cf. 23).

e) *Unit object.* Let $K = k[\varepsilon]$, $\varepsilon^2 = 0$. Prove that $K \bullet A = A$, $\mathbf{Hom}(A, K^!) = A$.

f) *Koszul complex* $K^\bullet(f)$. Substituting in d) $A = K$ associate with any morphism $f : B \to C$ the complex $K^\bullet(f) = (B^! \circ C, d_f)$, where d_f is the right multiplication by the image of $\varepsilon \in K$ in $B^! \circ C$ under the morphism corresponding to f. Compute $K^\bullet\left(\mathrm{id}_{S(V)}\right)$ explicitly (cf. Ex. I.7.5).

g) *Inner product and inner coproduct.* Define the inner product homomorphisms by

$$\mathbf{Hom}(B,C) \bullet \mathbf{Hom}(C,D) \longrightarrow \mathbf{Hom}(B,D),$$
$$B \bullet \mathbf{Hom}(B,C) \longrightarrow C.$$

Formulate and prove the associativity.

Let $\mathbf{hom}(B,C) = \mathbf{Hom}(B^!, C^!)^! = B^! \bullet C$. Define the inner coproduct homomorphisms by

$$\mathbf{hom}(B,D) \longrightarrow \mathbf{hom}(B,C) \circ \mathbf{hom}(B,D),$$
$$C \longrightarrow B \circ \mathbf{hom}(B,C).$$

Formulate and prove the associativity.

II.5 Additive and Abelian Categories

1. Abelian Variable of the Homology Theory

We have already mentioned in I.6.1 that any homology theory depends on two variables: abelian and nonabelian. An abelian variable is usually an object of an abelian category (while a nonabelian one is the category itself). The notion of abelian category is used to axiomatize main properties of the following categories.

a) Abelian groups.
b) Modules over a ring.
c) Systems of coefficients (I.4.8) and presheaves of abelian groups (I.5).
d) Sheaves of abelian groups (I.5).

We will formulate one by one axioms A1–A4 of an abelian category $\mathcal{C}$ (defining corresponding notions), check these axioms for the categories a)–d), and, finally, show how some axioms break up for certain similar looking but nonabelian categories such as, for example, the categories of topological abelian groups or of filtered groups.

2. Axiom A1

Any set $\mathrm{Hom}_\mathcal{C}(X,Y)$ is an abelian group (we will use additive notation); the composition of morphisms is bi-additive.

In other words, $\mathrm{Hom}_\mathcal{C}$ is a functor $\mathcal{C}^\circ \times \mathcal{C} \to \mathrm{Ab}$. Note also that this implies nonemptiness of each set $\mathrm{Hom}_\mathcal{C}(X,Y)$ because any group contains the zero element.

In all examples II.5.1a)–d) the group structure on any $\mathrm{Hom}_\mathcal{C}(X,Y)$ is clear.

3. Axiom A2

There exists a zero object $0 \in \mathrm{Ob}\,\mathcal{C}$, that is, an object such that $\mathrm{Hom}_\mathcal{C}(0,0)$ is the zero group. This implies that $\mathrm{Hom}_\mathcal{C}(0, X)$ and $\mathrm{Hom}_\mathcal{C}(X, 0)$ are also the zero groups for any $X \in \mathrm{Ob}\,\mathcal{C}$ and that any two zero objects are isomorphic.

In the examples II.5.1a)–d) zero objects are obvious ones.

4. Axiom A3

For any two objects X_1, X_2 there exist an object Y and morphisms

$$X_1 \underset{i_1}{\overset{p_1}{\rightleftarrows}} Y \underset{i_2}{\overset{p_2}{\leftrightarrows}} X_2 \tag{II.19}$$

such that

$$p_1 i_1 = \mathrm{id}_{X_1}, \quad p_2 i_2 = \mathrm{id}_{X_2}, \quad i_1 p_1 + i_2 p_2 = \mathrm{id}_Y$$
$$p_2 i_1 = p_1 i_2 = 0. \tag{II.20}$$

The following simple lemma clarifies the significance of this axiom.

5. Lemma. *The following two squares are respectively cartesian and co-cartesian:*

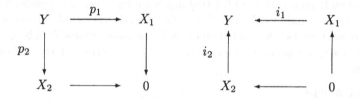

In other words, Y is both the direct sum and the direct product of X_1 and X_2. In particular, for given X_1 and X_2 any two diagrams of the form (II.19) are canonically isomorphic.

Proof. We shall show, that given a diagram $X_1 \xleftarrow{p'_1} Y' \xrightarrow{p'_2} X_2$ one can construct a unique morphism $\varphi: Y' \to Y$ such that

$$p'_1 = p_1 \circ \varphi, \quad p'_2 = p_2 \circ \varphi. \tag{II.21}$$

If such a φ exists, then, multiplying the first equality by i_1, the second one by i_2, and summing up the products we get (taking into account (II.20)) that $i_1 p'_1 + i_2 p'_2 = \varphi$. On the other hand, $\varphi = i_1 p'_1 + i_2 p'_2$ satisfies (II.21). This proves that the first square is cartesian; we leave the second verification to the reader. □

In categories II.5.1a)–d) the existence of direct sums and products is verified automatically.

Now we begin to analyse the least trivial property of categories II.5.1a)–d), namely the existence of exact sequences.

6. Kernel

Let a category $\mathcal{C}$ satisfy axioms A1 and A2, and let $\varphi : X \to Y$ be a morphism. Consider the following functor $\ker \varphi : \mathcal{C}^\circ \to Ab$:

$$(\ker \varphi)(Z) = \ker(X(Z) \longrightarrow Y(Z)),$$

$(\ker \varphi)(f) = $ is the restriction of $h_X(f)$ to $(\ker \varphi)(Z)$ (see II.3.1, II.3.2).

The inclusion $(\ker \varphi)(Z) \hookrightarrow X(Z)$ defines a morphism of functors $k : \ker \varphi \to h_X$. Assume that $\ker \varphi$ is represented by an object K. This object is defined together with a morphism $k : K \to X$ (theorem II.3.3) such that $\varphi \circ k = 0$. The diagram $K \xrightarrow{k} X \xrightarrow{\varphi} Y$ satisfies the following universal property: for any morphism $k' : K' \to X$ with $\varphi \circ k' = 0$ there exists a unique morphism $h : K' \to K$ with $k' = k \circ h$.

The morphism k, or the pair (K, k), is called the kernel of φ; where no confusion may arise we call the kernel the object K itself.

Let us prove that if the kernel (K, k) of the morphism φ exists, it is unique. First of all, applying the universal property of the kernel to $k : K \to X$ we see that the only morphism $\theta : K \to K$ with $k \circ \theta = k$ is $\theta = \mathrm{id}_K$. Let now $k_1 : K_1 \to X$ and $k_2 : K_2 \to X$ be two kernels of the morphism φ. From the universality of the diagrams $K_1 \xrightarrow{k_1} X \xrightarrow{\varphi} Y$ and $K_2 \xrightarrow{k_2} X \xrightarrow{\varphi} Y$ we get that there exist unique morphisms $\psi_1 : K_1 \to K_2$, $\psi_2 : K_2 \to K_1$ with $k_2 \circ \psi_1 = k_1$, $k_1 \circ \psi_2 = k_2$. Therefore $k_1 \circ \psi_2 \circ \psi_1 = k_1$ and cancelling, as mentioned above, by k_1, we get $\psi_2 \circ \psi_1 = \mathrm{id}_{K_1}$. Similarly, $\psi_1 \circ \psi_2 = \mathrm{id}_{K_2}$, and pairs (K_1, k_1) and (K_2, k_2) are isomorphic.

In categories a)–d) there is a set-theoretical construction of the kernel: $\varphi^{-1}(0)$ in groups and modules, the family $\varphi_x^{-1}(0)$ in the systems of coefficients (see I.5.8), the family $\varphi_U^{-1}(0)$ in sheaves (where a morphism of sheaves $\varphi : \mathcal{F} \to \mathcal{G}$ is given by morphisms $\varphi_U : \mathcal{F}(U) \to \mathcal{G}(U)$ for all open sets U).

7. Lemma. *a) In the categories of examples II.5.1a)–d) the set-theoretical kernel of a morphism $\varphi : X \to Y$ is an object K of the same category.*

b) The canonical embedding of K into X is the kernel in the corresponding category (in the sense of II.5.6).

We advise the reader to check at least that K so-defined in the category of sheaves is a sheaf itself (and not just a presheaf) (see I.5.4a)).

8. Cokernel

The naive definition of Coker φ as the object representing the functor $Z \mapsto \mathrm{Coker}(X(Z) \to Y(Z))$ is *wrong*. This functor is not isomorphic to the functor represented by the set-theoretical cokernel even in the category of abelian groups. Indeed, let $X = Y = \mathbb{Z}$, φ be the multiplication by an integer $n > 1$, $Z = \mathbb{Z}/n\mathbb{Z}$. Then $X(Z) = Y(Z) = 0$ so that

Coker $(X(Z) \to Y(Z)) = 0$ but $\text{Hom}(\mathbb{Z}/n\mathbb{Z}, \text{Coker}\,\varphi) \neq 0$ (here Coker $\varphi = \mathbb{Z}/n\mathbb{Z}$ is the the set-theoretic cokernel of ϕ). We strongly recommend to check that the functor $Z \mapsto \text{Coker}\,(X(Z) \to Y(Z))$ is not even representable.

The correct definition of Coker φ, provided that this functor is representable, requires the double dualization:

$$\text{Coker}\,\varphi = (\ker \varphi^\circ)^\circ,$$

where $^\circ$ denotes the duality functor (see II.1.7b)). The reader can check that this definition is equivalent to each of the following ones.

a) The cokernel of a morphism $\varphi : X \to Y$ is a morphism $c : Y \to K'$ such that for any $Z \in \text{Ob}\,\mathcal{C}$ the sequence of groups

$$0 \longrightarrow \text{Hom}_\mathcal{C}(K', Z) \longrightarrow \text{Hom}_\mathcal{C}(Y, Z) \longrightarrow \text{Hom}_\mathcal{C}(X, Z)$$

is exact (this means that $(K')^\circ$ represents $\ker \varphi^\circ$).

b) The cokernel of a morphism $\varphi : X \to Y$ is a morphism $c : Y \to K'$ such that $c \circ \varphi = 0$ and for any morphism $c_1 : Y \to K'_1$ with $c_1 \circ \varphi = 0$ there exists a unique morphism $h : K' \to K'_1$ with $c_1 = h \circ c$.

The cokernel $c : Y \to K'$ (similarly to the kernel), if it exists, is determined uniquely up to a unique isomorphism.

In categories II.5.1a)–c) one has the usual set-theoretic definition of the cokernel, and the analogue of Lemma II.5.7 is valid. For example, in the category Ab of abelian groups the cokernel of a homomorphism $\varphi : G \to H$ is the pair (K', c), where $K' = H/\varphi(G)$, $c : H \to K'$ is the factorization. Similarly one defines the cokernel of a morphism in the category of modules over a fixed ring. The cokernel of a morphism $\varphi : X \to Y$ in the category of presheaves is the pair (K', c), where K' is the presheaf $K'(U) = Y(U)/\varphi_U(X(U))$, $c_U : Y(U) \to K'(U)$ is the factorization. Similarly one defines the cokernel in the category of local systems on a fixed simplicial set.

The situation is more complicated in the category SAb of sheaves of abelian groups. The reason is that even if $\varphi : X \to Y$ is a morphism of sheaves $\{K'(U) = \text{Coker}\,\varphi_U\}$ might be a presheaf, but not a sheaf (see I.5.4b)). One can verify that $\{K'(U)\}$ is the cokernel in the category of presheaves.

In II.5.12–II.5.17 we shall construct cokernels of morphisms in the category of sheaves. Now we will state the last axiom.

9. Axiom A4

For any morphism $\varphi : X \to Y$ there exists a sequence

$$K \xrightarrow{k} X \xrightarrow{i} I \xrightarrow{j} Y \xrightarrow{c} K' \qquad (\text{II}.22)$$

with the following properties:

a) $j \circ i = \varphi$;
b) K is the kernel of φ, K' is the cokernel of φ;
c) I is both the cokernel of k and the kernel of c.

Any such sequence is called a *canonical decomposition*.

II.5 Additive and Abelian Categories

10. Definition. *An* additive category *is a category with axioms A1–A3; an* abelian category *is a category with axioms A1–A4.*

All the categories II.5.1a)–d) are additive. Moreover, they are abelian. The existence of the canonical decomposition of a morphism $\varphi : G \to H$ in the category Ab (i.e., the isomorphism $\operatorname{Im}\varphi \cong G/\ker\varphi$) is guaranteed by the homomorphism theorem for abelian groups; similarly one checks that the category of modules over a fixed ring is abelian. For categories II.5.1c) the canonical decomposition is constructed componentwise. For example, in the category PAb of presheaves of abelian groups the canonical decomposition of a morphism $\varphi : X \to Y$ is obtained from canonical decomposition of morphisms $\varphi_U : X(U) \to Y(U)$ in Ab,

$$K(U) \xrightarrow{k_U} X(U) \xrightarrow{i_U} I(U) \xrightarrow{j_U} Y(U) \xrightarrow{c_U} K'(U);$$

one can easily check that they are compatible with restrictions.

Below we prove that the category SAb of sheaves of abelian groups is abelian (proposition II.5.15).

11. Commentary on the Axiom A4

a) If a canonical decomposition of a morphism exists, then any other canonical decomposition is isomorphic to it, and this isomorphism is unique.

b) If one requires only the existence of kernels and cokernels, then for any morphism φ there exist two halves of the diagram (II.22),

$$K \xrightarrow{k} X \xrightarrow{i} I, \qquad I' \xrightarrow{j} Y \xrightarrow{c} K',$$

with $k = \ker\varphi$, $i = \operatorname{Coker} k$, $c = \operatorname{Coker}\varphi$, $j = \ker c$, as well as a morphism $l : I \to I'$ such that $\phi = j \circ l \circ i$. One can show that $\operatorname{Ker} l = \operatorname{Coker} l = 0$ (i.e., l is both a monomorphism and an epimorphism). The additional requirement of axiom A4 is that l is an isomorphism (i.e., it is invertible). Sometimes (I', j) is called the *image* of φ, and (I, i) is the *coimage* of φ.

c) The axiom A4 is self-dual in the following sense. Consider the diagram (II.22) in the dual category $\mathcal{C}^\circ$,

$$K'^\circ \xrightarrow{c^\circ} Y^\circ \xrightarrow{j^\circ} I^\circ \xrightarrow{i^\circ} X^\circ \xrightarrow{k^\circ} K^\circ. \qquad (\text{II.22}^\circ)$$

If (II.22) is the canonical decomposition of φ, then (II.22°) is the canonical decomposition of φ°.

Similar self-duality properties can be stated for axioms A1–A3. Hence, if we set $\operatorname{Hom}_\mathcal{C}(X, Y) = \operatorname{Hom}_{\mathcal{C}^\circ}(Y^\circ, X^\circ)$ as an abelian group, then the category dual to an additive one is additive, and that the category dual to an abelian one is abelian.

d) In an abelian category any morphism φ with $\ker\varphi = 0$ and $\operatorname{Coker}\varphi = 0$ is necessarily an isomorphism. Indeed, the cokernel of $0 \to X$ is isomorphic to $\operatorname{id}: X \to X$ and the kernel of $Y \to 0$ is isomorphic to $\operatorname{id}: Y \to Y$. Therefore, by axiom A4, i and j are isomorphisms, and $X \xrightarrow{\sim} Y$.

e) Morphisms φ with $\ker\varphi = 0$ are called *monomorphisms*, those with $\operatorname{Coker}\varphi = 0$ are called *epimorphisms*.

12. Sheaves and Presheaves

Let SAb (resp. PAb) be the additive category of sheaves (resp. presheaves) of abelian groups on a fixed topological space M. Any sheaf is a presheaf satisfying some additional conditions, so that one can define an inclusion functor $\iota: SAb \to PAb$. The main role in the proof that the category SAb is abelian is played by a functor in the opposite direction $s: PAb \to SAb$ and a canonical isomorphism of bifunctors

$$\operatorname{Hom}_{S\,Ab}(sX, Y) \xrightarrow{\sim} \operatorname{Hom}_{P\,Ab}(X, \iota Y). \qquad (\text{II}.23)$$

After the proof of (II.23) one checks that the cokernel of a morphism of sheaves $\varphi: X \to Y$ can be defined as $s(\tilde{K})$ where $\tilde{K}$ is the cokernel of φ in PAb, and then proves the existence of the canonical decomposition.

13. Proposition. *The functor* $\iota: SAb \to PAb$ *admits a left adjoint functor.*

Proof. We shall construct a functor $s: PAb \to SAb$ and a morphism of functors $\varepsilon: \operatorname{Id}_{P\,Ab} \to \iota \circ s$ such that the mapping (II.23) which associates with a morphism $\varphi: s(X) \to Y$ the composition $X \xrightarrow{\varepsilon_Y} \iota \circ s(X) \xrightarrow{\iota(\varphi)} \iota(X)$ determines the isomorphism (II.23) of bifunctors.

The sheaf $s(X)$ will coincide with the sheaf X^+ associated with a presheaf X (see I.5.6). Here we present another construction of $s(X)$ that is more appropriate in our situation. Let us recall that if a presheaf X is not a sheaf then either there exists a nonzero section $e \in X(U)$ that vanishes on elements of some covering of $U = \cup U_i$, or there exists a family of compatible sections $(e_i) \in \oplus X(U_i)$ that does not come from a section $e \in X(U)$.

So we construct sections of $s(X)$ over U as such collections. Namely, let

$s(X)(U)$ be the set of equivalence classes $(\{U_i\}, e_i)$, where $U = \cup U_i$, $e_i \in X(U_i)$ and $e_i|_{U_i \cap U_j} = e_j|_{U_i \cap U_j}$;

$(\{U_i\}, e_i) \sim (\{U'_j\}, e'_j)$ if there exists a covering $\{U''_k\}$ that refines $\{U_i\}$ and $\{U'_j\}$ and satisfies $e_i|_{U''_k} = e'_j|_{U''_k}$ for $U''_k \subset U_i \cap U'_j$.

II.5 Additive and Abelian Categories 115

Denote by $[\{U_i\}, e_i]$ the equivalence class of $(\{U_i\}, e_i)$. Define the restriction $r_{U,V} : s(X)(U) \to s(X)(V)$ for $V \subset U$ by the formula

$$r_{U,V}[\{U_i\}, e_i] = [\{U_i \cap V\}, e_i|_{U_i \cap V}].$$

Define also the addition in $s(X)(U)$ by

$$[\{U_i\}, e_i] + [\{V_j\}, f_j] = [\{U_i \cap V_j\}, e_i|_{U_i \cap V_j} + f_j|_{U_i \cap V_j}].$$

We leave to the reader the task to verify that these definitions are compatible with the equivalence relation and that they define on $s(X)$ the structure of a presheaf of abelian groups. Here we prove that $s(X)$ is a sheaf.

Indeed, let a section $[\{U_i\}, e_i] \in s(X)(U)$ become zero after being restricted to any element of the covering $\{V_j\}$. This means that for any j there exists a refinement $\{U'_{kj}\}$ of the covering $\bigcup_i (U_i \cap V_j) = V_j$ such that $e_i|_{U'_{kj}} = 0$ for any i, k with $U'_{kj} \subset U_i \cap V_j$. But then $[\{U_i\}, e_i] = [\{U'_{kj}\}, 0] = 0$.

Now let $U = \bigcup_j V_j$ and $g_j \in s(X)(V_j)$ be sections such that $r_{V_j, V_i \cap V_j}(g_i) = r_{V_i, V_i \cap V_j}(g_i)$. Let $[\{U_{jk}\}, g_{jk}]$ be a representative of the class g_j. The equality of restrictions to $V_j \cap V_i$ means that there exists a covering $\{U_{jil}\}$ of $V_j \cap V_i$ refining both $\{U_{jk} \cap V_i\}$ and $\{U_{ik} \cap V_j\}$ such that the restrictions of g_{jk} and $g_{ik'}$ to any U_{jil} contained in their definition regions coincide. Denote these coinciding restrictions by g_{jil}. Then

$$[\{U_{jil}\}, g_{jil}] \in s(X)(U)$$

being restricted to V_j coincides with g_j.

For a morphism of presheaves $\varphi : X \to Y$ define $s(\varphi)$ by

$$s(\varphi)[\{U_i\}, e_i] = [\{U_i\}, \varphi_{U_i}(e_i)].$$

This completes the construction of the functor s. The morphism of functors $\varepsilon : \mathrm{Id}_{P\,Ab} \to \iota \circ s$ consists of the maps $\varepsilon_X = \{\varepsilon_{X,U} : X(U) \to \iota \circ s(X)(U)\}$, where

$$\varepsilon_{X,U}(e) = [\{U\}, e].$$

Let us remark that if X is a sheaf, then ε is an isomorphism of sheaves X and $\iota \circ s(X)$. Sometimes we will identify X and $\iota \circ s(X)$ via this isomorphism.

Now let us prove that the map that sends $\varphi : s(X) \to Y$ to $\iota(\varphi) \circ \varepsilon_X : X \to \iota(Y)$ yields the isomorphism (II.23).

a) Suppose $\iota(\varphi) \circ \varepsilon_X = 0$, $e = [\{U_i\}, e_i] \in s(X)(U)$. Let $g = \varphi_U(e) \in Y(U)$. Then $g|_{U_i} = (\iota(\varphi) \circ \varepsilon_X)_{U_i}(e_i) = 0$ for any i; since Y is a sheaf, $g = 0$. Hence $\varphi = 0$.

b) Let $\psi : X \to \iota(Y)$ be a morphism of presheaves. We will find $\varphi : s(X) \to Y$ such that $\psi = \iota(\varphi) \circ \varepsilon_X$. Let $U \subset M$, $e = [\{U_i\}, e_i] \in s(X)(U)$ and let $g_i = \psi|_{U_i}(e_i) \in Y(U_i)$. Since ψ is a morphism of presheaves, the condition $e_i|_{U_i \cap U_j} = e_j|_{U_i \cap U_j}$ implies that $g_i|_{U_i \cap U_j} = g_j|_{U_i \cap U_j}$ for any nonempty intersection $U_i \cap U_j$. Next, since Y is a sheaf, there exists $g \in Y(U)$

with $g|_{U_i} = g_i$. We set $\varphi_U(e) = g$. The reader can check that $\varphi_U(e)$ does not depend on the choice of the representative $(\{U_i\}, e_i)$ of the section e, that $\varphi = \{\varphi_U\}$ is a morphism of sheaves, $s(X) \to Y$, and that $\psi = \iota(\varphi) \circ \varepsilon_X$. □

14. Example. For the presheaf X on $\mathbb{C} \setminus 0$ described in I.5.4b), $s(X)$ is the zero sheaf.

15. Proposition. *Let $\varphi : X \to Y$ be a morphism of sheaves in SAb and let*

$$K \xrightarrow{k} \iota X \xrightarrow{i} I \xrightarrow{j} \iota Y \xrightarrow{c} K' \qquad (II.24)$$

be the canonical decomposition of the morphism $\iota(\varphi)$ in the abelian category PAb. Then the diagram

$$sK \xrightarrow{s(k)} X = s\iota X \xrightarrow{s(i)} sI \xrightarrow{s(j)} Y = s\iota Y \xrightarrow{s(c)} sK' \qquad (II.25)$$

is the canonical decomposition of the morphism φ in the category SAb. In particular, SAb is an abelian category.

Proof. Identifying X with $s\iota X$ and Y with $s\iota Y$ we have $\varphi = s\iota(\varphi)$. Therefore $s(j) \circ s(i) = s(j \circ i) = s\iota(\varphi) = \varphi$ and (II.25) satisfies the property a) of the canonical decomposition of the morphism φ (see II.5.9). Two other properties are reduced to the corresponding properties of the decomposition (II.24) using the following result about the relation of kernels and cokernels in PAb and SAb. □

16. Lemma. *Let $\varphi : X \to Y$ be a morphism of presheaves of abelian groups, (K, k) and (K', c) be its kernel and cokernel in PAb. Then $(sK, s(k))$ and $(sK', s(c))$ are respectively the kernel and the cokernel of the morphism $s(\varphi) : sX \to sY$ in the category SAb.*

Proof. The statement about the cokernel follows immediately from Proposition II.5.13. Indeed, let $Z \in \mathrm{Ob}\,SAb$. Then the sequence of abelian groups

$$0 \longrightarrow \mathrm{Hom}_{P\,Ab}(K', \iota Z) \longrightarrow \mathrm{Hom}_{P\,Ab}(Y, \iota Z) \longrightarrow \mathrm{Hom}_{P\,Ab}(X, \iota Z)$$

is exact. By (II.23) the sequence

$$0 \longrightarrow \mathrm{Hom}_{S\,Ab}(sK', Z) \longrightarrow \mathrm{Hom}_{S\,Ab}(sY, Z) \longrightarrow \mathrm{Hom}_{S\,Ab}(sX, Z)$$

is isomorphic to it, and, therefore, also exact. Hence $(sK', s(c))$ is the cokernel of $s(\varphi)$ in SAb (see II.5.8a)).

Now we prove that $(sK, s(k))$ is the kernel of $s(\varphi)$. It is clear that $s(\varphi) \circ s(k) = s(\varphi \circ k) = 0$. Next, let $Z \in \mathrm{Ob}\,SAb$ and $\psi : Z \to sX$ be a morphism such that $s(\varphi) \circ \psi = 0$. We must prove that there exists a unique morphism $\theta : Z \to sK$ with $\psi = s(k) \circ \theta$.

Note first that if θ exists, it is unique. For this it is sufficient to check for any open $U \subset M$ that the morphism $s(k)_U : sK(U) \to sX(U)$ is an inclusion. Let $e = [\{U_i\}, e_i] \in sK(U)$. The equality $s(k)_U(e) = 0$ means that for any covering $\{V_j\}$ refining $\{U_i\}$ we have $k_{V_j}(e_i|_{V_j}) = 0$ for all $V_j \subset U_i$. But k_{V_j} being an inclusion, we have $e_i|_{V_j} = 0$, so that $e = [\{U_i\}, e_i] = [\{V_j\}, e_i|_{V_j}] = 0$.

To define θ we construct θ_U for all $U \subset M$.

Let $e \in Z(U)$ and $g = \psi_U(e)$. Since $s(\varphi)_U(g) = 0$, there exists a covering $\{U_i\}$ of U such that

$$\varphi_U \circ \psi_U(e)|_{U_i} = \varphi_{U_i} \circ \psi_{U_i}(e|_{U_i}) = 0. \qquad (\text{II}.26)$$

Next, for any i, $(K(U_i), k_{U_i})$ is the kernel of the homomorphism $\varphi_{U_i}: X(U_i) \to Y(U_i)$ of abelian groups. Therefore, for any i there exists a unique $h_i \in K(U_i)$ with $k_{U_i}(h_i) = \psi_{U_i}(e|_{U_i})$.

For any nonempty intersection $U_i \cap U_j$ we have

$$\begin{aligned} k_{U_i \cap U_j}(h_i|_{U_i \cap U_j}) &= \psi_{U_i \cap U_j}(e|_{U_i \cap U_j}) \\ &= k_{U_i \cap U_j}(h_j|_{U_i \cap U_j}). \end{aligned}$$

Since

$$k_{U_i \cap U_j} : K(U_i \cap U_j) \longrightarrow X(U_i \cap U_j)$$

is an inclusion, we have

$$h_i|_{U_i \cap U_j} = h_j|_{U_i \cap U_j}.$$

Therefore, the collection $(\{U_i\}, h_i)$ determines a section $h \in sK(U)$. We set $\theta_U(e) = h$.

The reader is requested to verify that:

a) The element $h \in sK(U)$ does not depend on the choice of the covering $\{U_i\}$ satisfying (II.26);
b) $\theta_U: Z(U) \to sK(U)$ is a homomorphism of abelian groups and $\{\theta_U, U \subset M\}$ is a morphism of sheaves $\theta: Z \to sK$.
c) $\psi = s(k) \circ \theta$. □

Now we give some examples of additive categories that are not abelian (because they do not satisfy the axiom A4).

17. Filtered Abelian Groups

An object of the category AbF of *filtered abelian groups* is an abelian group X together with an increasing sequence of subgroups $\ldots \subset F^i X \subset F^{i+1} X \subset \ldots \subset X$. Define $\text{Hom}_{AbF}(X, Y)$ by

$$\text{Hom}_{AbF}(X, Y) = \{\varphi \in \text{Hom}_{Ab}(X, Y) \mid \varphi(F^i X) \subset F^i Y \text{ for all } i\}.$$

Denote by $F^i \varphi : F^i X \to F^i Y$ the restriction of φ on $F^i X$.

The kernel of a morphism φ in AbF is the group $\ker \varphi$ (i.e., the kernel of φ in Ab); the filtration on $\ker \varphi$ is $\ker F^i \varphi$ (check!).

The cokernel of φ in AbF is the group Coker φ; the filtration is given by subgroups
$$F^i \operatorname{Coker}_{AbF}\varphi = F^i Y / F^i \cap \varphi(X)$$
(check!).

The next construction yields morphisms with zero kernel and cokernel that are not isomorphisms (i.e., that are not invertible in AbF). Consider a group X with two filtrations $\{F_1^i X\}$ and $\{F_2^i X\}$ such that $F_1^i X \subset F_2^i X$ for all i. Let φ be the identity morphism. If $F_1^i X \neq F_2^i X$ for at least one i, this morphism is not an isomorphism. On the other hand, it obviously has zero kernel and cokernel. Therefore, II.5.11d shows that AbF is nonabelian.

For a general morphism $\varphi : X \to Y$ we have (in the notations of II.5.11b)) $I = X/\ker \varphi$, $I' = \varphi(X)$ with filtrations
$$F^i I = F^n X / \ker F^i \varphi, \quad F^i I' = F^i Y \cap \varphi(X).$$

The canonical morphism $I \to I'$ is induced by φ; it is an isomorphism in Ab. But if the filtrations $\varphi(F^i X)$ and $F^i Y \cap \varphi(X)$ are different (as in the example above), φ has no canonical decomposition.

18. Topological Abelian Groups

Objects of the category AbT are abelian groups with a Hausdorff topology; morphisms in AbT are continuous group homomorphisms. Any morphism $\varphi : X \to Y$ in AbT has the kernel and the cokernel: $\ker \varphi$ is the kernel of φ in Ab with the induced topology; Coker φ is $Y/\overline{\varphi(X)}$, where $\overline{\varphi(X)}$ is the closure of $\varphi(X)$ in Y.

Let, for example, $\psi : Y \to Z$ be a morphism such that $\psi \circ \varphi = 0$. Then $\ker \psi = \psi^{-1}(0) \subset Y$ is a closed subgroup containing $\varphi(X)$ and hence $\overline{\varphi(X)}$ as well. Therefore, ψ can be factored through a morphism $Y/\overline{\varphi(X)} \to Z$.

In the notations of II.5.11b) we have $I = X/\ker \varphi$, $I' = \varphi(X)$ with the induced topology. If $\varphi(X)$ is not closed, then the canonical morphism $I \to I'$ is not an isomorphism. It may happen also that $\varphi(X)$ is closed but the topology induced from Y is weaker than the one induced from X. For example, the identity mapping of $\mathbb{R}$ with discrete topology into $\mathbb{R}$ with the usual topology has zero kernel and cokernel, but is not an isomorphism.

Exercises

In the following exercises we illustrate some methods allowing us to deal with objects of an arbitrary abelian category $\mathcal{A}$ as if they were just abelian groups. These methods are based on the following notion.

By an *element* y of an object Y (notation $y \in Y$) of an abelian category $\mathcal{A}$ we mean an equivalence class of pairs $(X, h), X \in \operatorname{Ob} \mathcal{A}, h : X \to Y$, by the equivalence relation

II.5 Additive and Abelian Categories

$$(X, h) \sim (X', h') \iff \begin{cases} \text{there exist } Z \in \mathrm{Ob}\,\mathcal{A} \text{ and epimorphisms} \\ h : Z \longrightarrow X, \; u' : Z \longrightarrow X' \text{ such that} \\ hu = h'u' \end{cases}. \quad (\mathrm{II}.27)$$

It is clear that the relation (II.27) is symmetric and reflexive.

1. Prove that $\sim$ is transitive.

   Hint. Prove first the following result that appears to be very useful in other problems too.

2. Let

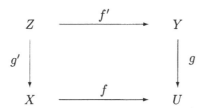

be a cartesian square in $\mathcal{A}$, so that $Z = X \underset{U}{\times} Y$ is the fibre product of X and Y over U. Prove that if f is an epimorphism, then f' is also an epimorphism. Furthermore, f' induces an epimorphism $\ker g \cong \ker g'$. More precisely, if (K', k') is the kernel of g', then $(K', f'k')$ is the kernel of g.

3. **Remarks.** Let us note here that the notion of element of an object Y of abelian category $\mathcal{A}$ is somewhat similar to the general notion of a Y-point (see II.3.2). The difference is that here we need additional factorization (II.27). The necessity for such a factorization can be seen from the properties of kernels and cokernels discussed in II.5.6–II.5.8. Indeed, since the functor $Y \mapsto \mathrm{Hom}_{\mathcal{A}}(X, Y)$ is left exact but not right exact (see II.6), the map $\mathrm{Hom}_{\mathcal{A}}(X, Y_1) \to \mathrm{Hom}_{\mathcal{A}}(X, Y_2)$ is an embedding if $Y_1 \to Y_2$ is a monomorphism but is not, in general, a surjection if $Y_1 \to Y_2$ is an epimorphism. The factorization (II.27) fixes this deficiency (see Ex. 5c)).

The main problem that occurs in attempts to describe morphisms in abelian categories by their actions on elements is that *a priori* we do not know which maps of sets {elements of Y_1} $\to$ {elements of Y_2} correspond to morphisms $Y_1 \to Y_2$ in $\mathcal{A}$. Therefore, usually morphisms are constructed by some methods that do not involve elements, and elements are used to verify the required properties.

4. Let $f : Y_1 \to Y_2$ be a morphism in $\mathcal{A}$, y an element of Y_1 and (X, h) a representative of y. Show that the formula $f(y) :$ {the class of $(X, f \circ h)$} determines a map {elements of Y_1} $\to$ {elements of Y_2} (which will be denoted by the same letter f).

Denote by 0 the class of the pair $(0,0)$, and by $-y$ for $y = (X, h)$ the class of the pair $((X, -h))$.

120   II. Main Notions of the Category Theory

5. Prove the following *diagram chase rules*:

a) $f : Y_1 \to Y_2$ is a monomorphism if and only if $f(y) = 0$ for $y \in Y_1$ implies $y = 0$.
b) $f : Y_1 \to Y_2$ is a monomorphism if and only if $f(y) = f(y')$ for $y, y' \in Y_1$ implies $y = y'$.
c) $f : Y_1 \to Y_2$ is an epimorphism if and only if for any $y \in Y_2$ there exists $y' \in Y_1$ such that $y = f(y')$.
d) $f : Y_1 \to Y_2$ is the zero morphism if and only if $f(y) = 0$ for all $y \in Y_1$.
e) A sequence $Y_1 \xrightarrow{f} Y \xrightarrow{g} Y_2$ is exact at Y if and only if $gf = 0$ and for any $y \in Y$ with $g(y) = 0$ there exists $y' \in Y_1$ with $f(y') = y$.
f) Assume we are given a morphism $g : Y_1 \to Y_2$ and elements $y, y' \in Y_1$ such that $g(y) = g(y')$. Then there exists $z \in Y_1$ such that $g(z) = 0$ and, moreover, for any $f : Y_1 \to Y$ with $f(y) = 0$ we have $f(z) = -f(y')$ and for any $f' : Y_1 \to Y$ with $f'(y') = 0$ we have $f'(z) = f'(y)$. (The element z is an analogue of the difference $y - y'$.)

6. Five-Lemma. Assume we are given a commutative diagram

$$\begin{array}{ccccccccc} X_1 & \to & X_2 & \to & X_3 & \to & X_4 & \to & X_5 \\ \downarrow f_1 & & \downarrow f_2 & & \downarrow f_3 & & \downarrow f_4 & & \downarrow f_5 \\ Y_1 & \to & Y_2 & \to & Y_3 & \to & Y_4 & \to & Y_5 \end{array}$$

with exact rows. Assume also that f_1 is an epimorphism, f_5 is a monomorphism, and f_2 and f_4 are isomorphisms. Then f_3 is also an isomorphism.

7. Snake-Lemma. Assume we are given a commutative diagram

$$\begin{array}{ccccccccc} 0 & \to & X_1 & \xrightarrow{g_1} & X_2 & \xrightarrow{g_2} & X_3 & \to & 0 \\ & & \downarrow f_1 & & \downarrow f_2 & & \downarrow f_3 & & \\ 0 & \to & Y_1 & \xrightarrow{h_1} & Y_2 & \xrightarrow{h_2} & Y_3 & \to & 0 \end{array} \quad \text{(II.28)}$$

with exact rows.

a) Prove that the sequences

$$0 \to \ker f_1 \xrightarrow{a_1} \ker f_2 \xrightarrow{a_2} \ker f_3,$$
$$\operatorname{Coker} f_1 \xrightarrow{b_1} \operatorname{Coker} f_2 \xrightarrow{b_2} \operatorname{Coker} f_3 \to 0. \quad \text{(II.29)}$$

where a_1, a_2 (resp. b_1, b_2) are induced by g_1, g_2 (resp. h_1, h_2), are exact.

b) Construct a natural morphism $\delta : \ker f_3 \to \operatorname{Coker} f_1$ that combines two exact sequences in (II.29) into one long exact sequence.

Hint. We present the construction of δ; all the required verifications are performed using Ex. 5.

To construct δ we extend (II.28) to the following commutative diagram with exact rows:

$$
\begin{array}{ccccccccc}
0 & \to & X_1 & \xrightarrow{t} & Z & \xrightarrow{s} & \ker f_3 & \to & 0 \\
& & \| & & \downarrow l & & \downarrow k & & \\
0 & \to & X_1 & \xrightarrow{g_1} & X_2 & \xrightarrow{g_2} & X_3 & \to & 0 \\
& & \downarrow f_1 & & \downarrow f_2 & & \downarrow f_3 & & \\
0 & \to & Y_1 & \xrightarrow{h_1} & Y_2 & \xrightarrow{h_2} & Y_3 & \to & 0 \\
& & \downarrow k' & & \downarrow l' & & \| & & \\
0 & \to & \operatorname{Coker} f_1 & \xrightarrow{s'} & Z' & \xrightarrow{t'} & Y_3 & \to & 0
\end{array}
$$
(II.30)

in which Z is the fibre product of X_2 and $\ker f_3$ over X_3, Z' is the cofibre product of Y_2 and $\operatorname{Coker} f_1$ over Y_1. By Ex. 2, s is an epimorphism with the kernel $t' : Z' \to Y_3$. Let us consider the morphism $\varepsilon = l' f_2 l : Z \to Z'$. It satisfies the conditions $\varepsilon t = s' k' f_1 = 0$ and $t' \varepsilon = f_3 k s = 0$. Since the top and the bottom rows of (II.30) are exact, there exists a unique morphism $\delta : \ker f_3 \to \operatorname{Coker} f_1$ such that $\varepsilon = s' \delta s$.

8. Jordan–Hölder Theorem. A nonzero object of an abelian category $\mathcal{A}$ is said to be *simple* it if has no proper subobjects. A *Jordan–Hölder series* of an object X is a filtration $\{0\} = X_0 \subset X_1 \subset \ldots \subset X_{n-1} \subset X_n = X$ such that all X_i/X_{i-1} are simple objects. If X admits such a series, it is said to have finite length. The Jordan–Hölder theorem says that any filtration $\{0\} = X'_0 \subset X'_1 \subset \ldots \subset X'_{m-1} \subset X'_m = X$ of a finite length object X such that all X'_i/X'_{i-1} are nonzero can be compressed to a Jordan–Hölder series, and any two Jordan–Hölder series of an object X are equivalent (i.e., determine the same, up to a permutation, collection of simple objects X_i/X_{i-1}). In particular, all Jordan–Hölder series have equal lengths.

9. Quotient Categories à la Serre. Let $\mathcal{A}$ be an abelian category, $\mathcal{B} \subset \mathcal{A}$ a full subcategory satisfying the following conditions:

(i) Together with any object B the subcategory $\mathcal{B}$ contains any subobject and any quotient object.

(ii) Together with any two objects B', B'' the category $\mathcal{B}$ contains any their extension (i.e., the middle term of any exact triple of the form $0 \to B' \to B \to B'' \to 0$).

For a fixed $A \in \operatorname{Ob} \mathcal{A}$ denote by $\operatorname{Sub}_{\mathcal{B}} A$ the partially ordered set consisting of all monomorphisms $\varphi : B \to A$ such that $\operatorname{Coker} \varphi \subset \operatorname{Ob} \mathcal{B}$ with the following ordering: $(\varphi : B \to A) < (\varphi' : B' \to A)$ if and only if φ can be factored through φ', i.e. $\varphi = \varphi' \theta$ for some (obviously unique) $\theta : B \to B'$.

Dual definition gives a partial ordering on the set $\operatorname{Qu}_{\mathcal{B}} A$ of all epimorphisms $\phi : A \to C$ with $\ker \phi \subset \operatorname{Ob} \mathcal{B}$.

For fixed $A, A' \in \operatorname{Ob} \mathcal{A}$ the map

$$(\varphi : B \longrightarrow A, \psi : A' \longrightarrow C) \mapsto \operatorname{Hom}_{\mathcal{A}}(B, C)$$

defines a functor
$$F_{A,A'} : \mathcal{C}(\mathrm{Sub}_\mathcal{B} A)^\circ \times \mathcal{C}(\mathrm{Qu}_\mathcal{B} A') \longrightarrow Ab$$
(here $\mathcal{C}(I)$ is the category associated with a partially ordered set I, see II.1.5d)).

Define the quotient category $\mathcal{A}/\mathcal{B}$ as follows:
$$\mathrm{Ob}\,\mathcal{A}/\mathcal{B} = \mathrm{Ob}\,\mathcal{A},$$
$$\mathrm{Hom}_{\mathcal{A}/\mathcal{B}}(A, A') = \varinjlim F_{A,A'}.$$

Define the composition and the identity morphisms in $\mathcal{A}/\mathcal{B}$.

Prove (or take for granted) that $\mathcal{A}/\mathcal{B}$ is an abelian category and that the natural functor $G : \mathcal{A} \to \mathcal{A}/\mathcal{B}$ (identity on objects) is exact. Prove that $G(A) = 0$ if and only if $A \in \mathrm{Ob}\,\mathcal{B}$. Prove that $\mathcal{A}/\mathcal{B}$ is the universal category with that property. More precisely, for any functor $F : \mathcal{A} \to \mathcal{C}$ from $\mathcal{A}$ into an abelian category $\mathcal{C}$ such that $F(A) = 0$ for $A \in \mathrm{Ob}\,\mathcal{B}$ there exists a unique functor $H : \mathcal{A}/\mathcal{B} \to \mathcal{C}$ such that $F = H \circ G$.

Standard examples of subcategories $\mathcal{B}$ satisfying the above conditions are $\mathcal{F}\,Ab$ (finite-abelian groups) in Ab, Vect_k^f (finite-dimensional vector spaces) in Vect_k, $2\text{-}Gr$ (2-groups) in $\mathcal{F}\,Gr$ (finite groups), etc. The existence and good properties of the quotient categories $\mathcal{A}/\mathcal{B}$ in the sense of Serre enable us to use arguments "modulo $\mathcal{B}$" (for example, to work in the category of abelian varieties up to isogeny).

II.6 Functors in Abelian Categories

1. Definition. *Let $\mathcal{C}, \mathcal{C}'$ be two additive categories. A functor $F : \mathcal{C} \to \mathcal{C}'$ is called* additive *if all maps*

$$F : \mathrm{Hom}_\mathcal{C}(X, Y) \longrightarrow \mathrm{Hom}_{\mathcal{C}'}(FX, FY), \quad X, Y \in \mathrm{Ob}\,\mathcal{C},$$

are homomorphisms of abelian groups.

In this book we will consider mostly additive functors. However, the reader should not think that all interesting functors are additive. Important non-additive functors are tensor powers and their canonical direct summands on the category A-mod over a commutative ring A,

$$T^n : A\text{-mod} \longrightarrow A\text{-mod}, \quad T^n(E) = E^{\otimes n}, \quad T^n(f) = f^{\otimes n}.$$

If A is a $\mathbb{Q}$-algebra then any idempotent α in the group algebra of the symmetric group S_n yields the functor

$$S^{(\alpha)}(E) = \mathrm{Im}\left(T^n(E) \xrightarrow{\alpha} T^n(E)\right).$$

In particular, in such a way we get the symmetric power $S^n(E)$ (corresponding to $\alpha = \frac{1}{n!}\sum_{s \in S_n} s$) and the exterior power $\Lambda^n(E)$ (corresponding to $\alpha = \frac{1}{n!}\sum_{s \in S_n} (\operatorname{sgn} s)s$).

The main characteristics of additive functors are related to whether or not they preserve kernels and cokernels. Let us introduce all the necessary definitions.

2. Definition. *a) A (cochain) complex in an additive category C is a sequence of objects and morphisms in C*

$$X^\bullet : \ldots \xrightarrow{d^{n-1}} X^n \xrightarrow{d^n} X^{n+1} \xrightarrow{d^{n+1}} \ldots$$

with the property $d^n \circ d^{n-1} = 0$ for all n.

b) If the category C is abelian, then the $(n+1)$-th cohomology of $X^\bullet$ is the object

$$H^{n+1}(X^\bullet) = \operatorname{Coker} a^n = \ker b^{n+1} \tag{II.31}$$

determined from the following commutative diagram

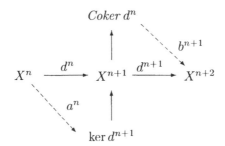

(The equality in (II.31) is a canonical isomorphism.)

c) A complex $X^\bullet$ in an abelian category is said to be acyclic *at the term X^n if $H^n(X^\bullet) = 0$.*

d) A complex $X^\bullet$ in an abelian category is said to be exact *(or* acyclic, *or an exact sequence) if it is acyclic at all terms.*

3. The Equivalence of Two Definitions of $H^n(X^\bullet)$

Here we will comment upon the definition of $H^n(X^\bullet)$ (Definition II.6.2b)) and, in particular, construct the canonical isomorphism in (II.31). The construction below gives an example of how to translate standard proofs in the category of abelian groups to an arbitrary abelian category (for another method, see Exs. II.5.1–II.5.7).

For convenience we will slightly change the notations. Let $X \xrightarrow{f} Y \xrightarrow{g} Z$ be a sequence of objects and morphisms in an abelian category C, with $gf = 0$. Let $(K, k) = \ker g$, $(K', k') = \operatorname{Coker} f$. By the definition of the kernel and the cokernel there exist morphisms $a : X \to K$, $b : K' \to Y$ such that the diagram

124   II. Main Notions of the Category Theory

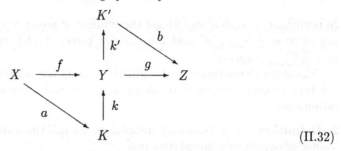

(II.32)

is commutative. We claim that *there exists a canonical isomorphism*

$$\text{Coker } a \xrightarrow{\sim} \ker b. \qquad (\text{II}.33)$$

The proof goes in several steps. At each step we shall construct some objects and morphisms, leaving to the reader the task to check all required properties.

a) First we state one general fact. Let $\varphi : X \to Y$ be a monomorphism (that is, $\ker \varphi = 0$) and $\psi : Y \to K'$ be the cokernel of φ. Then $(X, \varphi) = \ker \psi$.

The proof follows from the canonical decomposition of φ (see 5.9). Of course, one has also the dual fact: if $\psi : Y \to Z$ is an epimorphism then $\psi = \text{Coker} \ker \psi$.

b) Now we construct the required isomorphism (II.33) assuming that in (II.32) f is a monomorphism and g is an epimorphism.

Let $\varphi = k'k : K \to K'$; we prove that under the above assumption $(a, X) = \ker \varphi$, $(b, Z) = \text{Coker} \varphi$. Note first that $\varphi a = k'ka = k'f = 0$. Let now $\theta : U \to K$ be a morphism with $\varphi \theta = 0$, i.e., $k'k\theta = 0$. By a), $(f, X) = \ker k'$. Therefore, there exists a unique $\psi : U \to X$ with $k\theta = f\psi = ka\psi$. Since k is a monomorphism, $\theta = a\psi$, i.e., θ factors through X. It is easy to check that the decomposition $\theta = a\psi$ is unique. Therefore, $(a, X) = \ker \varphi$. Similarly, $(b, Z) = \text{Coker} \varphi$.

Now the required isomorphism (II.33) follows from the canonical decomposition of φ.

c) To reduce the general case to the one considered above, replace morphisms f and g in (II.32) by their canonical decompositions. We get the diagram

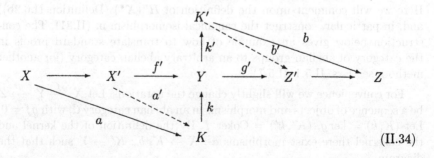

(II.34)

where

$$X' = \operatorname{Coker} \ker f = \ker \operatorname{Coker} f,$$
$$Z = \ker \operatorname{Coker} g = \operatorname{Coker} \ker g.$$

It is clear that in this diagram f' is a monomorphism and g' is an epimorphism. Applying several times part a) and the definitions of the kernel and the cokernel we see that there exist morphisms a' and b' such that the diagram (II.34) is commutative and $\operatorname{Coker} a' = \operatorname{Coker} a$, $\ker b' = \ker b$. Therefore, the general case is reduced to the one considered in b) and our claim is proved.

d) Let us note also that a) implies the following result:
Any short exact sequence

$$0 \longrightarrow X \xrightarrow{f} Y \xrightarrow{g} Z \longrightarrow 0$$

is canonically isomorphic to the sequence

$$0 \longrightarrow \ker g \longrightarrow Y \longrightarrow \operatorname{Coker} f \longrightarrow 0.$$

4. Definition. *Let $\mathcal{C}, \mathcal{C}'$ be two abelian categories. An additive functor $F : \mathcal{C} \to \mathcal{C}'$ is said to be* exact *if for any exact sequence*

$$0 \longrightarrow X \xrightarrow{f} Y \xrightarrow{g} Z \longrightarrow 0$$

in $\mathcal{C}$ the sequence

$$0 \longrightarrow F(X) \longrightarrow F(Y) \longrightarrow F(Z) \longrightarrow 0 \qquad (*)$$

in $\mathcal{C}'$ is also exact. F is said to be left exact *if the sequence $(*)$ is exact everywhere except, possibly, the term $F(Z)$. Similarly, F is said to be* right exact *if $(*)$ is exact everywhere except, possibly, the term $F(X)$.*

The interest in the exactness properties of an additive functor F can be explained as follows. In many situations we are interested (and historically this was the reason to study exactness properties) in the values of F on some specific objects. For example, the computation of the space of sections $\Gamma(\mathcal{F})$ for some sheaf $\mathcal{F}$ (or of its dimension) is the classical "Riemann–Roch-type problem". If the functor "sections of a sheaf" were exact, this problem would become much easier: a) for any short exact sequence $0 \longrightarrow \mathcal{F}_1 \longrightarrow \mathcal{F} \longrightarrow \mathcal{F}_2 \longrightarrow 0$ we would have $\dim \Gamma(\mathcal{F}) = \dim \Gamma(\mathcal{F}_1) + \dim \Gamma(\mathcal{F}_2)$; b) sheaves we are interested in, say on Riemann surfaces, can be easily represented as sequential extensions of some simple sheaves with known dimensions of $\dim \Gamma(\mathcal{F})$.

However, the functor Γ is not exact but only left exact, and the main task of the homological algebra is to overcome this nonexactness.

In this section we collect basic facts about exactness properties of some important functors.

5. Proposition. *Let C be an abelian category. The functors*

$$C \longrightarrow Ab : X \mapsto \mathrm{Hom}_C(Y, X) \tag{II.35}$$

(Y is fixed), and

$$C^\circ \longrightarrow Ab : X \mapsto \mathrm{Hom}_C(X, Y)$$

(Y is fixed) are left exact.

Proof. The left exactness of the functor (II.35) for a fixed object Y means that for any exact sequence

$$0 \longrightarrow X' \xrightarrow{f} X \xrightarrow{g} X'' \longrightarrow 0 \tag{II.36}$$

in C the following is true:

a) if $\varphi : Y \to X'$ satisfies $f \circ \varphi = 0$, then $\varphi = 0$.

b) for any $\varphi : Y \to X$ the equality $g \circ \varphi = 0$ is equivalent to the existence of $\psi : Y \to X'$ such that $\varphi = f \circ \psi$.

By II.6.3d), the sequence (II.36) can be replaced by the isomorphic sequence

$$0 \longrightarrow \ker g \xrightarrow{f} X \xrightarrow{g} X'' \longrightarrow 0.$$

Now a) follows from the fact that f is a monomorphism, so that $f \circ \varphi = f \circ 0 = 0$ if and only if $\varphi = 0$. Next, if $\varphi = f \circ \psi$ then $g \circ \varphi = g \circ f \circ \psi = 0$. Conversely, if $g \circ \varphi = 0$, then φ can be factored through $\ker g$, i.e., $\varphi = f \circ \psi$.

Similarly one proves the left exactness of the functor $X \mapsto \mathrm{Hom}_C(X, Y)$. □

6. Proposition. *Let A-mod (resp. mod-A) be the abelian category of left (resp. right) A-modules over a fixed ring A. The functor*

$$A\text{-mod} \longrightarrow Ab : X \mapsto Y \underset{A}{\otimes} X$$

(Y is a fixed object of mod-A) is right exact.

Proof. We must prove that for any exact sequence (II.36) of left A-modules and for any right A-module Y the sequence

$$Y \underset{A}{\otimes} X' \xrightarrow{f'} Y \underset{A}{\otimes} X \xrightarrow{g'} Y \underset{A}{\otimes} X'' \longrightarrow 0$$

is exact. Morphism g' is clearly an epimorphism: $y \otimes x'' = g'(y \otimes x)$, where $x \in X$ is an arbitrary element satisfying $g(x) = x''$. It is also clear that $g' \circ f' = 0$. Hence we must only prove that $\ker g' \subset \mathrm{Im}\, f'$. To do this we construct a morphism $\varphi : Y \underset{A}{\otimes} X'' \to \left(Y \underset{A}{\otimes} X\right)/\mathrm{Im}\, f'$ such that the composition $\varphi \circ g'$ coincides with the natural morphism $Y \underset{A}{\otimes} X \to \left(Y \underset{A}{\otimes} X\right)/\mathrm{Im}\, f'$ (this obviously suffices). Let $y \otimes x'' \in Y \underset{A}{\otimes} X''$ and let $x \in X$ satisfy $g(x) = x''$. It is clear that the image of $y \otimes x$ in $\left(Y \underset{A}{\otimes} X\right)/\operatorname{Im} f'$ does not depend on the choice of x (because, by the exactness of (II.36), x is unique modulo $\operatorname{Im} f$). It is easy to check also that the map $y \otimes x'' \to y \otimes x \mod \operatorname{Im} f'$ yields the required morphism $\varphi : Y \underset{A}{\otimes} X'' \to \left(Y \underset{A}{\otimes} X\right)/\operatorname{Im} f'$. □

Proposition 6 is true for sheaves of A-modules over a topological space, or over an arbitrary site. It remains true if one replaces A by a sheaf of rings over this site.

7. Proposition. *Let X be a topological space, $U \subset X$ an open set, SAb the category of sheaves of abelian groups on X. The functor*

$$SAb \longrightarrow Ab : \mathcal{F} \longrightarrow \mathcal{F}(U)$$

is left exact.

First Proof. Let PAb be the category of presheaves of abelian groups on X, $\iota : SAb \to PAb$ be the inclusion functor (see II.5.12). We prove first that ι is left exact. Let

$$0 \longrightarrow \mathcal{F}' \longrightarrow \mathcal{F} \longrightarrow \mathcal{F}'' \longrightarrow 0$$

be an exact sequence of sheaves. We can assume that $(\mathcal{F}', f) = \ker g$ (the kernel in SAb). According to II.5.7 we have $\ker(\iota g : \iota\mathcal{F} \to \iota\mathcal{F}'') = (\iota\mathcal{F}', \iota f)$ and the sequence

$$0 \longrightarrow \iota\mathcal{F}' \xrightarrow{\iota f} \iota\mathcal{F} \longrightarrow \iota\mathcal{F}''$$

is exact in PAb.

Now kernels and cokernels in the category PAb are defined separately on each open set (see II.5.5, II.5.8). Hence the functor

$$PAb \longrightarrow Ab : \mathcal{F} \longrightarrow \mathcal{F}(U)$$

is exact. So the sequence

$$0 \longrightarrow \iota\mathcal{F}'(U) \longrightarrow \iota\mathcal{F}(U) \longrightarrow \iota\mathcal{F}''(U) \longrightarrow 0$$

is exact. Next, for any sheaf $\mathcal{F}$ we have $\iota\mathcal{F}(U) = \mathcal{F}(U)$. Hence the functor $\mathcal{F} \mapsto \mathcal{F}(U)$ is left exact in the category SAb.

Second Proof. Define the constant presheaf $\mathbb{Z}_U$ with the fibre $\mathbb{Z}$ on U by

$$\begin{array}{rcll} \mathbb{Z}_U(V) &=& \mathbb{Z} & \text{if } V \cap U \text{ is nonempty,} \\ \mathbb{Z}_U(V) &=& \{0\} & \text{if } V \cap U \text{ is empty;} \\ r_{VV'} &=& \text{id} & \text{if } V' \subset V, V' \cap U \text{ is nonempty,} \\ r_{VV'} &=& 0 & \text{if } V' \subset V, V' \cap U \text{ is empty.} \end{array}$$

One can easily check that for any presheaf $\mathcal{F}$ on X we have

$$\operatorname{Hom}_{PAb}(\mathbb{Z}_U, \mathcal{F}) = \mathcal{F}(U).$$

Let now $s\mathbb{Z}_U \in \operatorname{Ob} SAb$ be the sheafification of $\mathbb{Z}_U$. Then for any sheaf $\mathcal{F}$ on X we have (using II.5.12)

$$\operatorname{Hom}_{S\,Ab}(s\mathbb{Z}_U, \mathcal{F}) = \operatorname{Hom}_{P\,Ab}(\mathbb{Z}_U, \iota\mathcal{F})$$
$$= (\iota\mathcal{F})(U) = \mathcal{F}(U)$$

and the left exactness of the functor $\mathcal{F} \mapsto \mathcal{F}(U)$ follows from the first part of Proposition II.6.5.

Let us remark that Proposition II.6.7 remains valid for the category of sheaves of abelian groups over an arbitrary site S.

The functors Hom and $\otimes$ in Propositions II.6.5 and II.6.6 depend on a fixed object; those objects for which these functors are exact (and not just exact from one side) are very important, and have special names.

8. Definition. *a) An object Y of an abelian category $\mathcal{A}$ is said to be* projective *if the functor $X \mapsto \operatorname{Hom}_{\mathcal{A}}(Y, X)$ is exact.*

b) An object Y of an abelian category $\mathcal{A}$ is said to be injective *if the functor $X \mapsto \operatorname{Hom}_{\mathcal{A}}(X, Y)$ is exact.*

c) A left A-module X (resp. a right A-module Y) is said to be flat *if the functor $Y \mapsto Y \underset{A}{\otimes} X$ (resp. $X \mapsto Y \underset{A}{\otimes} X$) is exact.*

Part c) of this definition can be applied to sheaves of modules on an arbitrary site.

Let us discuss various aspects of this definition.

9. Injectivity, Projectivity and Extension of Morphisms

Let us consider two diagrams in an abelian category $\mathcal{A}$:

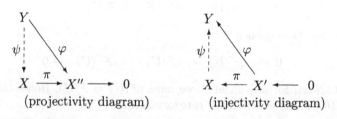

(projectivity diagram)   (injectivity diagram)

We will interpret these diagrams as the following properties of the object Y:

a) For any surjective morphism $X \to X''$ and for any morphism $Y \to X''$ there exists a morphism $Y \to X$ that makes the projectivity diagram commutative.

b) For any injective morphism $X' \to X$ and for any morphism $X' \to Y$ there exists a morphism $X \to Y$ that makes the injectivity diagram commutative.

(Main convention: a solid arrow (resp. a dashed arrow) means "for any" (resp. "there exists"); cf. the diagram in Definition II.6.2b)).

We claim that the property a) (resp. the property b)) is equivalent to the projectivity (resp. to the injectivity) of an object Y. Indeed, let, for example, an object Y satisfy a), and assume that we are given an exact triple

$$0 \longrightarrow X' \xrightarrow{f} X \xrightarrow{g} X'' \longrightarrow 0.$$

Let us consider the corresponding sequence of Homs,

$$0 \longrightarrow \mathrm{Hom}(Y, X') \xrightarrow{f^*} \mathrm{Hom}(Y, X) \xrightarrow{g^*} \mathrm{Hom}(Y, X'') \longrightarrow 0. \tag{II.37}$$

By Proposition II.6.5, it is exact in all terms except, possibly, $\mathrm{Hom}(Y, X'')$ and property a) means that g^* is surjective. Therefore the sequence (II.37) is exact and Y is projective.

On the other hand, let Y be projective. Since the surjection $X \to X''$ in the projectivity diagram can be completed to an exact triple, the exactness of (II.37) at $\mathrm{Hom}(Y, X'')$ yields property a).

10. Projective Modules and Free Modules

The previous statement easily gives a description of projective objects in the categories of modules (right or left ones): *a module over a ring is projective if and only if it is a direct summand of a free module.*

First, *a free module Y is projective*. Indeed, let $\{y_i\}$ be a family of free generators of Y. Suppose that we are given a projectivity diagram for Y. To construct ψ it suffices to define $\psi(y_i)$. We set $\psi(y_i) = x_i$ where $x_i \in X$ is an arbitrary element such that $\pi(x_i) = \varphi(y_i)$ (such an x_i exists because π is surjective).

Next, a *direct summand of a projective module is itself projective*. In fact, let $Y = Y_1 \oplus Y_2$ and we are given a projectivity diagram for Y_1

$$\tag{II.38}$$

Let $\varphi = (\varphi_i, 0) : Y \to X''$. Consider the corresponding projectivity diagram for Y. Since Y is projective, there exists a completing morphism $\psi = (\psi_1, \psi_2) : Y = Y_1 \oplus Y_2 \to X$. It is clear then that $\psi_1 : Y_1 \to X$ completes the diagram (II.38).

On the other hand, let Y be projective. There exists an exact sequence $\tilde{Y} \xrightarrow{\pi} Y \to 0$ with $\tilde{Y}$ being a free module (for $\tilde{Y}$ one can take, for example, the free module generated by all elements $y \in Y$, with the obvious π). Let us consider the following projectivity diagram:

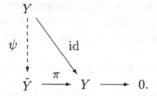

Y being projective, there exists a completing morphism ψ. It is clear that the pair (ψ, π) represents Y as a direct summand of the free module $\tilde{Y}$.

In the categories of sheaves of modules over ringed spaces there are, usually, very few projective objects. In some cases instead of projective objects one can consider locally free sheaves (i.e., sheaves that become free when restricted to elements of some appropriate covering or sieve).

11. Injective Modules and Divisibility

Any finite-dimensional linear space over a field k is both a projective and an injective object of the category Vect_f: it is projective because it is free, and injective because the category is equivalent to the dual one.

The category of modules over an arbitrary ring is, in general, very far from being equivalent to the dual category, and the classes of projective and of injective modules are very different from each other (see, however, IV.3). The projectivity of a module is related, roughly speaking, to the absence of relations among appropriate generators, while the injectivity is related to the possibility to divide by elements of the ring A. A typical obstruction to the injectivity is of the following kind: an element $x' \in X'$ can be divided by $a \in A$ in a bigger module $X \supset X'$ ($a\,x = x'$ for some $x \in X$) but not in X'. Then the identity morphism $\varphi = \text{id} : X' \to X'$ cannot be completed in the appropriate injectivity diagram.

Injective modules over a commutative ring are also closely related to prime ideals of this ring. As an example, we give a list of minimal injective abelian groups (i.e., of $\mathbb{Z}$-modules):

a) $\mathbb{Q}$ (the localization of $\mathbb{Z}$ by the complement to the prime ideal $\{0\}$),

b) $\mathbb{Q}_{(p)}/\mathbb{Z}$ (p is a prime, $\mathbb{Q}_{(p)}$ is an additive group of rational numbers with denominators p^n).

Let us check that these groups are injective. Consider an injectivity diagram

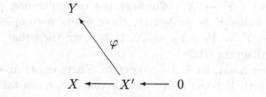

(so that X' is a subgroup of X), where Y is either $\mathbb{Q}$ or one of groups $\mathbb{Q}_{(p)}/\mathbb{Z}$. Let $\mathcal{E}$ be the set of pairs (X_1, φ_1), where X_1 is a subgroup of X containing X' (i.e., $X' \subset X_1 \subset X$) and $\varphi_1 : X_1 \to Y$ is an extension of φ (i.e., $\varphi_1 |_{X'} = \varphi$). Define on $\mathcal{E}$ a partial ordering: $(X_1, \varphi_1) \geq (X_2, \varphi_2)$ iff $X_1 \supseteq X_2$ and $\varphi_1 |_{X_2} = \varphi_1$. By the Zorn lemma, $\mathcal{E}$ contains at least one maximal element (X_1, φ_1). We show that this element is of the form (X, ψ), i.e. it gives a required extension of φ to X. Indeed, let $X_1 \neq X$ and $x \in X$, $x \notin X_1$. Let X_2 be the subgroup of X generated by X_1 and x. We extend $\varphi_1 : X_1 \to Y$ to $\varphi_2 : X_2 \to Y$. To do this we must define $\varphi_2(x)$. If $nx \notin X_1$ for all $n \in \mathbb{Z}$, $\varphi_2(x)$ can be any element of Y (for example, we can set $\varphi_2(x) = 0$). Otherwise, all $n \in \mathbb{Z}$ for which $nx \in X_1$ are multiples of some minimum element which we denote by n_0. Then $\varphi_2(x) \in Y$ should satisfy the condition $n_0 \varphi_2(x) = \varphi_1(n_0 x)$. Now it remains to verify that the equation $n_0 a = b$ (as an equation in a) can be solved in any of the groups $\mathbb{Q}$ or $\mathbb{Q}_{(p)}/\mathbb{Z}$. For $\mathbb{Q}$ this is clear, for $\mathbb{Q}_{(p)}/\mathbb{Z}$ one has to use the Euclid algorithm.

This example leads to an important observation: injective abelian groups are infinitely generated. This is a general phenomenon: we shall see in the next chapter (Sect. III.8.1) that to prove the existence of sufficiently many injective objects in, say, the category of sheaves of abelian groups one has to use infinite limits.

12. Flat Modules and Relations

Let us consider a left A-module X and a finite set of elements $x_1, \ldots, x_n \in X$. A family $(a_1, \ldots, a_n) \in A^n$ is called a *relation* among $x_1, \ldots, x_n$ if $\sum_{i=1}^n a_i x_i = 0$.

More generally, a family $(y_1, \ldots, y_n) \in Y^n$, where Y is a right A-module, is called a *relation among* $x_1, \ldots, x_n$ *in* Y if $\sum_{i=0}^n y_i \otimes x_i = 0$ in $Y \underset{A}{\otimes} X$.

Relations in Y can be obtained from relations in A as follows. Let $\left(a_1^{(j)}, \ldots, a_n^{(j)}\right) \in A^n, j = 1, \ldots, m$, be a family of relations among $x_1, \ldots, x_n$, and let $y^{(j)}, j = 1, \ldots, n$, be elements of Y. Then $\left(y_i = \sum_{j=1}^m y^{(j)} a_i^{(j)}\right) \in Y^n$ is a relation in Y:

$$\sum_{i=1}^n y_i \otimes x_i = \sum_{i=1}^n \sum_{j=1}^m y^{(j)} a_i^{(j)} \otimes x_i$$
$$= \sum_{i=1}^n \sum_{j=1}^m y^{(j)} \otimes a_i^{(j)} x_i = \sum_{j=1}^m y^{(j)} \otimes \sum_{i=1}^n a_i^{(j)} x_i = 0.$$

Now we can characterize flat right A-modules Y by the following property: *for any left A-module X and for any finite set of elements $x_1, \ldots, x_n \in X$ all relations among x_i with coefficients in Y are consequences of relations among x_i with coefficients in A.*

Indeed, let Y be a flat right A-module, X a left A-module, and $x_1, \ldots x_n \in X$. Define $\varphi : A^n \to X$ by $\varphi(a_1, \ldots, a_n) = \sum a_i x_i$. It is clear that $R = \ker \varphi \subset A^n$ consists of all relations among $x_1, \ldots, x_n$ in A. Tensoring the exact sequence
$$0 \longrightarrow R \overset{\iota}{\longrightarrow} A^n \overset{\varphi}{\longrightarrow} X$$
with the flat module Y we get the exact sequence
$$0 \longrightarrow Y \otimes R \overset{\iota'}{\longrightarrow} Y^n \overset{\varphi'}{\longrightarrow} Y \otimes X. \tag{II.39}$$

Now let $(y_1, \ldots, y_n) \in Y^n$ be a relation among $x_1, \ldots, x_n$ in Y. This means that $\varphi'(y_1, \ldots, y_n) = \sum y_i \otimes x_i = 0$. The exactness of (II.39) implies that there exist $y^{(j)} \in Y$ and $r^{(j)} \in \left(a_1^{(j)}, \ldots, a_n^{(j)}\right) \in R$, $j = 1, \ldots, m$, such that

$$(y_1, \ldots, y_n) = \sum_{j=1}^{m} y^{(j)} \otimes r^{(j)} = \left(\sum_{j=1}^{m} y^{(j)} a_1^{(j)}, \ldots, \sum_{j=1}^{m} y^{(j)} a_n^{(j)}\right).$$

Hence the relation $(y_1, \ldots, y_n)$ with coefficients in Y is a consequence of relations $r^{(j)}$ with coefficients in A.

On the other hand, let Y satisfy the above property, and let

$$0 \longrightarrow X' \overset{f}{\longrightarrow} X \overset{g}{\longrightarrow} X'' \longrightarrow 0$$

be an exact sequence of left A-modules. Since the functor $X \mapsto Y \underset{A}{\otimes} X$ is right exact, we must only check that $f' : Y \underset{A}{\otimes} X' \longrightarrow Y \underset{A}{\otimes} X$ is an injection. Let $\sum_{i=1}^{n} y_i \otimes x_i' \in Y \underset{A}{\otimes} X'$ be such that $f'\left(\sum y_i \otimes x_i'\right) = \sum y_i \otimes f(x_i') = 0$ in $Y \underset{A}{\otimes} X$. This means that $(y_1, \ldots, y_n)$ is a relation among $(f(x_1'), \ldots, f(x_n'))$ in Y. It is a consequence of relations in A, so that there exist $y^{(j)} \in Y$, $a_i^{(j)} \in A$; $j = 1, \ldots, m$, $i = 1, \ldots, n$, such that $y_i = \sum_{j=1}^{m} y^{(j)} a_i^{(j)}$ and $\sum_{i=1}^{n} a_i^{(j)} f(x_i') = 0$. Then $\sum_{i=1}^{n} a_i^{(j)} f(x_i') = f\left(\sum_{i=1}^{n} a_i^{(j)} x_i'\right) = 0$, and since f is an injection, $\sum_{i=1}^{n} a_i^{(j)} x_i' = 0$ for any $j = 1, \ldots, m$. Therefore $\sum_{i=1}^{n} y_i \otimes x_i' = \sum_{i=1}^{n} \sum_{j=1}^{m} y^{(j)} a_i^{(j)} \otimes x_i' = 0$ in $Y \underset{A}{\otimes} X'$.

13. Flat Modules and Projective Modules

The following properties of flat modules are almost obvious:

a) Free modules are flat.
b) Direct summands of flat modules are flat.
c) Inductive limits of families of flat modules are flat.

(To check this property one must use the fact that inductive limits commute with tensor products and preserve exactness. See some details in the next section.)

Properties a) and b) imply that projective modules are flat, and property c) implies that inductive limits of projective modules are flat. The theorem proved independently by Govorov and Lazard says that, conversely, any flat A-module is isomorphic to an inductive limit of free modules of finite type.

14. Acyclic Objects of a Site

Continuing the analysis of objects generating exact functors we return to the situation of Proposition II.6.6. Let $\mathcal{A}$ be an abelian category of sheaves on a site $\mathcal{S}$. An object $U \in \mathrm{Ob}\,\mathcal{S}$ is said to be $\mathcal{A}$-acyclic if the functor $\mathcal{A} \to \mathrm{Ab}$, $F \mapsto F(U)$, is exact.

Here are three important examples of acyclic objects. In all these examples $\mathcal{S}$ is the topology on a topological space X.

a) $\mathcal{A}$ is the category of constant sheaves of abelian groups on a topological space X, $U \subset X$ is an open subset homeomorphic to a ball.

b) $\mathcal{A}$ is the category of sheaves of modules on a scheme X that are locally isomorphic to cokernels of morphisms of $\mathcal{O}_X$-modules $\mathcal{O}_X^I \to \mathcal{O}_X^J$ (such sheaves are called quasicoherent; for schemes see II.4.10 and Ex. I.5.3). An open (in Zariski topology) subset $U \subset X$ is $\mathcal{A}$-acyclic iff $(U, \mathcal{O}_{X|U})$ is isomorphic to an affine scheme as a ringed space (the Serre theorem).

c) $\mathcal{A}$ is the category of modules on an analytic space that are locally (in the usual Hausdorff topology) isomorphic to cokernels of $\mathcal{O}_X$-modules $\mathcal{O}_X^I \to \mathcal{O}_X^J$ (quasicoherent analytic sheaves). The class of $\mathcal{A}$-acyclic subsets $U \subset X$ coincides with the class of the so-called Stein manifolds (the Cartan theorem).

Finally, we consider the exactness properties of direct and inverse images of sheaves.

15. Definition. *Let $f : M \to N$ be a continuous map of topological spaces, $\mathcal{F}$ a sheaf of sets on M. Its* direct image $f_\bullet \mathcal{F}$ *is defined as a sheaf on N whose sections over any open $U \subset X$ are given by*

$$f_\bullet(\mathcal{F})(U) = \mathcal{F}\left(f^{-1}(U)\right),$$

and the restriction map from U to $V \subset U$ is determined by the restriction map from $f^{-1}(U)$ to $f^{-1}(V)$.

Clearly, $f_\bullet(\mathcal{F})$ is a presheaf; one can check directly that it is a sheaf.

16. Properties of the Direct Image

a) The construction of $f_\bullet$ preserves additional structures that may exist on a sheaf $\mathcal{F}$: the direct image of a sheaf of groups is a sheaf of groups, and so on. If $\Phi = (f, \theta)$ is a morphism of ringed spaces (see II.4.5) and $\mathcal{F}$ is a sheaf of $\mathcal{O}_M$-modules, then $f_\bullet(\mathcal{F})$ has a natural structure of a sheaf of $\mathcal{O}_N$-modules. Indeed, we must define an action of $\mathcal{O}_N(U)$ on $f_\bullet(\mathcal{F})(U) = \mathcal{F}\left(f^{-1}(U)\right)$. But the morphism Φ of ring spaces contains a (compatible with restrictions) family of ringed homomorphisms $\varphi_U : \mathcal{O}_N(U) \to \mathcal{O}_M\left(f^{-1}(U)\right)$ and $\mathcal{F}\left(f^{-1}(U)\right)$ is a $\mathcal{O}_M\left(f^{-1}(U)\right)$-module. Therefore $f_\bullet(\mathcal{F})(U)$ becomes an $\mathcal{O}_N(U)$-module. This also shows that $\{\varphi_U\}$ can be considered as a morphism of sheaves of rings $\mathcal{O}_N \to f_\bullet(\mathcal{O}_M)$.

b) Let $f : M \to \{\bullet\}$ (one point space). Then $f_\bullet(\mathcal{F}) = \Gamma(M, \mathcal{F})$ (global sections).

More generally, if $f : M \to N$ is a locally trivial fibration, then $f_\bullet(\mathcal{F})$ is the sheaf of "sections of $\mathcal{F}$ along the fibres of f".

Let $i : M \to N$ be a closed embedding. The sheaf $i_\bullet(\mathcal{F})$ is sometimes called "the extension of $\mathcal{F}$ by zero": if $U \subset N \backslash M$, then $i_\bullet(\mathcal{F})(U)$ is the one-point set (if $\mathcal{F}$ is a sheaf of sets) or $i_\bullet(\mathcal{F})(U) = \{0\}$ (if $\mathcal{F}$ is a sheaf of abelian groups).

The use of the same name for the sheaf $i_\bullet(\mathcal{F})$ in the case when $i : M \to N$ is a general (not necessarily closed) embedding is safe if one remembers that the stalk of $i_\bullet(\mathcal{F})$ at a point of the boundary of $i(M)$ might be nontrivial.

c) Any morphism of sheaves $\varphi : \mathcal{F} \to \mathcal{G}$ on M induces (in an obvious way) the morphism of sheaves $f_\bullet(\varphi) : f_\bullet(\mathcal{F}) \to f_\bullet(\mathcal{G})$ on N. Therefore, $f_\bullet$ is a functor from the category of sheaves on M to the category of sheaves on N. The same is true for categories of abelian groups, of modules over ringed spaces, and so on.

We mention also two functorial properties of $f_\bullet$ with respect to f:

$$(fg)_\bullet = f_\bullet g_\bullet, \quad \mathrm{id}_\bullet = \mathrm{id}.$$

Now let us consider inverse images. Let again $f : M \to N$ be a morphism of topological spaces, and $\mathcal{F}$ a sheaf of sets on N.

The main property of the inverse image $f^\bullet(\mathcal{F})$ is that the functor $f^\bullet$ is right adjoint to the functor $f_\bullet$ (see II.3.23). Therefore, we give the definition of $f^\bullet(\mathcal{F})$ as an existence theorem. Such a choice is motivated also by the fact that when we consider, say, sheaves of modules on ringed spaces the construction of $f^\bullet(\mathcal{F})$ changes but preserves the adjunction property.

17. Proposition (– Definition). *There exists a functor* $f^\bullet : S\operatorname{Set}_N \to S\operatorname{Set}_M$ *and an isomorphism of bifunctors*

$$\operatorname{Hom}(f^\bullet(\mathcal{F}), \mathcal{G}) \xrightarrow{\sim} \operatorname{Hom}(\mathcal{F}, f_\bullet(\mathcal{G})). \qquad (\mathrm{II}.40)$$

II.6 Functors in Abelian Categories

Proof. We give an explicit construction of the functor $f^\bullet : S\operatorname{Set}_N \to S\operatorname{Set}_M$. Let $\mathcal{F}$ be a sheaf on N. For any open set $U \subset M$ we set

$$f^\bullet(\mathcal{F})(U) = \{\text{families } \{s'_x\}_{x \in U} \text{ where } s'_x \in \mathcal{F}_{f(x)}, \text{ and for any } \\ x \in U \text{ there exist: a neighbourhood } V \text{ of the point} \\ f(x) \text{ in } N, \text{ a neighbourhood } W \subset f^{-1}(V) \cap U \text{ of} \\ \text{the point } x \text{ in } M, \text{ and a section } s \in \mathcal{F}(V) \text{ such} \\ \text{that } s'_x = s_{f(z)} \text{ for all } z \in W \}.$$

Restrictions $f^\bullet(\mathcal{F})(U_1) \to f^\bullet(\mathcal{F})(U_2)$ for $U_2 \subset U_1$ are defined in the natural way. It is easy to see that $f^\bullet(\mathcal{F})$ is a sheaf of sets on M.

Next, for a morphism of sheaves $\varphi : \mathcal{F}_1 \to \mathcal{F}_2$ on N we define $f^\bullet(\varphi) : f^\bullet(\mathcal{F}_1) \to f^\bullet(\mathcal{F}_2)$ by the formula

$$f^\bullet(\varphi)(U)(\{s'_x\}) = \{\varphi_{f(x)}(s'_x)\}$$

for $U \subset M$, $s'_x \in (\mathcal{F}_1)_x$, $x \in U$.

So we have constructed a functor $f^\bullet : S\operatorname{Set}_N \to S\operatorname{Set}_M$. Now we check the adjunction property (II.40).

First assume that we have a morphism of sheaves $\psi : \mathcal{F} \to f_\bullet(\mathcal{G})$. Let s' be a section of $f^\bullet(\mathcal{F})$ over a neighbourhood U of a point $x \in M$, V be a neighbourhood of the point $f(x)$ in N, and suppose that $s \in \mathcal{F}(V)$ satisfies the condition $s'_z = s_{f(z)}$ for all $z \in W \subset f^{-1}(V) \cap U$. Then $\psi(V)(s) \in f_\bullet(\mathcal{G})(V) = \mathcal{G}(f^{-1}(V))$ and the morphism $\alpha(\psi) : f^\bullet(\mathcal{F}) \to \mathcal{G}$ corresponding to ψ is given by

$$(\alpha(\psi)(W))(s'|_W) = (\psi(V)(s))|_W.$$

It is easy to check that $\alpha(\psi)$ is well defined (in particular, $\alpha(\psi)$ does not depend on the choice of V, W and s).

On the other hand, let a morphism of sheaves $\varphi : f^\bullet(\mathcal{F}) \to \mathcal{G}$ be given. Let V be an open set in N and $s \in \mathcal{F}(V)$. We define the section $s' \in f^\bullet(\mathcal{F})(f^{-1}(V))$ by setting $s'_x = s_{f(x)}$, $x \in f^{-1}(V)$. Then $\tilde{s} = \varphi(f^{-1}(V))(s') \in \mathcal{G}(f^{-1}(V)) = f_\bullet(\mathcal{G})(V)$ and we set

$$\beta(\varphi)(V)(s) = \tilde{s}.$$

Again it is easy to check that $\beta(\varphi)$ is a morphism of sheaves.

We leave it to the reader to check that α and β are mutually inverse maps establishing the isomorphism of bifunctors (II.40). $\square$

18. Properties of Inverse Image

a) The sheaf $f^\bullet(\mathcal{F})$ can be defined also in two other ways. According to the first one, $f^\bullet(\mathcal{F})$ is the sheaf on M associated with the presheaf

$$U \mapsto \mathcal{F}(f(U)), \quad U \subset M.$$

Since $f(U)$ is, in general, not open in N, this definition of $f^\bullet(\mathcal{F})$ requires two limits, one to define $\mathcal{F}(f(U))$ (see I.5.5), and another to define the associated sheaf.

The second definition is given in terms of the total space F of the sheaf $\mathcal{F}$ (see I.5.5): sections of the sheaf $f^\bullet(\mathcal{F})$ over an open set $U \in M$ are continuous maps $s : U \to F$ such that $s(x) \in \mathcal{F}_{f(x)} \subset F$ for all $x \in U$.

An easy verification (using any of the two definitions) shows that for any $x \in M$ we have $f^\bullet(\mathcal{F})_x = \mathcal{F}_{f(x)}$.

b) The construction of $f^\bullet(\mathcal{F})$ used in the proof of Proposition II.6.17 makes sense also in the case when $\mathcal{F}$ is a presheaf (and not a sheaf) of sets on N. In any case $f^\bullet(\mathcal{F})$ will be a sheaf. For example, if $f : M \to M$ is the identity map, $f^\bullet(\mathcal{F})$ is the sheaf associated with a presheaf $\mathcal{F}$.

c) The construction of $f^\bullet$ preserves the inner composition laws: if $\mathcal{F}$ is a sheaf of abelian groups, of modules, etc., then $f^\bullet(\mathcal{F})$ belongs to the same category. However, if $\Phi = (f, \theta)$ is a morphism of ringed spaces and $\mathcal{F}$ is a sheaf of $\mathcal{O}_N$-modules, then $f^\bullet(\mathcal{F})$ does not carry, in general, a natural structure of sheaf of $\mathcal{O}_M$-modules.

It is quite instructive to see why the argument similar to II.6.16a) breaks. According to Proposition II.6.17, the morphism Φ determines also a morphism of sheaves of rings $f^\bullet(\mathcal{O}_N) \to \mathcal{O}_M$. But now $f^\bullet(\mathcal{F})$ is a sheaf of modules not over the target, but over the beginning of this arrow. Algebraic wisdom says us that the necessary base change in this case is given by the tensor product, so that it is natural to set

$$f^*(\mathcal{F}) = \mathcal{O}_M \underset{f^\bullet(\mathcal{O}_N)}{\otimes} f^\bullet(\mathcal{F})$$

(or $f^\bullet(\mathcal{F}) \underset{f^\bullet(\mathcal{O}_N)}{\otimes} \mathcal{O}_M$ depending on whether we work with left or right modules). One can check that

$$\operatorname{Hom}_{\mathcal{O}_M}(f^*(\mathcal{F}), \mathcal{G}) = \operatorname{Hom}_{\mathcal{O}_N}(\mathcal{F}, f_\bullet(\mathcal{G})),$$

so that $f^*(\mathcal{F})$ is a good definition of the direct image in the category of sheaves of modules.

d) Let $f : M \to \{\cdot\}$, and let $\mathcal{F}$ be a set (or a group, etc.) considered as a sheaf over the point. Then $f^\bullet(\mathcal{F})$ is the constant sheaf with fibre $\mathcal{F}$.

If $i : M \to N$ is an embedding, then $i^\bullet(\mathcal{F})$ is the restriction of $\mathcal{F}$ to M.

e) Similarly to the direct image, we have

$$(fg)^\bullet = g^\bullet f^\bullet, \quad \mathrm{id}^\bullet = \mathrm{id}.$$

Exactness properties of $f_\bullet$, $f^\bullet$ and f^* are described by the following proposition.

19. Proposition. *a) In the category of sheaves of abelian groups the functor $f_\bullet$ is left exact and the functor $f^\bullet$ is exact.*

b) In the category of sheaves of modules over ringed spaces the functor $f_\bullet$ is left exact and the functor f^* is right exact.

We give the proof of this proposition based on general exactness properties of adjoint functors.

20. Adjunction and Exactness

Exactness properties of adjoint functors are given by the following statement.

Let $\mathcal{C}, \mathcal{D}$ be two abelian categories, $F : \mathcal{C} \to \mathcal{D}$, $G : \mathcal{D} \to \mathcal{C}$ be two additive functors, and assume we are given an isomorphism of bifunctors

$$\mathrm{Hom}_{\mathcal{D}}(FX, Y) = \mathrm{Hom}_{\mathcal{C}}(X, GY) \tag{II.41}$$

so that F is left adjoint to G, and G is right adjoint to F. Then F is right exact and G is left exact.

Indeed, let

$$0 \longrightarrow Y' \xrightarrow{f} Y \xrightarrow{g} Y'' \longrightarrow 0$$

be an exact sequence in $\mathcal{D}$. Then, using (II.41) and the left exactness of the functor $Y \mapsto \mathrm{Hom}_{\mathcal{D}}(FX, Y) : \mathcal{D} \to \mathrm{Ab}$ (Proposition II.6.5), we see that the sequence

$$0 \to \mathrm{Hom}_{\mathcal{C}}(X, GY') \to \mathrm{Hom}_{\mathcal{C}}(X, GY) \to \mathrm{Hom}_{\mathcal{C}}(X, GY'')$$

is exact for any $X \in \mathrm{Ob}\,\mathcal{C}$. Hence GY' is the object of $\mathcal{C}$ representing the functor

$$X \mapsto \ker\left(\mathrm{Hom}_{\mathcal{C}}(X, GY) \longrightarrow \mathrm{Hom}_{\mathcal{C}}(X, GY'')\right),$$

so that $GY' = \ker G(g)$ (see II.5.6). Hence the sequence

$$0 \to GY' \to GY \to GY''$$

is exact, and G is left exact. Similarly one can prove that F is right exact.

Proof (of Proposition II.6.19). Propositions II.6.6 and II.6.17, Remark II.6.18c), and the statement from II.6.20 prove all the required properties, except the exactness of $f^\bullet$. The simplest way to prove the exactness of $f^\bullet$ is to note that the sequence

$$0 \longrightarrow \mathcal{F}' \longrightarrow \mathcal{F} \longrightarrow \mathcal{F}'' \longrightarrow 0$$

of sheaves of abelian groups on N is exact iff for any $y \in N$ the sequence

$$0 \longrightarrow \mathcal{F}'_y \longrightarrow \mathcal{F}_y \longrightarrow \mathcal{F}''_y \longrightarrow 0$$

of stalks at y is exact, and then to use the equality $f^\bullet(\mathcal{F})_x = \mathcal{F}_{f(x)}$, $x \in M$ (see II.6.18a)). $\square$

Exercises

1. Limits In Abelian Categories. a) Prove that the equalizer of morphisms $f, g : X \to Y$ in an abelian category $\mathcal{A}$ coincides with the kernel of $f - g$. Similar results hold for cokernels and coequalizers. Conclude that finite limits and colimits exist in any abelian category.

The existence of infinite limits (and colimits) imposes additional restrictions on an abelian category $\mathcal{A}$. Following Grothendieck [1], we formulate some standard axioms.

AB3). *The direct sum $\oplus X_i$ of any family $(X_i)_{i \in I}$ of objects in $\mathcal{A}$ exists.*

b) Let I be some set of indices. Define the category $\mathcal{A}^I$ whose objects are families $(X_I)_{i \in I}$, $X_i \in \mathrm{Ob}\,\mathcal{A}$ (with obvious morphisms), prove that it is abelian, and, assuming AB3, prove that $(X_i) \to \oplus X_i$ determines a functor $\mathcal{A}^I \to \mathcal{A}$. Prove that this functor is exact for finite I and is right exact for an arbitrary I. Give an example showing that, in general, it is not left exact (i.e., that the direct sum of monomorphisms might fail to be a monomorphism). Thus,

AB4). *AB3 is satisfied and the direct sum of monomorphisms is a monomorphism.*

c) Formulate dual axioms AB3*, AB4* (about direct products).

d) Prove that the category Ab (and, more generally, the category R-mod for any ring R) satisfies AB4, AB4*, and the category of sheaves of abelian groups on a topological space satisfies AB4 and AB3*, but does not satisfy, in general, AB4*.

2. Categories That Are Equivalent to Categories of Modules.

Let $\mathcal{A}$ be an abelian category satisfying AB3. An object P of $\mathcal{A}$ is said to be *strictly projective* (or a *projective generator*) if the functor $h' : X \mapsto \mathrm{Hom}_{\mathcal{A}}(P, X)$ from $\mathcal{A}$ to Ab is exact (projectivity of P), strict (i.e., $h'(X) = 0 \Rightarrow X = 0$), and commutes with direct sums. For such an object P we set $R = \mathrm{Hom}_{\mathcal{A}}(P, P)$. Any group $h'(X) = \mathrm{Hom}_{\mathcal{A}}(P, X)$ carries a natural structure of a right R-module. Prove that h' determines an equivalence of $\mathcal{A}$ and mod-R.

b) Let $\mathcal{A}$ be a Noetherian category (i.e., any increasing chain of subobjects stabilizes), and let P be a projective object in $\mathcal{A}$ such that h' is a strict functor. Then the ring $R = \mathrm{Hom}_{\mathcal{A}}(P, P)$ is right Noetherian and h' determines an equivalence between $\mathcal{A}$ and the category of finitely generated R-modules.

c) Prove that conditions from b) are satisfied in the case when $\mathcal{A}$ has only a finite number of pairwise nonisomorphic simple objects $X_1, \ldots, X_n$ such that for any i there exists an epimorphism $P_i \to X_i$ with a projective P_i and $P = \oplus P_i$. Categories with these (or similar) properties often appear in group representation theory.

III. Derived Categories and Derived Functors

III.1 Complexes as Generalized Objects

1. Generators and Relations

Homological algebra was founded by D. Hilbert. He considered, in particular, the following problem. Let $\sum_{j=1}^{m} a_{ij}x_j = 0$, $i = 1,\ldots,n$, $a_{ij} \in k[t_1,\ldots,t_r]$, be a system of linear homogeneous equations with coefficients lying in the polynomial ring over a field. All polynomial solutions are linear combinations (with polynomial coefficients) of a finite subset of solutions. However, in general there exists no basis of solutions that are linearly independent over $k[t_1,\ldots,t_r]$. Linear relations among elements of a generating system of solutions are, in turn, linear combinations of some finite set of relations, and again it might happen that there exists no free system of generators for relations.

Hilbert had called such linear relations among solutions, among relations and so on, *syzygies*. This name came from celestian mechanics where integral relations among periods of solar system planets are closely related to stability and resonance phenomena. Hilbert's fundamental theorem states that for the ring $k[t_1,\ldots,t_r]$ the chain of syzygies ends after $r+1$ steps; this is a generalization of the fact that the space of solution of a linear system over a field ($r = 0$) has a basis.

Using a more modern language, let us consider a ring A (not necessarily commutative), a left A-module E, and an exact sequence in the category of left A-modules

$$\ldots \longrightarrow F^{-n} \xrightarrow{d^{-n}} F^{-n+1} \xrightarrow{d^{-n+1}} F^{-n+2} \longrightarrow \ldots \longrightarrow F^0 \xrightarrow{\varepsilon} E \longrightarrow 0 \quad \text{(III.1)}$$

with F^i being free modules.

The complex $F^\bullet$ together with the augmentation morphism $\varepsilon : F^\bullet \to E$ is called a *free (left) resolution* of the module E. Choosing generators $(f_1^i,\ldots,f_{k_i}^i)$ of the free module F^i we can describe the resolution (III.1) in Hilbert's terms:

- $\varepsilon\left(f_1^0\right),\ldots,\varepsilon\left(f_{k_0}^0\right)$ are generators of the A-module E (because ε is a surjection);

140   III. Derived Categories and Derived Functors

– $d^{-1}\left(f_1^{-1}\right), \ldots, d^{-1}\left(f_{k_{-1}}^{-1}\right)$ are generators of the A-module of first syzygies, that is, of relations among the above generators (because d^{-1} maps F^{-1} onto the kernel of ε which consists just of these relations);

– $d^{-2}\left(f_1^{-2}\right), \ldots, d^{-2}\left(f_{k_{-2}}^{-2}\right)$ are generators of the A-module of second syzygies, and so on.

We shall return to the Hilbert theorem in Sect. III.5, and now we shall look at resolutions.

One of the main techniques in homological algebra is to replace E by its resolution $F^\bullet$. Let is discuss first the uniqueness of such a resolution.

We start with some general definitions. Let $\mathcal{A}$ be an abelian category. The category $\operatorname{Kom}(\mathcal{A})$ of complexes over $\mathcal{A}$ is defined in an obvious way (see II.1.4e) for $\mathcal{A} = Ab$).

2. Lemma (– Definition). *a) Let $K^\bullet$ and $L^\bullet$ be two complexes over $\mathcal{A}$ and let $k = (k^i)$, $k^i : K^i \to L^{i-1}$ be a collection of morphisms in $\mathcal{A}$. Then the maps*
$$h = kd + dk : K^\bullet \longrightarrow L^\bullet,$$
i.e.,
$$h^i = k^{i+1}d_K^i + d_L^{i-1}k^i : K^i \longrightarrow L^i$$

$$\begin{array}{ccccccc}
\ldots K^{i-1} & \longrightarrow & K^i & \longrightarrow & K^{i+1} & \longrightarrow & K^{i+2} \ldots \\
& {}^{k^i}\swarrow & \downarrow h^i & {}^{k^{i+1}}\swarrow & \downarrow h^{i+1} & {}^{k^{i+2}}\swarrow & \\
\ldots L^{i-1} & \longrightarrow & L^i & \longrightarrow & L^{i+1} & \longrightarrow & L^{i+2} \ldots
\end{array}$$

form a morphism of complexes.

The morphism $h : K^\bullet \to L^\bullet$ is said to be homotopic to 0 (notation $h \sim 0$).

b) Morphisms homotopic to 0 form an ideal in $\operatorname{Mor} \operatorname{Kom}(\mathcal{A})$ in the following sense: if $h_1, h_2 : K^\bullet \to L^\bullet$ and $h_1 \sim 0$, $h_2 \sim 0$, then $h_1 + h_2 \sim 0$; similarly, $fh_1 \sim 0$ and $h_2 g \sim 0$ whenever these compositions exist.

Morphisms $f, g : K^\bullet \to L^\bullet$ are said to be homotopic (to each other) if $f - g = kd + dk \sim 0$ (notation $f \sim g$); k is the corresponding homotopy.

c) If $f \sim g$, then $H^\bullet(f) = H^\bullet(g)$ where $H^\bullet(\bullet)$ is the mapping induced on the homology of a complex.

Proof. a) We have to verify that $dh = hd$. Formally, $dh = dkd = hd$ because $d^2 = 0$. The reader can insert appropriate indices in this formula.

b) If $h_i = dk_i + k_i d$, $i = 1, 2$, then $h_1 + h_2 = d(k_1 + k_2) + (k_1 + k_2)d$. Similarly, $fh_1 = d(fk_1) + (fk_1)d$, $h_2 g = d(k_2 g) + (k_2 g)d$, because f and g commute with d.

c) It suffices to verify that if $h : K^\bullet \to L^\bullet$ is homotopic to 0, then $H^\bullet(h) = 0$. Let $k = (k^i)$, $k^i : K^i \to L^{i-1}$, be the corresponding homotopy, $a \in H^i(K^\bullet)$ and $\bar{a} \in K^i$ a cocycle representing a. Then $H^i(h)a \in H^i(L^\bullet)$ can be represented by the cocycle

III.1 Complexes as Generalized Objects   141

$$h^i \bar{a} = k^{i+1} d_K^i \bar{a} + d_L^{i-1} k^i \bar{a} = d_L^{i-1}(k^i \bar{a})$$

(because $\bar{a}$ is a cocycle). Hence $H^i(h)a = 0$. □

Now we can formulate a theorem about the uniqueness of a resolution. Compared with the subsection III.1.1, we generalize the situation in two respects: instead of A-modules we consider an arbitrary abelian category $\mathcal{A}$, and instead of free resolutions we consider projective ones (so that F^i are projective in $\mathcal{A}$, see II.6.8a)). Finally, we study the behaviour of resolutions under morphisms.

3. Theorem. *Let $X, Y \in \operatorname{Ob} \mathcal{A}$, $P^\bullet \xrightarrow{\varepsilon_X} X$, $Q^\bullet \xrightarrow{\varepsilon_Y} Y$ projective resolutions of X, Y, $f : X \to Y$ a morphism. Then*

a) there exists a morphism of resolutions $R(f) : P^\bullet \to Q^\bullet$ extending f in the sense that the diagram

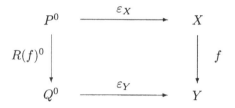

commutes;

b) any two such extensions $R(f)$ and $R'(f)$ are homotopic.

Proof. Construction of $R(f)^0$. $Q^0 \xrightarrow{\varepsilon_Y} Y$ being epimorphic, we get a projectivity diagram (see III.6.9)

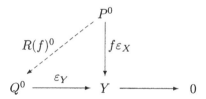

Since P^0 is projective, there exists a morphism $R(f)^0$ making this diagram commutative.

Inductive construction of $R(f)^{-i} : P^{-i} \to Q^{-i}$, $i > 0$. We need the following generalization of the projectivity diagram. Assume that in the diagram

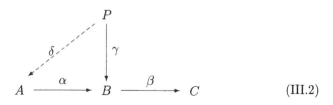   (III.2)

the lower row is exact, the object P is projective, and $\beta \circ \gamma = 0$. Then there exists a morphism $\delta : P \to Q$ making this diagram commutative. This generalization is a reformulation of the exactness of the functor $X \mapsto \mathrm{Hom}(P, X)$ for the projective object P (see II.6.8).

Now we can construct $R(f)^{-i}$. Let us assume that all $R(f)^{-j} : P^{-j} \to Q^{-j}$, $0 \leq j < i$, are already constructed and

$$d_Q^{-j} \circ R(f)^{-j} = R(f)^{-j+1} \circ d_P^{-j}. \tag{III.3}$$

Let us consider the following diagram:

$$\begin{array}{c} P^{-i} \\ R(f)^{-i} \swarrow \quad \downarrow R(f)^{-i+1} \circ d_P^{-i} \\ Q^{-i} \xrightarrow{d_Q^{-i}} Q^{-i+1} \xrightarrow{d_Q^{-i+1}} Q^{-i+2} \end{array} \tag{III.4}$$

Using (III.3) with $j = i - 1$ we get

$$d_Q^{-i+1} \circ R(f)^{-i+1} \circ d_P^{-i} = R(f)^{-i+2} \circ d_P^{-i+1} \circ d_P^{-i} = 0.$$

Since $Q^\bullet$ is a resolution, the lower row in (III.4) is exact and we get a diagram of the form (III.2) with a projective P^{-i}. Hence there exists $R(f)^{-i} : P^{-i} \to Q^{-i}$ and the commutativity of the diagram (III.4) is equivalent to (III.3) with $j = i$. Part a) is proved.

Construction of a Homotopy. Assume that we have now two morphisms of resolutions $R(f), R'(f) : P^\bullet \to Q^\bullet$, both extending $f : X \to Y$. To construct a homotopy k between these morphisms we proceed inductively constructing $k^{-i} : P^{-i} \to Q^{-i-1}$ one by one, similarly to the construction of $R(f)^{-i}$ above. We describe here only the general step, leaving the start of the induction to the reader.

Thus, assume that for all $j, 0 \leq j < i$, we have morphisms $k^{-j} : P^{-j} \to Q^{-j-1}$ such that

$$d_Q^{-j-1} \circ k^{-j} + k^{-j+1} \circ d_P^{-j} = R(f)^{-j} - R'(f)^{-j} \tag{III.5}$$

(we assume in this formula that $k^1 = 0$). We must construct k^{-i}. Consider the diagram

$$\begin{array}{c} P^{-i} \\ k^{-i} \swarrow \quad \downarrow \alpha \\ Q^{-i-1} \xrightarrow{d_Q^{-i-1}} Q^{-i} \xrightarrow{d_Q^{-i}} Q^{-i+1} \end{array} \tag{III.6}$$

III.1 Complexes as Generalized Objects   143

with $\alpha = R(f)^{-i} - R'(f)^{-i} - k^{-i+1} \circ d_P^{-i}$. Using the formula (III.5) with $j = -i + 1$ and the fact that $R(f)$, $R'(f)$ are morphisms of complexes, we easily see that $d_Q^{-i} \circ \alpha = 0$. Since $Q^\bullet$ is a resolution, we get again a diagram of the form (III.2). Now, P^{-i} being projective, there exists $k^{-i} : P^{-i} \to Q^{-i-1}$ making the diagram (III.6) commutative, which is equivalent to (III.5) with $j = i$. □

4. Remarks

a) Take $X = Y$ and $f = \mathrm{id}_X$ in Theorem III.1.3. Then Theorem III.1.3 claims that for any two resolutions of the same object there exists a morphism between them extending the identity morphism, and this morphism is unique modulo homotopy. This means that a projective resolution "is unique modulo homotopic equivalence".

b) The proof of the theorem shows that it remains valid even if we do not assume that Q^{-i} are projective; we need only that $Q^\bullet \xrightarrow{\varepsilon_Y} Y \longrightarrow 0$ is exact.

c) Reversing arrows and replacing projective resolutions by injective ones we obtain a dual version of Theorem III.1.3, as well as a dual version of its generalization from the previous remark.

5. Definition. *A morphism $f : K^\bullet \to L^\bullet$ of complexes in an abelian category $\mathcal{A}$ is said to be a* quasi-isomorphism *if the corresponding homology morphism $H^n(f) : H^n(K^\bullet) \to H^n(L^\bullet)$ is an isomorphism for any n.*

Earlier we have met the following examples of quasi-isomorphisms:

a) There exists a quasi-isomorphism among any two projective (injective) resolutions of one object (Theorem III.1.3a) and Lemma III.1.2c)).

b) Any object X of an abelian category $\mathcal{A}$ can be considered as a complex $\ldots \to 0 \to 0 \to X \to 0 \to 0 \to \ldots$ (with X at the 0-th place). This complex is acyclic outside zero, and its 0-th cohomology is isomorphic to X; such a complex will be called a 0-complex. The augmentation ε_X of a left resolution $P^\bullet \xrightarrow{\varepsilon_X} X$ determines a quasi-isomorphism of complexes

$$
\begin{array}{ccccccccc}
\ldots & \to & P^{-1} & \to & P^0 & \to & 0 & \to & 0 & \ldots \\
& & \downarrow & & \downarrow \varepsilon_X & & \downarrow & & \\
\ldots & \to & 0 & \to & X & \to & 0 & \to & 0 & \ldots
\end{array}
$$

In this way the notion of a resolution is a special case of the notion of a quasi-isomorphism.

c) Let $0^\bullet$ be the complex with all terms equal to the zero object of $\mathcal{A}$. Then the unique morphism $K^\bullet \to 0^\bullet$ (and $0^\bullet \to K^\bullet$) is a quasi-isomorphism if and only if $K^\bullet$ is acyclic.

6. What is the Derived Category?

The idea of the derived category, as we understand it today, can be formulated as follows.

a) An object X of an abelian category should be identified with all its resolutions.

b) The main reason for such an identification is that some most important functors, such as Hom, tensor products, Γ, should be redefined. Their "naive" definitions (see II.6) should be applied only to some special objects, which are acyclic with respect to this functor. If, say, X is a flat module and Y is an arbitrary one, then $X \otimes Y$ is the correct definition of the tensor product. But in general case in order to get a correct definition one has to replace $X \otimes Y$ by $P^\bullet \otimes Y$, where $P^\bullet$ is a flat resolution of the module X. Similarly, to get the correct definition of the group $\Gamma(\mathcal{F})$ of section of a sheaf $\mathcal{F}$ one has to take the complex $\Gamma(\mathcal{I}^\bullet)$, where $\mathcal{I}^\bullet$ is an injective resolution of $\mathcal{F}$.

c) To adopt this point of view we must consider from the very beginning not only objects of an abelian category and their resolutions, but arbitrary complexes. One of the reasons why we have to do this is that $P^\bullet \otimes Y$ and $\Gamma(\mathcal{I}^\bullet)$ in the above examples would usually have nontrivial cohomology in several degrees and not only in degree 0. (Some readers can recall that in the classical homological algebra the corresponding cohomology groups are called derived functors $\mathrm{Tor}_i(X, Y)$ and $H^i(\mathcal{F})$ respectively.) Hence the relation that enables us to identify an object and its resolution should be generalized to arbitrary complexes. The appropriate generalization is given by the notion of a quasi-isomorphism (Definition III.1.5).

d) The equivalence relation between complexes generated by quasi-isomorphisms is rather complicated, and what happens after the factorization by this equivalence relation is difficult to trace. The technique that enables us to do it form the core of the theory of derived categories; it will be discussed in the present chapter.

e) The above (see b)) redefinition of such functors as $\otimes$, Γ and others makes semi-exact functors into "exact" ones, in some sense. The very notion of exactness in a derived category is by no means obvious; see the discussion in Sect. III.3. The origin of this notion in the classical homological algebra is the exact sequence of higher derived functors, which is invariant under the change of a resolution.

III.2 Derived Categories and Localization

1. Definition (– Theorem). *Let $\mathcal{A}$ be an abelian category, $\mathrm{Kom}(\mathcal{A})$ the category of complexes over $\mathcal{A}$. There exists a category $D(\mathcal{A})$ and a functor $Q : \mathrm{Kom}(\mathcal{A}) \to D(\mathcal{A})$ with the following properties:*

a) $Q(f)$ is an isomorphism for any quasi-isomorphism f.

b) Any functor $F : \mathrm{Kom}(\mathcal{A}) \to \mathcal{D}$ transforming quasi-isomorphisms into isomorphisms can be uniquely factorized through $D(\mathcal{A})$, i.e., there exists a unique functor $G : D(\mathcal{A}) \to \mathcal{D}$ with $F = G \circ Q$.

The category $D(\mathcal{A})$ is called the derived category of the abelian category *$\mathcal{A}$.*

2. A Simple Proof of the Existence: Localization of a Category

Let $\mathcal{B}$ be an arbitrary category and S an arbitrary class of morphisms in $\mathcal{B}$. We show that there exists a universal functor transforming elements of S into isomorphisms. More precisely, we construct a category $\mathcal{B}[S^{-1}]$ and a "localization by S" functor $Q : \mathcal{B} \to \mathcal{B}[S^{-1}]$ with the universality property similar to that of III.2.1b) above.

To do this we set first $\operatorname{Ob} \mathcal{B}[S^{-1}] = \operatorname{Ob} \mathcal{B}$ and define Q to be the identity on objects.

To construct morphisms in $\mathcal{B}[S^{-1}]$ we proceed in several steps.

a) Introduce variables x_s, one for every morphism $s \in S$.

b) Construct an oriented graph Γ as follows:

> vertices of Γ = objects of $\mathcal{B}$;
> edges of Γ = {morphisms in $\mathcal{B}$} $\cup \{x_s, s \in S\}$;
> the edge $X \to Y$ is oriented from X to Y;
> the edge x_s has the same vertices as the edge s but the opposite orientation.

c) A *path* in Γ is a finite sequence of edges such that the end of any edge coincides with the beginning of the next one.

d) A morphism in $\mathcal{B}[S^{-1}]$ is an equivalence class of paths in Γ with the common beginning and the common end. Two paths are equivalent if they can be joined by a chain of elementary equivalences of the following type:

- two consecutive arrows in a path can be replaced by their composition;
- arrows $X \xrightarrow{s} Y \xrightarrow{x_s} X$ (resp. $Y \xrightarrow{x_s} X \xrightarrow{s} Y$) can be replaced by $X \xrightarrow{\mathrm{id}} X$ (resp. $Y \xrightarrow{\mathrm{id}} Y$).

Finally, the composition of two morphisms is induced by the conjunction of paths and the functor $Q : \mathcal{B} \to \mathcal{B}[S^{-1}]$ maps a morphism $X \to Y$ into the class of corresponding path (of length 1). For any $s \in S$ the morphism $Q(s)$ is clearly an isomorphism in $\mathcal{B}[S^{-1}]$, the inverse being the class of the path x_s.

Given another functor $\mathcal{B} \to \mathcal{B}'$ transforming morphisms from S into isomorphisms, the associated functor $G : \mathcal{B}[S^{-1}] \to \mathcal{B}'$ with the condition $F = G \circ Q$ is constructed as follows:

$$G(X) = F(X), \quad X \in \operatorname{Ob} \mathcal{B} = \operatorname{Ob} \mathcal{B}[S^{-1}];$$

$$G(f) = F(f), \quad f \in \operatorname{Mor} \mathcal{B},$$

$$G(\text{class of } x_s) = F(s)^{-1}, \quad s \in S.$$

The reader can easily verify that all definitions are unambiguous and that the functor G is unique.

3. Splitting and Derived Categories

As a first insight into the structure of the derived category we consider the following construction. A complex $K^\bullet$ is said to be *cyclic* if all its differentials are zero (so that all chains are cycles). Cyclic complexes form a complete subcategory $\text{Kom}_0(\mathcal{A}) \subset \text{Kom}(\mathcal{A})$. The structure of $\text{Kom}_0(\mathcal{A})$ is obvious: it is isomorphic to the category $\prod_{n=-\infty}^{\infty} \mathcal{A}[n]$, where $\mathcal{A}[n]$ is the "n-th copy of $\mathcal{A}$". Let i be the inclusion functor $\text{Kom}_0(\mathcal{A}) \to \text{Kom}(\mathcal{A})$ and h be the cohomology functor,

$$h : \text{Kom}(\mathcal{A}) \to \text{Kom}_0(\mathcal{A}), \quad h((K^n, d^n)) = (H^n(K^\bullet), 0),$$
$$h(f : K^\bullet \to L^\bullet) = (H^n(f)).$$

Since h transforms quasi-isomorphisms into isomorphisms, it can be factors through $D(\mathcal{A})$ so that for any $\mathcal{A}$ we have a functor

$$k : D(\mathcal{A}) \longrightarrow \text{Kom}_0(\mathcal{A}).$$

An abelian category $\mathcal{A}$ is said to be *semisimple* if any exact triple in $\mathcal{A}$ splits, i.e., is isomorphic to a triple of the form $0 \longrightarrow X \xrightarrow{(\text{id},0)} X \oplus Y \longrightarrow Y \longrightarrow 0$. For example, the category of linear spaces over a field or the category of finite-dimensional linear representations of a finite group over a field of characteristic zero is semisimple. The category of abelian groups is not semisimple: the sequence $0 \longrightarrow \mathbb{Z} \xrightarrow{2} \mathbb{Z} \longrightarrow \mathbb{Z}/2 \longrightarrow 0$ does not split.

4. Proposition. *If an abelian category $\mathcal{A}$ is semisimple then the functor $D(\mathcal{A}) \to \text{Kom}_0(\mathcal{A})$ is an equivalence of categories.* (The opposite is also true, see Ex. IV.1.1)

Proof. Let $K^\bullet$ be an arbitrary complex. We construct first two morphisms

$$f_K : (K^n, d^n) \longrightarrow (H^n(K^\bullet), 0), \quad g_K : (H^n(K^\bullet), 0) \longrightarrow (K^n, d^n),$$

both inducing identity maps on cohomology. To do this we set $B^n = \text{Im}\, d^{n-1}$, $Z^n = \ker d^n$, $H^n = H^n(K')$. We have two exact sequences in $\mathcal{A}$:

$$0 \longrightarrow Z^n \longrightarrow K^n \longrightarrow B^{n+1} \longrightarrow 0$$

(the second morphism is induced by d^n) and

$$0 \longrightarrow B^n \longrightarrow Z^n \longrightarrow H^n \longrightarrow 0.$$

Since $\mathcal{A}$ is semisimple, we have

$$K^n = B^n \oplus H^n \oplus B^{n+1}$$

and the morphism

$$d^n : B^n \oplus H^n \oplus B^{n+1} \longrightarrow B^{n+1} \oplus H^{n+1} \oplus B^{n+2}$$

is given by

$$d^n(b^n, h^n, b^{n+1}) = (b^{n+1}, 0, 0).$$

Now we define f_K and g_K by
$$f_K^n(b^n, h^n, b^{n+1}) = h^n, \quad g_K^n(h^n) = (0, h^n, 0).$$

Let us prove that k determines an equivalence of categories. To do this we define the functor $l : \mathrm{Kom}_0(\mathcal{A}) \to D(\mathcal{A})$ as the composition of the embedding $\mathrm{Kom}_0(\mathcal{A}) \to \mathrm{Kom}(\mathcal{A})$ with the localization $Q : \mathrm{Kom}(\mathcal{A}) \to D(\mathcal{A})$, and prove that k and l are quasi-inverse to each other.

First, it is clear that the composition $k \circ l$ is isomorphic to the identity functor in $\mathrm{Kom}_0(\mathcal{A})$. On the other hand, the functor $l \circ k : D(\mathcal{A}) \to D(\mathcal{A})$ maps a complex $(K^n, d^n) \in D(\mathcal{A})$ to $(H^n(K^\bullet), 0) \in D(\mathcal{A})$. The above morphisms of complexes $f_K : K^\bullet \to l \circ k(K^\bullet)$ and $g_K : l \circ k(K^\bullet) \to K^\bullet$ clearly give mutually inverse isomorphisms in $D(\mathcal{A})$, so that $\{f_K\}$ and $\{g_K\}$, $K^\bullet \in D$ are mutually inverse isomorphisms of functors $\mathrm{id}_{D(\mathcal{A})}$ and $l \circ k$. □

5. Variants

In applications it is often useful to consider complexes with various finiteness conditions. In particular, let

$$\mathrm{Kom}^+(\mathcal{A}) : K^i = 0 \quad \text{for} \quad i \leq i_0(K^\bullet),$$
$$\mathrm{Kom}^-(\mathcal{A}) : K^i = 0 \quad \text{for} \quad i \geq i_0(K^\bullet),$$
$$\mathrm{Kom}^b(\mathcal{A}) = \mathrm{Kom}^+(\mathcal{A}) \cap \mathrm{Kom}^-(\mathcal{A}).$$

These are full subcategories in $\mathrm{Kom}(\mathcal{A})$ and localizing by quasi-isomorphisms we can form the corresponding derived categories $D^+(\mathcal{A})$, $D^-(\mathcal{A})$, $D^b(\mathcal{A})$. For example, left projective resolutions lie in $\mathrm{Kom}^-(\mathcal{A})$, while right injective resolutions lie in $\mathrm{Kom}^+(\mathcal{A})$.

Now we see that we could define, say $D^+(\mathcal{A})$, either as the localization of $\mathrm{Kom}^+(\mathcal{A})$ by quasi-isomorphisms or as the full subcategory of $D(\mathcal{A})$ consisting of complexes $K^\bullet$ with $H^i(K^\bullet) = 0$ for $i < i_0$. We would like to be certain that these two constructions coincide. However, we do not have enough technique to prove this.

The problem is, that morphisms in $\mathcal{S}[S^{-1}]$ constructed by the method of III.2.2 are just formal expressions of the form

$$f_1 \circ s_1^{-1} \circ f_2 \circ s_2^{-1} \circ \ldots \circ s_k^{-1} \circ f_k \quad \text{where} \quad f_i \in \mathrm{Mor}\,\mathcal{B}, s_i \in S \quad (*)$$

and to manipulate such an expression we need some algebraic identities like "finding the common denominator". For example, in some cases it is rather hard even to figure out whether or not $\mathcal{B}[S^{-1}]$ is equivalent to the trivial category with one object and one morphism.

All the required algebraic identities are provided by the following definition.

6. Definition. *A class of morphisms $S \subset \mathrm{Mor}\,\mathcal{B}$ is said to be* localizing *if the following conditions are satisfied:*

a) S is closed under compositions: $\mathrm{id}_X \in S$ for any $X \in \mathrm{Ob}\,\mathcal{B}$ and $s \circ t \in S$ for any $s, t \in S$ whenever the composition is defined.

b) *Extension conditions:* for any $f \in \text{Mor}\,\mathcal{B}, s \in S$ there exist $g \in \text{Mor}\,\mathcal{B}, t \in S$ such that the following square

$$\left(\begin{array}{ccc} W & \xrightarrow{g} & Z \\ {\scriptstyle t}\downarrow & & \downarrow{\scriptstyle s} \\ X & \xrightarrow{f} & Y \end{array} \quad \text{resp.} \quad \begin{array}{ccc} W & \xleftarrow{g} & Z \\ {\scriptstyle t}\uparrow & & \uparrow{\scriptstyle s} \\ X & \xleftarrow{f} & Y \end{array} \right) \qquad \text{(III.7)}$$

is commutative.

c) *Let f, g be two morphisms from X to Y; the existence of $s \in S$ with $sf = sg$ is equivalent to the existence of $t \in S$ with $ft = gt$.*

7. Remarks

a) Let us consider in the left square in (III.7) the paths $x_s f$ and gx_t from X to Z. We claim that they represent the same morphism $X \to Z$ in $\mathcal{C}[S^{-1}]$. Indeed, the commutativity of the square means that $ft = sg$ in $\text{Mor}\,\mathcal{B}$ which implies the equivalence of paths $x_s ftx_t$ and $x_s sgx_t$ and hence the equivalence of paths $x_s f$ and gx_t. Hence in $\mathcal{B}[S^{-1}]$ we have (in clear notations) $s^{-1}f = gt^{-1}$ so that if S satisfies conditions a) and b) in III.2.6, we can move all denominators in (∗) to the right. Similarly, the second square in (III.7) enables us to move all denominators to the left. These two properties make the study of the localized categories much simpler.

The role of the condition III.2.3c) will become clear in Lemma III.2.8.

b) Unfortunately, quasi-isomorphisms in $\text{Kom}(\mathcal{A})$ do not form, in general, a localizing class. This obstacle can be bypassed as follows: first we use Lemma III.1.2 to construct the category $K(\mathcal{A})$ of complexes modulo homotopy equivalence and then verify that quasi-isomorphisms in this new category do form a localizing class of morphisms. This will be done in Sect. III.4 after we present some useful techniques in Sect. III.3.

8. Lemma. *Let S be a localizing class of morphisms in a category $\mathcal{B}$. Then $\mathcal{B}[S^{-1}]$ can be described as follows:*

$$\text{Ob}\,\mathcal{B}[S^{-1}] = \text{Ob}\,\mathcal{B}, \quad \text{and then}$$

a) *A morphism $X \to Y$ in $\mathcal{B}[S^{-1}]$ is the equivalence class of "roofs", i.e., of diagrams (s, f) in $\mathcal{B}$ of the form*

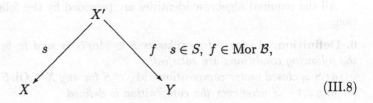

(III.8)

with two roofs being equivalent, $(s, f) \sim (t, g)$, if and only if there exists a third roof forming a commutative diagram of the form

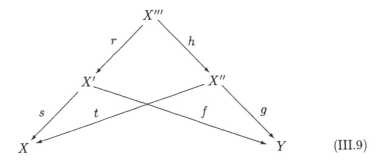

(III.9)

The identity morphism id : $X \to X$ is a class of the roof $(\mathrm{id}_X, \mathrm{id}_X)$.

b) The composition of morphisms represented by the roofs (s, f) and (t, g) is the class of the roof (st', gf') obtained using the first square in (III.7):

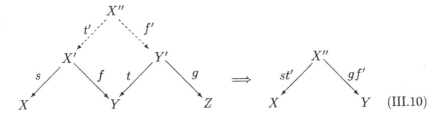

(III.10)

Proof. a) Let us prove first that the relation $(s, f) \sim (t, g)$ defined by (III.9) is an equivalence relation. It is obviously reflexive and symmetric. To prove transitivity let us assume that $(s, f) \sim (t, g)$, $(t, g) \sim (u, e)$ so that we have a commutative diagram

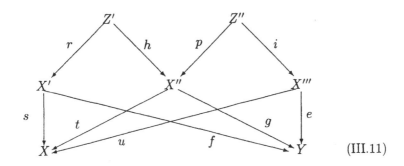

(III.11)

with s, t, u, sr, tp all belonging to S. We must construct a commutative diagram

150   III. Derived Categories and Derived Functors

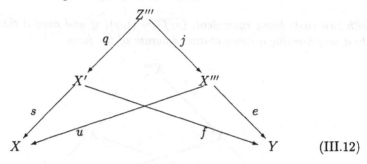
(III.12)

with $sq \in S$. To do this we first complete the diagram

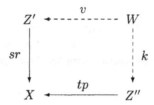

using the condition b) of Definition III.2.6, so that $v \in S$. Next, the commutativity of (III.11) implies that two morphisms $f_1 = hv$ and $f_2 = pk$ from W to X'' satisfy $tf_1 = tf_2$. Hence the condition III.2.6c) yields a morphism $w : Z''' \to W$, $w \in S$, such that $f_1 w = f_2 w$. The reader can easily prove that setting $q = rvw : Z''' \to X'$, $f = ikw : Z''' \to X'''$ we obtain the required diagram of the form (III.12).

b) The verification of the fact that the *composition of morphisms* (diagram (III.10)) *is well defined*, i.e., it does not depend on the choice of representatives in equivalence classes of roofs, is left to the reader.

c) Let us denote for a moment by $\tilde{\mathcal{B}}$ the category of roofs, i.e., we set $\mathrm{Ob}\,\tilde{\mathcal{B}} = \mathrm{Ob}\,\mathcal{B}$, $\mathrm{Mor}\,\tilde{\mathcal{B}} = \{$equivalence classes of roofs$\}$ with the composition given by the diagram (III.10). Lemma III.2.8 claims that $\tilde{\mathcal{B}} = \mathcal{B}[S^{-1}]$. We prove it by verifying that $\tilde{\mathcal{B}}$ satisfies the required universality property.

d) To construct the functor $F : \tilde{\mathcal{B}} \to \mathcal{B}$ we define it to be identical on objects and, for a morphism $f : X \to Y$,

$F(f) =$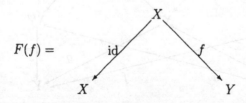

It is clear that if $s \in S$, then $F(s)$ is invertible in $\tilde{\mathcal{B}}$: the morphism $(F(s))^{-1}$ gets represented by a roof,

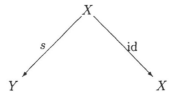

e) Let now $T : \mathcal{B} \to \mathcal{D}$ be a functor such that $T(s)$ is an isomorphism in $\mathcal{D}$ for any $s \in S$. We claim that T can be uniquely factored through F, i.e., there exists a unique functor $\tilde{\mathcal{G}} : \tilde{\mathcal{B}} \to \mathcal{D}$ with $T = G \circ F$.

Let us show first that if such a G exists, it is unique. Applying the equality $T = G \circ F$ to an object X we get

$$G(X) = T(X), \quad X \in \mathrm{Ob}\,\tilde{\mathcal{B}} = \mathrm{Ob}\,\mathcal{B}. \tag{III.13}$$

Next, let a morphism φ in $\tilde{\mathcal{B}}$ be represented by a roof (s, f). Then in $\mathrm{Mor}\,\tilde{\mathcal{B}}$ we have $\varphi \circ (\mathrm{id}, s) = (\mathrm{id}, f)$ so that $\varphi \circ F(s) = F(f)$. Applying G to this equality and recalling that $T = G \circ F$ we get $G(\varphi) \circ T(s) = T(f)$ and, $T(s)$ being invertible,

$$G(\varphi) = T(f) \circ (T(s))^{-1} \quad \text{for} \quad \varphi = (s, f). \tag{III.14}$$

To prove the existence of $G : \tilde{\mathcal{B}} \to \mathcal{D}$ we define it by (III.13) and (III.14). The reader can easily verify that $G(\varphi)$ does not depend on the choice of a representative (s, f) of φ and that $G(\mathrm{id}) = \mathrm{id}$, $G(\varphi_1) \circ G(\varphi_2) = G(\varphi_1 \circ \varphi_2)$, so that G is indeed a functor from $\tilde{\mathcal{B}}$ to $\mathcal{D}$.

Recalling the definition of F from d) we immediately see that $T = G \circ F$.
□

9. Remark. A diagram of the form (III.8) will be called a left S-roof. There exists a variant of Lemma III.2.8 in which instead of left S-roofs one uses right S-roofs:

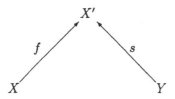

The composition of morphisms represented by right S-roofs is constructed using the second square in (III.7).

According to these two possibilities we could introduce the notions of left- and right-localizing classes of morphisms satisfying only half of the conditions III.2.6b) and III.2.6c). We leave it to an interested reader.

10. Proposition. Let $\mathcal{C}$ be a category, S a localizing system of morphisms in $\mathcal{C}$, and $\mathcal{B} \subset \mathcal{C}$ a full subcategory. Let a) and either b_1) or b_2) be satisfied, where a), b_1) and b_2) are the following conditions:

a) $S_{\mathcal{B}} = S \cap \operatorname{Mor} \mathcal{B}$ is a localizing system in $\mathcal{B}$.

b_1) For any $s : X' \to X$ with $s \in S$, $X \in \operatorname{Ob} \mathcal{B}$ there exists $f : X'' \to X'$, such that $sf \in S$, $X'' \in \operatorname{Ob} \mathcal{B}$.

b_2) The same as b_1) but with all arrows reversed.

Then $\mathcal{B}[S_{\mathcal{B}}^{-1}]$ is a full subcategory in $\mathcal{C}[S^{-1}]$. More precisely, the canonical functor $\mathcal{B}[S_{\mathcal{B}}^{-1}] \to \mathcal{C}[S^{-1}]$ is fully faithful.

Proof. By Lemma III.2.8, morphisms in $\mathcal{C}[S^{-1}]$ and in $\mathcal{B}[S_{\mathcal{B}}^{-1}]$ are equivalence classes of roofs (for $\mathcal{B}[S_{\mathcal{B}}^{-1}]$ this follows from the condition a)). The claimed property of the functor $I : \mathcal{B}[S_{\mathcal{B}}^{-1}] \to \mathcal{C}[S^{-1}]$ is that for any X, Y the map

$$I : \operatorname{Hom}_{\mathcal{B}[S_{\mathcal{B}}^{-1}]}(X, Y) \longrightarrow \operatorname{Hom}_{\mathcal{C}[S^{-1}]}(IX, IY) \qquad (\text{III}.15)$$

is a bijection. We verify it assuming that the condition b_1) is satisfied.

First let us check *that* (III.15) *is injective*. We must prove that if two $S_{\mathcal{B}}$-roofs (s, f) and (t, g) are equivalent in $\mathcal{C}$, then they are equivalent in $\mathcal{B}$ as well. Their equivalence in $\mathcal{C}$ is given by a diagram

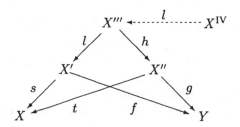

Applying b_1) to the morphism $sr : X''' \to X$ from S we find $X^{IV} \in \operatorname{Ob} \mathcal{B}$ and $l : X^{I)} \to X'''$ such that $srl \in S$. Since $\mathcal{B}$ is a full subcategory, we have $srl \in S \cap \operatorname{Mor} \mathcal{B} = S_{\mathcal{B}}$, $hl \in \operatorname{Mor} \mathcal{B}$ so that X^{IV}, together with morphisms srl and ghl, give the required $\mathcal{B}$-equivalence of $S_{\mathcal{B}}$-roofs (s, f) and (t, g).

Now let us verify that (III.15) is surjective. To do this we must check that any S-roof (s, f) with X, Y from $\mathcal{B}$ is equivalent to some $S_{\mathcal{B}}$-roof. We again use the condition b_1): the vertex X' is replaced by $X'' \in \mathcal{B}$,

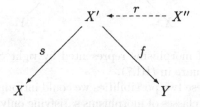

in such a way that $sr \in S \cap \operatorname{Mor} \mathcal{B} = S_{\mathcal{B}}$, $fr \in \operatorname{Mor} \mathcal{B}$. It is clear that $(s, f) \sim (sr, fr)$.

In the case of the condition b_2) instead of b_1) the proof is similar; one must use right S-roofs instead of left ones (see Remark III.2.9). □

Exercises

1. Bounded Complexes. Using Proposition III.2.10 prove that all arrows in the diagram

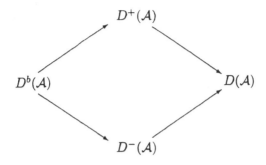

are embeddings of full subcategories.

2. Localizing Subcategories. A full abelian subcategory $\mathcal{B}$ of an abelian category $\mathcal{A}$ is said to be thick if it is closed under extensions, i.e., any extension of any two objects from $\mathcal{B}$ belongs to $\mathcal{B}$ (examples: the subcategory consisting of the single object 0, finite abelian groups in Ab, finitely generated modules in R-mod). Let $\text{Kom}_\mathcal{B}(\mathcal{A})$ be a full subcategory of $\text{Kom}(\mathcal{A})$ consisting of complexes K' such that $H^n(K^\bullet) \in \text{Ob}\,\mathcal{B}$ for all n. Define $D_\mathcal{B}(\mathcal{A})$ to be $\text{Kom}_\mathcal{B}(\mathcal{A})[S^{-1}]$, where S is the class of all quasi-isomorphisms. Define similarly $D_\mathcal{B}^{\pm,b}(\mathcal{A})$. Prove that $D_\mathcal{B}(\mathcal{A})$ is a full subcategory in $D(\mathcal{A})$.

Construct a natural functor $D(\mathcal{B}) \to D_\mathcal{B}(\mathcal{A})$ (identity on objects). In general this functor is neither full, nor faithful. Prove, however, that if any object of $\mathcal{B}$ is a subobject of an object in $\mathcal{B}$ which is injective in $\mathcal{A}$, then $D^+(\mathcal{B}) \to D_\mathcal{B}^+(\mathcal{A})$ is an equivalence of categories.

III.3 Triangles as Generalized Exact Triples

1. The Problem

In this section we introduce some diagrams in derived categories – called *distinguished triangles* – that may be considered as analogues of exact triples in abelian categories. The definition of such a generalization is not at all obvious. First of all, we do not even know that the category $D(\mathcal{A})$ is additive: to add morphisms we have, in a sense, to find "a common denominator". Next, although $D(\mathcal{A})$ will happen to be additive, it will never be abelian (see Proposition III.2.4) so that we cannot apply to $D(\mathcal{A})$ the usual definition of exactness.

However, although $D(\mathcal{A})$ is not abelian, distinguished triangles in $D(\mathcal{A})$ give some remarkable structure, which reflects the main homological properties of the initial abelian category $\mathcal{A}$.

2. Translation, Cylinder, Cone

Let $\mathcal{A}$ be an abelian category. All complexes below are assumed to be objects of Kom($\mathcal{A}$).

In what follows we will often give definitions of various morphisms and then verify identities among them by applying them to "elements" as if we were working with abelian groups or modules; see also exercises to Sect. II.5.

a) Fix an integer n and for any complex $K^\bullet = (K^i, d_K^i)$ define a new complex $K[n]^\bullet$ by $(K[n])^i = K^{n+i}$, $d_{K[n]} = (-1)^n d_K$. For a morphism of complexes $f : K^\bullet \to L^\bullet$ let $f[n] : K[n]^\bullet \to L[n]^\bullet$ coincide with f componentwise.

It is clear that $T^n : \text{Kom}(\mathcal{A}) \to \text{Kom}(\mathcal{A})$, $T^n(K^\bullet) = K[n]^\bullet$, $T^n(f) = f[n]$, is a functor, called the *translation by* n functor, which is an autoequivalence of Kom($\mathcal{A}$). It induces an autoequivalence of any of the categories $\text{Kom}^+(\mathcal{A})$, $\text{Kom}^-(\mathcal{A})$, $\text{Kom}^b(\mathcal{A})$ and of the corresponding derived categories.

b) Let $f : K^\bullet \to L^\bullet$ be a morphism of complexes. The *cone* of f is the following complex $C(f)$:

$$C(f)^i = K[1]^i \oplus L^i, \quad d_{C(f)}(k^{i+1}, l^i) = (-d_K k^{i+1}, f(k^{i+1}) + d_L l^i).$$

It is convenient to write elements of $C(f)$ as columns of height 2 and morphisms as matrices, so that

$$d_{C(f)} = \begin{pmatrix} d_{K[1]} & 0 \\ f[1] & d_L \end{pmatrix}.$$

One easily verifies that $d_{C(f)}^2 = 0$. In I.4.5d) we have defined the cone of a triangulated space by a transparent geometrical construction and have computed the chain complex of this cone. In our present language this cone may be considered (up to the change of chains to cochains) as the cone of the identity map.

If f is a morphism of 0-complexes (see III.1.4b)), then $C(f)$ is a complex $\ldots \to 0 \to K_{-1}^0 \to L_0^0 \to 0 \ldots$ In particular,

$$H^{-1}(C(f)) = \ker f, \quad H^0(C(f)) = \text{Coker } f.$$

c) In the same notations the *cylinder* $\text{Cyl}(f)$ of a morphism f is the following complex:

$$\text{Cyl}(f) = K^\bullet \oplus K[1]^\bullet \oplus L^\bullet,$$
$$d^i_{\text{Cyl}(f)}(k^i, k^{i+1}, l^i) = (d_K k^i - k^{i+1}, -d_K k^{i+1}, f(k^{i+1}) + d_L l^i).$$

Let us remark that if both complexes $K^\bullet$, $L^\bullet$ are bounded either from the left, or from the right, or from both sides, then both $C(f)$ and $\text{Cyl}(f)$ satisfy the same boundness condition. Moreover, if $K^\bullet$ and $L^\bullet$ lie in Kom($\mathcal{B}$), where $\mathcal{B} \subset \mathcal{A}$ is an additive subcategory, then $C(f)$ and $\text{Cyl}(f)$ also lie in Kom($\mathcal{B}$).

Main properties of the cone and of the cylinder are summarized in the following lemma.

3. Lemma. *For any morphism $f : K^\bullet \to L^\bullet$ there exists the following commutative diagram in $\mathrm{Kom}(\mathcal{A})$ with exact rows:*

$$
\begin{array}{ccccccccc}
 & & 0 & \longrightarrow & L^\bullet & \xrightarrow{\bar{\pi}} & C(f) & \xrightarrow{\delta = \delta(f)} & K[1]^\bullet & \longrightarrow & 0 \\
 & & & & \downarrow{\alpha} & & \| & & & & \\
0 & \longrightarrow & K^\bullet & \xrightarrow{\hat{f}} & \mathrm{Cyl}(f) & \xrightarrow{\pi} & C(f) & \longrightarrow & 0 & & \\
 & & \| & & \downarrow{\beta} & & & & & & \\
 & & K^\bullet & \xrightarrow{f} & L^\bullet & & & & & &
\end{array}
$$

(III.16)

It is functorial in f and has the following property:

α *and* β *are quasi-isomorphisms; moreover* $\beta\alpha = \mathrm{id}_L$ *and* $\alpha\beta$ *is homotopic to* $\mathrm{id}_{\mathrm{Cyl}(f)}$ *so that* $L^\bullet$ *and* $\mathrm{Cyl}(f)$ *are canonically isomorphic in the derived category.*

Proof. a) The definition of morphisms in the first row and the verification that they commute with d:

$$
\begin{array}{ccc}
l^i & \xrightarrow{\bar{\pi}} & (0, l^i) \\
\downarrow{d_L} & & \downarrow{d_{C(f)}} \\
d_L l^i & \xrightarrow{\bar{\pi}} & (0, d_L l^i)
\end{array}
$$

$$
\begin{array}{ccc}
(k^{i+1}, l^i) & \xrightarrow{\delta} & k^{i+1} \\
\downarrow{d_{C(f)}} & & \downarrow{d_{K[1]}} \\
(-d_K k^{i+1}, f(k^{i+1}) + d_L l^i) & \xrightarrow{\delta} & -d_K k^{i+1}
\end{array}
$$

The exactness of the first row is clear.

b) The definition of morphisms in the second row and the verification that they commute with d:

$$
\begin{array}{ccc}
(k^i, k^{i+1}, l^i) & \xrightarrow{\pi} & (k^{i+1}, l^i) \\
\downarrow{d_{\mathrm{Cyl}(f)}} & & \downarrow{d_{C(f)}} \\
(d_K k^i - k^{i+1}, -d_K k^{i+1}, f(k^{i+1}) + d_L l^i) & \xrightarrow{\pi} & (-d_K k^{i+1}, f(k^{i+1}) + d_L l^i)
\end{array}
$$

156   III. Derived Categories and Derived Functors

$$
\begin{array}{ccc}
k^i & \xrightarrow{\bar{f}} & (k^i, 0, 0) \\
\downarrow{\scriptstyle d_K} & & \downarrow{\scriptstyle d_{\mathrm{Cyl}(f)}} \\
d_K k^i & \xrightarrow{\bar{f}} & (d_K k^i, 0, 0)
\end{array}
$$

The exactness of the second row is clear.

c) *The definition of morphisms α and β and the verification that they commute with d:*

$$
\begin{array}{ccc}
l^i & \xrightarrow{d_L} & d_L l^i \\
\downarrow{\scriptstyle \alpha} & & \downarrow{\scriptstyle \alpha} \\
(0, 0, l^i) & \xrightarrow{d_{\mathrm{Cyl}(f)}} & (0, 0, d_L l^i)
\end{array}
$$

$$
\begin{array}{ccc}
(k^i, k^{i+1}, l^i) & \xrightarrow{d_{\mathrm{Cyl}(f)}} & (d_K k^i - k^{i+1}, -d_K k^{i+1}, f(k^{i+1}) + d_L l^i) \\
\downarrow{\scriptstyle \beta} & & \downarrow{\scriptstyle \beta} \\
f(k^i) + l^i & \xrightarrow{d_L} & f(d_K k^i) + d_L l^i
\end{array}
$$

The commutativity of the squares $\pi\alpha = \bar{\pi}$, $\beta f = \bar{f}$ is clear.

d) The formula $\beta\alpha = \mathrm{id}_L$ is clear. Define $h^i : \mathrm{Cyl}(f)^i \to \mathrm{Cyl}(f)^{i-1}$ by the formula

$$h^i\left(k^i, k^{i+1}, l^i\right) = (0, k^i, 0).$$

Let us verify that $\alpha\beta = \mathrm{id}_{\mathrm{Cyl}(f)} - (dh + hd)$. We have

$$
\begin{aligned}
\alpha\beta\left(k^i, k^{i+1}, l^i\right) &= (0, 0, f(k^i) + l^i), \\
d_{\mathrm{Cyl}(f)} h^i\left(k^i, k^{i+1}, l^i\right) &= (-k^i, -d_K k^i, f(k^i)), \\
h^i d_{\mathrm{Cyl}(f)}\left(k^i, k^{i+1}, l^i\right) &= (0, d_K k^i - k^{i+1}, 0).
\end{aligned}
$$

Since $\alpha\beta$ and $\beta\alpha$ induce the identity mappings on cohomology, α and β are quasi-isomorphisms. □

4. Definition. *a) A triangle in some category of complexes (Kom, D, D^+, ...) is a diagram of the form*

$$K^\bullet \xrightarrow{u} L^\bullet \xrightarrow{v} M^\bullet \xrightarrow{w} K[1]^\bullet.$$

III.3 Triangles as Generalized Exact Triples

b) *A morphism of triangles is a commutative diagram of the form*

$$\begin{array}{ccccccc} K^\bullet & \xrightarrow{u} & L^\bullet & \xrightarrow{v} & M^\bullet & \xrightarrow{w} & K[1]^\bullet \\ \downarrow f & & \downarrow g & & \downarrow h & & \downarrow f[1] \\ K_1^\bullet & \xrightarrow{u_1} & L_1^\bullet & \xrightarrow{v_1} & M_1^\bullet & \xrightarrow{w_1} & K_1[1]^\bullet \end{array}$$

Such a morphism is said to be an isomorphism if f, g, h are isomorphisms in the corresponding category.

c) *A triangle is said to be distinguished if it is isomorphic to the middle row*

$$K^\bullet \xrightarrow{\bar{f}} \mathrm{Cyl}(f) \xrightarrow{\pi} C(f) \xrightarrow{\delta} K[1]^\bullet.$$

of some diagram of the form (III.16).

The next proposition shows that any exact triple can be completed to a distinguished triangle.

5. Proposition. *An exact triple of complexes in $\mathrm{Kom}(\mathcal{A})$ is quasi-isomorphic to the middle row of an appropriate diagram of the form (III.16).*

Proof. Let $0 \longrightarrow K^\bullet \xrightarrow{f} L^\bullet \xrightarrow{g} M^\bullet \longrightarrow 0$ be an exact triple. Consider the following diagram:

$$\begin{array}{ccccccccc} 0 & \longrightarrow & K^\bullet & \longrightarrow & L^\bullet & & \longrightarrow & M^\bullet & \longrightarrow & 0 \\ & & \parallel & & \uparrow \beta & & & \uparrow \gamma & & \\ 0 & \longrightarrow & K^\bullet & \xrightarrow{\bar{f}} & \mathrm{Cyl}(f) & \xrightarrow{\pi} & C(f) & \longrightarrow & 0 \end{array} \qquad (\mathrm{III}.17)$$

where β is taken from (III.16) and γ is defined by $\gamma(k^{i+1}, l^i) = g(l^i)$.

Let us verify next that γ is a morphism of complexes:

$$\begin{array}{ccc} (k^{i+1}, l^i) & \xrightarrow{d_{C(f)}} & (-d_K k^{i+1}, f(k^{i+1}) + d_L l^i) \\ \downarrow \gamma & & \downarrow \gamma \\ g(l^i) & \xrightarrow{d_L} & d_L g(l^i) = g(f(k^{i+1})) + g(d_L l^i) \end{array}$$

Let us verify that the right square in (III.17) is commutative:

$$\begin{array}{ccc} (k^i, k^{i+1}, l^i) & \xrightarrow{\pi} & (k^{i+1}, l^i) \\ \downarrow \beta & & \downarrow \gamma \\ f(k^i) + l^i & \xrightarrow{g} & g(f(k^i)) + g(l^i) = g(l^i) \end{array}$$

To complete the proof we have to verify that γ is a quasi-isomorphism. Since g is an epimorphism of complexes, γ is an epimorphism as well, and $\ker \gamma$ is the following complex:

$$K[1]^\bullet \oplus \ker g = K[1]^\bullet \oplus \operatorname{Im} f, \quad d\left(k^{i+1}, k^i\right) = \left(-d_K k^{i+1}, k^{i+1} + d_K k^i\right).$$

This complex has zero cohomology because its identity mapping is homotopic to the zero one: $\chi d + d\chi = \operatorname{id}$ where $\chi = \{\chi^i : K^{i+1} \oplus K^i \to K^i \oplus K^{i-1}\}$, $\chi^i(k^{i+1}, k^i) = (k^i, 0)$. Hence the long exact cohomology sequence corresponding to the exact sequence of complexes

$$0 \longrightarrow \ker \gamma \longrightarrow C(f) \xrightarrow{\gamma} M^\bullet \longrightarrow 0$$

implies that γ is a quasi-isomorphism. $\square$

The next theorem shows that cohomological properties of distinguished triangles in $D(\mathcal{A})$ (or in $D^+(\mathcal{A})$, or ...) are quite similar to those of exact triples of complexes.

6. Theorem. *Let*

$$K^\bullet \xrightarrow{u} L^\bullet \xrightarrow{v} M^\bullet \xrightarrow{w} K[1]^\bullet \qquad (\text{III}.18)$$

be a distinguished triangle in $D(\mathcal{A})$. Then the sequence

$$\cdots \longrightarrow H^i(K^\bullet) \xrightarrow{H^i(u)} H^i(L^\bullet) \xrightarrow{H^i(v)} H^i(M^\bullet) \xrightarrow{H^i(w)}$$
$$\longrightarrow H^i(K[1]^\bullet) = H^{i+1}(K^\bullet) \longrightarrow \cdots$$

is exact.

Proof. It is sufficient to consider the distinguished triangle

$$K^\bullet \xrightarrow{\bar{u}} \operatorname{Cyl}(u) \xrightarrow{\pi} C(u) \xrightarrow{\delta} K[1]^\bullet$$

from (III.16), which is quasi-isomorphic to (III.18). According to Lemma 3, the sequence of complexes

$$0 \longrightarrow K^\bullet \xrightarrow{\bar{u}} \operatorname{Cyl}(u) \xrightarrow{\pi} C(u) \xrightarrow{\delta} 0$$

is exact. Let

$$\cdots \longrightarrow H^i(K^\bullet) \xrightarrow{H^i(\bar{u})} H^i(\operatorname{Cyl}(u)) \xrightarrow{H^i(\pi)} H^i(C(u)) \xrightarrow{\delta^i(\bar{u},\pi)}$$
$$\longrightarrow H^{i+1}(K^\bullet) \longrightarrow \cdots \qquad (\text{III}.19)$$

be the corresponding exact sequence of cohomology. To prove the theorem it suffices to prove that the connecting homomorphism $\delta^i(\bar{u}, \pi)$ in (III.19) coincides with $H^i(w)$. This can be done, for example, by direct computation using the decomposition

$$\operatorname{Cyl}(u)^i = K^i \oplus C(u)^i$$

and recalling the explicit construction of $\delta^i(\bar{u}, \pi)$ (see 1). Details are left to the reader. $\square$

III.4 Derived Category as the Localization of Homotopy Category

1. Definition. *Let $\mathcal{A}$ be an abelian category.* The homotopy category $K(\mathcal{A})$ is defined as follows:

$$\text{Ob } K(\mathcal{A}) = \text{Ob Kom}(\mathcal{A}),$$
$$\text{Mor } K(\mathcal{A}) = \text{Mor Kom}(\mathcal{A}) \quad \textit{modulo homotopy equivalence}$$

(see Lemma III.1.2). By $K^+(\mathcal{A})$, $K^-(\mathcal{A})$, $K^b(\mathcal{A})$ we denote full subcategories of $K(\mathcal{A})$ formed by complexes satisfying the corresponding boundness conditions.

Lemma III.1.2 shows that $K(\mathcal{A})$ is an additive category on which the functors H^i are well defined. Hence the definition III.1.5 of quasi-isomorphism can be literally applied to $K(\mathcal{A})$.

2. Proposition. *The localization of $K(\mathcal{A})$ by quasi-isomorphisms is canonically isomorphic to the derived category $D(\mathcal{A})$. The same is true for $K^*(\mathcal{A})$ and $D^*(\mathcal{A})$ where $* = +, -,$ or b.*

Proof. Denote temporarily by $\tilde{D}(\mathcal{A})$ the localization of $K(\mathcal{A})$ by quasi-isomorphisms. The composition $\text{Kom}(\mathcal{A}) \to K(\mathcal{A}) \to \tilde{D}(\mathcal{A})$ maps quasi-isomorphisms into isomorphisms; therefore, it factors through a functor $G : D(\mathcal{A}) \to \tilde{D}(\mathcal{A})$. By construction, G is a bijection on objects. The interpretation of morphisms in a localized category as classes of paths (see III.2.2) implies that G is surjective on morphisms, because each morphism in $K(\mathcal{A})$ can be lifted to a morphism in $\text{Kom}(\mathcal{A})$. □

It remains to show that G is injective on morphisms. From the description of morphisms in III.2.2 one can easily see that the injectivity follows from Lemma III.4.3 below:

3. Lemma. *Let $f, g : K^\bullet \to L^\bullet$ be homotopic to each other. Then $Q(f) = Q(g)$ in $D(\mathcal{A})$.*

Proof. Let $f = g + dh + hd$. Define the morphism $c(h) : C(f) \to C(g)$ by

$$C(h)\left(k^{i+1}, l^i\right) = \left(k^{i+1}, l^i + h\left(k^{i+1}\right)\right).$$

One can easily see that $c(h)$ commutes with differentials, so that it is really a morphism of complexes.

Similarly, define the morphism of complexes $\text{cyl}(h) : \text{Cyl}(f) \to \text{Cyl}(g)$ by

$$\text{cyl}(h)\left(k^i, k^{i+1}, l^i\right) = \left(k^i, k^{i+1}, l^i + h\left(k^{i+1}\right)\right).$$

Let us consider the diagram formed by the first rows of diagrams (III.16) from Sect. III.3 for f and g respectively:

$$0 \longrightarrow L^\bullet \xrightarrow{\bar\pi} C(f) \xrightarrow{\delta(f)} K^\bullet[1] \longrightarrow 0$$
$$\Big\| \qquad \Big\downarrow c(h) \qquad \Big\|$$
$$0 \longrightarrow L^\bullet \xrightarrow{\bar\pi} C(g) \xrightarrow{\delta(g)} K^\bullet[1] \longrightarrow 0$$

This diagram is commutative. Applying the five-lemma to the corresponding exact cohomology sequences,

$$H^{i-1}(K[1]^\bullet) \longrightarrow H^i(L^\bullet) \longrightarrow H^i(C(f)) \longrightarrow H^i(K[1]^\bullet) \longrightarrow H^{i+1}(L^\bullet)$$
$$\Big\| \qquad \Big\| \qquad \Big\downarrow H^i(c(h)) \qquad \Big\| \qquad \Big\|$$
$$H^{i-1}(K[1]^\bullet) \longrightarrow H^i(L^\bullet) \longrightarrow H^i(C(f)) \longrightarrow H^i(K[1]^\bullet) \longrightarrow H^{i+1}(L^\bullet)$$

we see that $c(h)$ is a quasi-isomorphism.

Using the middle rows of the diagram (III.16) in Sect. III.3 one can show similarly that $\mathrm{cyl}(h)$ is an isomorphism.

To complete the proof let us consider the following diagram:

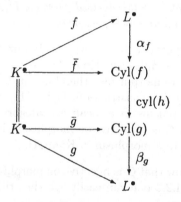

In $\mathrm{Kom}(\mathcal{A})$ the upper triangle of this diagram does not commute, while the square and the lower triangle commute. But in $D(\mathcal{A})$ the diagram becomes commutative in all parts. Indeed, by Lemma III.3.3, $Q(\alpha_f)$ and $Q(\beta_f)$ are mutually inverse, so that $f = \beta_f \circ \bar f$ implies $Q(f) = Q(\beta_f) \circ Q(\bar f)$ and hence $Q(\bar f) = Q(\alpha_f) \circ Q(f)$. Finally, the direct computation shows that $\beta_g \circ \mathrm{cyl}(h) \circ \alpha_f = \mathrm{id}_L$. Hence $Q(f) = Q(g)$ and the proof is complete. □

The role of Proposition III.4.2 is shown by the following theorem.

4. Theorem. *The class of quasi-isomorphisms in categories $K^*(\mathcal{A})$, $* = +, -, b, \emptyset$, is localizing.*

Proof. We must prove that conditions a) to c) of Definition III.2.6 are satisfied. For condition a), this is obvious.

III.4 Derived Category as the Localization of Homotopy Category

Let us verify the first extension condition. We must embed the diagram $K^\bullet \xrightarrow[\text{qis}]{f} L^\bullet \xleftarrow{g} M^\bullet$ into a commutative square (qis denotes quasi-isomorphisms)

$$\begin{array}{ccc} N^\bullet & \xdashrightarrow{k}_{\text{qis}} & M^\bullet \\ {\scriptstyle h}\downarrow & & \downarrow{\scriptstyle g} \\ K^\bullet & \xrightarrow[\text{qis}]{f} & L^\bullet \end{array}$$

The required square is a part of the following diagram, which is commutative in $K^*(\mathcal{A})$:

$$\begin{array}{ccccccc} C(\pi g)[-1] & \xrightarrow{k} & M^\bullet & \xrightarrow{\pi g} & C(f) & \longrightarrow & C(\bar\pi g) \\ {\scriptstyle h}\downarrow & & \downarrow{\scriptstyle g} & & \| & & \downarrow{\scriptstyle h[1]} \\ K^\bullet & \xrightarrow{f} & L^\bullet & \xrightarrow{\pi} & C(f) & \longrightarrow & K[1]^\bullet \end{array} \qquad \text{(III.20)}$$

The morphism k is $\delta(\pi g)[-1]$ in the notation of Lemma III.3.3. The morphism h we will construct explicitly. An element of $C(\pi g)[-1]^i = C(\pi g)^{i-1}$ is a triple (m^i, k^i, l^{i-1}), $m^i \in M^i$, $k^i \in K^i$, $l^{i-1} \in L^{i-1}$. Define $h(m^i, k^i, l^{i-1}) = -k^i$ and verify that the left square in (III.20) is commutative modulo homotopy. We have

$$k : (m^i, k^i, l^{i-1}) \longrightarrow m^i,$$
$$g \circ k - f \circ h : (m^i, k^i, l^{i-1}) \longrightarrow g(m^i) + f(k^i).$$

The last difference equals $\chi d_{C(\pi g)[-1]} + d_L \chi$, where $\chi = \{\chi^i\}$ and $\chi^i : C(\pi g)[-1]^i = C(\pi g)^{i-1} \to L^{i-1}$ is given by

$$\chi^i(m^i, k^i, l^{i-1}) = -l^{i-1}.$$

(One has to use the following easily verified formula:

$$d_{C(\pi g)[-1]}(m^i, k^i, l^{i-1}) = (d_M m^i, d_K k^i, -d_L l^{i-1} - f(k^i) - g(m^i)).)$$

It remains to prove that k is a quasi-isomorphism. Since f is a quasi-isomorphism, $C(f)$ is acyclic. But the upper row in (III.20) is a distinguished triangle and hence k is a quasi-isomorphism.

The second extension condition is proved in the same way: $M^\bullet \xleftarrow{g} K^\bullet \xrightarrow[\text{qis}]{f} L^\bullet$ can be embedded into a commutative square using the diagram

$$\begin{array}{ccccccc} C(f)[-1] & \xrightarrow{\tau} & K^\bullet & \xrightarrow{f} & L^\bullet & \longrightarrow & C(f) \\ \| & & \downarrow{\scriptstyle g} & & \downarrow{\scriptstyle h} & & \| \\ C(f)[-1] & \xrightarrow{g\tau} & M^\bullet & \xrightarrow{k} & C(g\tau) & \longrightarrow & C(f) \end{array}$$

III. Derived Categories and Derived Functors

Finally, let us prove the condition c) of III.2.6. Let $f : K^\bullet \to L^\bullet$ be a morphism in $K^*(\mathcal{A})$. We show that the existence of a quasi-isomorphism $s : L^\bullet \to \bar{L}^\bullet$ with $sf = 0$ in $K^*(\mathcal{A})$ implies the existence of a quasi-isomorphism $t : \bar{K}^\bullet \to K^\bullet$ with $ft = 0$ (the formula $sf = sg$ in III.2.6 is equivalent to $s(f - g) = 0$).

Let $\{h^i : K^i \to \bar{L}^{i-1}\}$ be a homotopy between sf and the zero morphism. Let us consider the following diagram:

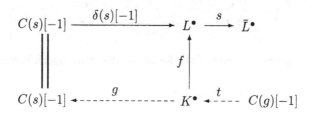

Define $g^i : K^i \to C(s)[-1]^i = L^i \oplus \bar{L}^{i-1}$ by $g^i(k^i) = \big(f^i(k^i), -h^i(k^i)\big)$. One can easily see that $g = \{g^i\}$ is a morphism of complexes. The commutativity of the square is clear. Next, let $t = \delta(g)[-1]$. Then $tf = \delta(s)[-1]gt = 0$ because $gt = 0$. Finally, t is a quasi-isomorphism because s is a quasi-isomorphism so that $C(s)[-1]$ is an acyclic complex. In the same way one proves that $ft = 0$ implies the existence of an s with $sf = 0$. □

5. Additivity of $D(\mathcal{A})$

To define the addition of morphisms in $D(\mathcal{A})$ we will show that for any two morphisms $\varphi, \varphi' : X \to Y$ in $D(\mathcal{A})$ one can find "a common denominator", i.e., represent them by roofs

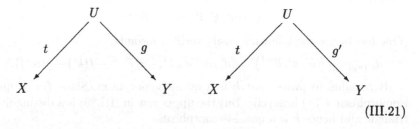

(III.21)

with the common top and the common left side.

Let φ, φ' be represented by some roofs

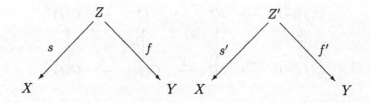

Let us extend the following diagram:

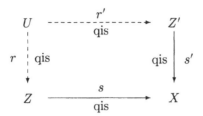

to a commutative square in $K(\mathcal{A})$. Since s, s', r are quasi-isomorphisms, r' is a quasi-isomorphism as well. Hence φ and φ' can be represented by the roofs of the form (III.21) with $t = sr = s'r'$, $g = fr$, $g' = f'r'$.

Now we define the sum $\varphi + \varphi'$ in $D(\mathcal{A})$ as the class of the roof $X \xleftarrow{t} U \xrightarrow{g+g'} Y$. The verification of the axioms of additive category (see II.6) for $D(\mathcal{A})$ is left to the reader.

Exercise

1. Morphisms in $D(\mathcal{A})$. Let $\mathcal{A}$ be an abelian category, $f : K^\bullet \to L^\bullet$ a morphism in Kom($\mathcal{A}$). By the definition of morphisms in $D(\mathcal{A})$, $f = 0$ in $D(\mathcal{A})$ iff there exists a quasi-isomorphism $s : L^\bullet \to M^\bullet$ such that sf is homotopic to zero (equivalently, there exists a quasi-isomorphism $t : N^\bullet \to K^\bullet$ such that ft is homotopic to zero).

If $f = 0$ in $D(\mathcal{A})$ then, clearly, $H^n(f) = 0$ for all n. Check that the morphism of complexes for the length 2 (for $\mathcal{A} = Ab$)

$$\begin{array}{ccccccccc} \cdots & \to & 0 & \to & \mathbb{Z} & \xrightarrow{a} & \mathbb{Z} & \to & 0 & \to & \cdots \\ & & & & {\scriptstyle b}\downarrow & & {\scriptstyle d}\downarrow & & & & \\ \cdots & \to & 0 & \to & \mathbb{Z} & \xrightarrow{c} & \mathbb{Z}/3 & \to & 0 & \to & \cdots \end{array}$$

(both b and c map the generator into the generator, a and d multiply the generator by 2) shows the converse is false. So we see that in the chain of implications

$$\{f = 0 \text{ in Kom}(\mathcal{A})\} \Rightarrow \{f = 0 \text{ in } K(\mathcal{A})\}$$
$$\Rightarrow \{f = 0 \text{ in } D(\mathcal{A})\} \Rightarrow \{H^n(f) = 0 \text{ for all } n\}$$

all implications are strict.

According to Lemma III.2.8, to prove that $(a \circ b)(\tilde{f}) = \tilde{f}$ we must construct a commutative diagram:

III.5 The Structure of the Derived Category

1. Objects Considered as Complexes

We shall say that $K^\bullet$ is an H^0-complex if $H^i(K^\bullet) = 0$ for $i \neq 0$. We note that this definition can be applied to any of the categories $\text{Kom}^*(\mathcal{A})$, $K^*(\mathcal{A})$, $D^*(\mathcal{A})$ of complexes, because H^i transforms quasi-isomorphisms to isomorphisms.

The notion of an H^0-complex is a generalization of the notion of 0-complex from III.1.5b).

We shall show that the full subcategory of $D^*(\mathcal{A})$ formed by H^0-complexes is equivalent to $\mathcal{A}$.

2. Proposition. *The functor $Q : \mathcal{A} \to D^*(\mathcal{A})$ yields an equivalence of $\mathcal{A}$ with the full subcategory of $D^*(\mathcal{A})$ formed by H^0-complexes.*

Proof. It is clear that the functor $\mathcal{A} \to K^*(\mathcal{A})$ sending an object of $\mathcal{A}$ to the corresponding 0-complex, is fully faithful, because the only homotopy between morphisms of 0-complexes is the zero homotopy. So we may (and will) consider $\mathcal{A}$ as a full subcategory of $K^*(\mathcal{A})$.

Let us prove that for 0-complexes the canonical mapping

$$a : \text{Hom}_{K^*(\mathcal{A})}(X, Y) \longrightarrow \text{Hom}_{D^*(\mathcal{A})}(Q(X), Q(Y))$$

is an isomorphism. The inverse mapping b is given by the functor $H^0 : D^*(\mathcal{A}) \to \mathcal{A}$. It is clear that $b \circ a = \text{id}$ on 0-complexes. To prove that $a \circ b = \text{id}$ consider a morphism $\tilde{f}$ of 0-complexes in $D^*(\mathcal{A})$ represented by a left roof:

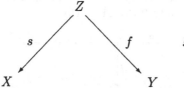

s is a quasi-isomorphism.

Let $g = H^0(f) \circ H^0(s)^{-1} : X \to Y$. The morphism $(a \circ b)(\tilde{f}) : X \to Y$ in $D^*(\mathcal{A})$ is represented by

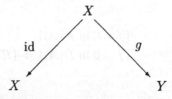

According to Lemma III.2.8, to prove that $(a \circ b)(\tilde{f}) = \tilde{f}$ we must construct a commutative diagram:

III.5 The Structure of the Derived Category   165

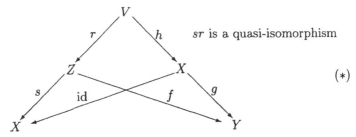

sr is a quasi-isomorphism

(∗)

Define V as follows: $V^i = Z^i$ for $i < 0$, $V^0 = \ker d_Z^0$, $V^i = 0$ for $i > 0$, d_V is induced by d_Z, $r : V \to Z$ is a natural embedding, $h : V \to X$ is given by

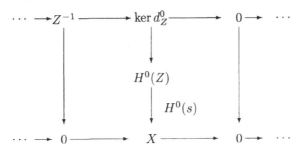

Necessary verifications: r is a quasi-isomorphism (obvious); the commutativity of the diagram: the least trivial equality $f \circ r = g \circ h$ follows from $g = H^0(f) \circ H^0(s)^{-1}$.

So the functor $Q : \mathcal{A} \to D^*(\mathcal{A})$ is fully faithful. Its image clearly lies in the subcategory of 0-complexes. It remains to prove that any H^0-complex Z is isomorphic in $D^*(\mathcal{A})$ to some 0-complex. The isomorphism is given by the upper part of the diagram (∗)

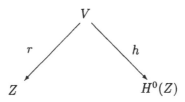

where both r and h are quasi-isomorphisms. □

Together with 0- and H^0-complexes we can consider i- and H^i-complexes for any i. The structure of morphisms of such complexes in $D^*(\mathcal{A})$ supplies us with the information about the derived category that may be considered as the information of the next complexity level.

We shall denote the i-complex corresponding to an object X of $\mathcal{A}$ by $X[-i]$ (X sits in degree i); earlier in this section we had written X instead of $X[0]$.

3. Definition. $\operatorname{Ext}_{\mathcal{A}}^i(X,Y) = \operatorname{Hom}_{D^*(\mathcal{A})}(X[0], Y[i])$.

4. Remarks

a) The meaning of $*$ in Definition III.5.3 does not matter because any i-complex is bounded, and the inclusions of the corresponding derived categories are fully faithful. Below we shall write simply $D(\mathcal{A})$.

b) Using the functor T^k we can identify $\operatorname{Ext}_{\mathcal{A}}^i(X,Y)$ with $\operatorname{Hom}_{D(\mathcal{A})}(X[k], Y[i+k])$ for any k. The multiplication of morphisms in $D(\mathcal{A})$ gives the composition law

$$
\begin{array}{rcl}
\operatorname{Ext}_{\mathcal{A}}^i(X,Y) & = & \operatorname{Hom}_{D(\mathcal{A})}(X[k], Y[i+k]) \\
\times & & \times \\
\operatorname{Ext}_{\mathcal{A}}^j(Y,Z) & = & \operatorname{Hom}_{D(\mathcal{A})}(Y[i+k], Z[i+j+k]) \\
\downarrow & & \downarrow \\
\operatorname{Ext}_{\mathcal{A}}^{i+j}(X,Z) & = & \operatorname{Hom}_{D(\mathcal{A})}(X[k], Z[i+j+k]).
\end{array}
$$

This composition law for Exts does not depend on the choice of k in the right column.

c) $\operatorname{Ext}_{\mathcal{A}}^i(X,Y)$ are abelian groups (since $D(\mathcal{A})$ is an additive category). The multiplication is bilinear. Moreover, $\operatorname{Ext}_{\mathcal{A}}^i$ gives the functor $\mathcal{A}^\circ \times \mathcal{A} \to Ab$.

Considering an exact sequence

$$0 \longrightarrow X' \longrightarrow X \longrightarrow X'' \longrightarrow 0 \quad (\text{resp. } 0 \longrightarrow Y' \longrightarrow Y \longrightarrow Y'' \longrightarrow 0)$$

in $\mathcal{A}$ as a distinguished triangle (see III.3.5), we get an exact sequence (compare with Sect. IV.1)

$$\ldots \longrightarrow \operatorname{Ext}_{\mathcal{A}}^i(X'',Y) \longrightarrow \operatorname{Ext}_{\mathcal{A}}^i(X,Y) \longrightarrow \operatorname{Ext}_{\mathcal{A}}^i(X',Y)$$
$$\longrightarrow \operatorname{Ext}_{\mathcal{A}}^{i+1}(X'', \mathcal{A}) \longrightarrow \ldots$$

(resp.

$$\ldots \longrightarrow \operatorname{Ext}_{\mathcal{A}}^i(X,Y') \longrightarrow \operatorname{Ext}_{\mathcal{A}}^i(X,Y) \longrightarrow \operatorname{Ext}_{\mathcal{A}}^i(X,Y'')$$
$$\longrightarrow \operatorname{Ext}_{\mathcal{A}}^{i+1}(X,Y) \longrightarrow \ldots).$$

d) Following Yoneda, let us consider the following construction of elements from $\operatorname{Ext}_{\mathcal{A}}^i(X,Y)$, $i > 0$. Let $K^\bullet$ be an acyclic complex of the form

$$
\begin{array}{rl}
K^\bullet : \ldots \longrightarrow 0 \longrightarrow K^{-i} & = Y \longrightarrow K^{-i+1} \longrightarrow \ldots \longrightarrow K^0 \longrightarrow X \\
& = K^1 \longrightarrow 0 \longrightarrow \ldots.
\end{array}
\quad \text{(III.22)}
$$

It determines the left roof

III.5 The Structure of the Derived Category   167

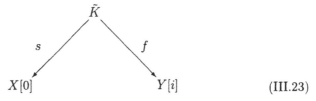
(III.23)

with $\tilde{K}^l = K^l$ for $l \neq 1$, $\tilde{K}^1 = 0$, $s^0 = d_K^0$, $f^{-i} = \mathrm{id}_Y$. Let $y(K^\bullet) : X[0] \to Y[i]$ be the morphism in the derived category corresponding to this roof.

Next, let $K^\bullet$ and $L^\bullet$ be two finite acyclic complexes such that the left element of $K^\bullet$ (i.e., Y in (III.22)) coincides with the right element of $L^\bullet$. Then we can form a new acyclic complex by chaining $L^\bullet$ and $K^\bullet$:

$$L^\bullet \circ K^\bullet : \ldots \longrightarrow 0 \longrightarrow Z = L^{-j} \longrightarrow \ldots \longrightarrow L^{-1} \xrightarrow{d_L} L^0 \xrightarrow{f} \\ \longrightarrow K^{-i+1} \longrightarrow \ldots \longrightarrow K^0 \longrightarrow K^1 = X \longrightarrow 0 \longrightarrow \ldots ; \quad \text{(III.24)}$$

here f is the composite morphism

$$f : L^0 \xrightarrow{d_L^0} L^1 = Y = K^{-i} \xrightarrow{d_K^{-i}} K^{-i+1}$$

(the acyclity of $L^\bullet \circ K^\bullet$ is easily verified).

5. Theorem. *a)* $\mathrm{Ext}_\mathcal{A}^i(X, Y) = 0$ for $i < 0$.

b) $\mathrm{Ext}_\mathcal{A}^0(X, Y) = \mathrm{Hom}_\mathcal{A}(X, Y)$.

c) Any element of $\mathrm{Ext}_\mathcal{A}^i(X, Y)$ is of the form $y(K^\bullet)$ for some complex $K^\bullet$ as in (III.22); for $y(K^\bullet) \in \mathrm{Ext}_\mathcal{A}^i(X, Y)$, $y(L^\bullet) \in \mathrm{Ext}_\mathcal{A}^j(Y, Z)$ we have

$$y(L^\bullet \circ K^\bullet) = y(L^\bullet) y(K^\bullet) \quad \text{(III.25)}$$

Proof. a) Let a morphism $\varphi : X[0] \to Y[-i]$, $i > 0$, in $D(\mathcal{A})$ be represented by a roof $X[0] \xleftarrow{s} K^\bullet \longrightarrow Y[-i]$. We shall construct a new complex $L^\bullet$ and a commutative diagram,

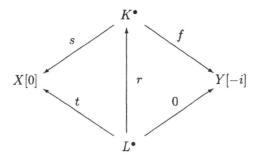

with r, t quasi-isomorphisms. This would imply that $\varphi = 0$ in $D(\mathcal{A})$.

We set
$$L^i = \begin{cases} K^j & \text{for } j < i-1, \\ \text{Ker } d_K^{i-1} & \text{for } j = i-1, \\ 0 & \text{for } j \geq i, \end{cases}$$

d_L is induced by d_K, r is the natural inclusion, $t^0 = s^0$. Since s is a quasi-isomorphism, we have $H^0(K) = X$, $H^j(K) = 0$ for $j \neq 0$. Therefore, r and t are quasi-isomorphisms. The commutativity is clear.

b) This is proved in Proposition III.5.2.

c) Let a morphism $X[0] \to Y[i], i > 0$, in $D(\mathcal{A})$ be represented by a roof $X[0] \xleftarrow{t} L^\bullet \xrightarrow{g} Y[i]$. Using it, we shall construct a new roof $X[0] \xleftarrow{s} K^\bullet \xrightarrow{f} Y[i]$ of the form (III.23) representing the same morphism in $D(\mathcal{A})$.

It will be convenient to construct this new roof in four steps. Each step starts from the roof obtained at the previous step (to simplify notations, we shall denote it by the same symbols $(L^\bullet, t, g)$; the result will be then a new roof $(L'^\bullet, t', g')$ together with a commutative diagram of the form

$$\begin{array}{ccccc} X[0] & \xleftarrow{t} & L^\bullet & \xrightarrow{g} & Y[i] \\ & & \updownarrow r & & \\ X[0] & \xleftarrow{t'} & L'^\bullet & \xrightarrow{g'} & Y[i] \end{array}$$

where r is a quasi-isomorphism between $L^\bullet$ and $L'^\bullet$ (sending either $L^\bullet$ to $L'^\bullet$ or $L'^\bullet$ to $L^\bullet$).)

Step 1. (Truncation from the left.) Let
$$(L')^j = \begin{cases} 0 & \text{for } j < -i, \\ \text{Coker } d_L^{-i-1} & \text{for } j = -i, \\ L^j & \text{for } j \geq -i+1. \end{cases}$$

It is clear how to construct the differential in $L'^\bullet$ and the morphisms t', g'. Let $r: L^\bullet \to L'^\bullet$ be the natural factorization; r is a quasi-isomorphism, because $L^\bullet$ is an H^0-complex and $i > 0$.

Step 2. (Truncation from the right.) Recall that now $L^\bullet$ is $L'^\bullet$ from Step 1. Let
$$(L')^j = \begin{cases} L^j & \text{for } j < 0, \\ \ker d_L^0 & \text{for } j = 0, \\ 0 & \text{for } j > 0. \end{cases}$$

The differential in $L'^\bullet$ and morphisms t', g', are the obvious ones; $r: L'^\bullet \to L^\bullet$ is the natural injection.

Step 3. (Double cone.) Now let
$$(L')^j = \begin{cases} 0 = L^j & \text{for } j < -i, \\ Y \oplus L^j & \text{for } j = -i, -i+1, \\ L^j & \text{for } j > -i+1. \end{cases}$$

The differentials are

$$d_{L'}^{-i}(y,l) = (y - g^{-i}(l), d_L^{-i}(l)),$$
$$d_{L'}^{-i+1}(y,l) = d_L^{-i+1}(l).$$

Other differentials are the same as in $L^\bullet$. The morphism $r : L^\bullet \to L'^\bullet$ is defined by

$$r^j(l) = \begin{cases} (g^{-i}(l), l) & \text{for } j = -i, \\ (0, l) & \text{for } j = -i+1, \\ l & \text{for } j > -i+1. \end{cases}$$

The definition of $(t')^0 : (L')^0 \to X$ depends on whether $-i+1$ equals to 0, or not:

$$\begin{array}{ll} \text{for } i \neq 1 : & (t')^0 = t^0, \\ \text{for } i = 1 : & (t')^0(y,l) = t^0(l). \end{array}$$

The morphism $(g')^{-i} : Y \oplus L^{-i} \to Y$ is the projection. For the reader's convenience we list all the necessary verifications:

a) $d_{L'} \circ d_{L'} = 0$ (easy);
b) $d_L \circ r = r \circ d_{L'}$ (easy);
c) $g' \circ r = g$ (easy);
d) $t' \circ r = t$ (easy);
e) t' is a quasi-isomorphism (easy); finally
f) r is a quasi-isomorphism.

This last statement is not completely obvious so we shall show how to prove it. It suffices to prove the acyclity of the cone $C(r)$. To do this we shall show that its identity map is homotopic to 0. The "nonstable" part of the corresponding homotopy h is shown on the diagram:

$$\begin{array}{c}
\to L^{-i} \to Y \oplus L^{-i} \oplus L^{-i+1} \to Y \oplus L^{-i+1} \oplus L^{-i+2} \to L^{-i+2} \oplus L^{-i+3} \to L^{-i+3} \oplus L^{-i+4} \to \cdots \\
(y,l,l') \quad (y,l,l') \quad (l,l') \quad (l,l') \\
l \quad (y,0,l) \quad (0,0,l) \quad (0,l) \\
\to L^{-i} \to Y \oplus L^{-i} \oplus L^{-i+1} \to Y \oplus L^{-i+1} \oplus L^{-i+2} \to L^{-i+2} \oplus L^{-i+3} \to \cdots
\end{array}$$

We highly recommend to the reader to write down the differentials in $C(r)$ and to verify the equality $dh + hd = \mathrm{id}_{C(r)}$.

Step 4. (Making g^{-i} an isomorphism.) At this moment the roof $(L'^\bullet, t', g')$ from Step 3 differs from (III.22) only in that $g^{-i} : (L')^{-i} \to Y$ might not be an isomorphism. To improve this we factorize $L'^\bullet$ by the two-term acyclic subcomplex

170   III. Derived Categories and Derived Functors

$$
\begin{array}{ccccccccc}
\cdots \to & 0 & \to & L^{-i} & \to & L^{-i} & \to & 0 & \to \cdots \\
& & & \downarrow l & & \downarrow l & & & \\
& & & \downarrow {\scriptstyle (0,l)} & & \downarrow {\scriptstyle (-g^{-i}(l),\, d_L^{-i}(l))} & & & \\
\cdots \to & 0 & \to & Y \oplus L^{-i} & \to & Y \oplus L^{-i+1} & \to & & \cdots
\end{array}
$$

Vertical arrows are injections because g^{-i} is an injection. Let $r : L'^\bullet \to K^\bullet$ be the factorization; it is a quasi-isomorphism due to Theorem I.6.8 (about exact sequence).

The morphism $f^{-i} : K^{-i} \to Y$ is induced by the projection $Y \oplus L^{-i} \to Y$. Finally, the construction of $s^0 : K^0 \to X$ is clear if $i \neq 1$, and when $i = 1$ s^0 it is induced by $(t')^0 : Y \oplus L^0 \to X$; this makes sense because $(t')^0 \circ (-g^{-i}, d_L) = 0$.

So, we have shown that our morphism $X[0] \to Y[i]$ is of the form $y(K^\bullet)$. It remains to verify (III.25). Let, as in III.5.4d),

$$K^\bullet : \ldots \quad \to \quad 0 \to K^{-i} =$$
$$= \quad Y \to K^{-i+1} \to \cdots \to K^0 \to K^1 = X \to 0 \to \cdots,$$

$$L^\bullet : \ldots \quad \to \quad 0 \to L^{-j} =$$
$$= \quad Z \to L^{-j+1} \to \cdots \to L^0 \to L^1 = Y \to 0 \to \cdots,$$

be two acyclic complexes. Let $M^\bullet = L^\bullet \circ K^\bullet$ (see (III.24)) so that

$$M^\bullet : \ldots \quad \to \quad 0 \to M^{-i-j} = Z \to$$
$$\to \quad M^{-i-j+1} \to \cdots \to M^0 \to M^1 = X \to 0 \to \cdots,$$

with $M^l = K^l$ for $-i+1 \leq l \leq 1$, $M^l = L^{l+i}$ for $-i-j \leq l \leq -i$. Let left roofs

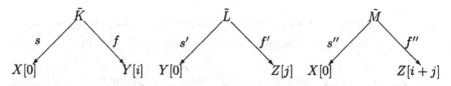

represent the morphisms $y(K^\bullet) : X[0] \to Y[i]$, $y(L^\bullet) : Y[0] \to Z[j]$, $y(L^\bullet \circ K^\bullet) : X[0] \to Z[i+j]$. The morphism $y(L^\bullet)[i] : Y[i] \to Z[i+j]$ is represented by the roof

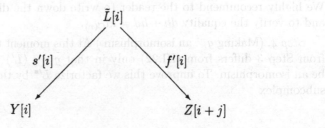

III.5 The Structure of the Derived Category   171

So to prove (III.25) we must construct the commutative diagram

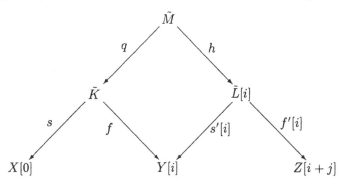

with $s \circ q = s''$, $f'[i] \circ g = f''$. Morphisms q and g are given by natural formulas:

$$q^l = \mathrm{id}_{K^l} \quad \text{for} \quad -i+1 \leq l \leq 0, \quad q^0 = d_L^0 : L^0 \to Y,$$

$$g^l = \mathrm{id}_{L^{l+i}} \quad \text{for} \quad -i-j \leq l \leq -i. \qquad \square$$

6. Homological Dimension

Theorem III.5.5 shows that, to some extent, the complexity of the derived category $D(\mathcal{A})$ can be measured by the following parameter.

Homological dimension $dh(\mathcal{A})$ is the maximum p such that there exist two objects $X, Y \in \mathrm{Ob}\,\mathcal{A}$ with $\mathrm{Ext}^p_{\mathcal{A}}(X,Y) \neq 0$ (or ∞ if there is no such p).

Obviously, $dh(\mathcal{A}) \geq 0$. Categories with $dh(\mathcal{A}) = 0$ are quite simple.

7. Proposition. *The following statements are equivalent:*
  a) $dh(\mathcal{A}) = 0$.
  b) $\mathrm{Ext}^1_{\mathcal{A}}(X,Y) = 0$ for all $X, Y \in \mathrm{Ob}\,\mathcal{A}$.
  c) The category $\mathcal{A}$ is semi-simple (see III.2.3).

Proof. a) $\Rightarrow$ b) is obvious.
  b) $\Rightarrow$ c) is proved as follows. Let

$$K^\bullet : 0 \longrightarrow Y \longrightarrow Z \longrightarrow X \longrightarrow 0$$

be an exact triple. By Theorem III.5.5a), b) and by 4c) we get the exact sequence of abelian groups

$$0 \longrightarrow \mathrm{Hom}(X,Y) \longrightarrow \mathrm{Hom}(X,Z)$$
$$\longrightarrow \mathrm{Hom}(X,X) \longrightarrow \mathrm{Ext}^1(X,Y) = 0.$$

Hence there exists $h : X \to Z$ whose image in $\mathrm{Hom}(X,X)$ is id_X. It is easy to see that $K^\bullet$ is isomorphic to the following split exact triple:

$$0 \longrightarrow Y \xrightarrow{(\mathrm{id}_Y, 0)} Y \oplus \mathrm{Im}\, h \xrightarrow{(0, k^{-1})} X \to 0$$

where $k : X \to \mathrm{Im}\, h$ is the canonical isomorphism.

c) ⇒ b) follows from Theorem III.5.5c) because a split exact triple defines the zero element in Ext^1 (see Ex. 3).

b) ⇒ a) To prove that $\mathrm{Ext}^p(X,Y) = 0$ for any $p \geq 2$ we represent any element $e \in \mathrm{Ext}^p(X,Y)$ as a product $e' \circ e''$ with $e' \in \mathrm{Ext}^{p-1}(Y,Z)$, $e'' \in \mathrm{Ext}^1(X,Y)$; this would suffice because $e'' = 0$.

Let $e = y(K^\bullet)$ with $K^\bullet$ as in (III.22). We set $e' = y(K'^\bullet)$, $e'' = y(K''^\bullet)$ where

$$K'^\bullet : \ldots \longrightarrow 0 \longrightarrow Y \longrightarrow K^{-p+1} \longrightarrow \ldots \longrightarrow K^{-1} \longrightarrow \mathrm{Im}\, d^{-1} \longrightarrow 0 \longrightarrow \ldots$$

$$K''^\bullet : \ldots \longrightarrow 0 \longrightarrow \mathrm{Im}\, d^{-1} \longrightarrow K^0 \longrightarrow X \longrightarrow 0 \longrightarrow \ldots . \qquad \square$$

8. Categories of Dimension 1

a) *The category Ab is one-dimensional.*
b) *The category $K[x]$-mod (where K is a field) is one-dimensional.*

It is easy to prove that these categories have a dimension ≥ 1; indeed, there are indecomposable exact triples

$$0 \longrightarrow \mathbb{Z} \xrightarrow{m} \mathbb{Z} \longrightarrow \mathbb{Z}/m\mathbb{Z} \longrightarrow 0, \quad m > 1,$$

$$0 \longrightarrow K[x] \xrightarrow{x} K[x] \longrightarrow K \longrightarrow 0.$$

To prove that the dimension is exactly 1, we develop some techniques.

9. Homological Dimension of an Object

For $X \in \mathrm{Ob}\, \mathcal{A}$ let

$$\mathrm{dhp}\, X = \sup(n \mid \exists Y \in \mathrm{Ob}\, \mathcal{A}, \mathrm{Ext}^n_\mathcal{A}(X,Y) \neq 0),$$
$$\mathrm{dhi}\, X = \sup(n \mid \exists Y \in \mathrm{Ob}\, \mathcal{A}, \mathrm{Ext}^n_\mathcal{A}(Y,X) \neq 0).$$

Here "p" and "i" are abbreviations for "projective" and "injective" respectively. This is justified by the following lemma.

10. Lemma. *The following properties of an object X are equivalent:*

$\mathrm{a_p})$ $\mathrm{dhp}\, X = 0$;
$\mathrm{b_p})$ $\mathrm{Ext}^1(X,Y) = 0$ *for all Y;*
$\mathrm{c_p})$ X *is a projective object.*

Similarly, the following properties of X are equivalent:

$\mathrm{a_i})$ $\mathrm{dhi}\, X = 0$;
$\mathrm{b_i})$ $\mathrm{Ext}^1(Y,X) = 0$ *for all Y;*
$\mathrm{c_i})$ X *is an injective object.*

Proof. $\mathrm{a_p}) \Rightarrow \mathrm{b_p})$ is obvious.

$\mathrm{b_p}) \Rightarrow \mathrm{c_p})$. Proposition III.5.7 implies that any exact triple with X as the third object splits. To prove the projectivity of X we must deduce from this the existence of the dashed arrow in the following diagram:

III.5 The Structure of the Derived Category

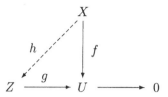

with the exact row. Let us form the fibre product $Z \underset{U}{\times} X$. By Ex. III.5.2, the projection $Z \underset{U}{\times} X \to X$ is surjective, so that there exists a splitting $k : X \to Z \underset{U}{\times} X$ and h is its composition with the projection $Z \underset{U}{\times} X \to Z$.

$c_p) \Rightarrow b_p)$. X being projective, any exact triple with the third object X splits, so $\mathrm{Ext}^1(X, Y) = 0$.

$b_p) \Rightarrow a_p)$. The proof repeats the last part of the proof of Proposition III.5.7.

The proof of equivalences $a_i) \Leftrightarrow b_i) \Leftrightarrow c_i)$ is similar. □

11. Proposition. *a) Let*

$$\ldots \longrightarrow 0 \longrightarrow X' \longrightarrow P^{-k} \longrightarrow \ldots \longrightarrow P^{-1} \longrightarrow P^0 \longrightarrow X \longrightarrow 0 \longrightarrow \ldots$$
(III.26)

by an acyclic complex with all P^{-i} projective. Then

$$\mathrm{dhp}\, X' = \max(\mathrm{dhp}\, X - k - 1, 0).$$

b) Let

$$\ldots \longrightarrow 0 \longrightarrow X \longrightarrow I^0 \longrightarrow I^1 \longrightarrow \ldots \longrightarrow I^k \longrightarrow X' \longrightarrow 0 \longrightarrow \ldots$$

be an acyclic complex with all I^i injective. Then

$$\mathrm{dhi}\, X' = \max(\mathrm{dhi}\, X - k - 1, 0).$$

12. Corollary. *a) If the category $\mathcal{A}$ has enough projective objects (that is, any object is a quotient of a projective one) then the condition $\mathrm{dhp}\, X \le k$ is equivalent to the existence of a projective resolution of X of the length $\le k + 1$.*

b) If the category $\mathcal{A}$ has enough injective objects (that is, any object is a subobject of an injective one), then the condition $\mathrm{dhi}\, X \le k$ is equivalent to the existence of an injective resolution of X of length $k + 1$.

Proof (of the proposition). Using the complex (III.26) we define for any object Y and for any $d \ge 0$ a map

$$\mathrm{Ext}^d(X', Y) \longrightarrow \mathrm{Ext}^{d+k+1}(X, Y) \qquad (\text{III.27})$$

and prove that this map is an isomorphism for $d \ge 1$ and a surjection for $d = 0$. This would immediately imply that $\mathrm{dhp}\, X = \mathrm{dhp}\, X' + k + 1$ if $\mathrm{dhp}\, X' \ge 1$ and $\mathrm{dhp}\, X \le k + 1$ if $\mathrm{dhp}\, X' = 0$ which is equivalent to the statement a).

We use the induction on k. For $k = 0$, (III.26) becomes an exact triple and we define the mapping (III.27) to be the connecting homomorphism in the long exact sequence of Exts corresponding to this exact triple:

$$\operatorname{Ext}^d(P^0, Y) \longrightarrow \operatorname{Ext}^d(X', Y) \longrightarrow \operatorname{Ext}^{d+1}(X, Y)$$
$$\longrightarrow \operatorname{Ext}^{d+1}(P^0, Y) = 0.$$

Our statement for $k = 0$ follows immediately as, by Lemma 10,

$$\operatorname{Ext}^d(P^0, Y) = 0 \quad \text{for} \quad d \geq 1.$$

To perform the induction step we split (III.26) into two parts:

$$0 \longrightarrow X' \longrightarrow P^{-k} \longrightarrow X'' \longrightarrow 0 \quad (X'' = \operatorname{Im} d^{-k} = \operatorname{Ker} d^{-k+1}) \quad \text{(III.28)}$$

$$0 \longrightarrow X'' \longrightarrow P^{-k+1} \longrightarrow \ldots \longrightarrow P^{-1} \longrightarrow P^0 \longrightarrow X \longrightarrow 0 \quad \text{(III.29)}$$

and define (III.27) as the composition of two mappings

$$\operatorname{Ext}^d(X', Y) \xrightarrow{\delta'} \operatorname{Ext}^{d+1}(X'', Y) \xrightarrow{\delta''} \operatorname{Ext}^{d+k+1}(X, Y)$$

where δ' is the connecting homomorphism of the exact triple (III.28) and δ'' is assumed to be constructed in the previous step using the complex (III.29). It is clear that this composition satisfies the required properties.

The proof of b) is similar. □

The corollary follows immediately.

13. Remark. The mapping (III.27) we have just constructed coincides up to a sign, with the multiplication by the Yoneda class of the complex (III.26) in $\operatorname{Ext}^{k+1}(X, X')$.

14. Modules Over a Principal Ideal Ring

Let A be a commutative ring with unity without zero divisors; we assume that any ideal in A is principal (for example, A may be $\mathbb{Z}$ or $K[x]$). Let $\mathcal{C}$ be the category of A-modules of finite type. The following three classes of objects in $\mathcal{C}$ coincide:

a) free modules; b) projective modules; c) torsion-free modules.

Hence any object has a projective resolution of length ≤ 2 so that $dh\, \mathcal{C} \leq 1$. If A is not a field, and $a \in A \setminus \{0\}$ is not invertible, then the class of the exact triple

$$0 \longrightarrow A \xrightarrow{a} A \longrightarrow A/a \longrightarrow 0$$

determines a nonzero element of $\operatorname{Ext}^1(A/a, A)$, so that $dh\, \mathcal{C} = 1$.

We have proved a somewhat weaker result than the one stated in III.5.8. However, the assumption of the finiteness of type can be dropped.

15. Hilbert Theorem

In III.1.1 we have stated the Hilbert syzygie theorem for modules over polynomial rings $k[t_1,\ldots,t_r]$. In the next subsection we reformulate this theorem in the category theory language. Being applied to modules over $k[t_1,\ldots,t_r]$, this theorem gives a somewhat weaker result than the classical syzygie theorem, as we get a bound for the length of a projective resolution, and not of a free one.

In fact, any projective module of finite type over a ring of polynomials with coefficients in a field is free. This highly nontrivial result was conjectured by Serre and proved independently by Quillen [5] and Suslin [1].

Let $\mathcal{A}$ be an abelian category. Denote by $\mathcal{A}[T]$ the following category:

$$\text{Ob } \mathcal{A}[T] = \{\text{pairs } (X, t) \text{ with } X \in \text{Ob } \mathcal{A},\, t \in \text{Hom}_\mathcal{A}(X, X)\}.$$

A morphism $(X, t) \to (X', t')$ in $\mathcal{A}[T]$ is a morphism $f : X \to X'$ in $\mathcal{A}$ such that $t' \circ f = f \circ t$.

16. Theorem. *a) $\mathcal{A}[T]$ is an abelian category.*

b) Let us assume that $\mathcal{A}$ has sufficiently many projective objects as well as all infinite direct sums. Then for any $(X, t) \in \text{Ob } \mathcal{A}[T]$ we have

$$dhp_{\mathcal{A}[T]}(X, t) \leq dhp_\mathcal{A}(X) + 1.$$

c) Under the above assumptions

$$dhp_{\mathcal{A}[T]}(X, 0) = dhp_\mathcal{A}(X) + 1.$$

This theorem will be proved in III.5.18–III.5.19.

17. The Classical Hilbert Theorem

Let $\mathcal{A}_r$ be the category of modules over the ring $k[t_1,\ldots,t_r]$. Define the functor

$$F_r : \mathcal{A}_{r-1}[T] \longrightarrow \mathcal{A}_r$$

by

$$F_r(X, t) = X \quad \text{with the previous action of } k[t_1,\ldots,t_{r-1}] \text{ and}$$
$$\text{with } t_r \text{ acting as } T.$$

One can easily verify that F_r is an equivalence of categories. If k is a field, we can start the induction on r from $r = 0$ using $dh\,\mathcal{A}_0 = 0$ and obtain from Theorem III.5.16 that $dh\,\mathcal{A}_r = r$.

One can of course take for k an arbitrary (not necessarily commutative) ring; in the definition of $k[t_1,\ldots,t_r]$ one always assumes that $t_1,\ldots,t_r$ commute with k and with each other.

18. $\mathcal{A}[T]$ is an Abelian Category

We must verify axioms A1–A4 from Sect. II.5. Axioms A1 and A2 are clear because $\mathrm{Hom}_{\mathcal{A}[T]}((X,t),(X',t'))$ is a subgroup of $\mathrm{Hom}_{\mathcal{A}}(X,X')$. Next, let $(X,t),(X',t')$ be two objects from $\mathcal{A}[T]$. Define their direct sum as $(X,t) \oplus (X',t') = (X \oplus X', t \oplus t')$. The construction of morphisms (II.19) and the verification of (II.20) from Sect. II.5.4 is standard.

Let us show now that $\mathcal{A}[T]$ contains kernels and cokernels of all morphisms. Below we will denote by the same letter morphism $(X,t) \to (X',t')$ in $\mathcal{A}[T]$ and the corresponding morphism $X \to X'$ in $\mathcal{A}$. Let (K,k) be the kernel of f in $\mathcal{A}$. Then $ftk = t'fk = 0$ so that there exists a unique morphism $s : K \to K$ with $tk = ks$ (see the diagram below),

$$\begin{array}{ccccccc}
K & \xrightarrow{k} & X & \xrightarrow{f} & X' & \xrightarrow{c} & C \\
s\downarrow & & t\downarrow & & t'\downarrow & & \downarrow v \\
K & \xrightarrow{k} & X & \xrightarrow{f} & X' & \xrightarrow{c} & C
\end{array}$$

Hence $(K,s) \in \mathrm{Ob}\,\mathcal{A}[T]$ and $k : (K,s) \to (X,t)$ is a morphism in $\mathcal{A}[T]$. We leave to the reader an easy verification that $((K,s),k)$ is the kernel of f in $\mathcal{A}[T]$. Similarly one can construct the cokernel $((C,v),c)$ of the morphism f in $\mathcal{A}[T]$.

Let us prove now the existence of the canonical decomposition of a morphism $f : (X,t) \to (X',t')$ in $\mathcal{A}[T]$. Let $(I,u) = \mathrm{Coker\,ker}\,f$, $(I',u') = \mathrm{ker\,Coker}\,f$ and $i : (X,t) \to (I,u)$, $j : (I',u') \to (X',t')$ be the corresponding morphisms in $\mathcal{A}[T]$. By II.5.11b) there exists a unique morphism $l : (I,u) \to (I',u')$ with $f = j \circ l \circ i$. As $I = \mathrm{Coker\,ker}\,f$ and $I' = \mathrm{ker\,Coker}\,f$ in $\mathcal{A}$, $l : I \to I'$ is an isomorphism in $\mathcal{A}$. Therefore $l : (I,u) \to (I',u')$ is an isomorphism in $\mathcal{A}[T]$ and A4 is satisfied for $\mathcal{A}[T]$.

19. Projective Objects in $\mathcal{A}[T]$

Let us assume now that $\mathcal{A}$ contains infinite direct sums (i.e., the axiom AB3 from Ex. II.6.2 is satisfied). For any $Y \in \mathrm{Ob}\,\mathcal{A}$ denote $Y^\infty = Y \oplus Y \oplus \ldots$ and define the shift morphism $s_Y : Y^\infty \to Y^\infty$ by $s_Y(y_1, y_2, \ldots) = (0, y_1, \ldots)$. We claim that *if Y is projective in $\mathcal{A}$, then (Y^∞, s_Y) is projective in $\mathcal{A}[T]$*.

Indeed, let

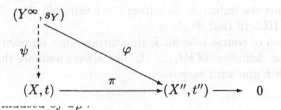

III.5 The Structure of the Derived Category

be a projectivity diagram in $\mathcal{A}[T]$. In $\mathcal{A}$ this diagram is described by morphisms $\varphi_i : Y \to X''$, $i = 1, 2, \ldots$ and an epimorphism $\pi : X \to X''$ such that $t''\varphi_i = \varphi_{i+1}$, $t''\pi = \pi t$. To construct a required morphism ψ in $\mathcal{A}[T]$ we have to find morphisms $\psi_i : Y \to X$, $i = 1, 2, \ldots$ in $\mathcal{A}$ such that $\pi \psi_i = \varphi_i$, $t\psi_i = \psi_{i+1}$.

As Y is projective in $\mathcal{A}$ and π is an epimorphism, there exists $\psi_1 : Y \to X$ in $\mathcal{A}$ with $\pi\psi_1 = \varphi_1$. Defining ψ_{i+1} inductively by $\psi_{i+1} = t\psi_i$ we have $\pi\psi_{i+1} = \pi t\psi_i = t''\pi\psi_i = t''\varphi_i = \varphi_{i+1}$.

The above claim implies, in particular, that if $\mathcal{A}$ has sufficiently many projective objects, then the same holds for $\mathcal{A}[T]$. Moreover, as the functor $X \mapsto (X, 0)$ represents $\mathcal{A}$ as a full subcategory of $\mathcal{A}[T]$, X is projective in $\mathcal{A}$ for any projective (X, t) in $\mathcal{A}[T]$.

Proof (of III.5.16b)). Let $(X, t) \in \mathrm{Ob}\,\mathcal{A}[T]$ and $dhp_{\mathcal{A}} X = k$. Let

$$0 \longrightarrow (Y, r) \longrightarrow (P^{-k}, t_{-k}) \longrightarrow \cdots \longrightarrow (P^0, t_0) \longrightarrow (X, t) \longrightarrow 0$$

be an acyclic complex with all (P^{-i}, t_{-i}), $0 \le i \le k$, projective in $\mathcal{A}[T]$. Then the complex

$$0 \longrightarrow Y \longrightarrow P^{-k} \longrightarrow \cdots \longrightarrow P^0 \longrightarrow X \longrightarrow 0$$

in $\mathcal{A}$ is also acyclic and all P^{-i} are projective in $\mathcal{A}$ by the last remark in the previous subsection. By III.5.10a) and III.5.11a), Y is projective in $\mathcal{A}$. Therefore (Y^∞, s_Y) is projective in $\mathcal{A}[T]$ and to prove that $dhp_{\mathcal{A}[T]}(X, t) \le k + 1$ it suffices to construct an exact sequence of the form

$$0 \longrightarrow (Y^\infty, s_Y) \xrightarrow{l} (Y^\infty, s_Y) \xrightarrow{q} (Y, r) \longrightarrow 0 \qquad (\mathrm{III}.30)$$

in $\mathcal{A}[T]$. Define l by the matrix of morphisms $l_{ij} : Y \to Y$, $i, j = 1, 2, \ldots$ in $\mathcal{A}$, with $l_{jj} = -r$, $l_{j,j-1} = \mathrm{id}_Y$, $l_{ij} = 0$ for all other pairs (i, j). Define also q by the family $q_j : Y \to Y$ of morphisms in $\mathcal{A}$, where $q_j = r^j$. Obviously, q and r are morphisms in $\mathcal{A}[T]$ (i.e., the necessary commutation relations hold). An easy verification of the exactness of (III.30) is left to the reader. □

Proof (of III.5.16c)). The required statement follows from the fact that

$$\mathrm{Ext}^n_{\mathcal{A}[T]}\left((X, 0), (Y, 0)\right) = \mathrm{Ext}^n_{\mathcal{A}}(X, Y) \oplus \mathrm{Ext}^{n-1}_{\mathcal{A}}(X, Y) \qquad (\mathrm{III}.31)$$

for any $X, Y \in \mathrm{Ob}\,\mathcal{A}$. The easiest way to prove it is to compute Exts using projective resolutions (see Ex. 1). Let

$$\cdots \longrightarrow P^{-n} \xrightarrow{d_P^{-n}} P^{-n+1} \longrightarrow \cdots \longrightarrow P^{-1} \xrightarrow{d_P^{-1}} P^0 \xrightarrow{\varepsilon} X \longrightarrow 0$$

be a projective resolution of X in $\mathcal{A}$, i.e., an exact sequence in $\mathcal{A}$ with projective P^{-i}. Then $\mathrm{Ext}^n_{\mathcal{A}}(X, Y) = H^n\left(\mathrm{Hom}(P^\bullet, Y)\right)$ is the cohomology of the complex

$$\cdots \longrightarrow \mathrm{Hom}_{\mathcal{A}}\left(P^{-n+1}, Y\right) \longrightarrow \mathrm{Hom}_{\mathcal{A}}\left(P^{-n}, Y\right) \longrightarrow \cdots$$

with differentials induced by d_P^{-i}.

Let us consider, for any $n \geq 0$, the exact sequence (III.30) in $\mathcal{A}[T]$ corresponding to $(P^{-n}, 0)$:

$$0 \longrightarrow ((P^{-n})^\infty, s_{P^{-n}}) \xrightarrow{l} ((P^{-n})^\infty, s_{P^{-n}}) \xrightarrow{q} (P^{-n}, 0) \longrightarrow 0.$$

The morphism $l = l^{-n}$ is given by $l^{-n}(p_1^{-n}, p_2^{-n}, \ldots) = (0, p_1^{-n}, p_2^{-n}, \ldots)$.

Let the complex

$$\ldots \longrightarrow Q^{-n} \xrightarrow{d_Q^{-n}} Q^{-n+1} \longrightarrow \ldots \longrightarrow Q^0 \xrightarrow{\varepsilon_Q} (X, 0) \longrightarrow 0 \qquad (\text{III.32})$$

in $\mathcal{A}[T]$ be given by

$$Q^{-n} = ((P^{-n})^\infty, s_{P^{-n}}) \oplus ((P^{-n+1})^\infty, s_{P^{-n+1}}) \quad \text{for } n > 0,$$

$$Q^0 = ((P^0)^\infty, s_{P^0}),$$

$$d_Q^{-n}(p_1^{-n}, p_2^{-n}, \ldots; p_1^{-n+1}, p_2^{-n+1}, \ldots) =$$
$$(d_P^{-n} p_1^{-n}, d_P^{-n} p_2^{-n} + (-1)^n p_1^{-n+1}, d_P^{-n} p_3^{-n} + (-1)^n p_2^{-n+1},$$
$$\ldots; d_P^{-n+1} p_1^{-n+1}, d_P^{-n+1} p_2^{-n+1}, \ldots),$$

$$\varepsilon_Q(p_1^0, p_2^0, \ldots) = \varepsilon_P(p_1^0).$$

The reader can easily verify that (III.32) is an exact sequence in $\mathcal{A}[T]$ (either directly or using properties of double complexes, see III.7.8). Moreover, by III.5.19, (III.32) is a projective resolution of $(X, 0)$ in $\mathcal{A}[T]$. To use it for the computation of $\text{Ext}_{\mathcal{A}[T]}((X, 0), (Y, 0))$ we note first that for any $Z \in \text{Ob } \mathcal{A}$ there exists a natural isomorphism

$$\text{Hom}_{\mathcal{A}[T]}((Z^\infty, s_Z), (Y, 0)) \cong \text{Hom}_{\mathcal{A}}(Z, Y) \qquad (\text{III.33})$$

given by the formula

$$\{f : (Z^\infty, s_Z) \longrightarrow (Y, 0)\} \mapsto \{\varphi : Z \longrightarrow Y, \varphi(z) = f(z, 0, 0, \ldots)\}.$$

Hence

$$\text{Hom}_{\mathcal{A}[T]}(Q^{-n}, (Y, 0)) \cong \text{Hom}_{\mathcal{A}}(P^{-n}, Y) \oplus \text{Hom}_{\mathcal{A}}(P^{-n+1}, Y).$$

Moreover, (III.33) implies that this isomorphism is compatible with differentials, and we get an isomorphism of complexes of abelian groups

$$\text{Hom}_{\mathcal{A}[T]}(Q^\bullet, (Y, 0)) \cong \text{Hom}_{\mathcal{A}}(P^\bullet, Y) \oplus \text{Hom}_{\mathcal{A}}(P^\bullet, Y)[1],$$

which implies (III.31). □

20. Derived Category and Injective Resolutions

The last topic of this section is the characterization of the derived category in terms of injective resolutions.

Let $\mathcal{I}$ be the full subcategory of an abelian category formed by all injective objects. Let $K^+(\mathcal{I})$ be the category whose objects are bounded from the left complexes of injective objects and whose morphisms are morphisms of complexes modulo homotopic equivalence. Let $K^+(\mathcal{I}) \to D^+(\mathcal{A})$ be the natural functor.

21. Theorem. *a) The above functor is an equivalence of $K^+(\mathcal{I})$ with a full subcategory of $D^+(\mathcal{A})$.*

b) If $\mathcal{A}$ has sufficiently many injective objects, then the functor yields an equivalence between $K^+(\mathcal{I})$ and $D^+(\mathcal{A})$.

22. Plan of the Proof

First we show that one can apply Proposition III.2.10 to the pair of categories $K^+(\mathcal{I}) \subset K^+(\mathcal{A})$ and to the class of all quasi-isomorphisms in $K^+(\mathcal{A})$. To do this we have to verify that:

A. Quasi-isomorphisms in $K^+(\mathcal{I})$ form a localizing system $S_\mathcal{I}$.
B. The condition b_2 of Proposition III.2.10 is satisfied.
   After that we get from Proposition III.2.10 that $K^+(\mathcal{I})[S_\mathcal{I}^{-1}]$ is a full subcategory of $K^+(\mathcal{A})[S^{-1}] = D^+(\mathcal{A})$ (the last equality is proved in Proposition 2).
C. Any element of $S_\mathcal{I}$ is an isomorphism. Hence
$$K^+(\mathcal{I})[S_\mathcal{I}^{-1}] = K^+(\mathcal{I}).$$
D. If $\mathcal{A}$ has sufficiently many injective objects, then any object of $D^+(\mathcal{A})$ is isomorphic to one from $K^+(\mathcal{I})$.

23. Proof of 22A

It was proved in Theorem III.5.4 that quasi-isomorphisms form a localizing system in $K^+(\mathcal{A})$. That proof was based on the construction of new complexes by completing certain commutative diagrams. The construction was explicit, with all the new complexes being just the cones of morphisms of the previous ones. As direct sums of injective objects are injective, this construction applied to complexes in the category $K^+(\mathcal{I})$ will produce new complexes which belong to the same category $K^+(\mathcal{I})$. Therefore, axioms of the localizing system are satisfied for quasi-isomorphisms in $K^+(\mathcal{I})$.

24. Proof of 22B,C

We prove the following statement (which is stronger than the condition b_2 of Proposition III.2.10):

(∗) *Let $s : I \to K$ be a quasi-isomorphism between an object from $K^+(\mathcal{I})$ and an object from $K^+(\mathcal{A})$. Then there exists a morphism of complexes $t : K \to I$ such that $t \circ s$ is homotopic to id_I.*

Using Lemma III.3.3 we construct the distinguished triangle $I \xrightarrow{s} K \xrightarrow{\delta} C(s) \to I[1]$ in $K^+(\mathcal{A})$. Since s is a quasi-isomorphism, its cone $C(s)$ is acyclic. We prove now that any morphism from an acyclic complex to a left-bounded complex formed by injective elements is homotopic to the zero morphism. The homotopy is constructed by induction starting from the left; one may assume that we begin with C^1:

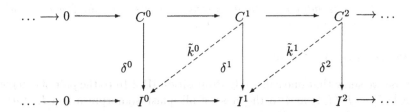

The existence of $\tilde{k}^0$ is the extension condition for injective objects. The existence of $\tilde{k}^1$ can be reduced to this condition as follows. Let us consider the diagram

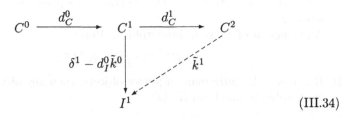

(III.34)

It differs from the extension diagram for injective objects in that $\operatorname{Ker} d_C^1$ may be nonzero. However, $(\delta^1 - d_I^0 \tilde{k}^0) \circ d_C^0 = 0$. So, instead of (III.34) one can consider the diagram

$$0 \to \operatorname{Coker} d_C^0 \xrightarrow{d'} C^2$$
$$\downarrow \qquad \swarrow \tilde{k}^1$$
$$I^1$$

that gives the required morphism $\tilde{k}^1$. The general induction step is of the same structure.

Now we write down this homotopy componentwise:
$$\tilde{k} : C = I[1] \oplus K \longrightarrow I[1], \quad \tilde{k} = (k, t').$$
We have $\delta = \tilde{k}d + d\tilde{k}$. On the other hand, by Lemma III.3.3, we have $\delta = (\mathrm{id}_{I[1]}, 0)$. So
$$(\mathrm{id}_{I[1]}, 0) = (k, t')d_C + d_{I[1]}(k, t'), \quad t' : K \longrightarrow I[1],$$
$$\mathrm{id}_{I[1]} = kd_{I[1]} + d_{I[1]}k + t's, \quad 0 = t'd_C - d_{I[1]}t'.$$
Hence t' is a morphism of complexes and $t's$ is homotopic to $\mathrm{id}_{I[1]}$.

25. Proof of 22D

We prove that for any complex $C^\bullet \in D^+(\mathcal{A})$ there exist a complex $I^\bullet \in K^+(\mathcal{I})$ and a quasi-isomorphism $t : C^\bullet \to I^\bullet$. One can assume that $C^i = 0$ for $i < 0$. We construct I^i, d_I^i and $t^i : C^i \to I^i$ by induction on i.

Steps 0 and 1:

$$\begin{array}{ccccc}
0 & \longrightarrow & C^0 & \xrightarrow{d_C^0} & C^1 \\
& & \downarrow t^0 & & \downarrow a \\
0 & \longrightarrow & I^0 & \xdashrightarrow{b} & I^0 \coprod_{C^0} C^1 \xdashrightarrow{c} I^1
\end{array}$$

Here we consequently construct: an injection $t^0 : C^0 \to I^0$ with $I^0 \in \mathcal{I}$ (it exists because $\mathcal{A}$ has sufficiently many injectives); the cocartesian square with the vertex $I^0 \coprod_{C^0} C^1$; an injection c of this node into $I^1 \in \mathcal{I}$. Then we define $d_I^0 = c \circ b$, $t^1 = c \circ a$.

Step $(i+1)$:

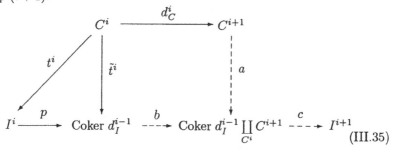

(III.35)

Here we construct the cocartesian square with the vertex $\mathrm{Coker}\, d_I^{i-1} \coprod_{C^i} C^{i+1}$ and an injection of this vertex into $I^{i+1} \in \mathcal{I}$. Then we define $d_I^{i+1} = c \circ b \circ p$, $t^{i+1} = c \circ a$.

182   III. Derived Categories and Derived Functors

It is clear that this construction yields a complex $I^\bullet \in K^+(\mathcal{I})$ and a morphism of complexes $t = (t^i) : C^\bullet \to I^\bullet$. We must prove only that t is a quasi-isomorphism, i.e., that $H^i(t) : H^i(C^\bullet) \to H^i(I^\bullet)$ is an isomorphism for each $i \geq 0$. To do this we use the language of elements (see Ex. II.5). First, using the exact sequence

$$0 \longrightarrow \operatorname{Im} d_C^{i-1} \longrightarrow \operatorname{Ker} d_C^i \longrightarrow H^i(C^\bullet) \longrightarrow 0$$

and Ex. II.5.4a)–c) it is easy to verify that elements of $H^i(C^\bullet)$ are equivalence classes of pairs (X, h), $X \in \operatorname{Ob} \mathcal{A}$, $h : X \to C^i$ with $d_C^i \circ h = 0$, modulo the relation

$(X, h) \sim (X', h') \Leftrightarrow \{\text{there exist } V \in \operatorname{Ob} \mathcal{A}, \text{ epimorphisms } u : V \to X, u' : V \to X', \text{ and a morphism } v : V \to C^{i-1} \text{ such that } hu - h'u' = d_C^{i-1}v \}.$

We need also the following property of a cocartesian square

$$\begin{array}{ccc} U & \xrightarrow{f} & X \\ g \downarrow & & \downarrow g' \\ Y & \xrightarrow{f'} & Z \end{array}$$

with $Z = X \coprod_U Y$. Let $X \oplus Y$ be the direct sum of X and Y and $\pi : X \oplus Y \to Z$ the natural morphism. Then the sequence

$$U \xrightarrow{(f,g)} X \oplus Y \xrightarrow{\pi = (g', f')} Z \longrightarrow 0$$

is exact.

Now let us prove that t is a quasi-isomorphism. Using diagram (III.35) we check that $H^i(t)$ is an epimorphism and $H^{i+1}(t)$ is a monomorphism. Our arguments can be easily extended to cases $i = 0, 1$.

a) $H^i(t)$ *is an epimorphism.* We use Ex. II.5.4c). Let $x = (X, h)$ be an element of $H^i(I)$ so that $X \in \operatorname{Ob} \mathcal{A}$, $h : X \to I^i$, $d_I^i \circ h = 0$. The last equality implies that $b \circ p(x) = 0$ so that the morphism $\pi = (b, a)$: $\operatorname{Coker} d_I^{i-1} \oplus C^{i+1} \to \operatorname{Coker} d_I^{i-1} \coprod_{C^i} C^{i+1}$ transforms the element $(p(x), 0) \in \operatorname{Coker} d_I^{i-1} \oplus C^{i+1}$ into the zero element. From the exactness of the sequence

$$C^i \xrightarrow{(\tilde{t}^i, -d_C^i)} \operatorname{Coker} d_I^{i-1} \oplus C^{i+1} \xrightarrow{(b,a)} \operatorname{Coker} d_I^{i-1} \coprod_{C^i} C^{i+1} \longrightarrow 0 \quad \text{(III.36)}$$

we get, using Ex. II.5.4e), that

$$(p(x), 0) = (\tilde{t}, -d_C^i) \tilde{x}$$

for some $\tilde{x} = (\tilde{X}, \tilde{h}) \in C^i$. This means that $d_C^i \circ \tilde{h} = 0$ and $\tilde{x}$ determines an element of $H^i(C)$ such that $x = H^i(t)\tilde{x}$.

b) $H^{i+1}(t)$ *is a monomorphism.* Let $h : X \to C^{i+1}$ be a representative of an element $x \in H^{i+1}(C)$ and $H^{i+1}(t)x = 0$ in $H^{i+1}(I)$. We prove that $x = 0$ in $H^{i+1}(C)$. The equality $t^{i+1} = c \circ a$, where c is an injection (see (III.35)), implies that there exist an object $V \in \mathrm{Ob}\,\mathcal{A}$, an epimorphism $u : V \to X$, and a morphism $v : V \to \mathrm{Coker}\, d_I^{i-1}$ such that $a \circ h \circ u = b \circ v$. Hence the morphism
$$(bv, -ahu) : V \longrightarrow \mathrm{Coker}\, d_I^{i-1} \coprod_{C^i} C^{i+1}$$
equals 0, i.e.,
$$(b, a)(v, -hu) = 0$$
where $(v, -hu)$ is the morphism $V \to \mathrm{Coker}\, d_I^{i-1} \oplus C^{i+1}$. From the exactness of the sequence (III.36) and from Ex. II.5.4f) we get that there exist an object $W \in \mathrm{Ob}\,\mathcal{A}$, an epimorphism $w : W \to V$, and a morphism $z : W \to C^i$ such that
$$(v, -hu)w = (\tilde{t}^i, -d_C^i)z.$$
This equality implies, in particular, that
$$huw = d_C^i z.$$
As $uw : W \to X$ is an epimorphism, the element $x = (X, h)$ equals 0 in $H^{i+1}(C)$.

Exercises

1. Ext and Resolutions. a) Verify that the statement (∗) proved in III.5.24 and the dual statement for complexes of projective objects imply that the natural homomorphism
$$\mathrm{Hom}_{K(\mathcal{A})}(X^\bullet, Y^\bullet) \longrightarrow \mathrm{Hom}_{D(\mathcal{A})}(X^\bullet, Y^\bullet)$$
is an isomorphism in each of the following cases:

(i) $Y^\bullet \in \mathrm{Ob}\, Kom^+(\mathcal{I})$;
(ii) $X^\bullet \in \mathrm{Ob}\, Kom^-(\mathcal{P})$ ($\mathcal{P}$ is the class of projective objects in $\mathcal{A}$).

Deduce that $\mathrm{Ext}_{\mathcal{A}}(X, Y)$ for $X, Y \in \mathrm{Ob}\,\mathcal{A}$ can be computed as follows. Let $\ldots \to P^{-1} \xrightarrow{d_P^{-1}} P^0 \to X \to 0$ be a projective resolution of X (resp. $0 \to Y \to I^0 \xrightarrow{d_I^0} I^1 \to \ldots$ be an injective resolution of Y). Then $\mathrm{Ext}_{\mathcal{A}}^n(X, Y)$ is the n-th cohomology group of the complex
$$0 \longrightarrow \mathrm{Hom}(P^0, Y) \longrightarrow \mathrm{Hom}(P^{-1}, Y) \longrightarrow \mathrm{Hom}(P^{-2}, Y) \longrightarrow \ldots$$
(resp. of the complex
$$0 \longrightarrow \mathrm{Hom}(X, I^0) \longrightarrow \mathrm{Hom}(X, I^1) \longrightarrow \mathrm{Hom}(X, I^2) \longrightarrow \ldots)$$
with the differentials induced by d_P (resp. by d_I).

2. Ext^1 and Extensions. Let $X, Y \in \text{Ob } \mathcal{A}$. Denote by $E(X, Y)$ the set of equivalence classes of exact triples

$$E: 0 \longrightarrow Y \longrightarrow Z \longrightarrow X \longrightarrow 0$$

by the following equivalence relation: $E \sim E'$ if an only if there exists a commutative diagram

$$\begin{array}{ccccccccc} E: & 0 & \longrightarrow & Y & \longrightarrow & Z & \longrightarrow & X & \longrightarrow & 0 \\ & & & \| & & \downarrow \alpha & & \| & & \\ E': & 0 & \longrightarrow & Y & \longrightarrow & Z' & \longrightarrow & X & \longrightarrow & 0 \end{array}$$

(in which α is necessarily an isomorphism by the five-lemma).

a) Prove that $\sim$ is an equivalence relation.

b) For $\varphi: Y \to Y'$, $E \in E(X, Y)$ define φE as the class of the triple

$$0 \longrightarrow Y' \longrightarrow Z' \longrightarrow X \longrightarrow 0$$

where $Z' = Y' \coprod_Y Z$. Show that φE is well defined. Show that this is an action of $\text{Hom}(Y, Y')$ on $E(X, Y)$. Define similarly $E\psi$ for $\psi: X' \to X$. Prove the following associativity relations: $(\varphi_1 \varphi_2) E = \varphi_1(\varphi_2 E)$, $E(\psi_1 \psi_2) = (E\psi_1)\psi_2$, $(\varphi E)\psi = \varphi(E\psi)$.

c) Let $E, E' \in E(X, Y)$. Define $E \oplus E' \in E(X \oplus X', Y \oplus Y')$ as the class of the triple

$$0 \longrightarrow Y \oplus Y' \longrightarrow Z \oplus Z' \longrightarrow X \oplus X' \longrightarrow 0.$$

Let $\Delta_X : X \to X \oplus X$, $\nabla_Y : Y \oplus Y \to Y$ be the natural diagonal and codiagonal morphisms. Prove that the operation $(E, E') \mapsto \nabla_Y(E \oplus E')\Delta_X \in E(X, Y)$ (Baer sum) makes $E(X, Y)$ an abelian group with the class of split triples

$$0 \longrightarrow Y \longrightarrow X \oplus Y \longrightarrow X \longrightarrow 0$$

as the zero element.

d) Prove that $E(X, Y)$ is isomorphic to $\text{Ext}^1(X, Y)$.

3. Ext^2 and Filtrations. Let $X \subset Y \subset Z$ be a three-term filtration of an object Z of an abelian category $\mathcal{A}$. Let us construct the exact sequence

$$0 \longrightarrow X \longrightarrow Y \xrightarrow{f} Z/X \longrightarrow Z/Y \longrightarrow 0$$

where f is the composition of the embedding $Y \to Z$ and the surjection $Z \to Z/X$.

a) Prove that the element $\gamma \in \text{Ext}^2(Z/Y, X)$ corresponding to this exact sequence is the product $\gamma_1 \gamma_2$, where $\gamma_1 \in \text{Ext}^1(Y/X, X)$, $\gamma_2 \in \text{Ext}^1(Z/Y, Y/X)$ correspond to exact triples

$$0 \longrightarrow X \longrightarrow Y \longrightarrow Y/X \longrightarrow 0$$
$$0 \longrightarrow Y/X \longrightarrow Z/X \longrightarrow Z/Y \longrightarrow 0.$$

b) Prove that $\gamma = 0$ in $\mathrm{Ext}^2(Z/Y, X)$.

c) Prove the converse. Namely, let A, B, C be three objects of $\mathcal{A}$, $\gamma_1 \in \mathrm{Ext}^1(B, C)$, $\gamma_2 \in \mathrm{Ext}^1(A, B)$ and $\gamma_1\gamma_2 = 0$ in $\mathrm{Ext}^2(A, C)$. Prove that there exists an object Z of $\mathcal{A}$ and a filtration $X \subset Y \subset Z$ such that $A = Z/Y$, $C = X$ and γ_1, γ_2 correspond to exact triples from a).

4. The following exercise generalizes Ex. 1a).

a) Let $\mathcal{A}$ be an abelian category, $X \in \mathrm{Ob}\,\mathcal{A}$. Denote by add X the full subcategory of $\mathcal{A}$ consisting of finite direct sums of direct summands of X. It is clear that add X is an additive subcategory of $\mathcal{A}$. We have also the natural functor $\varphi_X : K^b(\mathrm{add}\ X) \to D^b(\mathcal{A})$ (composition of the inclusion $K^b(\mathrm{add}\ X) \to K^b(\mathcal{A})$ with the localization $Q : K^b(\mathcal{A}) \to D^b(\mathcal{A})$).

Prove that if $\mathrm{Ext}^i(X, X) = 0$ for all $i > 0$, then φ_X is a fully faithful functor. (Verify the equality $\mathrm{Hom}_{K^b(\mathrm{add}\ X)}(K^\bullet, L^\bullet) = \mathrm{Hom}_{D^b(\mathcal{A})}(K^\bullet, L^\bullet)$ first in the case when $K^\bullet$, $L^\bullet$ are 0-complexes and then use III.6.14a) or IV.1.3 to proceed by induction on the number of nonzero terms in complexes $K^\bullet$ and $L^\bullet$.)

b) Prove that if any object Y of $\mathcal{A}$ has a finite resolution by objects from add X (i.e., if there exists a complex from $\mathrm{Kom}^b(\mathrm{add}\ X)$ which is quasi-isomorphic to the 0-complex Y), then φ_X is an equivalence of categories.

III.6 Derived Functors

1. Motivations

We have already mentioned in Sect. III.1.6 that some of the most important additive functors F in abelian categories, such as Hom, $\otimes$, Γ, are not exact, and to restore their exactness we have to redefine them. More explicitly, let $F : \mathcal{A} \to \mathcal{B}$ be a left (resp. right) exact functor between abelian categories. In this section we define and study its extension $RF : D^+(\mathcal{A}) \to D^+(\mathcal{B})$ (resp. $LF : D^-(\mathcal{A}) \to D^-(\mathcal{B})$) which will be called *right* (resp. *left*) *derived functor of F*.

The functors RF (resp. LF) will be *exact* in the following sense: *they map distinguished triangles into distinguished ones*.

In particular, if we define classical derived functors by

$$R^i F = H^i(RF),\quad L^i F = H^i(LF)$$

then any exact triple $0 \to A \to B \to C \to 0$ in $\mathcal{A}$ yields a long exact cohomology sequence

$$\cdots \longrightarrow R^i F(A) \longrightarrow R^i F(B) \longrightarrow R^i F(C) \longrightarrow R^{i+1} F(A) \longrightarrow \cdots$$

and similarly for L.

But what should be an appropriate way to extend F to complexes? The first and most obvious idea is to make F act on complexes componentwise. In any case, such an extension transforms homotopic morphisms into homotopic ones, so that we obtain functors $K^*(F) : K^*(\mathcal{A}) \to K^*(\mathcal{B})$, $* = \emptyset, +, -, b$.

2. Proposition. *Assume F to be exact.*

a) $K^*(F)$ transforms quasi-isomorphisms into quasi-isomorphisms so that it induces a functor $D^*(F) : D^*(\mathcal{A}) \to D^*(\mathcal{B})$.

b) $D^*(F)$ is an exact functor, i.e., it transforms distinguished triangles into distinguished triangles.

Proof. a) Let us prove first that $K^*(F)$ maps an acyclic (i.e., quasi-isomorphic to 0) complex $K^\bullet$ into an acyclic one. Let $B^i = \ker d_K^i = \operatorname{Im} d_K^{i+1}$. The functor F maps the exact sequence

$$0 \longrightarrow B^i \xrightarrow{e^i} K^i \xrightarrow{p^i} B^{i+1} \longrightarrow 0$$

into the exact sequence

$$0 \longrightarrow F(B^i) \xrightarrow{F(e^i)} F(K^i) \xrightarrow{F(p^i)} F(B^{i+1}) \longrightarrow 0.$$

From $d^i = e^{i+1} \circ p^i$ we get $F(d^i) = F(e^{i+1}) \circ F(p^i)$. Moreover, $F(p^i)$ is an epimorphism, while $F(e^{i+1})$ is a monomorphism. Hence $F(B^i)$ is isomorphic to the image of $F(d^i)$ and $F(B^{i+1})$ is isomorphic to the cokernel of $F(d^{i+1})$ so that $F(K^\bullet)$ is an acyclic complex.

Let us note that if $f : K^\bullet \to L^\bullet$ is a morphism of complexes then there exists a canonical isomorphism of $F(C(f))$ onto $C(F(f))$ (see III.3.2). By III.3.6, f is a quasi-isomorphism if and only if $C(f)$ is acyclic. In this case, as we have just proved, $F(C(f))$ is also acyclic and $F(f)$ is a quasi-isomorphism.

b) Looking through the definitions in III.3.2 and III.3.3 we see that any additive functor F maps the cylinder of f into the cylinder of $F(f)$, and the main diagram of Lemma III.3.3 into a similar diagram. Hence any additive functor F maps a triangle of the form

$$K^\bullet \xrightarrow{\bar{f}} \operatorname{Cyl}(f) \xrightarrow{\pi} C(f) \xrightarrow{\delta} K^\bullet[1]$$

into a triangle of the same form (see III.3.4). If F is exact, then, by a), $K^*(F)$ maps distinguished (i.e., quasi-isomorphic to the above) triangles into distinguished triangles. □

3. Adapted Classes of Objects

The basic idea underlying the construction of derived functors RF and LF in the general case is that we apply F term by term, but not to every complex. Instead we select some special representatives in the equivalence classes of quasi-isomorphic complexes. For example, as we will see later on, to compute

$R\Gamma$ we have to apply Γ to (bounded from the left) complexes of injective sheaves, and to compute $L(M \otimes \bullet)$ we have to use (bounded from the right) complexes of flat modules.

To proceed in the general case we introduce the following definition. A class of objects $\mathcal{R} \subset \mathrm{Ob}\,\mathcal{A}$ is said to be *adapted* to a (left or right) exact functor F if it is stable under finite direct sums and satisfies the following two conditions:

a) For a left exact functor F: F maps any acyclic complex from $\mathrm{Kom}^+(\mathcal{R})$ into an acyclic complex.
   For a right exact functor F: F maps any acyclic complex from $\mathrm{Kom}^-(\mathcal{R})$ into an acyclic complex.

b) For a left exact functor F: any object from $\mathcal{A}$ is a subobject of an object from $\mathcal{R}$.
   For a right exact functor F: any object from $\mathcal{A}$ is a quotient of object from $\mathcal{R}$.

(If this last condition is satisfied we will say that $\mathcal{R}$ is sufficiently large, or that there is sufficiently many elements in $\mathcal{R}$.)

Let us note that if F is exact, then the first part of the proof of Proposition III.6.2 implies that any class $\mathcal{R}$ satisfying the condition b) (in particular, the class of all objects of $\mathcal{A}$) is adapted to F.

4. Proposition. *Let $\mathcal{R}$ be a class of objects adapted to a left exact functor $F : \mathcal{A} \to \mathcal{B}$ and $S_\mathcal{R}$ a class of quasi-isomorphisms in $K^+(\mathcal{R})$. Then $S_\mathcal{R}$ is a localizing class of morphisms in $K^+(\mathcal{R})$ and the canonical functor*

$$K^+(\mathcal{R})[S_\mathcal{R}^{-1}] \longrightarrow D^+(\mathcal{A})$$

is an equivalence of categories.

A similar statement (with K^+ and D^+ replaces by K^- and D^- respectively) holds for a right exact functor F.

Proof. To prove that $S_\mathcal{R}$ is a localizing system we can argue similarly to Sect. III.5.23.

Next, repeating the proof of Sect. III.5.22D (see Sect. III.5.25) with $\mathcal{I}$ replaced by $\mathcal{R}$ (this is possible because in III.5.25 we have used only the fact that any object of $\mathcal{A}$ can be embedded into an object from $\mathcal{I}$, but not that objects from $\mathcal{I}$ are injective) we see that:

(∗) For any complex $K^\bullet \in K^+(\mathcal{A})$ there exists a quasi-isomorphism $t : K^\bullet \to R^\bullet$ with $R^\bullet \in K^+(\mathcal{R})$.

Hence to prove that the functor $K^+(\mathcal{R})[S_\mathcal{R}^{-1}] \to D^+(\mathcal{A})$ is fully faithful we can use Proposition III.2.10 checking condition b_2): if $s : L^\bullet \to K^\bullet$ is a quasi-isomorphism then, constructing a quasi-isomorphism $t : K^\bullet \to R^\bullet$, we get $t \circ s \in S_\mathcal{R}$. The same property (∗) implies that any complex from $D^+(\mathcal{A})$ is quasi-isomorphic to a complex from $K^+(\mathcal{R})$.

For a left exact functor F the proof is similar (one must use condition b_1). □

5. Construction of the Derived Functor

Under the assumptions of Proposition III.6.4 we define the derived functor RF of a left exact functor F on objects of the category $K^+(\mathcal{R})[S_\mathcal{R}^{-1}]$ term by term:
$$RF(K^\bullet)^i = F(K^i) \quad \text{for} \quad K^\bullet \in \text{Ob } K^+(\mathcal{R}).$$

As F maps acyclic objects from $K^+(\mathcal{R})$ into acyclic ones, arguments similar to those used in the proof of Proposition III.6.2 show that quasi-isomorphisms in $K^+(\mathcal{R})$ are mapped into quasi-isomorphisms. So we can consider RF as a functor from $K^+(\mathcal{R})[S_\mathcal{R}^{-1}]$ to $D^+(\mathcal{B})$.

To complete the definition of RF we have to choose an equivalence $\Phi : D^+(\mathcal{A}) \to K^+(\mathcal{R})[S_\mathcal{R}^{-1}]$ that is inverse to the natural embedding, and define
$$RF(K^\bullet) = RF(\Phi(K^\bullet)).$$

Similarly, we can construct the derived functor LF for a right exact functor F (replacing everywhere K^+ and D^+ by K^- and D^-).

This construction contains some ambiguity: the choice of Φ and, what is more important, the choice of $\mathcal{R}$ are nonunique. It is rather clear in what sense the construction does not depend on Φ. To state and to prove the independence on the choice of $\mathcal{R}$ we need a formal definition of the derived functor via a universal property.

6. Definition. *The* derived functor *of an additive left exact functor* $F : \mathcal{A} \to \mathcal{B}$ *is a pair consisting of an exact functor* $RF : D^+(\mathcal{A}) \to D^+(\mathcal{B})$ *and a morphism of functors* $\varepsilon_F : Q_\mathcal{B} \circ K^+(F) \to RF \circ Q_\mathcal{A}$:

$$\begin{array}{ccc}
 & D^+(\mathcal{A}) & \\
Q_\mathcal{A} \nearrow & & \searrow RF \\
K^+(\mathcal{A}) & & D^+(\mathcal{B}) \\
K^+(F) \searrow & & \nearrow Q_\mathcal{B} \\
 & K^+(\mathcal{B}) &
\end{array}$$

satisfying the following universal property: for any exact functor $G : D^+(\mathcal{A}) \to D^+(\mathcal{B})$ *and any morphism of functors* $\varepsilon : Q_\mathcal{B} \circ K^+(F) \to G \circ Q_\mathcal{A}$ *there exists a unique morphism of functors* $\eta : RF \to G$ *making the diagram*

$$\begin{array}{ccc}
 & Q_\mathcal{B} \circ K^+(F) & \\
\varepsilon_F \swarrow & & \searrow \varepsilon \\
RF \circ Q_\mathcal{A} & \xrightarrow{\eta \circ Q_\mathcal{A}} & G \circ Q_\mathcal{A}
\end{array} \qquad (\text{III.37})$$

commutative.

Similarly, the derived functor of a right exact functor $F : \mathcal{A} \to \mathcal{B}$ is a pair consisting of an exact functor $LF : D^-(\mathcal{A}) \to D^-(\mathcal{B})$ and a morphism of functors $\varepsilon_F : LF \circ Q_\mathcal{A} \to Q_\mathcal{B} \circ K^-(F)$ satisfying the universal property similar to (III.37) (with a morphism of functors $\eta : G \to LF$).

Let us remark that if F is exact then, by Proposition III.6.2b) both RF and LF coincide with the result of the term-by-term application of F to complexes.

7. Uniqueness of the Derived Functor

Let (RF, ε_F) and $(\tilde{R}F, \tilde{\varepsilon}_F)$ be two derived functors for F. By definition, there exist unique morphisms $\tilde{R}F \xrightarrow{\eta'} RF \xrightarrow{\eta} \tilde{R}F$ with the required commutativity properties. These commutativity properties imply, in particular, that $\eta' \circ \eta$ and $\eta \circ \eta'$ are automorphisms of functors RF and $\tilde{R}F$ respectively. Hence, by the uniqueness, they are identity isomorphisms, so that η and η' are mutually inverse isomorphisms of functors which, moreover, are uniquely defined.

8. Theorem. *Assume that a left exact functor F admits an adapted class of objects $\mathcal{R}$. Then the derived functor RF exists and can be defined by the construction from III.6.5.*

A similar statement holds for right exact functors.

The proof occupies III.6.8–III.6.11. We will consider only left exact functors F. First, we repeat the constructions from III.6.5 introducing necessary additional notations. Let

$$\Psi : K^+(\mathcal{R})[S_\mathcal{R}^{-1}] \longrightarrow D^+(\mathcal{A})$$

be a natural embedding which, under the conditions of Theorem III.6.8, is an equivalence of categories. Let

$$\Phi : D^+(\mathcal{A}) \longrightarrow K^+(\mathcal{R})[S_\mathcal{R}^{-1}]$$

be a quasi-inverse to Ψ so that we have the following isomorphisms of functors:

$$\alpha : \mathrm{Id}_{K^+(\mathcal{R})[S_\mathcal{R}^{-1}]} \longrightarrow \Phi \circ \Psi$$

$$\beta : \mathrm{Id}_{D^+(\mathcal{A})} \longrightarrow \Psi \circ \Phi.$$

By the property III.6.3a) of the class $\mathcal{R}$ the term-by-term application of F to complexes from $\mathcal{R}$ determines a functor

$$\bar{F} : K^+(\mathcal{R})[S_\mathcal{R}^{-1}] \longrightarrow D^+(\mathcal{B})$$

with the property

$$\bar{F} \circ Q_\mathcal{R} = Q_\mathcal{B} \circ K^+(F)$$

190   III. Derived Categories and Derived Functors

(here $Q_\mathcal{R} : K^+(\mathcal{R}) \to K^+(\mathcal{R})[S_\mathcal{R}^{-1}]$ is the localization functor). The functor $\bar{F}$ is obviously exact (i.e., it transforms distinguished triangles into distinguished ones).

Let us define RF by the formula
$$RF = \bar{F} \circ \Phi : D^+(\mathcal{A}) \longrightarrow D^+(\mathcal{B}).$$

To prove Theorem III.6.8 we must:

a) prove that RF is an exact functor;
b) construct a morphism of functors $\varepsilon_F : Q_\mathcal{B} \circ K^+(F) \to RF \circ Q_\mathcal{A}$ and prove its uniqueness;
c) verify the universality property from III.6.6.

Let us prove a). It is clear that Φ commutes with the translation functor T. Therefore, it suffices to verify that Φ maps distinguished triangles in $D^+(\mathcal{A})$ into distinguished triangles in $K^+(\mathcal{R})[S_\mathcal{R}^{-1}]$. Since Φ is an equivalence of categories, the following lemma yields the required property.

9. Lemma. *Let*
$$\Delta = \left\{ X \xrightarrow{f} Y \longrightarrow Z \longrightarrow X[1] \right\}$$
be a triangle (not necessarily distinguished) in $K(\mathcal{R})[S_\mathcal{R}^{-1}]$ that becomes distinguished when considered as a triangle in $D(\mathcal{A})$. Then Δ is isomorphic in $D(\mathcal{A})$ to a standard triangle in $K(\mathcal{R})$.

Proof. Since $\Psi : K(\mathcal{R})[S_\mathcal{R}^{-1}] \to D(\mathcal{A})$ is an equivalence of categories, the morphism $f : X \to Y$ on $D(\mathcal{A})$ comes from a morphism in $K(\mathcal{R})[S_\mathcal{R}^{-1}]$, i.e., it can be represented by a roof

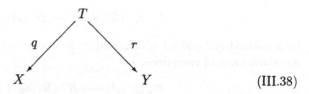

(III.38)

with $T \in \mathrm{Kom}(\mathcal{R})$, $q \in S_\mathcal{R}$. We prove that Δ is isomorphic in $D(\mathcal{A})$ to the following standard triangle:

$$T \xrightarrow{r} Y \longrightarrow C(r) \longrightarrow T(1)$$

in $K(\mathcal{R})$, where $C(r)$ is the cone of r. Indeed, consider the diagram

$$\begin{array}{ccccccc}
T & \xrightarrow{r} & Y & \longrightarrow & C(r) & \longrightarrow & T(1) \\
{\scriptstyle q}\downarrow & & {\scriptstyle \mathrm{id}}\downarrow & & {\scriptstyle q[1]}\downarrow & & \\
X & \xrightarrow{f} & Y & \longrightarrow & Z & \longrightarrow & X[1]
\end{array} \qquad (\mathrm{III.39})$$

In $D(\mathcal{A})$ both rows are distinguished triangles and the left square is commutative. Therefore, there exists a morphism $v : C(r) \to Z$ in $D(\mathcal{A})$ that completes the diagram to a morphism of triangles. Since q and id are isomorphisms in $D(\mathcal{A})$, v is also an isomorphism and (q, id, v) is the required isomorphisms of triangles in $D(\mathcal{A})$. □

10. The Construction of the Functor Morphism ε_F

Let $X \in \mathrm{Ob}\, K^+(\mathcal{A}) = \mathrm{Ob}\, D^+(\mathcal{A})$ and $Y = \Phi \circ Q_\mathcal{A}(X) \in \mathrm{Ob}\, K^+(\mathcal{R})[S_\mathcal{R}^{-1}] = \mathrm{Ob}\, K^+(\mathcal{R})$. Applying to $X \in D^+(\mathcal{A})$ the functor morphism β we get an isomorphism

$$\beta(X) : X \longrightarrow \Psi \circ \Phi(X) = \Psi(Y)$$

in $D^+(\mathcal{A})$, which is given by a diagram

$$X \xrightarrow{s} Z \xleftarrow{t} Y \qquad (\mathrm{III}.40)$$

in $K^+(\mathcal{A})$ with $s, t \in \mathrm{Qis}$. Moreover, by the property $(*)$ from the proof of Proposition III.6.4, we can assume that $Z \in \mathrm{Ob}\, K^+(\mathcal{R})$. Applying $K^+(F)$ to this diagram we get the diagram

$$K^+(F)(X) \xrightarrow{K^+(F)(s)} K^+(F)(Z) \xleftarrow{K^+(F)(t)} K^+(F)(Y) \qquad (\mathrm{III}.41)$$

in $K^+(\mathcal{B})$. By the property III.6.3a) of the class $\mathcal{R}$ we see that $K^+(F)(s)$ is a quasi-isomorphism and (III.41) yields a morphism in $D^+(\mathcal{B})$:

$$\varepsilon_F(X) : Q_\mathcal{B} K^+(F)(X) \longrightarrow Q_\mathcal{B} K^+(F)(Y) = \bar{F} Q_\mathcal{R}(Y)$$
$$= \bar{F} \Phi Q_\mathcal{A}(X) = RF Q_\mathcal{A}(X).$$

Let us prove that $\varepsilon_F(X)$ does not depend on the choice of diagram (III.40). Two such diagrams with the middle objects Z_1, Z_2 can be completed to a commutative diagram of the form

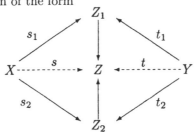

and the reader can easily verify that the morphisms $Q_\mathcal{B} K^+(F)(X) \to Q_\mathcal{B} K^+(F)(Y)$ corresponding to (s_1, t_1) and to (s_2, t_2) coincide with the morphism corresponding to (s, t).

Let us prove that the family $\varepsilon_F(X)$, $X \in K^+(\mathcal{A})$, determines a morphism of functors $\varepsilon_F(X) : Q_\mathcal{B} \circ K^+(F) \to RF \circ Q_\mathcal{A}$. Let $\varphi : X_1 \to X_2$ be a morphism in $K^+(\mathcal{A})$, $Y_1 = \Phi Q_\mathcal{A}(X_1)$, $Y_2 = \Phi Q_\mathcal{A}(X_2)$. From the fact that β is a morphism of functors we get the following commutative diagram in $D^+(\mathcal{A})$:

$$\begin{array}{ccc} X_1 & \xrightarrow{\beta(X_1)} & \Psi(Y_1) \\ \varphi \downarrow & & \downarrow \Psi\Phi(\varphi) \\ X_2 & \xrightarrow{\beta(X_2)} & \Psi(Y_2) \end{array} \qquad (\text{III.42})$$

Representing each of three morphisms $\beta(X_1)$, $\beta(X_2)$, $\Psi\Phi(\varphi)$ in this diagram by a roof in $K^+(\mathcal{A})$ and recalling the property (*) from the proof of Proposition 4 we can assume that all objects in these roofs are from $K^+(\mathcal{R})$. Moreover, $Y_1, Y_2 \in \mathrm{Ob}\, K^+(\mathcal{R})$. Applying to these roofs the functor $K^+(F)$ and using the property a) of the class $\mathcal{R}$ we obtain roofs in $K^+(\mathcal{B})$. These roofs determine some morphisms in $D^+(\mathcal{B})$ and the commutative diagram (III.42) yields the commutative diagram

$$\begin{array}{ccc} Q_\mathcal{B} K^+(F)(X_1) & \xrightarrow{\varepsilon_F(X_1)} & RF \circ Q_\mathcal{A}(X_1) \\ Q_\mathcal{B} K^+(F)(\varphi) \downarrow & & \downarrow RF \circ Q_\mathcal{A}(\varphi) \\ Q_\mathcal{B} K^+(F)(X_2) & \xrightarrow{\varepsilon_F(X_2)} & RF \circ Q_\mathcal{A}(X_2) \end{array}$$

which shows that ε_F is a morphism of functors.

The uniqueness of ε_F follows from the universal property of the localization functor $Q_\mathcal{A}: K^+(\mathcal{A}) \to D^+(\mathcal{A})$.

11. The Universal Property of RF

Let an exact functor $G: D^+(\mathcal{A}) \to D^+(\mathcal{B})$ and a morphism of functors $\varepsilon: Q_\mathcal{B} \circ K^+(F) \to G \circ Q_\mathcal{A}$ be given. We construct a morphism of functors $\eta: RF \to G$ for which the diagram

$$\begin{array}{ccc} & Q_\mathcal{B} \circ K^+(F) & \\ \varepsilon_F \swarrow & & \searrow \varepsilon \\ RF \circ Q_\mathcal{A} & \xrightarrow{\eta \circ Q_\mathcal{A}} & G \circ Q_\mathcal{A} \end{array} \qquad (\text{III.43})$$

of functor morphisms commutes.

To construct η we must define for each $X \in \text{Ob } K^+(\mathcal{A}) = \text{Ob } D^+(\mathcal{A})$ a morphism $\eta(X) : RF(X) \to G(X)$ in $D^+(\mathcal{B})$.

The morphism of functors ε determines a morphism $\varepsilon(X) : F(X) \to G(X)$ in $D^+(\mathcal{B})$ for each $X \in \text{Ob } K^+(\mathcal{A})$. Similarly, a morphism of functors $\beta : \text{Id}_{D^+(\mathcal{A})} \to \Psi \circ \Phi$ (see 8) determines, for each $X \in \text{Ob } K^+(\mathcal{A})$, an isomorphism $\beta(X) : X \to \Psi \circ \Phi(X)$ in $D^+(\mathcal{A})$.

As in III.6.10, let us represent this isomorphism by the roof (III.40). Since ε is a morphism of functors $K^+(\mathcal{A}) \to D^+(\mathcal{B})$, we have a commutative diagram in $D^+(\mathcal{B})$:

$$\begin{array}{ccc}
F(X) & \xrightarrow{\varepsilon(X)} & G(X) \\
{\scriptstyle Q_\mathcal{B} \circ F(s)} \downarrow & & \downarrow {\scriptstyle G \circ Q_\mathcal{A}(s)} \\
F(Z) & \xrightarrow{\varepsilon(Z)} & G(Z) \\
{\scriptstyle Q_\mathcal{B} \circ F(t)} \uparrow & & \uparrow {\scriptstyle G \circ Q_\mathcal{A}(t)} \\
F(Y) & \xrightarrow{\varepsilon(Y)} & G(Y)
\end{array}$$

In this diagram, $Q_\mathcal{A}(s)$ and $Q_\mathcal{A}(t)$ are isomorphisms because s and t are quasi-isomorphisms. Hence, $G \circ Q_\mathcal{A}(t)$ and $Q_\mathcal{B} \circ F(t)$ are isomorphisms. Inverting these two isomorphisms, we obtain the commutative diagram

$$\begin{array}{ccc}
Q_\mathcal{B} \circ K^+(F)(X) & \xrightarrow{\varepsilon(X)} & G(X) \\
{\scriptstyle \varepsilon_F(X)} \downarrow & & \downarrow {\scriptstyle G(\beta(X))} \\
RFQ_\mathcal{A}(X) & \xrightarrow{\varepsilon(\Psi \circ \Phi(X))} & G(\Psi \circ \Phi(X))
\end{array} \qquad (\text{III.44})$$

(cf. arguments after (III.41)).

Since $\beta(X)$ is an isomorphism in $D^+(\mathcal{A})$ and G is a functor from $D^+(\mathcal{A})$ to $D^+(\mathcal{B})$, the right arrow in this diagram is also an isomorphism. Let us define

$$\eta(X) = G(\beta(X))^{-1} \circ \varepsilon(\Psi \circ \Phi(X)) : RF(X) \longrightarrow G(X).$$

From the fact that β is a morphism of functors we easily deduce that the family of morphisms $\eta(X)$ determines a morphism of functors $\eta : RF \to G$. The commutativity of (III.43) and the uniqueness of $\eta(X)$ follow from the commutativity of (III.44) and the fact that $G(\beta(X))$ is an isomorphism. □

In application it is sometimes useful to know that some classes of objects are adapted to *all functors* that are exact from a fixed side.

In Theorem III.6.12 below we will consider left exact functors F. The reader can easily prove a similar result for right exact functors.

12. Theorem. *If $\mathcal{A}$ contains sufficiently many injective objects, then the class $\mathcal{I}$ of all injective objects is adapted to any left exact functor F.*

Proof. We have to verify that F maps acyclic complexes from $\text{Kom}^+(\mathcal{I})$ into acyclic ones. Let $I^\bullet$ be such a complex. The zero morphism $0 : I^\bullet \to I^\bullet$ is a quasi-isomorphism. By Sect. III.5.24, it is homotopic to $\text{id}_{I^\bullet}$. Hence the zero morphism of $F(I^\bullet)$ is homotopic to $\text{id}_{F(I^\bullet)}$, so that $F(I^\bullet)$ is acyclic. □

13. Classical Derived Functors

a) A functor H from a derived (or homotopy) category to an abelian category is said to be *cohomological* if for any distinguished triangle $X \to Y \to Z \to X[1]$ the sequence

$$\cdots \longrightarrow H(T^i X) \longrightarrow H(T^i Y) \longrightarrow H(T^i Z) \longrightarrow H(T^{i+1} X) \longrightarrow \cdots$$

is exact. For example, $H = H^0$ is a cohomological functor (see III.3.5 and I.6.8). Another example: $H = \text{Hom}(U, \bullet)$ (see IV.1.3).

b) Let F be a left exact (resp. right exact) functor between two abelian categories. Then $R^i F = H^0(T^i(RF)) = H^i(RF)$ (resp. $L^i F = H^i(LF)$) is called *the classical i-th derived functor* for F. One can easily see that $R^i F = 0$ for $i < 0$, $R^0 F = F$ (resp. $L^i F = 0$ for $i > 0$, $L^0 F = F$).

Let us give some examples.

14. Ext^i as a Derived Functor

Let $\mathcal{A}$ be an abelian category with sufficiently many injective objects. Fix an object X and consider the functor $\text{Hom}(X, \bullet)$ from $\mathcal{A}$ to the category Ab of abelian groups. This functor is left exact and we can consider the derived functor $R\text{Hom}(X, \bullet)$. Let us show that there exists an isomorphism of functors

$$\text{Ext}^i(X, \bullet) \cong R^i \text{Hom}(X, \bullet).$$

Indeed, to compute $\text{Ext}^i(X, Y)$ we replace Y by its injective resolution $Y \to I_Y$ so that

$$\text{Ext}^i(X, Y) = \text{Hom}_{D(\mathcal{A})}(X, I_Y[i]).$$

By the injectivity of I_Y we have (see Ex. III.5.1i))

$$\text{Hom}_{D(\mathcal{A})}(X, I_Y[i]) = \text{Hom}_{K(\mathcal{A})}(X, I_Y[i]).$$

On the other hand, by the construction of the derived functor,
$$R^i \operatorname{Hom}(X, Y) = H^i(\operatorname{Hom}(X, I_Y)).$$
Now we must use the following general construction. Define "inner Hom" in the category Kom($\mathcal{A}$) by the formula
$$\operatorname{Hom}^n(A^\bullet, B^\bullet) = \prod_{i \in \mathbb{Z}} \operatorname{Hom}(A^i, B^{i+n})$$
$$df = d_B \circ f - (-1)^n f \circ d_A, \quad f \in \operatorname{Hom}^n(A^\bullet, B^\bullet).$$
Then we have the following canonical isomorphisms:
$$Z^i \operatorname{Hom}^\bullet(A^\bullet, B^\bullet) = \operatorname{Hom}_{\operatorname{Kom}(\mathcal{A})}(A^\bullet, B^\bullet[i]),$$
$$B^i \operatorname{Hom}^\bullet(A^\bullet, B^\bullet) = \{\text{ morphisms homotopic to the zero morphism}\},$$
so that
$$H^i \operatorname{Hom}^\bullet(A^\bullet, B^\bullet) = \operatorname{Hom}_{K(\mathcal{A})}(A^\bullet, B^\bullet[i]).$$

Similarly, one can prove that if $\mathcal{A}$ contains sufficiently many projective objects then $\operatorname{Ext}^i(\bullet, Y) \cong R^i \operatorname{Hom}(\bullet, Y)$ as functors of the first argument (they are contravariant so that we have to consider them as being defined on $\mathcal{A}^0$).

To formulate the functoriality properties of Ext^i and $R^i \operatorname{Hom}$ in both arguments simultaneously one should develop an appropriate formalism of derived multifunctors.

15. Functors Tor$_i$

Let R be an associative ring with unity. Fix a left R-module N and consider the functor $\bullet \otimes N$ (tensor product over R) on the category of right R-modules. This functor is right exact and flat modules form an adapted class for it. Using this class we can construct the left derived functor, which is usually denoted $\bullet \overset{L}{\otimes} N$. The corresponding classical derived functors are denoted by Tor$_i$:
$$\operatorname{Tor}^R_i(M, N) = H^{-i}(M \overset{L}{\otimes} N).$$

Earlier in this section we used the existence of an adapted class to prove the existence of the derived functor. A partial inversion of this result is formulated as follows.

Let, say, RF exists. An object X is said to be F-*acyclic* if $R^i F(X) = 0$ for all $i \neq 0$.

16. Theorem. *a) An adapted to F class of objects exists if and only if the class $\mathcal{Z}$ of all F-acyclic objects is sufficiently large, i.e., if any object is a subobject of an F-acyclic one.*

b) If $\mathcal{Z}$ is sufficiently large then it contains any adapted to F class and any sufficiently large subclass of $\mathcal{Z}$ is adapted.

c) If $\mathcal{Z}$ is sufficiently large then it contains all injective objects.

Proof. Assume that there exists a class of objects $\mathcal{R}$ adapted to F. By the construction from III.6.5, $RF(X[0])$ is quasi-isomorphic to $F(X)[0]$ for any $X \in \mathcal{R}$. So $\mathcal{R} \subset \mathcal{Z}$ and $\mathcal{Z}$ is sufficiently large.

Conversely, let $\mathcal{R}$ be a sufficiently large subclass in $\mathcal{Z}$. To prove that it is adapted to F we must prove that F maps acyclic complexes from $\text{Kom}^+(\mathcal{R})$ into acyclic ones. If this acyclic complex is a triple $0 \to K^0 \to K^1 \to K^2 \to 0$ then the exactness of $0 \to F(K^0) \to F(K^1) \to F(K^2) \to 0$ follows from the equality $R^1 F(K^0) = 0$. In the general case we can consecutively split off exact triples. Letting $X^i = \text{Im}\, d^i$, $X^0 = K^0$ we get exact triples

$$0 \longrightarrow K^0 \longrightarrow K^1 \longrightarrow X^1 \longrightarrow 0,$$
$$0 \longrightarrow X^1 \longrightarrow K^2 \longrightarrow X^2 \longrightarrow 0,$$

and so on, with $X^{i+1} \in \mathcal{Z}$ (because $X^i, K^{i+1} \in \mathcal{Z}$). Hence the triples $0 \to F(X^i) \to F(K^{i+1}) \to F(X^{i+1}) \to 0$ are exact and $F(K^\bullet)$ is acyclic.

Finally, let $\mathcal{Z}$ be sufficiently large. Embedding an injective object I into an object X from $\mathcal{Z}$ and looking at the injectivity diagram (II.6.9)

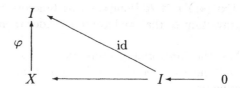

we see that φ splits off I as a direct summand of X. Since F is an additive functor, $R^i F(I)$ is a direct summand of $R^i F(X) = 0$ for $i \neq 0$. □

Exercises

In Exs. 1–5 we give another construction of derived functors. This construction guarantees the existence of, say, RF in a much more general situation; however now RF will take values not in the derived category itself but in some extension of it.

1. Category of Coindices. A category $\mathcal{I}$ is said to be a category of *coindices* if it is small, nonempty, and satisfies the following conditions:

a) $\mathcal{I}$ is connected, i.e., any two objects can be joint by a sequence of morphisms (directions are irrelevant).

b) Any pair of morphisms $j' \leftarrow i \to j$ can be embedded into a commutative square

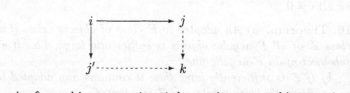

c) For any pair of morphisms $u, v : i \to j$ there exists a morphism $w : j \to k$ such that $w \circ u = w \circ v$.

A category $\mathcal{J}$ is said to be a category of *indices* if $\mathcal{J}^\circ$ is a category of coindices.

2. Inductive Limits. Let $\mathcal{I}$ be a category of coindices, $F: \mathcal{I} \to \mathcal{C}$, $i \to X_i$, a functor. It determines a functor $\hat{F}: \mathcal{I} \to \hat{\mathcal{C}} = Funct(\mathcal{C}^\circ, Set)$ by

$$\hat{F}(i) = \operatorname{Hom}_{\mathcal{C}}(Y, F(i)), \quad Y \in \operatorname{Ob}\mathcal{C}^\circ = \operatorname{Ob}\mathcal{C}, \quad i \in \operatorname{Ob}\mathcal{I}.$$

Define an object L of the category $\hat{\mathcal{C}}$ as the inductive limit $L = \varinjlim \hat{F}$ (see II.3.19), so that for any $Y \in \operatorname{Ob}\mathcal{C}^\circ$

$$L(Y) = \varinjlim F_Y$$

where $F_Y : \mathcal{I} \to Set$, $F_Y(i) = \operatorname{Hom}_{\mathcal{C}}(Y, F(i))$. The existence of L follows from the existence, for any Y, of the inductive limit $L = \varinjlim F_Y$ in the category Set; this is proved similarly to Theorem II.3.20.

Such functors L (for all possible $\mathcal{I}$ and F) form a full subcategory $\operatorname{Ind} \mathcal{C}$ of $\hat{\mathcal{C}}$, which is called the category of inductive limits in $\mathcal{C}$.

Define the category of projective limits in $\operatorname{Pro}\mathcal{C}$ in $\mathcal{C}$ as $(\operatorname{Ind}\mathcal{C}^\circ)^\circ$. This is a subcategory of $((\mathcal{C}^\circ)^\wedge)^\circ$. In $\operatorname{Pro}\mathcal{C}$ one can take limits of functors of the form $\mathcal{J} \to \mathcal{C}$, where $\mathcal{J}$ is a category of indices.

3. Inductive Limits and Localization. Let S be a localizing class of morphisms in a category $\mathcal{C}$ (see III.2.6). It is said to be *saturated* if any morphism which is both a right divisor of some morphism in S and a left divisor of some morphism in S, itself belongs to S.

For any object $X \in \operatorname{Ob}\mathcal{C}$ the category $\mathcal{I}_X$ of morphisms $s: X \to X'$, $s \in S$, is a category of coindices, and the category $\mathcal{J}_X$ of morphisms $s: X' \to X$, $s \in S$, is a category of indices. Let

$$X^+ = \varinjlim_{\mathcal{I}_X} X', \quad X^- = \varprojlim_{\mathcal{J}_X} X'.$$

Assume that S is saturated. Then the mappings $X \to X^+$, $X \to X^-$ can be extended to functors $\mathcal{C} \to \operatorname{Ind}\mathcal{C}$ and $\mathcal{C} \to \operatorname{Pro}\mathcal{C}$ that map morphisms from S into isomorphisms. Hence they determine canonical functors $\mathcal{C}[S^{-1}] \to \operatorname{Ind}\mathcal{C}$ and $\mathcal{C}[S^{-1}] \to \operatorname{Pro}\mathcal{C}$.

4. Weak Derived Functors. Let $\mathcal{A}$, $\mathcal{B}$ be two abelian categories, $F: \mathcal{A} \to \mathcal{B}$ an additive functor. Denote also by F the term by term extension of F to a functor from $K^*(\mathcal{A})$ to $K^*(\mathcal{B})$ (where $* = \emptyset, +, -, b$). Let $S_\mathcal{A}$ (resp. $S_\mathcal{B}$) be the class of quasi-isomorphisms in $K^*(\mathcal{A})$ (resp. $K^*(\mathcal{B})$). Define the weak right derived functor

$$R_w F : K^*(\mathcal{A})[S_\mathcal{A}^{-1}] = D^*(\mathcal{A}) \longrightarrow \operatorname{Ind} K^*(\mathcal{B})[S_\mathcal{B}^{-1}] = \operatorname{Ind} D^*(\mathcal{B})$$

as a unique functor for which the diagram

$$
\begin{array}{ccc}
K^*(\mathcal{A}) & \xrightarrow{X \mapsto F(X^+)} & \operatorname{Ind} K^*(\mathcal{B}) \\
\downarrow & & \downarrow \\
K^*(\mathcal{A})[S_\mathcal{A}^{-1}] & \xrightarrow{R_w F} & \operatorname{Ind} K^*(\mathcal{B})[S_\mathcal{B}^{-1}]
\end{array}
$$

(where $F(X^+) = \varinjlim_{\mathcal{I}_X} F(X')$) is commutative.

If $R_w F$ takes values in the subcategory of Ind $D^*(\mathcal{B})$ formed by representable objects then it "coincides" with RF.

An object $X \in K^*(\mathcal{A})$ is said to be *F-acyclic from the right* if the canonical morphism $F(X) \to R_w F(X)$ is an isomorphism.

The weak left derived functor $L_w F$ is defined similarly using the diagram

$$
\begin{array}{ccc}
K^*(\mathcal{A}) & \xrightarrow{X \mapsto F(X^-)} & \operatorname{Pro} K^*(\mathcal{B}) \\
\downarrow & & \downarrow \\
K^*(\mathcal{A})[S_\mathcal{A}^{-1}] & \xrightarrow{L_w F} & \operatorname{Pro} K^*(\mathcal{B})[S_\mathcal{B}^{-1}]
\end{array}
$$

5. Derived Functors. Keeping the assumptions of the last exercise, let us assume also that any object of $\mathcal{A}$ is a quotient object (resp. a subobject) of a left (resp. right) F-acyclic object. Then the functor $L_w F$ on $D^-(\mathcal{A})$ (resp. the functor $R_w F$ on $D^+(\mathcal{A})$) takes values in $D^-(\mathcal{B})$ (resp. in $D^+(\mathcal{B})$) and, therefore, "coincides" with LF (resp. with RF).

6. Exactness of Limits. Let $\mathcal{A}$ be an abelian category with countable direct products, $\mathcal{C}(\mathbb{Z}^+)$ the category corresponding to the ordered set of positive integers (see II.1.5d)).

a) Prove that $\mathcal{A}^{\mathbb{Z}^+} = \operatorname{Funct}(\mathcal{C}(\mathbb{Z}^+), \mathcal{A}^\circ)$ is an abelian category.

b) Prove that the functor $\varprojlim_{\mathbb{Z}^+} : \mathcal{A}^{\mathbb{Z}^+} \to \mathcal{A}$ is left exact.

c) An object $X = (X_i, p_{ij}) \in \mathcal{A}^{\mathbb{Z}^+}$ is said to satisfy the condition ML (Mittag–Leffler) if for any i there exists $j > i$ such that $p_{ij} : X_j \to X_i$ is an epimorphism.

Prove that the class of all objects X satisfying the condition ML is adapted to the functor $\varprojlim$ (use the description of $\varprojlim$ in Ex. II.3.10).

d) Prove that if $0 \to X \to S \to Y \to 0$ is an exact sequence in $\mathcal{A}^{\mathbb{Z}^+}$ and S satisfies the condition ML, then Y also satisfies this condition. Deduce that the right derived functors $R^i \varprojlim$ are zero for $i \geq 2$.

In Exs. 7–10 we present some results of Spaltenstein [1] that enable us to work with unbounded complexes in derived categories.

7. Projective and Injective Resolutions. Let $\mathcal{A}$ be an abelian category. By a *left projective resolution* of a complex $A^\bullet \in \operatorname{Ob} \operatorname{Kom}(\mathcal{A})$ we mean a quasi-isomorphism $P^\bullet \to A^\bullet$ with all P^i's projective in $\mathcal{A}$. Similarly, we define a *right injective resolution*.

Theorem III.4.4 (and its analogue for projective resolutions) claims that any object $A^\bullet \in \operatorname{Ob} \operatorname{Kom}^+(\mathcal{A})$ (resp. $A^\bullet \in \operatorname{Ob} \operatorname{Kom}^-(\mathcal{A})$) has a right injective (resp. left projective) resolution which is unique up to a homotopy equivalence.

Without the boundedness assumptions the uniqueness may fail: for $\mathcal{A} = (\mathbb{Z}/4)$-mod the complex

$$P^\bullet : \ldots \longrightarrow \mathbb{Z}/4 \xrightarrow{2} \mathbb{Z}/4 \xrightarrow{2} \mathbb{Z}/4 \longrightarrow \ldots$$

is acyclic and consists of free $\mathbb{Z}/4$-modules. Hence it is a left projective resolution of the zero complex. However, the morphism $P^\bullet \to 0^\bullet$ is not a homotopy equivalence: tensoring with $\mathbb{Z}/2$ we obtain the complex

$$P^\bullet \otimes \mathbb{Z}/2 : \ldots \longrightarrow \mathbb{Z}/2 \xrightarrow{0} \mathbb{Z}/2 \xrightarrow{0} \mathbb{Z}/2 \longrightarrow \ldots$$

which has nonzero cohomology and, therefore, is not homotopic to $0^\bullet$.

8. A complex $P^\bullet$ is said to be K-projective if for any acyclic complex $A^\bullet$ the complex of abelian groups $\operatorname{Hom}^\bullet(P^\bullet, A^\bullet)$ is acyclic. Similarly, one can define K-injective complexes $I^\bullet$ (using $\operatorname{Hom}^\bullet(A^\bullet, I^\bullet)$). Prove the following properties of K-projective complexes.

a) Let $A^\bullet$ be a 0-complex (i.e., $A^i = 0$ for $i \neq 0$). Then $A^\bullet$ is K-projective if and only if A^0 is projective in $\mathcal{A}$.

b) If any two vertices of a distinguished triangle in $D(\mathcal{A})$ are K-projective, so is the third.

c) For $P^\bullet \in \operatorname{Ob} \operatorname{Kom}(\mathcal{A})$ the following conditions are equivalent:

(i) $P^\bullet$ is K-projective.
(ii) For any $A^\bullet \in \operatorname{Ob} \operatorname{Kom}(\mathcal{A})$ the natural homomorphism

$$\operatorname{Hom}_{K(\mathcal{A})}(P^\bullet, A^\bullet) \longrightarrow \operatorname{Hom}_{D(\mathcal{A})}(P^\bullet, A^\bullet)$$

is an isomorphism.
(iii) Any quasi-isomorphism $s : A^\bullet \to P^\bullet$ admits a right inverse $t : P^\bullet \to A^\bullet$ in $K(\mathcal{A})$.

d) By a K-projective (left) resolution of a complex $A^\bullet$ we mean a quasi-isomorphism $P^\bullet \to A^\bullet$ with a K-projective $P^\bullet$. Prove that a K-projective resolution, whenever it exists, is unique up to a homotopy equivalence. If $A^\bullet \in \operatorname{Ob} \operatorname{Kom}^-(\mathcal{A})$ and $\mathcal{A}$ has sufficiently many projective objects then a K-projective resolution is a left projective resolution of $A^\bullet$.

Similar results hold for K-injective (right) resolutions.

9. In view of the property 8d) we can use K-projective and K-injective resolutions to compute values of derived functors on unbounded complexes. In particular, to compute $R \operatorname{Hom}(A^\bullet, B^\bullet)$ we can use either a K-projective resolution of $A^\bullet$ or a K-injective resolution of $B^\bullet$.

As to the existence of K-resolutions, the following results are proved by Spaltenstein [1].

a) Let R be an associative ring with unity and $\mathcal{A} = R$-mod. Then any complex $A^\bullet \in \mathrm{Ob}\,\mathrm{Kom}(\mathcal{A})$ admits a K-projective and a K-injective resolution.

b) Let $\mathcal{O}$ be a sheaf of rings on a topological space X and $\mathcal{A}$ the category of sheaves of $\mathcal{O}$-modules. Then any complex $A^\bullet \in \mathrm{Ob}\,\mathrm{Kom}(\mathcal{A})$ admits a K-projective resolution.

10. Similarly, one can define and prove the existence of K-flat resolutions (used to compute derived functors of the tensor product), K-soft resolutions (used to compute $Rf_!$, see Sect. III.8), and so on.

III.7 Derived Functor of the Composition. Spectral Sequence

1. Theorem. *Let $\mathcal{A}, \mathcal{B}, \mathcal{C}$ be three abelian categories, $F : \mathcal{A} \to \mathcal{B}$, $G : \mathcal{B} \to \mathcal{C}$ two additive left exact functors. Let $\mathcal{R}_\mathcal{A} \subset \mathrm{Ob}\,\mathcal{A}$ (resp. $\mathcal{R}_\mathcal{B} \subset \mathrm{Ob}\,\mathcal{B}$) be a class of objects adapted to F (resp. to G). Assume that $F(\mathcal{R}_\mathcal{A}) \subset \mathcal{R}_\mathcal{B}$. Then the derived functors RF, RG, $R(G \circ F) : D^+(\bullet) \to D^+(\bullet)$ exist and the natural morphism of functors $R(G \circ F) \to RG \circ RF$ is an isomorphism.*

Proof. The definition of an adapted class in III.6.3 and the conditions of the theorem show that $\mathcal{R}_\mathcal{A}$ is adapted not only to F, but to $G \circ F$ as well. Hence, RF, RG and $R(G \circ F)$ exist and to compute them we can use the construction from III.6.6.

Next, RF and RG are exact functors. Hence $R(G \circ F)$ is also exact and the morphism $E : R(G \circ F) \to RG \circ RF$ is defined by the universality property.

For $K^\bullet \in \mathrm{Ob}\,\mathrm{Kom}^+(\mathcal{R}_\mathcal{A})$ the morphism $E(K^\bullet) : R(G \circ F)(K^\bullet) \to RG \circ RF(K^\bullet)$ is an isomorphism. Since any object of $D^+(\mathcal{A})$ is isomorphic to such an object $K^\bullet$, E is an isomorphism of functors. □

A similar result holds for right exact functors.

2. Example. Let $\mathcal{A}, \mathcal{B}, \mathcal{C}$ be the categories of sheaves of abelian groups on topological spaces U, V, W respectively, and let $F = f_\bullet$, $G = g_\bullet$, where $f : U \to V$, $g : V \to W$ are continuous maps. In the classical situation W is a point, so that G and $G \circ F$ are functors of sections on V and U respectively. Instead of $R(G \circ F)$ and $RG \circ RF$, the classical theory considers $R^i(G \circ F)(\mathcal{F}) = H^i(U, \mathcal{F})$ and $R^p G(R^q F(\mathcal{F})) = H^k(V, R^j f_\bullet \mathcal{F})$. These groups are connected by the Leray spectral sequence, which is a rather complicated algebraic structure (for the Leray spectral sequence see Sect. III.8, in particular Theorem III.8.3d)).

In our more general setting we can also construct a spectral sequence that would encode, with some loss of information, the functor isomorphism E.

To be more precise, let us consider the action of E on 0-complexes, that is, on objects $X \in \mathrm{Ob}\,\mathcal{A}$. Then E yields an isomorphism in $\mathcal{B}$:

$$E^n(X) : R^n(G \circ F)(X) \cong R^n G(RF(X)).$$

The idea that leads to spectral sequences is that we would like to approximate the complex $RF(X)$ by complexes with smaller number of nontrivial cohomology objects. In the simplest case, when $RF(X)$ has only one nontrivial cohomology $R^k F(X)$, we have $RF(X) \cong R^k F(X)[-k]$ in the derived category (see Proposition III.5.2) and since RG is exact, we have an isomorphism

$$E^n(X) : R^n(G \circ F)(X) \cong R^{n-k}G(R^k F(X)).$$

In the general case the properties of $E^n(X)$ are described in the following manner:

a) We define some filtration on $R^n(G \circ F)(X)$, such that

b) the consecutive quotients of this filtration are obtained from $R^p G(R^q F(X))$, $p + q = n$, by a series of steps, each consisting of taking a subquotient.

It is this algebraic structure that is called a spectral sequence. We give its formal definition consisting of two main parts: a) abstract spectral sequence; b) construction of the spectral sequence for the composition of two functors.

3. Spectral Sequence

Let $\mathcal{A}$ be an abelian category. A *spectral sequence* in $\mathcal{A}$ is a family of objects in $\mathcal{A}$ of the form $E = (E_r^{p,q}, E^n)$, $p, q, r, n \in \mathbb{Z}$, $r \geq 1$, and a family of morphisms between these objects with some properties that we will describe shortly. But first we will say a few words about how one may try to look at all these data.

The reader is advised to imagine a stack of square-lined sheets of paper with each square numbered by a pair of integers $(p, q) \in \mathbb{Z}^2$. An object $E_r^{p,q}$ is assumed to sit in the (p, q)-th square at the r-th sheet. Objects E^n sit at the last, "transfinite" sheet, and occupy the whole diagonal $p + q = n$.

Now we describe morphisms and the conditions they should satisfy.

a) On the r-th sheet we have morphisms $d_r^{p,q} : E_r^{p,q} \to E_r^{p+r,q-r+1}$. For $r = 1$ they act from a square to its right neighbour, for $r = 2$ they act by a chess springer move (one square down and two squares to the right). For $r \geq 3$ we get a generalized springer move.

Condition: $d_r^2 = 0$; more explicitly, $d_r^{p+r,q-r+1} \circ d_r^{p,q} = 0$ for all p, q, r. Using $(E_r^{p,q}, d_r^{p,q})$ we can construct the cohomology of the r-th sheet:

$$H^{p,q}(E_r) = \ker d_r^{p,q} / \mathrm{Im}\, d_r^{p+r,q-r+1}.$$

The following data are included in the definition of E:

b) Isomorphisms $\alpha_r^{p,q} : H^{p,q}(E_r) \to E_{r+1}^{p,q}$.

Usually we will assume that on the $(r+1)$-th sheet we have just cohomology of the r-th sheet and $\alpha_r^{p,q}$ are identities.

The main condition on $\alpha_r^{p,q}$ is the existence of the limit objects $E_\infty^{p,q}$. The simplest way to ensure this, which usually goes well in applications, is the following:

c) For any pair (p,q) there exists r_0 such that $d_r^{p,q} = 0$, $d_r^{p+r,q-r+1} = 0$ for $r \geq r_0$. In this case $\alpha_r^{p,q}$ identify all $E_r^{p,q}$ for $r \geq r_0$ and we will denote this object by $E_\infty^{p,q}$.

At this moment on the transfinite sheet ($r = \infty$) we have objects $E_\infty^{p,q}$ and objects E^n along the diagonal $p + q = n$. The last collection of data connects these two types of objects.

d) A decreasing regular filtration $\ldots \supset F^p E^n \supset F^{p+1} E^n \supset \ldots$ on each E^n and isomorphisms $\beta^{p,q} : E_\infty^{p,q} \to F^p E^{p+q}/F^{p+1} E^{p+q}$ are given. (Recall that a decreasing filtration is said to be *regular* if $\bigcap_p F^p E^n = \{0\}$, $\bigcup_p F^p E^n = E^n$.)

If these conditions are satisfied we say that the spectral sequence $(E_r^{p,q})$ *converges* to (E^n) or that (E^n) is the *limit* of $(E_r^{p,q})$.

Let us stress once more that the components of one spectral sequence E are all objects $(E_r^{p,q}, E^n)$, all morphisms $(d_r^{p,q}, \alpha_r^{p,q}, E^{p,q})$, and all filtrations on E^n.

A morphism $f : E \to E'$ of spectral sequences is a family of morphisms $f_r^{p,q} : E_r^{p,q} \to E_r'^{p,q}$, $f^n : E^n \to E'^n$ that commute with structural morphisms and are compatible with filtrations. With this definition spectral sequences in a category themselves form a category. It is clearly additive, but usually nonabelian.

4. Remarks About Boundness Conditions

a) Working with complexes that are bounded from one side we usually get spectral sequences with nonzero objects placed in only one quadrant (the region of the form, say $p \geq p_0$, $q \geq q_0$).

Let the only nonzero $E_r^{p,q}$ be those in the quadrant I. Then the condition III.7.3c) is automatically satisfied, because for $r > r_0(p,q)$ either the beginnings or the ends of arrows $d_r^{p,q}$, $d_r^{p+r,q-r+1}$ lie outside the quadrant. Moreover, filtrations on E^n are automatically finite: $F^p E^n = 0$ for $p < p_-(n)$, $F^p E^n = E^n$ for $p > p_+(n)$. The same holds, of course, for spectral sequences in quadrant III.

b) The larger is the number of zero objects $E_r^{p,q}$ and of zero morphisms $d_r^{p,q}$, the better a spectral sequence may serve as a computational tool. One special case has its own name: E is said to be *degenerate* at E_r, if $d_{r'}^{p,q} = 0$ for $r' \geq r$ and for all p,q. In this case we have obviously $E_\infty^{p,q} = E_r^{p,q}$.

c) Traditionally, people use spectral sequences when they wish to learn something about E^n while knowing $E_1^{p,q}$ or $E_2^{p,q}$.

Let us show that invariants similar to the Euler characteristic can be computed explicitly even in the case when we know nothing about differentials $d_r^{p,q}$.

Let C be an abelian group and $\chi : \mathrm{Ob}\,\mathcal{A} \to C$ an additive function, i.e., a map satisfying the conditions $\chi(X) = \chi(Y) + \chi(X/Y)$ for any pair consisting of an object X and its subobject Y, and $\chi(X) = \chi(X')$ for isomorphic objects X and X'. For a finite complex $K^\bullet$ let $\chi(K^\bullet) = \sum(-1)^i \chi(K^i)$. Then it is easy to show that $\chi(K^\bullet) = \sum(-1)^i \chi(H^i(K^\bullet))$.

To apply this fact to E let us consider complexes $(E_r^\bullet, d_r^\bullet)$, where $E_r^n = \bigoplus_{p+q=n} E_r^{p,q}$, $d_r^n = \bigoplus_{p+q=n} d_r^{p,q}$. Let us assume that for some r_0 the direct sums are finite and the complex $E_{r_0}^\bullet$ is bounded. Then the same is true for all $r \geq r_0$ and

$$\chi(E_r^\bullet) = \sum (-1)^i \chi\left(H^i(E_r^\bullet)\right) = \chi(E_{r+1}^\bullet) = \chi(E_\infty^\bullet).$$

Moreover,

$$\chi(E^n) = \sum_p \chi\left(F^p E^n / F^{p+1} E^n\right) = \sum_p \chi\left(E_\infty^{p,n-p}\right),$$

and finally

$$\sum_n (-1)^n \chi(E^n) = \chi(E_r^\bullet).$$

5. Spectral Sequence of a Filtered Complex

Let $K^\bullet$ be an object of the category $\mathrm{Kom}(\mathcal{A})$ given together with a decreasing filtration by its subcomplexes $F^p K^\bullet$. This means that in K^n we have subobjects $\ldots \supset F^p K^n \supset F^{p+1} K^n \supset \ldots$ and $d^n(F^p K^n) \subset F^p K^{n+1}$.

Let us give two useful examples.
Let

$$(F^p K^\bullet)^n = \tau_{\leq -p}(K^\bullet)^n = \begin{cases} K^n & \text{for } n < -p, \\ \ker d^{-p} & \text{for } n = -p, \\ 0 & \text{for } n > -p. \end{cases}$$

This is obviously a filtration which is called *canonical*. It kills cohomology of $K^\bullet$ one by one:

$$H^n(F^p K^\bullet) = \begin{cases} H^n(K^\bullet) & \text{for } n \leq -p, \\ 0 & \text{for } n > -p. \end{cases}$$

Next, let

$$(\tilde{F}^p K^\bullet) = \sigma_{\geq p}(K^\bullet)^n = \begin{cases} 0 & \text{for } n < p, \\ K^n & \text{for } n \geq p. \end{cases}$$

This filtration is called *stupid*: it also kills cohomology one by one, but doing so it damages them before killing:

$$H^n(\tilde{F}^p K^\bullet) = \begin{cases} 0 & \text{for } n < p, \\ \ker d^p & \text{for } n = p, \\ H^n(K^\bullet) & \text{for } n > p. \end{cases}$$

Now, given a filtered complex, we construct a spectral sequence. We will work as if $\mathcal{A}$ were the category of abelian groups.

a) *Construction of $E_r^{p,q}$ and of $d_r^{p,q}$.* Let

$$Z_r^{p,q} = d^{-1}\left(F^{p+r}K^{p+q+1}\right) \cap \left(F^p K^{p+q}\right). \qquad \text{(III.45)}$$

This group "bounds from above" cycles in K^{p+q} that belong to the p-th filtration subgroup: the differential d does not necessarily map them to 0, but increases the filtration index by at least r.

$Z_r^{p,q}$ contains a trivial part which is the sum of the following two subgroups:

$$Z_{r-1}^{p+1,q-1} = d^{-1}\left(F^{p+r}K^{p+q+1}\right) \cap \left(F^{p+1}K^{p+q}\right),$$
$$dZ_{r-1}^{p-r+1,q+r-2} = d\left(F^{p-r+1}K^{p+q-1}\right) \cap \left(F^p K^{p+q}\right).$$

Let

$$E_r^{p,q} = Z_r^{p,q} \Big/ \left(Z_{r-1}^{p+1,q-1} + dZ_{r-1}^{p-r+1,q+r-2}\right). \qquad \text{(III.46)}$$

We claim that d induces a differential

$$d_r^{p,q} : E_r^{p,q} \longrightarrow E_r^{p+r,q-r+1}.$$

Indeed, d clearly maps $Z_r^{p,q}$ into $E_r^{p+r,q-r+1}$ (because $d^2 = 0$) and $Z_{r-1}^{p+1,q-1} + dZ_{r-1}^{p-r+1,q+r-2}$ into $dZ_{r-1}^{p+1,q-1}$ (again because $d^2 = 0$).

b) *Construction of $\alpha_r^{p,q}$.* We construct morphisms

$$\left(Z_{r+1}^{p,q} + Z_{r-1}^{p+1,q-1}\right) \Big/ \left(Z_{r-1}^{p+1,q-1} + dZ_{r-1}^{p-r+1,q+r-2}\right) \longrightarrow Z\left(E_r^{p,q}\right) \qquad \text{(III.47)}$$

$$\left(dZ_r^{p-r,q+r-1} + Z_{r-1}^{p+1,q-1}\right) \Big/ \left(Z_{r-1}^{p+1,q-1} + dZ_{r-1}^{p-r+1,q+r-2}\right) \longrightarrow B\left(E_r^{p,q}\right) \qquad \text{(III.48)}$$

(with cycles and boundaries of d_r on the right-hand sides of (III.47) and (III.48)) and show that they are in fact isomorphisms.

First of all, the reader can easily verify that the groups on the left-hand sides of (III.47) and (III.48) are well defined, i.e., we indeed take a quotient of a group by a subgroup. As the definition (III.46) of $E_r^{p,q}$ involves quotients by the same subgroup and $Z_{r+1}^{p,q} + Z_{r-1}^{p+1,q-1} \subset Z_r^{p,q}$, the morphism in (III.47) will be defined if we show that

$$d\left(Z_{r+1}^{p,q} + Z_{r-1}^{p+1,q-1}\right) \subset Z_{r-1}^{p+r+1,q-r} + dZ_{r-1}^{p+1,q-1}.$$

But this is clear because $dZ_{r+1}^{p,q} \subset F^{p+r+1}K^{p+q+1}$ by (III.45). Similarly, one can prove that the morphism (III.48) is well defined.

Now, (III.47) and (III.48) are clearly monomorphisms because they are obtained from monomorphisms by taking quotients by the same subgroup.

Let us verify that (III.47) is surjective. We have

$$Z\left(E_r^{p,q}\right) = Z_r^{p,q} \cap d^{-1}\left(Z_{r-1}^{p+r+1,q-r}\right.$$
$$\left. + dZ_{r-1}^{p+1,q-1}\right) \Big/ \left(Z_{r-1}^{p+1,q-1} + dZ_{r-1}^{p-r+1,q+r-2}\right).$$

III.7 Derived Functor of the Composition. Spectral Sequence

Next
$$d^{-1}\left(Z_{r-1}^{p+r+1,q-r} + dZ_{r-1}^{p+1,q-1}\right) = d^{-1}\left(Z_{r-1}^{p+r+1,q-r}\right) + Z_{r-1}^{p+1,q-1},$$

$$Z_r^{p,q} \cap \left(d^{-1}\left(Z_{r-1}^{p+r+1,q-r}\right) + Z_{r-1}^{p+1,q-1}\right) =$$
$$Z_r^{p,q} \cap d^{-1}\left(Z_{r-1}^{p+r+1,q-r}\right) + Z_{r-1}^{p+1,q-1}$$

because $Z_{r-1}^{p+1,q-1} \subset Z_r^{p,q}$. Finally,

$$Z_r^{p,q} \cap d^{-1}\left(Z_{r-1}^{p+r+1,q-r}\right) =$$
$$d^{-1}\left(F^{p+r}K^{p+q+1}\right) \cap d^{-1}\left(Z_{r-1}^{p+r+1,q-r}\right) \cap F^p K^{p+q}$$
$$\subset d^{-1}\left(F^{p+r+1}K^{p+q+1}\right) \cap F^p K^{p+q} = Z_{r+1}^{p,q}.$$

Comparing this with (III.47) we have the desired result. The surjectivity of (III.48) follows directly from definitions.

So, (III.47) and (III.48) yield the isomorphism
$$\left(Z_{r+1}^{p,q} + Z_{r-1}^{p+1,q-1}\right) \Big/ \left(dZ_r^{p-r,q+r+1} + Z_{r-1}^{p+1,q-1}\right) \longrightarrow H\left(E_r^{p,q}\right).$$

To complete the construction of $\alpha_r^{p,q}$ it remains to identify the left-hand side with
$$E_{r+1}^{p,q} = Z_{r+1}^{p,q} \Big/ \left(dZ_r^{p-r,q+r-1} + Z_r^{p+1,q-1}\right).$$

To do this we note that
$$\left(Z_{r+1}^{p,q} + Z_{r-1}^{p+1,q-1}\right) \Big/ \left(dZ_r^{p-r,q+r+1} + Z_{r-1}^{p+1,q-1}\right) =$$
$$Z_{r+1}^{p,q} \Big/ \left(Z_{r+1}^{p,q} \cap \left(dZ_r^{p-r,q+r+1} + Z_{r-1}^{p+1,q-1}\right)\right).$$

Next
$$dZ_r^{p-r,q+r+1} \subset Z_{r+1}^{p,q},$$
$$Z_{r+1}^{p,q} \cap Z_{r-1}^{p+1,q-1} = Z_r^{p+1,q-1}.$$

Hence
$$Z_{r+1}^{p,q} \cap \left(dZ_r^{p-r,q+r-1} + Z_{r-1}^{p+1,q-1}\right) = dZ_r^{p-r,q+r-1} + Z_r^{p+1,q-1}.$$

c) *Construction of E^n.* Let $E^n = H^n(K^\bullet)$ and $F^p E^n$ be the image of $H^n(F^p K^\bullet)$ under the natural morphism $F^p K^\bullet \to K^\bullet$.

Now we must verify that under some conditions that ensure the existence of E_∞ the spectral sequence we have constructed converges to E^n.

Let us assume that for each n the filtration F^p on K^n is finite and regular. This means that there exist $p_+(n)$ and $p_-(n)$ such that $F^{p_+(n)} K^n = K^n$, $F^{p_-(n)} K^n = 0$. Then the condition III.7.3c) is automatically satisfied with

$$r_0(p,q) = \max\ (p_+(p+q+1) - p_-(p+q) + 1,$$
$$p_+(p+q) - p_-(p+q-1) + 1). \qquad \text{(III.49)}$$

So the groups $E_\infty^{p,q}$ exist and, moreover, (III.46) shows that
$$E_\infty^{p,q} = Z_r^{p,q}/Z_{r-1}^{p+1,q-1} \qquad \text{for}\quad r \geq r_0(p,q)$$
and, by (III.45) and (III.49),
$$Z_r^{p,q} = Z\left(F^p K^{p+q}\right), \quad Z_{r-1}^{p+1,q-1} = Z\left(F^{p+1} K^{p+q}\right) \qquad \text{for}\quad r \geq r_0(p,q)+1.$$
Finally, we get an isomorphism
$$F^p E^n / F^{p+1} E^n = E_\infty^{p,q},$$
which establishes the convergence.

6. Example. Here we compute spectral sequences related to the stupid and to the canonical filtration of a complex $K^\bullet$ (see the beginning of III.7.5).

a) *Stupid filtration $\tilde{F}^p K^\bullet$.* We have
$$Z_r^{p,q} = \begin{cases} 0 & \text{for}\quad q < 0, \\ \ker d^{p+q} & \text{for}\quad q \geq 0,\ r < q+1, \\ K^{p+q} & \text{for}\quad q \geq 0,\ r \geq q+1. \end{cases}$$

$$E_r^{p,q} = \begin{cases} 0 & \text{for}\quad q \neq 0, \\ K^p & \text{for}\quad q = 0,\ r = 1, \\ H^p(K^\bullet) & \text{for}\quad q = 0,\ r \geq 2 \quad \text{and}\quad q = 0,\ r = \infty. \end{cases}$$

The differential $d_r^{p,q} : E_r^{p,q} \to E_r^{p+r,q-r+1}$ coincides with $d^p : K^p \to K^{p+1}$ for $q = 0,\ r = 1$ and is trivial in all other cases.

$E^n = H^n(K^\bullet)$; the filtration on E^n is trivial:
$$F^p E^n = \begin{cases} E^n & \text{for}\quad p \leq n, \\ 0 & \text{for}\quad p > n. \end{cases}$$

b) *Canonical filtration.* We have
$$E_1^{p,q} = \begin{cases} H^p(K^\bullet) & \text{for}\quad q = -2p, \\ 0 & \text{for}\quad q \neq -2p. \end{cases}$$
$$d_1^{p,q} = 0 \quad \text{for all}\quad p,q.$$

Hence $E_r^{p,q} = E_1^{p,q}$ and $d_r = 0$ for all $r \geq 1$. Next, $E^n = H^n(K^\bullet)$ and the filtration on E^n is trivial:
$$F^p E^n = \begin{cases} E^n & \text{for}\quad n \leq -p, \\ 0 & \text{for}\quad n > -p. \end{cases}$$

Let us note that the spectral sequence related to the canonical filtration yields a functor from the category $D(\mathcal{A})$ to the category of spectral sequences:

if $f : K^\bullet \to L^\bullet$ is a quasi-isomorphism, then the corresponding morphism $E(f) : E(K^\bullet) \to E(L^\bullet)$ of spectral sequences is an isomorphism (i.e., all $E_r^{p,q}(K^\bullet) \to E_r^{p,q}(L^\bullet)$, $E^n(K^\bullet) \to E^n(L^\bullet)$ are isomorphisms). On the other hand, the spectral sequence related to the stupid filtration is functorial in $\mathrm{Kom}(\mathcal{A})$ but not in $D(\mathcal{A})$.

We strongly recommend to the reader to perform all missing computations.

7. Theorem. *Assume that, under the conditions of Theorem III.7.1 above, the class $\mathcal{R}_\mathcal{A}$ coincides with the class $\mathcal{I}_\mathcal{A}$ of all injective objects in $\mathcal{A}$ and that the class $\mathcal{I}_\mathcal{B}$ is sufficiently large. Then for any $X \in \mathrm{Ob}\,\mathcal{A}$ there exists a spectral sequence with*
$$E_2^{p,q} = R^p G(R^q F(X))$$
converging to $R^n(G \circ F)(X)$. It is functorial in X.

Plan of the Proof. Objects $R^q F(X)$ are cohomology objects of the complex $RF(X) = F(I_X^\bullet)$, where $I_X^\bullet$ is an injective resolution of X. To compute $R(G \circ F)(X) = RG \circ RF(X)$ we have to apply G term by term to a complex from $\mathrm{Kom}^+(\mathcal{R}_\mathcal{B})$ which is quasi-isomorphic to $F(I_X^\bullet)$. However, for $F(\mathcal{I}_\mathcal{B}) \subset \mathcal{R}_\mathcal{B}$ we can take $F(I_X^\bullet)$ itself so that $R^n(G \circ F)(X)$ are cohomology objects of the complex $G \circ F(I_X^\bullet)$.

On the other hand, to compute $R^p G(R^q F(X))$ we must apply G term by term to resolutions in $\mathrm{Kom}^+(\mathcal{R}_\mathcal{B})$ of objects $R^q F(X)$.

These two resolutions are related via a double complex that is called *the Cartan–Eilenberg resolution* of the complex $K^\bullet = F(I_X^\bullet)$. This Cartan–Eilenberg resolution consists of the following data.

a) A double complex (L^{ij}) with differentials d_I, d_{II} of bidegree $(1,0)$ and $(0,1)$ respectively; it satisfies the conditions $L^{ij} = 0$ for $j < 0$ or $i \le 0$, L^{ij} are injective (the main definitions concerning double complexes are given in 8 below).

b) A morphism of complexes $\varepsilon : K^\bullet \to L^{\bullet,0}$.

To state the conditions imposed on the above data we note that (L^{ij}) gives rise to the following complexes:

$$
\begin{aligned}
0 &\longrightarrow K^i \xrightarrow{\varepsilon} L^{i,0} \longrightarrow L^{i,1} \longrightarrow \cdots \\
0 &\longrightarrow B^i(K^\bullet) \xrightarrow{\varepsilon} B_I^i(L^{\bullet,0}) \longrightarrow B_I^i(L^{\bullet,1}) \longrightarrow \cdots \\
0 &\longrightarrow Z^i(K^\bullet) \longrightarrow Z_I^i(L^{\bullet,0}) \longrightarrow Z_I^i(L^{\bullet,1}) \longrightarrow \cdots \\
0 &\longrightarrow H^i(K^\bullet) \longrightarrow H_I^i(L^{\bullet,0}) \longrightarrow H_I^i(L^{\bullet,1}) \longrightarrow \cdots
\end{aligned}
\qquad \text{(III.50)}
$$

We require that:

c) All these complexes are acyclic.

d) Exact triples

$$0 \longrightarrow B_I^i(L^{\bullet,j}) \longrightarrow Z_I^i(L^{\bullet,j}) \longrightarrow H_I^i(L^{\bullet,j}) \longrightarrow 0$$
$$0 \longrightarrow Z_I^i(L^{\bullet,j}) \longrightarrow L^{ij} \longrightarrow B_I^{i+1}(L^{\bullet,j}) \longrightarrow 0$$

split.

Under these conditions the objects B_I, Z_I, H_I are injective, so that the complexes (III.50) are injective resolutions of $K^\bullet$, $B^\bullet(K^\bullet)$, $Z^\bullet(K^\bullet)$, $H^\bullet(K^\bullet)$ respectively. In particular, $R^pG(R^qF(X))$ can be thought of as the p-th cohomology of the complex $G(H_I^q(L^{\bullet,\bullet}))$. On the other hand, $K^\bullet = F(I_X^\bullet)$ is quasi-isomorphic to the *diagonal complex SL* of L (see 8) so that $R^n(G \circ F)(X) = H^n(G(F(I_X^\bullet)))$ is isomorphic to $H^nG(SL)$. Moreover, L has a filtration $L^{\bullet, \geq k}$ and the term E_2 of the spectral sequence corresponding to this filtration is isomorphic to $R^pG(R^qF(X)) = G(H_I^q(L^{\bullet,\bullet}))$.

It will be convenient to perform the main steps of the proof in a somewhat more general situation. In particular, we will not assume that $K^\bullet \in \text{Kom}^+(\mathcal{B})$ is of the form $F(I_X^\bullet)$.

8. Double Complexes

A double complex $L = \left(L^{ij}, d_I^{ij}, D_{II}^{ij}\right)$ is a collection of objects L^{ij} and morphisms $d_I^{ij}: L^{ij} \to L^{i+1,j}$, $d_{II}^{ij}: L^{ij} \to L^{i,j+1}$ satisfying the relations

$$d_I^2 = 0, \quad d_{II}^2 = 0, \quad d_I d_{II} + d_{II} d_I = 0. \tag{III.51}$$

Let $(SL)^n = \bigoplus_{i+j=n} L^{ij}$ (in all situations we consider in this book these direct sums are finite; in the general case their existence has to be assumed separately). The condition (III.51) means that the operator

$$d = d_I + d_{II} : (SL)^n \longrightarrow (SL)^{n+1}$$

satisfies the condition $d^2 = 0$ so that $((SL)^\bullet, d)$ is a complex called the *diagonal complex* of L.

Morphisms of double complexes are defined in an obvious way. A homotopy between two morphisms $f, g : L^{\bullet\bullet} \to M^{\bullet\bullet}$ is a collection of morphisms $h_I^{ij}: L^{ij} \to M^{i-1,j}$, $h_{II}^{ij}: L^{ij} \to M^{i,j-1}$ such that

$$g - f = (d_I + d_{II})(h_I + h_{II}) + (h_I + h_{II})(d_I + d_{II})$$

as morphisms of bigraded objects. Writing down this equality componentwise we obtain on the right-hand side, for each pair (i,j), three morphisms $(ij) \mapsto (ij)$, $(ij) \mapsto (i+1, j-1)$, $(ij) \mapsto (i-q, j+1)$, the first of which should be equal to $g^{ij} - f^{ij}$, while the last two should be zero. In particular, $h_I + h_{II}$ gives a homotopy of morphisms $Sf, Sg : SL \to SM$.

Let L be a double complex. Define
$$Z_I^i(L^{\bullet,j}) = \ker d_I^{i,j},$$
$$B_I^i(L^{\bullet,j}) = \operatorname{Im} d_I^{ij},$$
$$H_I^{ij}(L^{\bullet,j}) = H_I^i(L^{\bullet,j}) = Z_I^i(L^{\bullet,j})/B_I^i(L^{\bullet,j}),$$

so that Z_I, B_I, H_I are cocycles, coboundary and cohomology of the ordinary complex $(L^{\bullet,j}, d^{\bullet,j})$ respectively. The differential d_{II}^{ij} induces the morphisms
$$Z_I^i(L^{\bullet,j}) \longrightarrow Z_I^i(L^{\bullet,j+1}),$$
$$B_I^i(L^{\bullet,j}) \longrightarrow B_I^i(L^{\bullet,j+1}),$$
$$H_I^i(L^{\bullet,j}) \longrightarrow H_I^i(L^{\bullet,j+1}),$$

which can be considered as differentials in the last three complexes in (III.50). The cohomology of the last complex in (III.50) will be denoted by $H_{II}^j\left(H_I^{i,\bullet}(L^{\bullet\bullet})\right)$. Similarly one can define cohomology $H_I^i\left(H_{II}^{\bullet,j}(L^{\bullet\bullet})\right)$.

The category of double complexes (resp. double complexes up to homotopy) is denoted by Kom^{**} (resp. by K^{**}).

9. Spectral Sequences of a Double Complex

Let $L = \left(L^{ij}, d_I^{ij}, d_{II}^{ij}\right)$ be a double complex and SL be the corresponding diagonal complex. On SL there exist two decreasing filtrations
$$F_I^p(SL)^n = \bigoplus_{\substack{i+j=n \\ i \geq p}} L^{ij}, \qquad F_{II}^q(SL)^n = \bigoplus_{\substack{i+j=n \\ j \geq q}} L^{ij}.$$

The construction from 5 yields two spectral sequences ${}^I E_r^{pq}$ and ${}^{II} E_r^{pq}$. If (L^{ij}) lies in the first quadrant, as it will be in our applications, both these filtrations are finite and regular on each $(SL)^n$, so that by 5c) both spectral sequences converge to the common limit $H^n(SL)$. On the other hand, their E_2 terms can be computed directly as follows.

10. Proposition. We have ${}^I E_2^{p,q} = H_I^p\left(H_{II}^{\bullet,q}(L^{\bullet\bullet})\right)$, ${}^{II} E_2^{p,q} = H_{II}^p\left(H_I^{q,\bullet}(L^{\bullet\bullet})\right)$.

Proof. First, let us compute ${}^I E_1^{pq}$. By (III.45),
$$
\begin{aligned}
{}^I Z_1^{pq} &= d^{-1}\left(F^{p+1}(SL)^{p+q+1}\right) \cap F^p(SL)^{p+q} \\
&= (d_I + d_{II})^{-1}\left(\bigoplus_{\substack{i+j=p+q+1 \\ i \geq p+1}} L^{ij}\right) \cap \bigoplus_{\substack{i+j=p+q \\ i \geq p}} L^{ij} \\
&= \ker d_{II}^{pq} \oplus \left(\bigoplus_{\substack{i+j=p+q \\ i \geq p+1}} L^{ij}\right).
\end{aligned}
$$

210   III. Derived Categories and Derived Functors

To use (III.46) we need also ${}^I Z_0^{p+1,q-1}$ and $d\left({}^I Z_0^{p,q-1}\right)$. We have

$${}^I Z_0^{p+1,q-1} = (d_I + d_{II})^{-1}\left(\bigoplus_{\substack{i+j=p+q+1\\i\geq p+1}} L^{ij}\right) \cap \bigoplus_{\substack{i+j=p+q\\i\geq p+1}} L^{ij} = \bigoplus_{\substack{i+j=p+q\\i\geq p+1}} L^{ij},$$

$${}^I Z_0^{p,q-1} = \bigoplus_{\substack{i+j=p+q-1\\i\geq p}} L^{ij}.$$

Hence

$${}^I Z_0^{p+1,q-1} + d\left({}^I Z_0^{p,q-1}\right) = \operatorname{Im} d_{II}^{p,q-1} \oplus \left(\bigoplus_{\substack{i+j=p+q\\i\geq p}} L^{ij}\right).$$

and

$${}^I E_1^{p,q} = {}^I Z_1^{p,q} / \left({}^I Z_0^{p+1,q-1} + d\left({}^I Z_0^{p,q-1}\right)\right) = H_{II}^{p,q}(L^{\bullet\bullet}).$$

We claim that under this identification the differential

$${}^I d_I^{p,q} : {}^I E_1^{p,q} \longrightarrow {}^I E_1^{p,q+1}$$

in the spectral sequence coincides with the differential

$$d_I^{p,q} : H_{II}^{p,q}(L^{\bullet\bullet}) \longrightarrow H_{II}^{p,q+1}(L^{\bullet\bullet})$$

which is induced by $d_I^{p,q}$ (see III.7.8). This is clear from the definitions because on H_{II} the differential $d = d_I + d_{II}$ induces the same map as d_I.

Thus, we obtain that ${}^I E_2^{p,q} = H_I^p(H_{II}^q(L^{\bullet\bullet}))$.

Similarly one computes ${}^{II} E_2^{p,q}$. □

11. Proposition. *Let $K^\bullet$ be in $\operatorname{Kom}^+(\mathcal{B})$ and let the class $\mathcal{I}_\mathcal{B}$ be sufficiently large. Then:*

a) $K^\bullet$ has a Cartan–Eilenberg resolution.

b) Any morphism $K^\bullet \to \tilde{K}^\bullet$ in $\operatorname{Kom}^+(\mathcal{B})$ can be extended to a morphism of arbitrary Cartan–Eilenberg resolutions of $K^\bullet$, $\tilde{K}^\bullet$ respectively, and this extension is unique up to a homotopy of double complexes.

c) If two morphisms $K^\bullet \to \tilde{K}^\bullet$ are homotopic, then any two their extensions to morphisms of Cartan–Eilenberg resolutions are also homotopic.

In other words, taking a Cartan–Eilenberg resolution determines a functor from $K^+(\mathcal{B})$ to $K^{++}(\mathcal{I}_\mathcal{B})$.

Proof. First let us prove one general statement. Let

$$0 \longrightarrow X' \longrightarrow X \longrightarrow X'' \longrightarrow 0 \tag{III.52}$$

be an exact sequence in $\mathcal{B}$, and let $I_{X'}^\bullet, I_X^\bullet, I_{X''}^\bullet$, be injective resolutions of X', X, X'' respectively. By an exact sequence of resolutions we mean the commutative diagram of the form

$$
\begin{array}{ccccccccc}
& & 0 & & 0 & & 0 & & \\
& & \downarrow & & \downarrow & & \downarrow & & \\
0 & \to & X' & \to & X & \to & X'' & \to & 0 \\
& & \downarrow & & \downarrow & & \downarrow & & \\
0 & \to & I^0_{X'} & \to & I^0_X & \to & I^0_{X''} & \to & 0 \\
& & \downarrow & & \downarrow & & \downarrow & & \\
0 & \to & I^1_{X'} & \to & I^1_X & \to & I^1_{X''} & \to & 0 \\
& & \downarrow & & \downarrow & & \downarrow & & \\
& & \vdots & & \vdots & & \vdots & &
\end{array}
\qquad (\text{III.53})
$$

with exact rows and columns. Such an exact sequence of resolutions will be denoted by

$$0 \to I^\bullet_{X'} \to I^\bullet_X \to I^\bullet_{X''} \to 0.$$

We leave to the reader a rather easy verification of the following statements.

A. Let an exact sequence (III.52) and injective resolutions $I^\bullet_{X'}$, $I^\bullet_{X''}$ of objects X', X'' be given. Then there exists at least one injective resolution $I^\bullet_X$ of the object X that can be included into an exact sequence of resolutions (III.53).

B. Let

$$
\begin{array}{ccccccccc}
0 & \to & X' & \to & X & \to & X'' & \to & 0 \\
& & \downarrow f' & & \downarrow f & & \downarrow f'' & & \\
0 & \to & Y' & \to & Y & \to & Y'' & \to & 0
\end{array}
$$

a commutative diagram with exact rows and

$$
\begin{array}{ccccccccc}
0 & \to & I^\bullet_{X'} & \to & I^\bullet_X & \to & I^\bullet_{X''} & \to & 0 \\
& & \downarrow F' & & \downarrow F & & \downarrow F'' & & \\
0 & \to & I^\bullet_{X'} & \to & I^\bullet_X & \to & I^\bullet_{X''} & \to & 0
\end{array}
\qquad (\text{III.54})
$$

be the corresponding exact sequence of injective resolutions. Then there exist morphisms of complexes F', F, F'' extending f', f, f'' and making the diagram of complexes (III.54) commutative. If G', G, G'' is another triple of morphisms of complexes making (III.54) commutative and h' (resp. h'') a homotopy between F' and G' (resp. between F'' and G'') then there exists a homotopy h between F and G such that for any n the diagram

$$
\begin{array}{ccccccccc}
0 & \to & I^n_{X'} & \to & I^n_X & \to & I^n_{X''} & \to & 0 \\
& & \downarrow h'^n & & \downarrow h^n & & \downarrow h''^n & & \\
0 & \to & I^{n-1}_{X'} & \to & I^{n-1}_X & \to & I^{n-1}_{X''} & \to & 0
\end{array}
$$

commutes.

Let us proceed now with the proof of Proposition III.7.11. Let us choose for any i projective resolutions $I^\bullet_{B^i(K^\bullet)}$ and $I^\bullet_{H^i(K^\bullet)}$ of objects $B^i(K^\bullet)$ and $H^i(K^\bullet)$ respectively. The exact sequence

$$0 \longrightarrow B^i(K^\bullet) \longrightarrow Z^i(K^\bullet) \longrightarrow H^i(K^\bullet) \longrightarrow 0$$

and the statement A above yield the existence of an injective resolution $I_{Z^i(K^\bullet)}$ of $Z^i(K^\bullet)$ and an exact sequence of resolutions

$$0 \longrightarrow I^\bullet_{B^i(K^\bullet)} \longrightarrow I^\bullet_{Z^i(K^\bullet)} \longrightarrow I^\bullet_{H^i(K^\bullet)} \longrightarrow 0.$$

Similarly, the exact sequence

$$0 \longrightarrow Z^i(K^\bullet) \longrightarrow K^i \longrightarrow B^{i+1}(K^\bullet) \longrightarrow 0$$

shows that there exist an injective resolution $I^\bullet_{K^i}$ of K^i and an exact sequence of resolutions

$$0 \longrightarrow I^\bullet_{Z^i(K^\bullet)} \longrightarrow I^\bullet_{K^i} \longrightarrow I^\bullet_{B^{i+1}(K^\bullet)} \longrightarrow 0.$$

Let us define now the double complex L as follows:

$$L^{ij} = I^j_{K^i},$$

d^{ij}_I is the composition of the following sequence of morphisms:
$$I^j_{K^i} \longrightarrow I^j_{B^{i+1}(K^\bullet)} \longrightarrow I^j_{Z^{i+1}(K^\bullet)} \longrightarrow I^j_{K^{i+1}},$$

$d^{ij}_{II} = (-1)^j d^j_{I_{K^i}}$ where $d^j_{I_{K^i}} : I^j_{K^i} \longrightarrow I^{j+1}_{K^i}$ is the differential in $I^\bullet_{K^i}$.

One can easily verify that $L = \left(L^{ij}, d^{ij}_I, d^{ij}_{II} \right)$ is a double complex which is a Cartan–Eilenberg resolution of the complex $K^\bullet$ with respect to morphisms

$$K^i \xrightarrow{\varepsilon^i} L^{i,0} = I^0_{K^i}.$$

This proves the Proposition III.7.11a).

To prove the first statement in part b) of the proposition it suffices to apply Theorem III.1.3 (or, more exactly, Remark III.1.4c)) to morphisms $B^i(K^\bullet) \longrightarrow B^i(\tilde{K}^\bullet)$, $H^i(K^\bullet) \longrightarrow H^i(\tilde{K}^\bullet)$ induced by the morphism $K^\bullet \longrightarrow \tilde{K}^\bullet$ and then to use several times the statement B above.

To prove the second statement in part b) of the proposition we must construct a homotopy between two extensions of a morphism $K^\bullet \to \tilde{K}^\bullet$ to Cartan–Eilenberg resolutions. This can be done similarly to the construction of a morphism of Cartan–Eilenberg resolutions above; the main difference is that we have to use the statement B instead of the statement A.

Finally, to prove part c) of the proposition let us consider two homotopic morphisms of complexes $f, g : K^\bullet \to \tilde{K}^\bullet$. Let h be a homotopy between morphisms f, g, and let $L^{\bullet\bullet}$, $\tilde{L}^{\bullet\bullet}$ be Cartan–Eilenberg resolutions of $K^\bullet$ and

$\tilde{K}^\bullet$ respectively. Any morphism $h^p : K^p \to \tilde{K}^{p-1}$ can be extended to at least one morphism of complexes $H^{p,\bullet} : L^{p,\bullet} \to \tilde{L}^{p-1,\bullet}$. These morphisms give a homomorphism $H : L^{\bullet\bullet} \to \tilde{L}^{\bullet\bullet}$ of bidegree $(-1,0)$, which commutes with morphisms $K^\bullet \to L^{\bullet,0}$ and $\tilde{K}^\bullet \to \tilde{L}^{\bullet,0}$, and anticommutes with differentials d_{II} in $L^{\bullet\bullet}$ and in $\tilde{L}^{\bullet\bullet}$. One can easily see that $Q = F + d_{I,\tilde{L}}H + Hd_{I,L}$ is a morphism of double complexes extending the morphism $g : K^\bullet \to \tilde{K}^\bullet$ and the pair $(H,0)$ gives a homotopy between F and Q. Thus part c) of the proposition follows from part b). □

12. Lemma. *Suppose we are given a complex* (K^i), *a bicomplex* (L^{ij}), *and a morphism of complexes* $\varepsilon : K^\bullet \to L^{\bullet,0}$. *Let us assume that* $d_{II} \circ \varepsilon = 0$ *and all complexes*

$$0 \longrightarrow K^i \xrightarrow{i} L^{i,0} \xrightarrow{d_{II}} L^{i,1} \xrightarrow{d_{II}} \ldots$$

are acyclic. Then the induced morphism of complexes

$$\tilde{\varepsilon} : K^\bullet \to (SL)^\bullet$$

is a quasi-isomorphism.

Proof. One can easily see that $\tilde{\varepsilon}$ is a morphism of complexes. By assumptions, $H_{II}^{pq}(L^{\bullet\bullet}) = 0$ for $q > 0$ so that the term ${}^I E_1^{pq}$ of the first spectral sequence of the double complex $L^{\bullet\bullet}$ is concentrated in the zeroth row and, moreover, ε indices an isomorphism of complexes $K^\bullet$ and $H_{II}^{p,0}(L^{\bullet\bullet}) = {}^I E_1^{p0}$. Hence ε induces a cohomology isomorphism $H^p(K^\bullet) \to {}^I E_2^{p0}$ and ${}^I E_2^{pq}$ are zero for $q > 0$ so that ${}^I E_\infty^{pq} = 0$ for $q > 0$ and

$$H^p(SL) = E^p = {}^I E_\infty^{p0} = {}^I E_2^{p0}.$$

□

13. Corollary. *Assume that* $K^\bullet \in \mathrm{Kom}^+(\mathcal{B})$, *the class* $\mathcal{I}_\mathcal{B}$ *is sufficiently large, and* $L^{\bullet\bullet}$ *is a Cartan–Eilenberg resolution of* $K^\bullet$. *Assume also that* $G : \mathcal{B} \to \mathcal{C}$ *is a left exact additive functor. Then* $RG(K^\bullet) = G(SL)$ *in* $D(\mathcal{C})$. *(Take into account the fact that, by Theorem III.6.9, $\mathcal{I}_\mathcal{B}$ is adapted to G.)*

14. Hypercohomology of a Functor with Respect to a Complex

Under the conditions of the previous corollary, the objects $R^i G(K^\bullet)$ (or $H^i(G(SL))$) are usually called *hypercohomologies* of the functor G with respect to the complex $K^\bullet$. These hypercohomology objects are the limit objects of the spectral sequence ${}^{II}E_r^{pq}$ corresponding to the filtration $F_{II}^q G(SL) = G(F_{II}^q(SL))$ (because G transforms direct sums into direct sums). By Proposition III.7.10 the second term of this spectral sequence is ${}^{II}E_2^{pq} = H_{II}^q\left(H_I^{q,\bullet}(G(L^{\bullet\bullet}))\right)$. Let us compute it.

By the definition of Cartan–Eilenberg resolution the complexes $(L^{\bullet,j}, d_I)$ are split in the sense of III.7.7d). Hence

$$H_I^{q,\bullet}(G(L^{\bullet\bullet})) = G\left(H_I^{q,\bullet}(L^{\bullet\bullet})\right)$$

(more precisely, some canonical morphism is an isomorphism). But the last exact sequence in (III.50) shows that $(H_I^{q,\bullet}, d_{II})$ is an injective resolution of $H^q(K^\bullet)$. Hence $G(H_I^{q,\bullet}) = RG(H^q(K^\bullet))$ and

$$H_{II}^q\left(H_I^{q,\bullet}(G(L^{\bullet\bullet}))\right) = R^p G\left(H^q(K^\bullet)\right). \tag{III.55}$$

15. Completion of the Proof of Theorem 7

Let $K^\bullet = F(I_X^\bullet) = RF(X)$, where $I_X^\bullet$ is an injective resolution of X. Then $H^q(K^\bullet)$ in (III.55) is $R^q F(X)$ and the spectral sequence $E_2^{pq} = R^q G\left(R^p F(X)\right)$ converges to $R^n G(SL)$, where $L^{\bullet\bullet}$ is a Cartan–Eilenberg resolution of $K^\bullet$. But $G(SL) = RG(K^\bullet) = RG \circ RF(X)$, which is isomorphic to $R(G \circ F)(X)$ by Theorem III.7.1.

The functoriality of the spectral sequence in X follows from the functoriality of $I_X^\bullet$ in X (Theorem III.1.3 and Remark III.1.4c)) and from functoriality of Cartan–Eilenberg resolution (Proposition III.7.11).

Exercises

1. Serre–Hochschild Spectral Sequence. a) Show that the cohomology $H^n(G, A)$ of a group G with coefficients in a G-module A is a right derived functor $R^n F$ of the functor

$$F(A) = H^0(G, A) = A^G \stackrel{\text{def}}{=} \{a \in A, \ ga = a \quad \text{for all} \quad g \in G\}$$

on the category, G-mod, of left G-modules. Deduce that $H^n(G, A) = \text{Ext}_{\mathbb{Z}[G]}^n(\mathbb{Z}, A)$, where $\mathbb{Z}[G]$ is the group ring of G and $\mathbb{Z}$ is the trivial $\mathbb{Z}[G]$-module ($gn = n$ for $g \in G$, $n \in \mathbb{Z}$).

b) Let H be a normal subgroup of G. Show that for any G-module A the group G acts on $H^n(H, A)$ (for any $g \in G$ the map $a \mapsto ga$ is an automorphism of the functor $A \mapsto A^H$), and the action of elements from H is trivial. Write down this action explicitly (on cocycles) using the definition of $H^n(H, A)$ from Chap. I (see I.2.8, I.4.9c), I.7.2, I.7.6).

c) Using the formula $A^G = (A^H)^{G/H}$ construct a functorial in $A \in \text{Ob } G$-mod spectral sequence (Serre–Hochschild spectral sequence) with

$$E_2^{pq} = H^p\left(G/H, H^q(H, A)\right), \quad E_\infty^n = H^n(G, A).$$

d) State and prove similar results for homology $H_n(G, A)$.

2. Spectral Sequence for Čech Cohomology. a) Let us recall that for any open covering $\mathcal{U} = (U_i)$ of a topological space X and for any presheaf $\mathcal{F}$ on X we denote by $C^\bullet(\mathcal{U}, \mathcal{F})$ the complex of Čech cochains of $\mathcal{F}$ with respect to the covering $\mathcal{U}$, and by $\hat{H}^p(\mathcal{U}, \mathcal{F})$ the cohomology groups of this complex. The mapping $\mathcal{F} \mapsto C^\bullet(\mathcal{U}, \mathcal{F})$ gives a functor $C_\mathcal{U}^\bullet$ from the category $\mathcal{S}Ab_X$ of sheaves of abelian groups on X to the category $\text{Kom}^{\geq 0}(Ab)$ of complexes of abelian groups concentrated in non-negative degrees. Let H^0:

$\text{Kom}^{\geq 0}(Ab) \to Ab$ be the functor of 0-cohomology. Show that the spectral sequence corresponding to the composition of functors $H^0 \circ C_{\mathcal{U}}^{\bullet}$ has

$$E_2^{pq} = \hat{H}^p(\mathcal{U}, \mathcal{H}^q(\mathcal{F}))$$

where $\mathcal{H}^q(\mathcal{F})$ is a presheaf on X given by $V \mapsto H^q(V, \mathcal{F})$ and

$$H_{\infty}^n = H^n(X, \mathcal{F}).$$

b) Construct functorial in $\mathcal{F} \in \mathcal{S}Ab_X$ homomorphisms

$$\hat{H}^p(\mathcal{U}, \mathcal{F}) \longrightarrow H^p(X, \mathcal{F}).$$

Show that if all nonempty intersections $U_{i_0} \cap \ldots \cap U_{i_p}$ are $\mathcal{F}$-acyclic (i.e., $H^q(U_{i_0} \cap \ldots \cap U_{i_p}, \mathcal{F}) = 0$ for all $q > 0$) then these homomorphisms are isomorphisms.

c) If a covering $\mathcal{U}'$ is a refinement of $\mathcal{U}$, then we have a natural morphism of functors $C_{\mathcal{U}}^{\bullet} \to C_{\mathcal{U}'}^{\bullet}$, so that homomorphisms $\hat{H}^p(\mathcal{U}, \mathcal{F}) \to \hat{H}^p(\mathcal{U}', \mathcal{F})$ are defined. Denoting by $\hat{H}^p(X, \mathcal{F})$ the inductive limits $\hat{H}^p(X, \mathcal{F}) = \varinjlim \hat{H}^p(\mathcal{U}, \mathcal{F})$ over all open coverings $\mathcal{U}$, construct a spectral sequence with $E_2^{pq} = \hat{H}^p(X, \mathcal{H}^q(\mathcal{F}))$, $E_{\infty}^n = \hat{H}^n(X, \mathcal{F})$. Show that the corresponding homomorphism $a_n : \hat{H}^n(X, \mathcal{F}) \to H^n(X, \mathcal{F})$ is an isomorphism for $n = 0, 1$ and a monomorphism for $n = 2$. Show that if X is paracompact, then a_n is an isomorphism for all n.

3. Exact Pairs and Spectral Sequences. Another (historically earlier) method to construct a spectral sequence is related to so-called exact pairs. We will work in a category of modules over a fixed ring R; however, all definitions and results can be easily generalized to an arbitrary abelian category $\mathcal{A}$.

a) An *exact pair* is a collection (D, E, i, j, k) consisting of two modules and three morphisms:

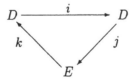

such that the sequence $D \xrightarrow{i} D \xrightarrow{j} E \xrightarrow{k} D \xrightarrow{i} D$ is exact, i.e., at any vertex of the above triangle the image equals the kernel.

In particular, $(jk)^2 = 0$ so that we can take the homology $H(E, jk) = \ker(jk)/\operatorname{Im}(jk)$.

The derived pair

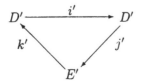

216   III. Derived Categories and Derived Functors

is defined as follows: $D' = \mathrm{Im}\, i$, $E' = H(E, jk)$, i', j', k' are induced by i, j, k respectively:
   i' is the restriction of i to $\mathrm{Im}\, i \subset D$;
   $j'(i(x))$ is the class of $j(x)$ in $H(E, jk)$, $x \in D$;
   k' (class of y) $= k(y)$, $y \in E$, $jk(y) = 0$.
   Prove that the derived pair of an exact pair is well defined and exact.

So, we can define the sequence of exact pairs $P_r = (D_r, E_r, i_r, j_r, k_r)$: for $r = 1$ it is the initial exact pair $P_1 = (D, E, i, j, k)$ and P_{r+1} for $r \geq 1$ is the derived pair of P_r.

b) Let us assume that the exact pair P_1 is bigraded; this means that D and E are bigraded, $D = \oplus D^{pq}$, $E = \oplus E^{pq}$ and i, j, k have the bidegree $(-1, 1)$, $(0, 0)$, $(1, 0)$ respectively.

Show that the derived exact pairs P_r are also bigraded and the morphisms i_r, j_r, k_r have the bidegree $(-1, 1)$, $(r - 1, -r + 1)$, $(1, 0)$ respectively.

In this case $d_r = j_r k_r$ is a differential of E_r of bidegree $(r, -r + 1)$ and the homology of d_r is isomorphic to E_{r+1} (as bigraded modules), so that (E_r^{pq}, d_r) is a part of a spectral sequence.

Construct the limit E_∞ of this spectral sequence. To do this it is convenient to represent the exact pair P_1 as an infinite diagram formed by coupled exact sequences:

$$\begin{array}{ccccccccc}
& & \uparrow & & & & \uparrow & & \\
\cdots \to & E^{p-2,q} & \xrightarrow{k} & D^{p-1,q} & \xrightarrow{j} & E^{p-1,q} & \xrightarrow{k} & D^{p,q} & \xrightarrow{j} \cdots \\
& & & \uparrow i & & & & \uparrow i & \\
\cdots \to & E^{p-1,q-1} & \xrightarrow{k} & D^{p,q-1} & \xrightarrow{j} & E^{p,q-1} & \xrightarrow{k} & D^{p+1,q-1} & \xrightarrow{j} \cdots \\
& & \uparrow & & & & \uparrow & &
\end{array}$$

Here each sequence formed by a step up i, two steps to the right j, k, a new step up i, and so on, is exact. In this description E_r^{pq} is a subquotient of E^{pq} obtained as a result of the factorization of $k^{-1}(\mathrm{Im}\, i^{r-1})$ by $j(\ker i^{r-1})$ (k^{-1} is a preimage under k).

c) Let $F^p K^\bullet$ be a decreasing filtration of a complex $K^\bullet$ as in III.7.5. Using the exact cohomology sequences corresponding to exact triples

$$0 \longrightarrow F^{p+1} K^\bullet \longrightarrow F^p K^\bullet \longrightarrow F^p K^\bullet / F^{p+1} K^\bullet \longrightarrow 0$$

construct a bigraded exact pair with $D^{pq} = H^{p+q}(F^p K^\bullet)$, $E^{pq} = H^{p+q}(F^p K^\bullet / F^{p+1} K^\bullet)$ and morphisms i, j, k of bidegrees $(-1, 1)$, $(0, 0)$, $(1, 0)$ respectively. Show that the spectral sequence corresponding to this pair coincides with the spectral sequence of the filtered complex $K^\bullet$ in III.7.5.

III.7 Derived Functor of the Composition. Spectral Sequence 217

4. More About the Spectral Sequence of a Filtered Complex. a) For any complex $K^\bullet$ define the filtration $F^p K^\bullet$ by

$$(F^p K^\bullet) = \begin{cases} 0 & \text{for } n < p, \\ \operatorname{Im} d^{p-1} & \text{for } n = p, \\ K^n & \text{for } n > p. \end{cases}$$

This filtration is similar to the canonical filtration in the sense that

$$H^n(F^p K^\bullet) = \begin{cases} 0 & \text{for } n < p, \\ H^n(K^\bullet) & \text{for } n \geq p. \end{cases}$$

Compute the spectral sequence associated with this filtration.

b) Verify that for any decreasing filtration $F^p K^\bullet$ the term E_1^{pq} of the associated spectral sequence is given by

$$E_1^{pq} = H^{p+q}(F^p K^\bullet / F^{p+1} K^\bullet).$$

In the next two exercises we consider the derived functor of the tensor product of modules. Let A be a ring, A-mod and mod-A the categories of left and right A-modules respectively.

5. Flat Modules. Prove that the class of flat modules (see II.6.8) is adapted to the functor $M \mapsto M \underset{A}{\otimes} N$ from A-mod to Ab. To do this prove first that if in an exact sequence

$$0 \longrightarrow M_1 \longrightarrow M_2 \longrightarrow M_3 \longrightarrow 0$$

of right A-modules M_2 and M_3 are flat, then M_1 is also flat.

6. Definition of $M^\bullet \overset{L}{\otimes} N^\bullet$. Let $M^\bullet \in \operatorname{Kom}^-(\text{mod-}A)$, $N^\bullet \in \operatorname{Kom}^-(A\text{-mod})$. We want to define an object $M^\bullet \overset{L}{\otimes} N^\bullet \in D^-(Ab)$ determining the functor

$$\bullet \overset{L}{\otimes} \bullet : D^-(\text{mod-}A) \times D^-(A\text{-mod}) \longrightarrow D^-(Ab).$$

a) Prove that $M^\bullet$ has a bounded from the right flat resolution $P^\bullet$ (i.e., $P^\bullet \in \operatorname{Kom}^-(\text{mod-}A)$ is quasi-isomorphic to $M^\bullet$, all P^i are flat). Similarly, $N^\bullet$ has a flat resolution $Q^\bullet \in \operatorname{Kom}^-(A\text{-mod})$.

b) Let $\alpha : \tilde{N}^\bullet \to N^\bullet$ be a quasi-isomorphism of bounded from the right complexes of left A-modules, $M^\bullet$ be a bounded from the right complex of right A-modules. Prove that $\beta = 1 \otimes \alpha : M^\bullet \otimes \tilde{N}^\bullet \to M^\bullet \otimes N^\bullet$ is a quasi-isomorphism. To do this consider two spectral sequences $\tilde{E}_r, E_r$ associated with filtrations on complexes $P^\bullet \otimes \tilde{N}^\bullet$ and $P^\bullet \otimes N^\bullet$ induces by the stupid filtration on $P^\bullet$ (see III.7.5). Prove that β induces a morphism of spectral sequences $\tilde{E}_r \to E_r$. Prove that the flatness of P^i implies

$$\tilde{E}_1^{ij} = P^i \otimes H^j(\tilde{N}^\bullet), \quad E_1^{ij} = P^i \otimes H^j(N^\bullet)$$

so that $\beta_1 : \tilde{E}_1^{ij} \to E_1^{ij}$ is an isomorphism. Hence
$$\beta_\infty : \tilde{E}_\infty^n = H^n(P^\bullet \otimes \tilde{N}^\bullet) \longrightarrow E_\infty^n = H^n(P^\bullet \otimes N^\bullet)$$
is also an isomorphism.

c) Define $M^\bullet \overset{L}{\otimes} N^\bullet$ for $M^\bullet$, $N^\bullet$ as in a) by the formula $M^\bullet \overset{L}{\otimes} N^\bullet = P^\bullet \otimes Q^\bullet = M^\bullet \otimes Q^\bullet = P^\bullet \otimes N^\bullet$ in $D^-(Ab)$. Prove that if $M^\bullet = \{M^0\}$ is a 0-complex, then $M^\bullet \overset{L}{\otimes} \bullet$ is the left derived functor for $M^0 \otimes \bullet$. A similar result holds for $\bullet \overset{L}{\otimes} N^\bullet$.

III.8 Sheaf Cohomology

1. Proposition. *Let X be a topological space, $\mathcal{R}$ a sheaf of unitary rings on X. Then any sheaf of $\mathcal{R}$-modules can be embedded into an injective sheaf of $\mathcal{R}$-modules.*

Proof. Let $\mathcal{F}$ be a sheaf of $\mathcal{R}$-modules (say, left). For any point $x \in X$ we can construct a monomorphism $\mathcal{F}_x \hookrightarrow I(x)$ of $\mathcal{R}_x$-modules, where $I(x)$ is injective over $\mathcal{R}_x$. Let us define now a sheaf of $\mathcal{R}$-modules $\mathcal{I}$ by
$$\mathcal{I}(U) = \prod_{x \in U} I(x), \qquad U \text{ open in } X$$
(with obvious restrictions). We have a canonical embedding $\mathcal{F} \to \mathcal{I}$. The injectivity of $\mathcal{I}$ would follow from the existence, for any sheaf $\mathcal{G}$ of $\mathcal{R}$-modules, of the canonical homomorphism
$$\text{Hom}_\mathcal{R}(\mathcal{G}, \mathcal{I}) = \prod_{x \in X} \text{Hom}_{\mathcal{R}_x}(\mathcal{G}_x, I(x)).$$
Denoting by $\mathcal{I}_x$ the stalk of $\mathcal{I}$ at a point x, we have the canonical homomorphism $v_x : \mathcal{I}_x \to I(x)$ of $\mathcal{R}_x$-modules, and the family $\{v_x\}$ determines a homomorphism of the left-hand side of the required identity into the right-hand side. We leave to the reader the proof that this homomorphism is an isomorphism. □

2. Direct Images and Cohomology

By Theorem III.6.12, Proposition III.8.1 enables us to construct the derived functor $R\text{Hom}(\mathcal{G}, \bullet) : D^+(\mathcal{R}\text{-mod}) \to D^+(Ab)$, as well as the derived functor for the direct image in the following situation. Let $(f, \varphi) : (X, \mathcal{R}_X) \to (Y, \mathcal{R}_Y)$ be a morphism of ringed spaces (II.6.16), where $\varphi : \mathcal{R}_Y \to f_\bullet(\mathcal{R}_X)$ is a morphism of sheaves of modules. Then for any $\mathcal{F} \in \mathcal{R}_X\text{-mod}$, φ determines the structure of an $\mathcal{R}_Y$-module on the sheaf $f_\bullet(\mathcal{F})$ and the functor $f_\bullet : \mathcal{R}_X\text{-mod} \to \mathcal{R}_Y\text{-mod}$ is left exact. Hence we can construct the right derived functor

III.8 Sheaf Cohomology

$$Rf_\bullet : D^+(\mathcal{R}_X\text{-mod}) \longrightarrow D^+(\mathcal{R}_Y\text{-mod}).$$

In particular, when Y is a point, we have $\mathcal{R}_Y = \mathbb{Z}$, $f_\bullet = \Gamma$, and the derived functor $R\Gamma : D^+(\mathcal{R}_X\text{-mod}) \to D^+(Ab)$ is the cohomology functor: $R^i\Gamma(\mathcal{F}) = H^i(X, \mathcal{F})$.

3. Theorem. *a) Let $\Phi : \mathcal{R}_X\text{-mod} \to SAb$ be the forgetful functor (of the structure of $\mathcal{R}_X$-module). Then the functors $R\Gamma$ and $R\Gamma \circ \Phi$ are naturally isomorphic. In other words, in computing $H^i(X, \mathcal{F})$ it does not matter whether we consider $\mathcal{F}$ as a $\mathcal{R}_X$-module or just a sheaf of abelian groups.*

b) Let $X = \cup U_i$ be an open covering with $H^q\left(U_{i_1} \cap \ldots \cap U_{i_p}, \mathcal{F}\right) = 0$ for all $q > 0$, $p \geq 1$ (such a covering is called $\mathcal{F}$-acyclic). Then $H^i(X, \mathcal{F})$ coincides with the i-dimensional cohomology of the Čech complex of this covering (see I.7.4).

c) $H^i(X, \mathcal{F}) = \mathrm{Ext}^i_{\mathcal{R}_X\text{-mod}}(\mathcal{R}_X, \mathcal{F})$.

d) Let $f : X \to Y$ be a mapping of topological spaces. Then $R^q f_\bullet(\mathcal{F})$ is naturally isomorphic to the sheaf associated with the presheaf $U \mapsto H^q\left(f^{-1}(U), \mathcal{F}\right)$.

e) Let $X \xrightarrow{f} Y \xrightarrow{g} Z$ be three topological spaces and two mappings, $\mathcal{F}$ be a sheaf of $\mathcal{R}_X$-modules. There exists a spectral sequence with $E_2^{pq} = R^p g_\bullet (R^q f_\bullet(\mathcal{F}))$, $E_\infty^{p+q} = R^{p+q}(gf)_\bullet(\mathcal{F})$; it is functorial in $\mathcal{F}$.

Proof (of Theorem III.8.3a). It is clear that $\Gamma = \Gamma \circ \Phi$ (as functors from $\mathcal{R}_X$-mod to Ab) so that $R\Gamma = R(\Gamma \circ \Phi)$. We will show that we can apply Theorem III.7.1 to get that $R(\Gamma \circ \Phi) = R\Gamma \circ R\Phi$. The required statement would follow because, by the exactness of Φ, $R\Phi$ coincides with the term-by-term action of Φ.

To apply Theorem III.7.1 we must prove that there exists a class of sheaves of $\mathcal{R}_X$-modules which is adapted to Φ and is transformed by Φ into a class of sheaves adapted to Γ. As the first class we choose injective $\mathcal{R}_X$-modules, and as the second class we choose flabby sheaves of abelian groups. Let us recall that a sheaf $\mathcal{F}$ is said to be *flabby* if the restriction maps $\Gamma(X, \mathcal{F}) \to \Gamma(U, \mathcal{F})$ are surjective for all *open* $U \subset X$ (see Ex. I.5.2). Let us verify all the required properties.

a) *Any sheaf of abelian groups is a subsheaf of a flabby sheaf. Any injective sheaf of $\mathcal{R}_X$-modules is flabby.*

Define a sheaf $\mathcal{CF}$ on X by $\mathcal{CF}(U) = \prod_{x \in U} \mathcal{F}_x$ (cf. the proof of Proposition III.8.1). It is clear that $\mathcal{CF}$ is a flabby sheaf and $\mathcal{F}$ is a subsheaf of $\mathcal{CF}$. If $\mathcal{F}$ is an injective sheaf of $\mathcal{R}_X$-modules then $\mathcal{CF}$ is also a sheaf of $\mathcal{R}_X$-modules and $\mathcal{F}$ splits off as a direct summand of $\mathcal{CF}$. By the definition of a flabby sheaf one immediately sees that a direct summand of a flabby sheaf is flabby.

b) Let

$$0 \longrightarrow \mathcal{F} \xrightarrow{\varphi} \mathcal{G} \xrightarrow{\psi} \mathcal{H} \longrightarrow 0 \tag{III.56}$$

be an exact sequence of sheaves of abelian groups with $\mathcal{F}$ being flabby. Then the sequence

$$0 \longrightarrow \Gamma(X,\mathcal{F}) \xrightarrow{\Gamma(\varphi)} \Gamma(X,\mathcal{G}) \xrightarrow{\Gamma(\psi)} \Gamma(X,\mathcal{H}) \longrightarrow 0 \qquad (\text{III.57})$$

is also exact.

The functor $\Gamma(X,\bullet)$ being left exact, it remains to prove that $\Gamma(\psi):\Gamma(X,\mathcal{G}) \to \Gamma(X,\mathcal{H})$ is an epimorphism. Let $s \in \Gamma(X,\mathcal{H})$ and let us consider the set E of pairs (U,t), where $U \subset X$ is an open set and $t \in \Gamma(U,\mathcal{G})$ is a section satisfying $\psi(t) = s|_U$. Define a partial ordering on E as follows: $(U',t') \leq (U'',t'')$ if $U' \subset U''$ and $t' = t''|_{U'}$. Let (U,t) be a maximum element in E. We prove that $U = X$. Indeed, let $U \neq X$ and $x \in X\setminus U$. By the surjectivity of ψ there exists a neighbourhood V of the point x and a section $t_1 \in \Gamma(V,\mathcal{G})$ such that $\psi(t_1) = s_V$. By the exactness of (III.56), on $U \cap V$ we have $t|_{U\cap V} - t_1|_{U\cap V} = \varphi(r)$ for some $r \in \Gamma(U \cap V,\mathcal{F})$. By the flabbiness of $\mathcal{F}$, there exists an extension r_1 of r to a section on X. Letting $t_2 = t_1 + \varphi(r_1|_V)$ we have $t|_{U\cap V} = t_2|_{U\cap V}$. Hence there exists $\bar{t} \in \Gamma(U \cup V,\mathcal{G})$ with $\bar{t}|_U = t$, $\bar{t}|_V = t_2$ so that $(U,t) < (U \cup V, \bar{t})$ and (U,t) is not a maximal element.

c) *If in (III.56) the sheaves $\mathcal{F}$ and $\mathcal{G}$ are flabby, then $\mathcal{H}$ is also flabby.*

By b), any section s of the sheaf $\mathcal{H}$ on an open set $U \subset X$ is of the form $s = \psi(t)$. As t can be extended to X, s can also be extended to X.

d) *Γ maps a bounded from the left acyclic complex of flabby sheaves into an acyclic complex of abelian groups.*

Let $0 \longrightarrow \mathcal{F}^0 \xrightarrow{d^0} \mathcal{F}^1 \longrightarrow \ldots$ be an acyclic complex of flabby sheaves. Let $\mathcal{Z}^i = \ker d^i = \operatorname{Im} d^{i-1}$. Then the sequences $0 \longrightarrow \mathcal{Z}^i \longrightarrow \mathcal{F}^i \longrightarrow \mathcal{Z}^{i+1} \longrightarrow 0$ are exact and the induction by i, together with c), shows that all $\mathcal{Z}^i$ are flabby (note that $\mathcal{Z}^0 = 0$ is obviously flabby). By b), the sequence

$$0 \longrightarrow \Gamma(X,\mathcal{Z}^i) \longrightarrow \Gamma(X,\mathcal{F}^i) \longrightarrow \Gamma(X,\mathcal{Z}^{i+1}) \longrightarrow 0$$

is also exact, so that $\Gamma(X,\mathcal{Z}^i) = \ker\bigl(\Gamma(d^i)\bigr) = \operatorname{Im}\bigl(\Gamma(d^{i-1})\bigr)$ and part d) is proved. □

Proof (of Theorem III.8.3b)). Let $\check{C}^\bullet(\mathcal{U},\mathcal{F})$ be the Čech complex associated with the sheaf $\mathcal{F}$ with respect to an open covering $\mathcal{U}$ of X, so that

$$\check{C}^k(\mathcal{U},\mathcal{F}) = \oplus_I \mathcal{F}(U_I),$$

where the sum is taken over all families of $k+1$ indices $I = \{i_0,\ldots,i_k\}$, $U_I = U_{i_0} \cap \ldots \cap U_{i_k}$.

For any open set $V \subset X$ denote by $\mathcal{U} \cap V$ the covering of V by the sets $U_i \cap V$. Let $\check{C}^\bullet(\mathcal{U} \cap V, \mathcal{F}|_V)$ be the Čech complex associated with the restriction of the sheaf $\mathcal{F}$ to V, and with the covering $\mathcal{U} \cap V$. There are natural restriction maps to smaller open sets $V' \subset V$ that commute with differentials in the Čech complexes and make $\check{C}^\bullet(\mathcal{U} \cap V, \mathcal{F}|_V)$ a complex of sheaves on X, which we denote by $\check{C}^\bullet(\mathcal{U},\mathcal{F})$.

First of all, we prove that a natural morphism of sheaves $\mathcal{F} \to \check{C}^0(\mathcal{U},\mathcal{F})$ makes $\check{C}^\bullet(\mathcal{U},\mathcal{F})$ into a resolution of $\mathcal{F}$. Indeed, we must prove that any $x \in X$ has an open neighbourhood V such that the sequence of abelian groups

$$\mathcal{F}(V) \longrightarrow \check{C}^0(\mathcal{U} \cap V, \mathcal{F}|_V) \longrightarrow \check{C}^1(\mathcal{U} \cap V, \mathcal{F}|_V) \longrightarrow \cdots \qquad (*)$$

is exact. This can be proved locally with respect to $x \in X$, so that we can assume that some element of the covering $\mathcal{U}$, say U_0, coincides with X. Then $U_{I \cup 0} = U_I$ for $0 \notin I$. For $p > 0$ define $h_p : \check{C}^p(\mathcal{U}, \mathcal{F}) \to \check{C}^{p-1}(\mathcal{U}, \mathcal{F})$ by the formula

$$(h_p(\varphi))_I = \begin{cases} \varphi_{I \cup 0} & \text{if } 0 \notin I, \\ 0 & \text{if } 0 \in I \end{cases}$$

for $\varphi = \oplus \varphi_I \in \check{C}^p(\mathcal{U}, \mathcal{F}) = \oplus_I \mathcal{F}(U_I)$. One can easily verify that $\{h_p\}$ is a homotopy between the zero and identity morphisms of the complex $\mathcal{F} \to \check{C}^\bullet(\mathcal{U}, \mathcal{F})$. This proves that $(*)$ is indeed a resolution of the sheaf $\mathcal{F}$.

The above result yields canonical maps of cohomology groups $\check{H}^q(X, \mathcal{F}) \to H^q(X, \mathcal{F})$ induced by a morphism of the resolution from $(*)$ to any injective resolution of $\mathcal{F}$. One can easily see that these maps commute with morphisms of sheaves $\mathcal{F} \to \mathcal{F}'$.

We prove that under the condition b) of Theorem III.8.3, these morphisms are isomorphisms. Let us consider a flabby resolution

$$\mathcal{F} \longrightarrow \mathcal{K}^0 \xrightarrow{d_K} \mathcal{K}^1 \longrightarrow \cdots \qquad (**)$$

of $\mathcal{F}$. Construct the double complex $L^{pq} = \check{C}^q(\mathcal{K}^p)$, with the differential d_I induced by differential d_K in the above flabby resolution and the differential d_{II} induced by the differential in the Čech complexes. Since all sheaves $\check{C}^q(\mathcal{K}^p)$ are flabby (this follows from Ex. I.5.2h)) and $\check{C}^\bullet(\mathcal{K}^p)$ is a resolution of $\mathcal{K}^p$ for any p, the complex of groups $\check{C}^\bullet(\mathcal{K}^p)$ is exact for any p. Therefore, the term $^{II}E_1^{pq}$ for the second spectral sequence of the bicomplex L^{pq} vanishes for $q > 0$. Hence, this spectral sequence degenerates and the cohomology groups of the complex associated with the bicomplex $\check{C}^q(\mathcal{K}^p)$ are isomorphic to $^{II}E_2^{n0}$. But $^{II}E_2^{n0}$ is the cohomology of the complex obtained by applying the global section functor $\Gamma(X, \bullet)$ to the flabby resolution $(**)$ of $\mathcal{F}$, i.e., to $H^n(X, \mathcal{F})$.

Now we consider the first spectral sequence of the bicomplex L^{pq}. Its term $^I E_1^{pq}$ is the row cohomology of our bicomplex, i.e.,

$$^I E_1^{pq} = \bigoplus_{I, |I|=q+1} H^p(U_1, \mathcal{F}).$$

By the condition of the theorem, $^I E_1^{pq} = 0$ for $p > 0$ and $^I E_1^{0q} = \mathcal{F}(U_I)$ (since each sheaf $\mathcal{F}|_{U_I}$ is flabby). Therefore, the first spectral sequence also degenerates, and $^I E_2^{0n} = \check{H}^q(X, \mathcal{F}) = E_\infty^q = H^q(X, \mathcal{F})$. $\square$

Proof (of Theorem III.8.3c)). By III.6.15,

$$\operatorname{Ext}^i_{\mathcal{R}_X\text{-mod}}(\mathcal{R}_X, \mathcal{F}) = R^i \operatorname{Hom}_{\mathcal{R}_X\text{-mod}}(\mathcal{R}_X, \mathcal{F})$$
$$= R^i \Gamma(X, \mathcal{F}) = H^i(X, \mathcal{F}). \qquad \square$$

Proof (of Theorem III.8.3d)). Let us consider an injective resolution $\mathcal{I}^\bullet$ of the sheaf $\mathcal{F}$. By definition, $R^q f_\bullet(\mathcal{F}) = H^q(f_\bullet(\mathcal{I}^\bullet))$. Next, by definition of $f_\bullet$, this sheaf is associated to the presheaf $U \mapsto H^q(f^{-1}(U), \mathcal{I}^\bullet)$. But $H^q(f^{-1}(U), \mathcal{I}^\bullet) = H^q(f^{-1}(U), \mathcal{F})$ because $\mathcal{I}^\bullet$ is an injective resolution. □

Proof (of Theorem III.8.3e)). By the statement a) of the theorem it suffices to prove the statement e) in the category SAb of sheaves of abelian groups. We apply Theorem III.7.7 to the pair of functors $F = f_\bullet$, $G = g_\bullet$ taking for $\mathcal{I}_B$ the class of all injective sheaves on Y. This is possible because $f_\bullet$ maps injective sheaves on X into injective sheaves on Y. Indeed, injective sheaves $\mathcal{F}$ can be characterized by the property that $\mathrm{Hom}(\mathcal{G}', \mathcal{F}) \to \mathrm{Hom}(\mathcal{G}'', \mathcal{F})$ is an epimorphism for any monomorphism of sheaves $\mathcal{G}'' \to \mathcal{G}'$. But, by Proposition II.6.17, $\mathrm{Hom}(\mathcal{G}', f_\bullet \mathcal{F}) = \mathrm{Hom}(f^\bullet \mathcal{G}, \mathcal{F})$ and, by II.6.19, $f^\bullet$ is an exact functor. So $f^\bullet$ maps monomorphisms into monomorphisms, and $f_\bullet \mathcal{F}$ is an injective sheaf whenever $\mathcal{F}$ is. □

4. Tensor Products and Flat Sheaves

Let $\mathcal{R} = \mathcal{R}_X$ be a sheaf of unitary rings on a topological space X. For any sheaf of left $\mathcal{R}$-modules $\mathcal{N}$ we can define a functor

$$\bullet \otimes \mathcal{N} : \mathrm{mod}\text{-}\mathcal{R} \longrightarrow SAb, \qquad \mathcal{F} \mapsto \mathcal{F} \underset{\mathcal{R}}{\otimes} \mathcal{N}.$$

Similarly to the corresponding statement for modules over a ring (see II.6.6), one can prove that this functor is right exact.

A sheaf $\mathcal{N}$ is said to be *flat* (over $\mathcal{R}$) if the functor $\bullet \otimes \mathcal{N}$ is exact. The reader can easily verify that $\mathcal{N}$ is flat if and only if its stalk $\mathcal{N}_x$ at any point $x \in X$ is a flat module over the ring $\mathcal{R}_x$. In particular, for any open $U \subset X$ the sheaf $\mathcal{R}_U$ defined by

$$(\mathcal{R}_U)_x = \mathcal{R}_x \quad \text{if} \quad x \in U,$$
$$(\mathcal{R}_U)_x = 0 \quad \text{if} \quad x \notin U$$

(see also II.6.7) is flat.

5. Proposition. *a) Any sheaf of left $\mathcal{R}$-modules is a quotient of a flat sheaf.*
b) The class of all flat sheaves is adapted to the functor

$$\mathcal{M} \otimes \bullet : \mathcal{R}\text{-mod} \longrightarrow SAb$$

of tensoring with an arbitrary sheaf $\mathcal{M}$ of right $\mathcal{R}$-modules.

Proof. We prove that for any sheaf of left S-modules $\mathcal{F}$ there exists a collection of open sets $(U_i)_{i \in I}$ and an epimorphism $\underset{i}{\oplus} \mathcal{R}_{U_i} \to \mathcal{F}$. This would imply part a) because a direct sum of flat sheaves is flat.

First of all, we have
$$\mathrm{Hom}_{\mathcal{R}\text{-mod}}(\mathcal{R}, \mathcal{F}) = \Gamma(X, \mathcal{F})$$
($\varphi : \mathcal{R} \to \mathcal{F}$ corresponds to $\varphi(\mathbf{1}) \in \Gamma(X, \mathcal{F})$, where $\mathbf{1}$ is the unit element of the ring $\Gamma(X, \mathcal{R})$). Similarly, for any open $U \subset X$ we have
$$\mathrm{Hom}_{\mathcal{R}\text{-mod}}(\mathcal{R}_U, \mathcal{F}) = \Gamma(U, \mathcal{F}).$$
Now let us fix a family of pairs $(U_i, a_i \in \Gamma(U_i, \mathcal{F}))$ with the property that for any point $x \in X$ the stalk $\mathcal{F}_x$ is generated (as a $\mathcal{R}_x$-module) by the images in $\mathcal{F}_x$ of those a_i that $x \in U_i$. It is clear that the morphism
$$\bigoplus_{i \in I} \mathcal{R}_{U_i} \longrightarrow \mathcal{F}$$
determined by the sections $a_i \in \Gamma(U_i, \mathcal{F})$ is an epimorphism.

Let us note also that we have actually proved the existence, for any $\mathcal{F} \in \mathcal{R}$-mod, of the resolution
$$\ldots \longrightarrow \mathcal{L}^{-n} \longrightarrow \ldots \longrightarrow \mathcal{L}^0 \longrightarrow \mathcal{F} \longrightarrow 0$$
in which all terms are direct sums of the form $\oplus \mathcal{R}_{U_i}$.

Part b) of the proposition can be reduced to the corresponding statement for modules over rings (see Ex. III.7.4) by using the characterization of flat sheaves in terms of their stalks, see III.8.4. □

Similar results are, of course, true for sheaves of right $\mathcal{R}$-modules.

6. Inverse Images and Tensor Products

By Proposition 5, we can construct the left derived functor
$$\mathcal{M} \overset{L}{\otimes} \bullet : D^-(\mathcal{R}\text{-mod}) \longrightarrow D^-(SAb).$$
Its cohomology sheaves are denoted by Tor:
$$\mathrm{Tor}_i(\mathcal{M}, \mathcal{N}) = H^i(\mathcal{M} \overset{L}{\otimes} \mathcal{N}).$$
We can construct also the functor
$$\mathcal{M}^\bullet \overset{L}{\otimes} \bullet : D^-(\mathcal{R}\text{-mod}) \longrightarrow D^-(SAb)$$
for $\mathcal{M}^\bullet \in D^-(\text{mod-}\mathcal{R})$.

In the similar way we can define the functors
$$\bullet \overset{L}{\otimes} \mathcal{N}, \quad \bullet \overset{L}{\otimes} \mathcal{N}^\bullet : D^-(\text{mod-}\mathcal{R}) \longrightarrow D^-(SAb)$$
for $\mathcal{N} \in \mathcal{R}\text{-mod}$, $\mathcal{N}^\bullet \in D^-(\mathcal{R}\text{-mod})$, and the bifunctor
$$\bullet \overset{L}{\otimes} \bullet : D^-(\text{mod-}\mathcal{R}) \times D^-(\mathcal{R}\text{-mod}) \longrightarrow D^-(SAb).$$

Similarly to Ex. III.7.6 one can prove that the sheaf $\text{Tor}_i(\mathcal{M},\mathcal{N})$ does not depend on whether we define it using $\mathcal{M}\overset{L}{\otimes}\bullet$, or $\bullet\overset{L}{\otimes}\mathcal{N}$, or $\bullet\overset{L}{\otimes}\bullet$. If $\mathcal{R}$ is a sheaf of commutative (or supercommutative) rings, then left modules can be identified with right modules and $\mathcal{M}\otimes\mathcal{N}$ has the structure of an $\mathcal{R}$-module, so that $\mathcal{M}\overset{L}{\otimes}\bullet$ takes values in $D^-(\mathcal{R}\text{-mod})$.

Let now $(f,\varphi):(X,\mathcal{R}_X)\to(Y,\mathcal{R}_Y)$ be a morphism of ringed spaces with (super)commutative structure sheaves. Then for any sheaf of $\mathcal{R}_Y$-modules $\mathcal{F}$ we can define a sheaf of $\mathcal{R}_X$-modules

$$f^*(\mathcal{F}) = \mathcal{R}_X \underset{f^\bullet(\mathcal{R}_Y)}{\otimes} f^\bullet(\mathcal{F}).$$

The corresponding left derived functor

$$Lf^*(\mathcal{F}) = \mathcal{R}_X \underset{f^\bullet(\mathcal{R}_Y)}{\overset{L}{\otimes}} f^\bullet(\mathcal{F})$$

provides higher inverse image functors:

$$L_i f^*(\mathcal{F}) = H^{-i}\left(\mathcal{R}_X \underset{f^\bullet(\mathcal{R}_Y)}{\overset{L}{\otimes}} f^\bullet(\mathcal{F})\right).$$

A morphism (f,φ) is said to be *flat* if $\mathcal{R}_X$ is a flat $f^\bullet(\mathcal{R}_Y)$-module. This property is one of the weakest and, at the same time, one of the most useful algebraic analogues of the geometrical notion of a "locally trivial fibration". It is widely used in algebraic and analytic geometry.

7. Higher Direct Images with Compact Support

Up to the end of this section we will consider only sheaves of abelian groups on locally compact topological spaces satisfying the first countability axiom. Later we will impose some additional conditions about the finiteness of the dimension of involved spaces. They will not, however, be too restrictive; in particular, they hold for all topological manifolds. In this situation, with any map $f:X\to Y$ we will associate functors $Rf_!$ and $f^!$ on appropriate categories.

8. Definition (– Lemma). *Let $f:X\to Y$ be a morphism of locally compact topological spaces and $\mathcal{F}$ a sheaf on X. For any open $U\subset X$ let*

$$f_!(\mathcal{F})(U) = \left\{s\in\Gamma\left(f^{-1}(U),\mathcal{F}\right),\ \text{supp}(s)\xrightarrow{f} U\ \text{is proper}\right\}.$$

(Recall that a morphism is said to be proper *if the preimage of any compact set is compact.) Then:*

a) $f_!(\mathcal{F})$ is a subsheaf of $f_\bullet(\mathcal{F})$.

b) The map $\mathcal{F}\to f_!(\mathcal{F})$ can be extended to a left exact functor called the direct image with compact support.

Proof. a) It is clear that $f_!(\mathcal{F})$ is a subpresheaf of the sheaf $f_\bullet(\mathcal{F})$. It is clear also that any set of compatible sections of $f_!(\mathcal{F})$ can be glued together into a unique section of $f_\bullet(\mathcal{F})$. It remains to prove that this section belongs to $f_!(\mathcal{F})$. In other words, we must verify that for any family (U_i) of open subsets of Y the following condition is satisfied: if $s \in \Gamma(U_i, \mathcal{F})$ and $V_i = \operatorname{supp} s_i \to U_i$ are proper maps, then $\cup V_i \to \cup U_i$ us also a proper map. To prove this let K be a compact set in $\cup U_i$. Let us choose a finite subcovering $K \subset \bigcup_{j \in J} U_j$ and consider a finite compact covering $K = \bigcup_{j \in J} K_j$ with $K_j \subset U_j$. Then $f^{-1}(K_j) \cap V_j$ are compact sets so that

$$f^{-1}(K) \cap \left(\bigcup_i V_i\right) = f^{-1}(K) \cap \left(\bigcup_{j \in J} V_j\right) = \bigcup_{j \in J} \left(f^{-1}(K) \cap V_j\right)$$

is also compact.

b) The functoriality of $f_!$ follows from the fact that under a morphism $\varphi : \mathcal{F} \to \mathcal{G}$ of sheaves the support of a section cannot increase. Finally, the left exactness of $f_!$ follows from the left exactness of $f_\bullet$ and from definitions. $\square$

9. Sections with Compact Support

An important particular case of the above situation appears when $f : X \to \mathrm{pt}$ is the mapping to a point. In this case $f_!(\mathcal{F})$ is an abelian group formed by all sections $s \in \Gamma(X, \mathcal{F})$ such that $\operatorname{supp} s$ is a compact in X. This group is called the group of sections of $\mathcal{F}$ with compact support and is denoted by $\Gamma_c(X, \mathcal{F})$.

For an arbitrary $f : X \to Y$ the sheaf $f_!(\mathcal{F})$ can be, in some sense, recovered from the groups of compactly supported sections of $\mathcal{F}$ over various subsets of X. More explicitly, we have the following:

10. Proposition. *The stalk of the sheaf $f_!(\mathcal{F})$ at a point $y \in Y$ is isomorphic to $\Gamma_c\left(f^{-1}(y), \mathcal{F}|_{f^{-1}(y)}\right)$.*

Proof. Let us construct, first of all, a homomorphism

$$\varphi : f_!(\mathcal{F})_y \longrightarrow \Gamma_c\left(f^{-1}(y), \mathcal{F}|_{f^{-1}(y)}\right).$$

Let $s \in f_!(\mathcal{F})_y$, U a neighbourhood of y, and $t \in \Gamma(U, f_!(\mathcal{F}))$ a representative of s, so that $t \in \Gamma\left(f^{-1}(U), \mathcal{F}\right)$ is a section of $\mathcal{F}$ over $f^{-1}(U)$ and the map $\operatorname{supp} t \to U$ is proper. It is clear that $t|_{f^{-1}(y)}$ lies in $\Gamma_c\left(f^{-1}(y), \mathcal{F}|_{f^{-1}(y)}\right)$ (because $\operatorname{supp}\left(t|_{f^{-1}(y)}\right) = (\operatorname{supp} t) \cap f^{-1}(y)$). One can easily verify that the resulting element of $\Gamma_c\left(f^{-1}(y), \mathcal{F}|_{f^{-1}(y)}\right)$ depends only on s and not on the choices involved in its construction. We define $\varphi(s) = t|_{f^{-1}(y)}$.

226   III. Derived Categories and Derived Functors

Let us prove that φ is injective. Let $\varphi(s) = 0$. Then $t|_{f^{-1}(y)} = 0$, i.e., (supp t) $\cap f^{-1}(y) = \emptyset$, so that $y \notin f(\text{supp } t)$. Moreover, $f|_{\text{supp } t}$ is a proper map of locally compact spaces, so that $f(\text{supp } t)$ is closed in Y. Hence $s = 0$.

Let us prove that φ is surjective. Let $U_1 \supset U_2 \subset \ldots$ be a sequence of open subsets in Y with $\cap \bar{U}_i = \{y\}$. Then $\cap f^{-1}(U_i) = f^{-1}(y)$ and, X being locally compact,
$$\Gamma_c\left(f^{-1}(y), \mathcal{F}|_{f^{-1}(y)}\right) = \varinjlim A_i,$$
where $A_i = \{$the group of sections $t \in \Gamma\left(f^{-1}(U_i), \mathcal{F}\right)$ with supp $t = K \cap f^{-1}(U_i)$ for some compact subset $K \subset X\}$ (verify this!). On the other hand,
$$(f_! \mathcal{F})_y = \varinjlim B_i$$
where $B_i = \{$the group of sections $t \in \Gamma\left(f^{-1}(U_i), \mathcal{F}\right)$ such that supp $t \to U_i$ is a proper map$\}$. It is clear that A_i is a subgroup of B_i for any i. Hence φ is surjective. □

11. Sheaves Adapted to $f_!$

A sheaf $\mathcal{F}$ on X is said to be *soft* if for any closed $K \subset X$ the restriction mapping $\Gamma(X, \mathcal{F}) \to \Gamma(K, \mathcal{F})$ is surjective (for more details about soft sheaves see Ex. I.5.2 c–e).

Since any injective sheaf is flabby (see III.8.3a)), and any flabby sheaf is obviously soft, the class of soft sheaves is sufficiently large.

12. Proposition. *The class of soft sheaves is adapted to the functor $f_!$.*

Proof. By the previous remark and Ex. I.5.2c) it suffices to prove the following statement: if
$$0 \longrightarrow \mathcal{F} \xrightarrow{\varphi} \mathcal{G} \xrightarrow{\psi} \mathcal{H} \longrightarrow 0 \tag{III.58}$$
be an exact sequence of soft sheaves, then the sequence
$$0 \longrightarrow f_! \mathcal{F} \longrightarrow f_! \mathcal{G} \longrightarrow f_! \mathcal{H} \longrightarrow 0$$
is also exact.

Since $f_!$ is left exact, we must only prove the surjectivity of the last morphism in this sequence, i.e., the surjectivity of the mapping $(f_! \mathcal{G})_y \to (f_! \mathcal{H})_y$ for any $y \in Y$. The restriction of an exact sequence of soft sheaves to $f^{-1}(y)$ is again an exact sequence of soft sheaves. By Proposition III.8.10, it suffices to prove the following statement. *In an exact sequence of soft sheaves* (III.58), *the mapping* $\Gamma_c(X, \mathcal{G}) \to \Gamma_c(X, \mathcal{H})$ *is surjective.*

So, let $s \in \Gamma_c(X, \mathcal{H})$ and let K be a compact set containing supp s. Let us cover K by a finite number of compacts $K_1, \ldots, K_n$ such that $s|_{K_i}$ is obtained (by ψ) from some section $t_i \in \Gamma(K_i, \mathcal{G})$. Let $L_i = K_1 \cup \ldots \cup K_i$; we prove by induction on i that there exists a section $r_i \in \Gamma(L_i, \mathcal{G})$ with $\psi(r_i) = s|_{L_i}$. Let us assume that r_{i-1} is already constructed. Let

$$v = r_{i-1}|_{L_{i-1} \cap K_i} - t_i|_{L_{i-1} \cap K_i}.$$

We have $\psi(v) = 0$, so that $v = \varphi(v')$ for some $v' \in \Gamma(L_{i-1} \cap K_i, \mathcal{F})$. Let us extend v' to a section v'' of the sheaf $\mathcal{F}$ over K_i (using the softness of $\mathcal{F}$). Then the restrictions of $t'_i = t_i + \varphi(v'')$ and of r_{i-1} to $L_{i-1} \cap K_i$ coincide, so that they can be glued together to the required section r of the sheaf $\mathcal{G}$ over $L_i = L_{i-1} \cup K_i$.

So we have constructed a section $r \in \Gamma(K, \mathcal{G})$ with $\psi(r) = s$. Let M be the boundary of K. Then $\psi(r|_M) = 0$ so that $r|_M = \varphi(u)$ for some $u \in \Gamma(M, \mathcal{F})$. Since $\mathcal{F}$ is a soft sheaf, u can be extended to a section $u' \in \Gamma(K, \mathcal{F})$. Then $r' = r - \varphi(u')|_M$, i.e., r' can be extended by zero outside K, yielding $s' \in \Gamma_c(X, \mathcal{F})$ with $\psi(s') = s$. □

Let us note that in the proof we have used only that $\mathcal{F}$ is a soft sheaf.

13. Higher Direct Images with Compact Support

The previous proposition enables us to define the right derived functor of $f_!$:

$$Rf_! : D^+(SAb_X) \longrightarrow D^+(SAb_Y).$$

Its cohomology sheaves are called *higher direct images with compact support* and denoted by $R^i f_!(\mathcal{F}) \in SAb_Y$ (for $\mathcal{F} \in SAb_X$).

In particular, for $f : X \to \text{pt}$ we get the functor

$$R\Gamma_c : D^+(SAb) \longrightarrow D^+(Ab)$$

and its cohomology $H^i_c(X, \mathcal{F})$ (cohomology of $\mathcal{F}$ with compact support).

Let us list some properties of $Rf_!$.

a) *The stalk of $R^i f_!(\mathcal{F})$ at a point $y \in Y$ is canonically isomorphic to $H^i_c(f^{-1}(y), \mathcal{F}|_{f^{-1}(y)})$*. This follows from Proposition III.8.10 and from the softness of the restriction of a soft sheaf to $f^{-1}(y)$.

b) *For continuous maps $f : X \to Y$, $g : Y \to Z$ we have*

$$R(gf)_! = Rg_! \circ Rf_!. \tag{III.59}$$

This follows from the fact that $f_!$ maps soft sheaves on X into soft sheaves on Y (see Ex. 1b)). Using the results of Sect. III.7 we can rewrite (III.59) as a spectral sequence relating $R^p f_!$, $R^q g_!$, and $R^{p+q}(gf)_!$.

14. Dimension

In general, the functor $Rf_!$ transforms bounded complexes of sheaves on X into unbounded complexes of sheaves on Y. However, for a wide class of spaces the boundedness property is preserved.

By the *dimension* $\dim_c X$ of a locally compact space X we shall mean the smallest n such that for any sheaf $\mathcal{F} \in SAb_X$ we have $H^i_c(X, \mathcal{F}) = 0$ for all $i > n$. We skip here proofs of the following properties of $\dim_c X$ (see Iversen [1]):

a) Let
$$0 \longrightarrow \mathcal{F} \longrightarrow \mathcal{L}^0 \longrightarrow \dots \longrightarrow \mathcal{L}^{n-1} \longrightarrow \mathcal{L}^n \longrightarrow 0$$
be an exact sequence of abelian groups on a locally compact topological space X with $\dim_c X \leq n$ and with $\mathcal{L}^0, \dots, \mathcal{L}^{n-1}$ soft. Then $\mathcal{L}^n$ is also soft.

b) $\dim_c \mathbb{R}^n = n$.

c) Let Y be an open or a closed subset of X. Then $\dim_c Y \leq \dim_c X$.

d) $\dim_c X$ can be computed locally: if each point $x \in X$ has a neighbourhood U with $\dim_c U \leq n$, then $\dim_c X \leq n$.

Properties b)–d) imply, in particular, that the majority of reasonable topological spaces X (in particular, topological manifolds and geometric realizations of finite-dimensional simplicial sets, see Sect. I.2) have finite dimension $\dim_c X$.

e) Let $f : X \to Y$ be a map of locally compact topological spaces with $\dim_c X \leq n$. Then $R^p f_! \mathcal{F} = 0$ for $p > n$ and for any $\mathcal{F} \in SAb_X$.

f) In the conditions of e) $Rf_!$ can be considered as a functor from $D^b(SAb_X)$ to $D^b(SAb_Y)$ and from $D^-(SAb_X)$ to $D^-(SAb_Y)$.

15. Inverse Image with Compact Support

By the general ideology (see II.6.17), for any map $f : X \to Y$ the inverse image $f^!$ with compact support should be defined as a functor from the category of sheaves on Y to the category of sheaves on X that is adjoint to $f_! : SAb_X \to SAb_Y$. However, for a general f the functor $f_!$ does not admit a right adjoint functor, and to define $f^!$ we have to pass to derived categories. Moreover, we have to assume X and Y to be finite-dimensional.

16. Theorem. *Let $f : X \to Y$ be a continuous map of locally compact finite-dimensional (in the sense of $\dim_c$) topological spaces. There exists a functor*
$$f^! : D^+(SAb_Y) \longrightarrow D^+(SAb_X)$$
and a functorial in $\mathcal{F}^\bullet \in (D^+(SAb_X))^0$, $\mathcal{G}^\bullet \in D^+(SAb_Y)$ isomorphism
$$R\operatorname{Hom}(Rf_! \mathcal{F}^\bullet, \mathcal{G}^\bullet) \cong R\operatorname{Hom}(\mathcal{F}^\bullet, f^! \mathcal{G}^\bullet) \qquad (\text{III.60})$$
in $D^+(Ab)$.

The proof of this theorem occupies almost all the remaining part of this section (III.8.18–III.8.24).

17. Corollary. *The functor $f^!$ is right adjoint to $Rf_!$.*

Proof. We have to apply $H^0 : D^+(Ab) \to Ab$ to both parts of (III.60). □

18. Comments on Theorem III.8.16 and the Plan of the Proof

For a sheaf $\mathcal{F}$ on X and an open set $U \subset X$ denote by $\mathcal{F}_U$ the extension of $\mathcal{F}$ by zero outside U (using the functors $j_!$, $j^\bullet$, where $j : U \to X$ is the embedding we can write $\mathcal{F}_U = j_! j^\bullet \mathcal{F}$). For two open sets $U \subset V$ we have a natural morphism of sheaves $\mathcal{F}_V \to \mathcal{F}_U$ which induces, for each sheaf $\mathcal{G}$ on Y, a homomorphism

$$\operatorname{Hom}(f_! \mathcal{F}_U, \mathcal{G}) \longrightarrow \operatorname{Hom}(f_! \mathcal{F}_V, \mathcal{G}). \tag{III.61}$$

It is clear that
$$U \mapsto \operatorname{Hom}(f_! \mathcal{F}_U, \mathcal{G}) \tag{III.62}$$

together with restriction mappings (III.61), defines a presheaf on X. If this presheaf were a sheaf, everything would be OK: denoting the sheaf $U \mapsto \operatorname{Hom}(f_! \mathbb{Z}_U, \mathcal{G})$ (where $\mathbb{Z}$ is the constant sheaf on X) by $f^? \mathcal{G}$, we would have

$$\operatorname{Hom}(f_! \mathcal{F}, \mathcal{G}) = \operatorname{Hom}(\mathcal{F}, f^? \mathcal{G})$$

(see Theorem III.8.19 below) and $f_!$ would possess the right adjoint functor. However, (III.62) is a sheaf only in some very special cases. More precisely,

$$U \mapsto \operatorname{Hom}(f_!(\mathcal{F}_U \otimes \mathcal{L}), \mathcal{G})$$

is a sheaf if $\mathcal{L}$ is a flat soft sheaf on X (Proposition III.8.20). This is the reason why we have to work in derived categories: we have to replace the constant sheaf $\mathbb{Z}$ by its resolution

$$0 \longrightarrow \mathbb{Z} \longrightarrow \mathcal{L}^\varkappa \longrightarrow \ldots \longrightarrow \mathcal{L}^{\varkappa-\varkappa} \longrightarrow \mathcal{L}^\varkappa \longrightarrow \varkappa$$

consisting of flat soft sheaves (see III.8.22), and then the functor $f^!$ can be rather easily constructed in the derived category.

As a first step in the proof of Theorem III.8.16 we describe representable functors on the category SAb_X of abelian groups on a topological space X.

19. Theorem. *A functor $F : SAb_X \to (Ab)^0$ is representable if and only if it transforms inductive limits in SAb_X into projective limits in Ab.*

Proof. The "only if" part of the theorem is true in the general situation: for an arbitrary abelian category $\mathcal{A}$ a functor of the form $X \mapsto \operatorname{Hom}_\mathcal{A}(X, Y)$ transforms inductive limits in $\mathcal{A}$ into projective limits in Ab. To prove the "if" part let us note that
$$U \mapsto F(\mathbb{Z}_U),$$

together with restriction mappings $F(\mathbb{Z}_U) \to F(\mathbb{Z}_V)$ for $U \subset V$ induced by embeddings $\varphi_{UV} : \mathbb{Z}_U \hookrightarrow \mathbb{Z}_V$, determines a presheaf $\mathcal{G}$ of abelian groups on X such that $\Gamma(U, \mathcal{G}) = F(\mathbb{Z}_U)$. Let us prove that $\mathcal{G}$ is a sheaf. Let (U_i) be a family of open subsets in X and $U = \cup U_i$. There exists an exact sequence of sheaves

$$\bigoplus_{i,j} \mathbb{Z}_{U_i \cap U_j} \xrightarrow{\alpha} \bigoplus_i \mathbb{Z}_{U_i} \xrightarrow{\beta} \mathbb{Z}_U \longrightarrow 0$$

where $\alpha = \bigoplus_{i,j} \left(\varphi_{U_i \cap U_j, U_i} - \varphi_{U_i \cap U_j, U_j}\right)$, $\beta = \bigoplus_i \varphi_{U_i, U}$. By the condition of the theorem the functor F is right exact and transforms direct sums in SAb_X into direct products in Ab. Hence it preserves the exactness of this sequence and we get the following exact sequence:

$$0 \longrightarrow F(\mathbb{Z}_U) \xrightarrow{F(\beta)} \prod_i F(\mathbb{Z}_{U_i}) \xrightarrow{F(\alpha)} \prod_{i,j} F(\mathbb{Z}_{U_i \cap U_j})$$

or

$$0 \longrightarrow \Gamma(U, \mathcal{G}) \longrightarrow \prod_i \Gamma(U_i, \mathcal{G}) \longrightarrow \prod_{i,j} \Gamma(U_i \cap U_j, \mathcal{G})$$

and this means that $\mathcal{G}$ is a sheaf.

Now let us construct a functorial in $\mathcal{F}$ isomorphism

$$\mathrm{Hom}(\mathcal{F}, \mathcal{G}) \cong F(\mathcal{F}) \tag{III.63}$$

To do this we will specify the element $e \in F(\mathcal{F})$ which would correspond to the identity morphism $\mathcal{G} \to \mathcal{G}$, and then map an arbitrary morphism $f: \mathcal{F} \to \mathcal{G}$ into $F(f)e \in F(\mathcal{F})$. It is clear that in such a way we get a homomorphism of the form (III.63) which is functorial in $\mathcal{F}$, and we will have to prove only that it is an isomorphism.

To construct e we introduce the following category I:

$$\mathrm{Ob}\, I \;=\; \{\text{pairs } (U, a_U),\; U \subset X,\; a_U \in F(\mathbb{Z}_U)\},$$

$$\mathrm{Hom}_I((U, a_U), (V, a_V)) \;=\; \{\text{morphisms of sheaves } f: \mathbb{Z}_U \to \mathbb{Z}_V \text{ such that } F(f)a_V = a_U\}.$$

Let $G: I \to SAb_X$ be a functor that associates with a pair (U, a_U) the sheaf $\mathbb{Z}_U$ (with natural action on morphisms). From $\mathrm{Hom}_{SAb_X}(\mathbb{Z}_U, \mathcal{G}) = \Gamma(U, \mathcal{G}) = F(\mathbb{Z}_U)$ we see that a_U determines a morphism of sheaves $a_U: \mathbb{Z}_U \to \mathcal{G}$. The family $\{a_U\}$ for all $(U, a_U) \in \mathrm{Ob}\, I$ forms a morphism of the functor G into the functor $\mathrm{Const}_\mathcal{G}: I \to SAb_X$ taking the constant value $\mathcal{G}$. This gives us a morphism

$$a: \varinjlim G \longrightarrow \varinjlim \mathrm{Const}_\mathcal{G} = \mathcal{G}$$

and one can easily see that it is an isomorphism.

Applying to a the functor F we get an isomorphism

$$F(a): F(\mathcal{G}) \longrightarrow F(\varinjlim G) = \varprojlim(F \circ G).$$

Let us construct now $e \in F(\mathcal{G})$ as follows. For $(U, a_U) \in \mathrm{Ob}\, I$ we have an element $a_U \in (F \circ G)(U, a_U) = F(\mathbb{Z}_U)$ and for $f: (U, a_U) \to (V, a_V)$ we have $(F \circ G)(f)(a_V) = a_U$. By the universal property of $\varprojlim$, the collection $\{a_U, (U, a_U) \in \mathrm{Ob}\, I\}$ determines a unique element $e \in \varprojlim(F \circ G) = F(\mathcal{G})$.

Let us show that the homomorphism $\operatorname{Hom}(\mathcal{F}, \mathcal{G}) \to F(\mathcal{F}) : f \mapsto F(f)(e)$ is an isomorphism for any $\mathcal{F}$. By the construction, e has the following property: for any U and any $f : \mathbb{Z}_U \to \mathcal{G}$ the element $F(f)(e) \in F(\mathbb{Z}_U)$ corresponds to f under the isomorphism $\operatorname{Hom}(\mathbb{Z}_U, \mathcal{G}) = \Gamma(U, \mathcal{G}) = F(\mathbb{Z}_U)$. Hence (III.63) is an isomorphism in the case when $\mathcal{F} = \mathbb{Z}_U$ for some open $U \subset X$. Next, any sheaf $\mathcal{F}$ on X is an inductive limit of sheaves of the form $\mathbb{Z}_U$ (more precisely, an inductive limit of a functor taking values $\mathbb{Z}_U$; this can be proved in the same way as in the construction of the morphism a above using an appropriate category $I(\mathcal{F})$). Since f transforms inductive limits into projective ones, (III.63) is an isomorphism for any $\mathcal{F}$. □

Let us show that this theorem can be applied to functors we are interested in.

20. Proposition. *Let $\mathcal{L}$ be a flat soft sheaf on X, $\mathcal{G}$ an arbitrary sheaf on Y. The functor*

$$\mathcal{F} \mapsto \operatorname{Hom}(f_!(\mathcal{L} \otimes \mathcal{F}), \mathcal{G}) \quad \text{from} \quad SAb_X \quad \text{to} \quad (Ab)^0$$

transforms inductive limits in SAb_X into inductive limits in $(Ab)^0$ (i.e., into projective limits in Ab).

Proof. We prove that the functor

$$f_!^{\mathcal{L}} : \mathcal{F} \mapsto f_!(\mathcal{L} \otimes \mathcal{F})$$

commutes with inductive limits. Proposition III.8.20 would follow from this fact and from the properties of the functor $\operatorname{Hom}(\bullet, \mathcal{G})$ (see the beginning of the proof of Theorem III.8.19). By Theorem II.3.20 it suffices to check that $f_!^{\mathcal{L}}$ maps cokernels into cokernels (i.e., it is right exact) and direct sums into direct sums. The second statement is clear because both the tensor product and $f_!$ commute with direct sums.

Let us prove that $f_!^{\mathcal{L}}$ is right exact. We claim that $f_!^{\mathcal{L}}$ is even exact. Indeed, $\mathcal{F}$ has a resolution

$$\ldots \xrightarrow{d^{-2}} \mathcal{E}^{-1} \xrightarrow{d^{-1}} \mathcal{E}^0 \longrightarrow \mathcal{F} \longrightarrow 0$$

with all terms of the form $\oplus \mathbb{Z}_{U_j}$ (see III.8.5). Tensoring this resolution with $\mathcal{L}$ we obtain the exact (because $\mathcal{L}$ is flat) sequence

$$\ldots \xrightarrow{d^{-2}} \mathcal{L} \otimes \mathcal{E}^{-1} \xrightarrow{d^{-1}} \mathcal{L} \otimes \mathcal{E}^0 \longrightarrow \mathcal{L} \otimes \mathcal{F} \longrightarrow 0,$$

with all terms of the form $\mathcal{L} \otimes (\oplus \mathbb{Z}_{U_j}) = \oplus \mathcal{L}_{U_j}$, so that all of them are soft (because $\mathcal{L}$ is soft). Let $n \geq \dim_c X$. Let us consider the following acyclic complex:

$$0 \longrightarrow \ker(I \otimes d^{-n}) \longrightarrow \mathcal{L} \otimes \mathcal{E}^{-n} \longrightarrow \ldots \longrightarrow \mathcal{L} \otimes \mathcal{E}^0 \longrightarrow \mathcal{L} \otimes \mathcal{F} \longrightarrow 0.$$

All $\mathcal{L} \otimes \mathcal{E}^{-i}$ in this complex are soft sheaves. Therefore, by III.8.14a), $\mathcal{L} \otimes \mathcal{F}$ is also soft. Now let

$$0 \longrightarrow \mathcal{F}' \longrightarrow \mathcal{F} \longrightarrow \mathcal{F}'' \longrightarrow 0$$

be an exact sequence of sheaves on X. Then ($\mathcal{L}$ being flat)

$$0 \longrightarrow \mathcal{L} \otimes \mathcal{F}' \longrightarrow \mathcal{L} \otimes \mathcal{F} \longrightarrow \mathcal{L} \otimes \mathcal{F}'' \longrightarrow 0$$

is an exact sequence of soft sheaves, so that the sequence

$$0 \longrightarrow f_!^{\mathcal{L}} \mathcal{F}' \longrightarrow f_!^{\mathcal{L}} \mathcal{F} \longrightarrow f_!^{\mathcal{L}} \mathcal{F}'' \longrightarrow 0$$

is also exact. □

21. Corollary. *For any soft flat sheaf $\mathcal{L}$ on X and for any sheaf $\mathcal{G}$ on Y there exists a sheaf $f^!(\mathcal{L}, \mathcal{G})$ and a functorial in $\mathcal{F}$ isomorphism*

$$\mathrm{Hom}\,(f_!(\mathcal{L} \otimes \mathcal{F}), \mathcal{G}) \cong \mathrm{Hom}\,(\mathcal{F}, f^!(\mathcal{L}, \mathcal{G})).$$

Moreover, if $\mathcal{G}$ is an injective sheaf, then $f^!(\mathcal{L}, \mathcal{G})$ is also injective.

Proof. The first statement follows from Theorem III.8.19 and Proposition III.8.20. To prove the second one, let us remark that, as it was established while proving Proposition III.8.20, the functor $f_!^{\mathcal{L}} : \mathcal{F} \mapsto f_!(\mathcal{L} \otimes \mathcal{F})$ is exact. By the injectivity of $\mathcal{G}$, the functor $\mathcal{F} \mapsto \mathrm{Hom}\,(f_!^{\mathcal{L}} \mathcal{F}, \mathcal{G}) = \mathrm{Hom}\,(\mathcal{F}, f^!(\mathcal{L}, \mathcal{G}))$ is also exact, i.e., the sheaf $f^!(\mathcal{L}, \mathcal{G})$ is injective. □

The next proposition shows that the constant sheaf $\mathbb{Z}_X$ on X has a resolution formed by sheaves of the above type.

22. Proposition. *Any flat sheaf $\mathcal{F}$ on X has a resolution*

$$0 \longrightarrow \mathcal{F} \longrightarrow \mathcal{L}^0 \longrightarrow \mathcal{L}^1 \longrightarrow \ldots \longrightarrow \mathcal{L}^n \longrightarrow 0$$

($n = \dim_c X$) formed by soft flat sheaves.

Proof. We construct a resolution of $\mathcal{F}$

$$0 \longrightarrow \mathcal{F} \xrightarrow{\varepsilon} \mathcal{C}^0 \xrightarrow{d^0} \mathcal{C}^1 \xrightarrow{d^1} \ldots$$

similar to the Godement's resolution. Namely, let

$$\Gamma(U, \mathcal{C}^0) = \prod_{x \in U} \mathcal{F}_x,$$

$\varepsilon : \mathcal{F} \longrightarrow \mathcal{C}^0$ be a natural monomorphism, and then inductively

$$\Gamma(U, \mathcal{C}^i) = \prod_{x \in U} \left(\mathcal{C}^{i-1} / \mathrm{Im}\, d^{i-2} \right)_x,$$

$d^i : \mathcal{C}^{i-1} \longrightarrow \mathcal{C}^i$ the composition of the projection
$$\mathcal{C}^{i-1} \longrightarrow \mathcal{C}^{i-1}/\operatorname{Im} d^{i-2} \text{ and the natural morphism}$$
$$\mathcal{C}^{i-1}/\operatorname{Im} d^{i-2} \longrightarrow \mathcal{C}^i.$$

Since the direct product of any family of flat $\mathbb{Z}$-modules is a flat module, $\mathcal{C}^0$ is a flat sheaf. Next, $\mathcal{F}_x$ is a direct summand of $(\mathcal{C}^0)_x$ for any $x \in X$, so that $\mathcal{C}^0/\operatorname{Im}\varepsilon$ is also flat. Similarly, $\mathcal{C}^i$ and $\ker d^i$ are flat sheaves for any i. It is clear also that all $\mathcal{C}^i$ are soft.

Let us consider now the resolution
$$0 \longrightarrow \mathcal{F} \longrightarrow \mathcal{L}^0 \longrightarrow \mathcal{L}^1 \longrightarrow \ldots \longrightarrow \mathcal{L}^n \longrightarrow 0$$
where $\mathcal{L}^i = \mathcal{C}^i$ for $i < n$ and $\mathcal{L}^n = \ker d^n = \operatorname{Im} d^{n-1}$. By the above, all $\mathcal{L}^i$, $i < n$, are both soft and flat, and $\mathcal{L}^n$ is flat. By III.8.14a), it is also soft. □

23. Construction of $f^!(\mathcal{G}^\bullet)$

Now let
$$0 \longrightarrow \mathbb{Z}_X \longrightarrow \mathcal{L}^0 \longrightarrow \mathcal{L}^1 \longrightarrow \ldots \longrightarrow \mathcal{L}^n \longrightarrow 0$$
be a bounded resolution of the constant sheaf $\mathbb{Z}_X$ on X formed by soft flat sheaves, and $\mathcal{G}^\bullet$ a bounded from the left complex of sheaves on Y. Let $A^{ij} = f^!(\mathcal{L}^{-i}, \mathcal{G}^j)$. Differentials in $\mathcal{L}^\bullet$ and in $\mathcal{G}^\bullet$ provide maps $d_I^{ij} : A^{ij} \to A^{i+1,j}$ and $d_{II}^{ij} : A^{ij} \to A^{i,j+1}$ making $\{A^{ij}\}$ a bicomplex. Denote by $f^!(\mathcal{L}^\bullet, \mathcal{G}^\bullet)$ the corresponding diagonal complex (see III.7.8). By the first part of Corollary III.8.21 there exists an isomorphism of complexes of abelian groups
$$\operatorname{Hom}^\bullet(f_!(\mathcal{L}^\bullet \otimes \mathcal{F}^\bullet), \mathcal{G}^\bullet) \cong \operatorname{Hom}^\bullet(\mathcal{F}^\bullet, f^!(\mathcal{L}^\bullet, \mathcal{G}^\bullet)), \tag{III.64}$$
which is functorial in $\mathcal{F}^\bullet \in \operatorname{Kom}^+(SAb_X)$, $\mathcal{G}^\bullet \in \operatorname{Kom}^+(SAb_Y)$ (here $\mathcal{L}^\bullet \otimes \mathcal{F}^\bullet$ is the complex associated with the bicomplex $\{\mathcal{L}^i \otimes \mathcal{F}^j\}$, $f_!$ acts on complexes term by term, and $\operatorname{Hom}^\bullet$ is the complex of abelian groups defines in III.6.15).

Now we define $f^!(\mathcal{G}^\bullet)$ as follows. Let $\mathcal{G}^\bullet \to \mathcal{I}^\bullet$ be a quasi-isomorphism of $\mathcal{G}^\bullet$ with a complex formed by injective sheaves on Y. Define
$$f^!(\mathcal{G}^\bullet) = f^!(\mathcal{L}^\bullet, \mathcal{I}^\bullet).$$
One can easily verify that $f^!(\mathcal{G}^\bullet)$ does not depend on the choices of resolutions $\mathcal{L}^\bullet$ and $\mathcal{I}^\bullet$ (up to a canonical isomorphism in $D^+(SAb_X)$).

24. Completion of the Proof of Theorem III.8.16

First of all, as $\mathcal{L}^\bullet$ is a flat resolution of the constant sheaf and $\mathcal{L}^\bullet \otimes \mathcal{F}^\bullet$ consist of soft sheaves, we have
$$f_!(\mathcal{L}^\bullet \otimes \mathcal{F}^\bullet) = Rf_!(\mathcal{F}^\bullet)$$
in $D^+(SAb_Y)$. Next, $\mathcal{I}^\bullet$ and $f^!(\mathcal{L}^\bullet, \mathcal{I}^\bullet)$ are injective and $\mathcal{I}^\bullet$ is isomorphic to $\mathcal{G}^\bullet$ in $D^+(SAb_Y)$, so that (III.60) follows from (III.64).

To complete the proof of Theorem III.8.16 we notice that the functoriality of the mapping $\mathcal{G}^\bullet \mapsto f^!(\mathcal{G}^\bullet)$ follows from the uniqueness of the representing object.

25. Properties of $f^!$

a) The construction of $f^!$ and Theorem III.8.16 can be generalized to the case where the category SAb is replaced by the category of sheaves of R-modules, where R is a fixed noetherian ring (in particular, a field).

b) Formula (III.59) (see III.8.13) implies that $(gf)^! = g^! \circ f^!$ for two continuous maps $f : X \to Y$, $g : Y \to Z$.

c) If $f : X \to Y$ is an embedding of an open or closed subset, then the right adjoint functor for $f_!$ exists already an the level of the category of sheaves, so that we do not need derived categories. Namely, if $f : U \to Y$ is an open embedding, then the right adjoint to $f_! : SAb_U \to SAb_X$ is the restriction $f^\bullet : SAb_X \to SAb_U$. On the other hand, if $f : X \to Y$ is a closed embedding, then the right adjoint to $f_! : SAb_X \to SAb_Y$ is the functor γ_X of "sections supported on X" defined as follows: let $U \subset X$, $V \subset Y$ be open sets with $U = V \cap Y$. Then

$$\Gamma(U, \gamma_X \mathcal{F}) = \{ s \in \Gamma(V, \mathcal{F}), \text{ supp } s \subset X \}.$$

We leave to the reader the verification of the facts that $\Gamma(V, \gamma_X \mathcal{F})$ is well defined and $\gamma_X \mathcal{F}$ is a sheaf, and of the required conjugacy properties.

To conclude this section we consider the situation which is, in some sense, opposite to the one in c) above. Namely, we assume that $f : X \to$ pt is the map to a point.

26. The Dualizing Complex

Let Y be a point. Then sheaves of Y are just abelian groups, and we will denote by $\mathbb{Z} \in D^+(SAb_Y) = D^+(Ab)$ the 0-complex with the zero component $\mathbb{Z}$. For any (finite-dimensional, locally compact) topological space X let $\mathfrak{D}_X^\bullet = f^!(\mathbb{Z})$, where $f : X \to$ pt. The complex $\mathfrak{D}_X^\bullet$ is called *the dualizing complex* on X. In this case Theorem III.8.16 takes the following form (Poincaré duality):

$$R\text{Hom}\,(R\Gamma_c(X, \mathcal{F}^\bullet), \mathbb{Z}) \cong R\text{Hom}(\mathcal{F}^\bullet, \mathfrak{D}_X^\bullet). \tag{III.65}$$

The complex $\mathfrak{D}_X^\bullet$ reflects some topological singularities of X. In particular, if X is a stratified space whose strata are topological manifolds, then the cohomology sheaves of $\mathfrak{D}_X^\bullet$ are constructible with respect to this stratification.

The structure of $\mathfrak{D}_X^\bullet$ is most simple in the case when X is nonsingular.

27. Corollary. *Let X be an n-dimensional topological manifold with boundary. Then $\mathfrak{D}_X^\bullet = \omega_X[n]$, where the sheaf ω_X is defined by*

$$\Gamma(U, \omega_X) = \text{Hom}_{Ab}\,(H_c^n(U, \mathbb{Z}), \mathbb{Z})$$

for any open $U \subset X$.

Proof. First of all, for $V \subset U \subset X$, the "extension by zero" functor determines a morphism $H^n_c(V, \mathbb{Z}) \to H^n_c(U, \mathbb{Z})$, so that ω_X is a presheaf on X. To verify that ω_X is a sheaf we can either use Theorem III.8.19 and Proposition III.8.20, or proceed directly: ω_X is the kernel of the morphism of sheaves $(\mathcal{S}^n)^* \to (\mathcal{S}^{n-1})^*$ where

$$0 \longrightarrow \mathbb{Z} \longrightarrow \mathcal{S}^0 \longrightarrow \mathcal{S}^1 \longrightarrow \ldots \longrightarrow \mathcal{S}^n \longrightarrow 0 \tag{III.66}$$

is a soft resolution of the constant sheaf, and $\mathcal{S}^*$ for a soft $\mathcal{S}$ is the sheaf defined by

$$U \mapsto \mathrm{Hom}_{Ab}\left(\Gamma_c(U, \mathcal{S}), \mathbb{Z}\right).$$

Now, substituting in (III.65) $\mathcal{F}^\bullet = \mathbb{Z}_U$ we get

$$R\mathrm{Hom}\left(R\Gamma_c(X, \mathbb{Z}_U), \mathbb{Z}\right) = R\mathrm{Hom}\left(\mathbb{Z}_U, \mathfrak{D}^\bullet_X\right)$$

so that (as $\mathrm{Hom}(\mathbb{Z}_U, \mathcal{E}) = \Gamma(U, \mathcal{E})$ for any sheaf $\mathcal{E}$ on X)

$$R\mathrm{Hom}\left(R\Gamma_c(U, \mathbb{Z}), \mathbb{Z}\right) = \Gamma(U, \mathcal{I}^\bullet),$$

where $\mathcal{I}^\bullet$ is a complex of injective sheaves which is quasi-isomorphic to $\mathfrak{D}^\bullet_X$ and bounded from the left.

Thus, the cohomology sheaf $H^{-i}(\mathfrak{D}^\bullet_X)$ of $\mathfrak{D}^\bullet_X$ is the sheaf associated with the presheaf

$$U \mapsto H^{-i}\left(R\mathrm{Hom}\left(R\Gamma_c(U, \mathbb{Z}), \mathbb{Z}\right)\right). \tag{III.67}$$

We have to prove that this sheaf is isomorphic to ω_X for $i = n$ and is the zero sheaf for other i's. Denote by N_U^{-i} the group on the right-hand side of (III.67). Using the resolution (III.66) to compute $R\Gamma_c$ and the injective resolution $0 \to \mathbb{Z} \to \mathbb{Q} \to \mathbb{Q}/\mathbb{Z} \to 0$ of the group $\mathbb{Z}$ in the category Ab to compute $R\mathrm{Hom}$, we can easily verify that N_U^{-i} enters the exact triple

$$0 \longrightarrow \mathrm{Ext}^1\left(H^{i+1}_c(U, \mathbb{Z}), \mathbb{Z}\right) \longrightarrow N_U^{-i} \longrightarrow \mathrm{Hom}\left(H^{-i}_c(U, \mathbb{Z}), \mathbb{Z}\right) \longrightarrow 0.$$

Any point of our n-dimensional topological manifold X has a fundamental system of neighbourhoods, each being homeomorphic either to $\mathbb{R}^n$ or to $\mathbb{R}_+ \times \mathbb{R}^{n-1}$. The required fact follows from the equalities

$$H_c(\mathbb{R}^n, \mathbb{Z}) = \begin{cases} 0 & \text{for } i \neq n, \\ \mathbb{Z} & \text{for } i = n. \end{cases}$$

$$H^i_c\left(\mathbb{R}_+ \times \mathbb{R}^{n-1}, \mathbb{Z}\right) = 0 \quad \text{for all } i,$$

$$\mathrm{Ext}^1(\mathbb{Z}, \mathbb{Z}) = 0. \qquad \square$$

28. Remarks

Replacing $\mathbb{Z}$ with an arbitrary noetherian ring R we can obtain analogues of the above corollary. In particular, if $R = k$ is a field and X is a topological manifold without boundary, then $\omega_X = \tau_X$ is the sheaf of k-orientations of X (in particular, the constant sheaf k if X is oriented or if char $k = 2$). If X is a manifold with boundary ∂X, then $\omega_X = i_!\tau$, where τ is the sheaf of k-orientations of $X - \partial X$ and $i: X - \partial X \to X$ is the inclusion.

Taking cohomology of both sides of (III.65) we can express the Poincaré duality in a more standard form (if k is a field) by saying that for any sheaf of k-modules $\mathcal{F}$ on X there exists a canonical isomorphism

$$\mathrm{Hom}_k\left(H^i_c(X,\mathcal{F}),k\right) \cong \mathrm{Ext}^{n-i}(\mathcal{F},\omega_X). \tag{III.68}$$

Denoting by $\int_X : H^n_c(X,\omega_X) \to k$ the fundamental class of X, i.e., the preimage of $\mathbf{1} \in \mathrm{Hom}(\omega_X,\omega_X)$ under the isomorphism (III.68) for $i = n = \dim X$, $\mathcal{F} = \omega_X$, we can express (III.68) as the composition of the canonical pairing $\mathrm{Ext}^{n-i}(\mathcal{F},\omega_X) \times H^i_c(X,\mathcal{F}) \to H^n_c(X,\omega_X)$ with $\int_X$.

Exercises

1. Softness Criterion. a) Let Y be a closed subset of a locally compact topological space X and $U = X\setminus Y$. Prove that for any sheaf of abelian groups $\mathcal{F}$ on X there exists the exact sequence

$$\cdots \longrightarrow H^n_c(U,\mathcal{F}) \xrightarrow{\alpha_n} H^n_c(X,\mathcal{F}) \xrightarrow{\beta_n} H^n_c(Y,\mathcal{F}) \longrightarrow H^{n+1}_c(U,\mathcal{F}) \longrightarrow \cdots$$

where α_n is induced by the mapping $\alpha_0 : \Gamma_c(U,\mathcal{F}) \to \Gamma_c(X,\mathcal{F})$ (extension by zero) and β_n is induced by the restriction $\beta_0 : \Gamma_c(X,\mathcal{F}) \to \Gamma_c(Y,\mathcal{F})$. To do this prove that if $\mathcal{F}$ is a soft sheaf, then the sequence

$$0 \longrightarrow \Gamma_c(U,\mathcal{F}) \xrightarrow{\alpha_0} \Gamma_c(X,\mathcal{F}) \xrightarrow{\beta_0} \Gamma_c(Y,\mathcal{F}) \longrightarrow 0$$

is exact, and use soft resolutions to compute H^n_c.

b) Deduce from a) that a sheaf $\mathcal{F}$ on X is soft if and only if $H^1_c(U,\mathcal{F}) = 0$ for all open $U \subset X$.

c) Prove that the functor $f_! : SAb_X \to SAb_Y$ for a continuous mapping $f : X \to Y$ transforms soft sheaves on X into soft sheaves on Y.

2. Mayer–Vietoris Theorems. Let X be a topological space which is the union of two closed subsets X_1, X_2 and $\mathcal{F}$ a sheaf of abelian groups on X. The Mayer–Vietoris exact sequence relates cohomology $\mathcal{F}$ on X_1, X_2, X, and $X_2 \cap X_2$.

a) Construct the long exact sequence

$$\cdots \longrightarrow H^{n-1}(X_1 \cap X_2, \mathcal{F}) \longrightarrow H^n(X,\mathcal{F}) \longrightarrow H^n(X_1,\mathcal{F}) \oplus H^n(X_2,\mathcal{F})$$

$$\longrightarrow H^n(X_1 \cap X_2, \mathcal{F}) \longrightarrow \cdots \tag{*}$$

To do this denote by $i_1 : X_1 \to X$, $i_2 : X_2 \to X$, $i : X_1 \cap X_2 \to X$ the natural embeddings and construct, for any $\mathcal{G} \in SAb_X$, the exact sequence of sheaves

$$0 \longrightarrow \mathcal{G} \xrightarrow{\alpha} i_{1\bullet}i_1^{\bullet}\mathcal{G} \oplus i_{2\bullet}i_2^{\bullet}\mathcal{G} \xrightarrow{\beta} i_{\bullet}i^{\bullet}\mathcal{G} \longrightarrow 0$$

where α is the sum of two adjunction morphisms (expressing the adjunction of direct and inverse images), and β is the difference of two adjunction morphisms. Then apply this exact sequence to an injective resolution of $\mathcal{F}$ and use the isomorphism

$$H^n(X, j_{\bullet}\mathcal{H}) \cong H^n(Y, \mathcal{H}) \qquad (**)$$

valid for any closed embedding $j : Y \to X$ and any SAb_Y.

b) Assuming X to be locally compact prove the existence of a long exact sequence similar to (*) for groups H_c^n. Use, instead of (**), the isomorphism

$$H_c^n(X, j_{\bullet}\mathcal{H}) \cong H_c^n(Y, \mathcal{H}).$$

The following series of exercises contains some results about the relation of various functors in derived categories of sheaves on topological spaces. Below, all spaces are assumed to be locally compact, paracompact, and finite-dimensional (in the sense of $\dim_c$), and all maps are assumed to be continuous. The equalities between objects of (derived) categories are functorial isomorphisms.

3. $f^{\bullet}$ and $\overset{L}{\otimes}$. For $f : X \to Y$, $\mathcal{F}^{\bullet}, \mathcal{G}^{\bullet} \in D^-(SAb_Y)$ prove that $f^{\bullet}(\mathcal{F}^{\bullet} \overset{L}{\otimes} \mathcal{G}^{\bullet}) = f^{\bullet}\mathcal{F}^{\bullet} \overset{L}{\otimes} f^{\bullet}\mathcal{G}^{\bullet}$. To do this replace $\mathcal{F}^{\bullet}$ and $\mathcal{G}^{\bullet}$ by their flat resolutions (quasi-isomorphic complexes formed by flat sheaves), and use the identity $f^{\bullet}(\mathcal{F} \otimes \mathcal{G}) = f^{\bullet}\mathcal{F} \otimes f^{\bullet}\mathcal{G}$ for $\mathcal{F}, \mathcal{G} \in SAb_Y$, which follows from $(\mathcal{F} \otimes \mathcal{G})_y = \mathcal{F}_y \otimes \mathcal{G}_y$.

4. $R\operatorname{Hom}$ and $\overset{L}{\otimes}$. For $\mathcal{F}^{\bullet}, \mathcal{G}^{\bullet} \in D^-(SAb_X)$, $\mathcal{H}^{\bullet} \in D^+(SAb_X)$ we have

$$R\operatorname{Hom}(\mathcal{F}^{\bullet} \otimes \mathcal{G}^{\bullet}, \mathcal{H}^{\bullet}) = R\operatorname{Hom}(\mathcal{F}^{\bullet}, R\operatorname{Hom}(\mathcal{G}^{\bullet}, \mathcal{H}^{\bullet})).$$

To prove this verify the corresponding statement for sheaves of abelian groups on X, and then replace $\mathcal{G}^{\bullet}$ with a flat resolution and $\mathcal{H}^{\bullet}$ with an injective resolution.

5. $Rf_{\bullet}$ and $R\operatorname{Hom}$. For $f : X \to Y$, $\mathcal{F}^{\bullet} \in D^-(SAb_Y)$, $\mathcal{G}^{\bullet} \in D^+(SAb_X)$ we have

$$Rf_{\bullet}R\operatorname{Hom}(f^{\bullet}\mathcal{F}^{\bullet}, \mathcal{G}^{\bullet}) = R\operatorname{Hom}(\mathcal{F}^{\bullet}, Rf_{\bullet}\mathcal{G}^{\bullet}).$$

Replacing $\mathcal{G}^{\bullet}$ with an injective resolution show that $f_{\bullet}\mathcal{G}^{\bullet}$ (with the term by term action) is formed by injective sheaves, and $\operatorname{Hom}(f^{\bullet}\mathcal{F}^{\bullet}, \mathcal{G}^{\bullet})$ is formed by soft sheaves. After that the required isomorphism follows from the isomorphism

$$f_{\bullet}\operatorname{Hom}(f^{\bullet}\mathcal{F}, \mathcal{G}) = \operatorname{Hom}(\mathcal{F}, Rf_{\bullet}\mathcal{G})$$

in SAb_Y, which follows from the adjointness of $f^{\bullet}$ and $f_{\bullet}$.

6. Base Change Formulas. Let

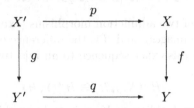

be a commutative diagram of spaces and continuous mappings. Then in $D^+(SAb_{Y'})$ we have

$$q^{\bullet}Rf_!\mathcal{F}^{\bullet} = Rg_!p^{\bullet}\mathcal{F}^{\bullet},$$

$$Rg_{\bullet}p^!\mathcal{F}^{\bullet} = q^!Rf_!\mathcal{F}^{\bullet}, \quad \mathcal{F}^{\bullet} \in D^+(SAb_X).$$

To prove the first formula verify the equality $q^{\bullet}f_!\mathcal{F} = g_!p^{\bullet}\mathcal{F}$, $\mathcal{F} \in SAb_X$, by computing the stalks of both sheaves at a point $y' \in Y'$ and then replace $\mathcal{F}^{\bullet}$ with its soft resolution. To prove the second formula interchange X and Y', apply the first formula, and use the fact that the functors in the following pairs are adjoint: $(f^{\bullet}, Rf_{\bullet})$, $(g^{\bullet}, Rg_{\bullet})$, $(Rq_!, q^!)$, $(Rp_!, p^!)$.

7. Projection formula. For $f: X \to Y$, $\mathcal{F}^{\bullet} \in D^-(SAb_X)$, $\mathcal{G}^{\bullet} \in D^-(SAb_Y)$ we have

$$Rf_!\left(\mathcal{F}^{\bullet} \overset{L}{\otimes} f^{\bullet}\mathcal{G}^{\bullet}\right) = Rf_!\mathcal{F}^{\bullet} \overset{L}{\otimes} \mathcal{G}^{\bullet}.$$

For the proof, verify the corresponding formula in SAb_Y and then replace $\mathcal{F}^{\bullet}$ by a soft resolution and $\mathcal{G}^{\bullet}$ by an injective resolution.

8. $f^!$ and $R\operatorname{Hom}$. For $f: X \to Y$, $\mathcal{F}^{\bullet}, \mathcal{G}^{\bullet} \in D^+(SAb_Y)$ we have

$$f^!R\operatorname{Hom}(\mathcal{F}^{\bullet}, \mathcal{G}^{\bullet}) = R\operatorname{Hom}(f^{\bullet}\mathcal{F}^{\bullet}, f^!\mathcal{G}^{\bullet}).$$

For the proof replace $\mathcal{F}^{\bullet}$ with a soft resolution, $\mathcal{G}^{\bullet}$ with an injective resolution, and use the explicit construction of $f^!$ (see III.8.23).

IV. Triangulated Categories

IV.1 Triangulated Categories

1. Axioms

Let $\mathcal{D}$ be an additive category. A structure of *triangulated category* on $\mathcal{D}$ is defined by specifying data a) and b) which satisfy axioms TR1–TR4 below.

a) *Additive automorphism* $T : \mathcal{D} \to \mathcal{D}$ called the translation functor.

As in 2, we will write $X[n]$ for $T^n(X)$ and $f[n]$ for $T^n(f)$. Now we can literally repeat parts a) and b) of definition III.3.4 introducing in $\mathcal{D}$ *triangles*

$$X \xrightarrow{u} Y \xrightarrow{v} Z \xrightarrow{w} X[1]$$

and *morphisms of triangles*

$$\begin{array}{ccccccc} X & \xrightarrow{u} & Y & \xrightarrow{v} & Z & \xrightarrow{w} & X[1] \\ \downarrow f & & \downarrow g & & \downarrow h & & \downarrow f[1] \\ X' & \xrightarrow{u'} & Y' & \xrightarrow{v'} & Z' & \xrightarrow{w'} & X'[1] \end{array}$$

The last part of Definition III.3.4 should be given axiomatically. Namely, among all triangles in $\mathcal{D}$ we should distinguish:

b) *The class of distinguished triangles.*

As we will see, the following axioms give a rather satisfactory axiomatic description of the properties of the triangles defined in III.3.4c).

TR1. a) $X \xrightarrow{\mathrm{id}} X \to 0 \to X[1]$ *is a distinguished triangle.*
b) *Any triangle isomorphic to a distinguished one is itself distinguished.*
c) *Any morphism* $X \xrightarrow{u} Y$ *can be completed to a distinguished triangle* $X \xrightarrow{u} Y \xrightarrow{v} Z \xrightarrow{w} X[1]$.

TR2. *A triangle* $X \xrightarrow{u} Y \xrightarrow{v} Z \xrightarrow{w} X[1]$ *is distinguished if and only if the triangle* $Y \xrightarrow{v} Z \xrightarrow{w} X[1] \xrightarrow{-u[1]} Y[1]$ *is distinguished.*

IV. Triangulated Categories

TR3. *Assume we are given two distinguished triangles and two morphisms f, g as in the diagram below:*

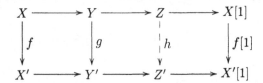

This diagram can be completed (not necessarily uniquely) to a morphism of triangles by a morphism $h : Z \to Z'$.

The last axiom deals with the rather big "octahedron diagram". One way to represent this diagram is to draw the two "caps" of the octahedron with a common brim:

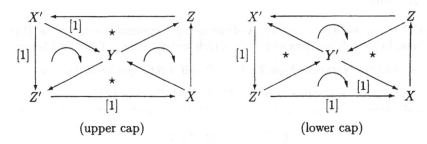

(upper cap)   (lower cap)

In these diagrams X, Y, etc. are objects of $\mathcal{D}$; arrows of the type $X' \xrightarrow{[1]} Z'$ represent morphisms $X' \longrightarrow Z'[1]$ in $\mathcal{D}$; triangles marked $\star$ are distinguished, and those marked $\frown$ are commutative. Finally, one requires that the two composite morphisms $Y \to Y'$ (through Z and through Z') coincide, and that the two composite morphisms $Y' \to Y[1]$ (through $X[1]$ and through X') also coincide.

Now we can formulate the last axiom.

TR4. *Any diagram of the type "upper cap" can be completed to an octahedron diagram.*

2. Remarks About the Formal Structure of the Axioms

a) Axiom TR2 implies that any distinguished triangle can be canonically included into a helix such that any three consecutive morphisms form a distinguished triangle.

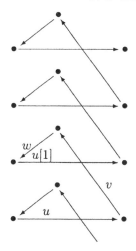

Any morphism of triangles generates two helices chained together by horizontal arrows. Axioms TR2 and TR3 imply that given two neighbouring horizontal arrows forming a commutative square, one can construct a morphism of helices.

b) One can consider a diagram of type "upper cap" as a morphism of distinguished triangles with the middle morphism id_Y. Including this morphism into the "double helix" as above, and looking at its consecutive full (360°) turns, one can see that this double helix contains diagrams of the type "lower cap". An equivalent (modulo axioms TR1–TR3) formulation of the axiom TR4 is as follows: any lower cap can be completed to an octahedron.

c) We will show in IV.1.4 below that any completions of a morphism $X \xrightarrow{u} Y$ to a distinguished triangle in TR1 are isomorphic, i.e., if we take $f = \text{id}_X$, $g = \text{id}_Y$, then any morphism h whose existence is guaranteed by TR3 is an isomorphism.

Therefore, any upper cap can be recovered uniquely up to an isomorphism from one of its commutative triangles, say $X \to Y \to Z$, by complementing $X \to Y$ and $Y \to Z$ to distinguished triangles.

Next we show that from TR1–TR3 imply the following property of distinguished triangles (cf. III.3.6).

3. Proposition. *Let $\mathcal{D}$ be a triangulated category and let $X \xrightarrow{u} Y \xrightarrow{v} Z \xrightarrow{w} X[1]$ be a distinguished triangle. Then for any object in $\mathcal{D}$ the following sequences are exact:*

$$\ldots \longrightarrow \text{Hom}(U, X[i]) \xrightarrow{u_*[i]} \text{Hom}(U, Y[i]) \xrightarrow{v_*[i]} \text{Hom}(U, Z[i])$$
$$\xrightarrow{w_*[i]} \text{Hom}(U, X[i+1]) \longrightarrow \ldots ,$$

$$\ldots \longrightarrow \text{Hom}(X[i+1], U) \xrightarrow{w^*[i]} \text{Hom}(Z[i], U) \xrightarrow{v^*[i]} \text{Hom}(Y[i], U)$$
$$\xrightarrow{u^*[i]} \text{Hom}(X[i], U) \longrightarrow \ldots$$

Proof. We give the proof for the first sequence; the second one is dealt with similarly. By remark IV.1.2a) it suffices to prove the exactness at the term $\mathrm{Hom}(U,Y)$. Check first that $vu = 0$ (and, therefore, the composition of any two consecutive morphisms in a distinguished triangle is zero). This follows from the axiom TR3 applied to $X \xrightarrow{\mathrm{id}} X \longrightarrow 0 \longrightarrow X[1]$ and our triangle $X \longrightarrow Y \longrightarrow Z \longrightarrow X[1]$:

$$\begin{array}{ccccccc}
X & \xrightarrow{\mathrm{id}} & X & \longrightarrow & 0 & \longrightarrow & X[1] \\
\downarrow \mathrm{id} & & \downarrow u & & \downarrow h & & \downarrow \\
X & \xrightarrow{u} & Y & \xrightarrow{v} & Z & \xrightarrow{w} & X[1]
\end{array}$$

The only possible h is $h = 0$, so the commutativity implies $vu = 0$.

Now let $f : U \to Y$ satisfy $vf = 0$. We want to prove that $f = ug$ for some $g : U \to X$. We obtain g from a morphism of distinguished triangles

$$\begin{array}{ccccccc}
U & \xrightarrow{\mathrm{id}} & U & \longrightarrow & 0 & \longrightarrow & U[1] \\
\downarrow g & & \downarrow f & & \downarrow & & \downarrow g[1] \\
X & \xrightarrow{u} & Y & \xrightarrow{v} & Z & \xrightarrow{w} & X[1]
\end{array}$$

This morphism g is constructed using TR2 and TR3: first TR3 is applied to

$$\begin{array}{ccccccc}
U & \longrightarrow & 0 & \longrightarrow & U[1] & \xrightarrow{-\mathrm{id}} & U[1] \\
\downarrow f & & \downarrow & & \downarrow g[1] & & \downarrow f[1] \\
Y & \xrightarrow{v} & Z & \xrightarrow{w} & X[1] & \xrightarrow{-u[1]} & Y[1]
\end{array}$$

and then g is recovered from $g[1]$ (cf. remark IV.1.2a)). □

4. Corollary. *a) If f and g in TR3 are isomorphisms, then h is also an isomorphism.*

b) The distinguished triangle in TR1c) is determined uniquely up to an isomorphism.

Proof. The diagram in TR3 induces the commutative diagram

$$\begin{array}{ccccccccc}
\mathrm{Hom}(Z',X) & \to & \mathrm{Hom}(Z',Y) & \to & \mathrm{Hom}(Z',Z) & \to & \mathrm{Hom}(Z',X[1]) & \to & \mathrm{Hom}(Z',Y'[1]) \\
\downarrow f_* & & \downarrow g_* & & \downarrow h_* & & \downarrow f[1]_* & & \downarrow g[1]_* \\
\mathrm{Hom}(Z',X') & \to & \mathrm{Hom}(Z',Y') & \to & \mathrm{Hom}(Z',Z') & \to & \mathrm{Hom}(Z',X'[1]) & \to & \mathrm{Hom}(Z',Y'[1])
\end{array}$$

whose rows are exact by Proposition 3. In this diagram f and g, so also $f[1]$ and $g[1]$, are isomorphisms. Therefore $f_*, g_*, f_*[1], g_*[1]$ are isomorphisms as well. By the five-lemma (Ex. II.5.6), h_* also is an isomorphism. This means that there exists $\varphi : Z' \to Z$ with $h\varphi = \mathrm{id}_{Z'}$. Using a similar diagram with $\mathrm{Hom}(Z, \bullet)$, we conclude that there exists $\psi : Z' \to Z$ with $\psi h = \mathrm{id}_Z$. Therefore $\varphi = \psi$, and h is an isomorphism.

Part b) follows immediately from a). □

Let us discuss now when a morphism between vertices of two distinguished triangles can be completed to a morphism of triangles

$$\begin{array}{ccccccc}
X & \xrightarrow{u} & Y & \xrightarrow{v} & Z & \xrightarrow{w} & X[1] \\
\downarrow f & \text{\textcircled{1}} & \downarrow g & & \downarrow h & \text{\textcircled{2}} & \downarrow f[1] \\
X' & \xrightarrow{u'} & Y' & \xrightarrow{v'} & Z' & \xrightarrow{w'} & X'[1]
\end{array}$$

We have $vu = 0$ so that if such a completion exists, then $v'gu = 0$. Let us show that this condition is also sufficient.

5. Corollary. *If $v'gu = 0$, then g can be completed to a morphism of triangles. If, moreover, $\mathrm{Hom}(X, Z'[-1]) = 0$, then this morphism of triangles is unique.*

Proof. Consider the exact sequence from Proposition IV.1.3 for morphisms from X to the lower triangle:

$$\cdots \longrightarrow \mathrm{Hom}(X, Z'[-1]) \longrightarrow$$
$$\longrightarrow \mathrm{Hom}(X, X') \longrightarrow \mathrm{Hom}(X, Y') \longrightarrow \mathrm{Hom}(X, Z') \longrightarrow \cdots$$
$$\cup \qquad\qquad \cup \qquad\qquad \cup$$
$$f \mapsto\!\!-\!\!-\!\!\to gu \qquad\qquad \mapsto 0$$

It is clear that a morphism f that makes the square ① commutative is constructed as a preimage of gu and is unique up to an element from the image of $\mathrm{Hom}(X, Z'[-1])$. By TR2, any choice of f gives a morphism of triangles, and, in particular, a morphism h that makes the square ② commutative. Similar arguments applied to the exact sequence of morphisms of the upper triangle to Z' shows that h is unique provided $\mathrm{Hom}(X[1], Z') = 0$. □

6. Cohomological Functors

A functor $H : \mathcal{D} \to \mathcal{A}$ from a triangulated category $\mathcal{D}$ into an abelian category $\mathcal{A}$ is called a *cohomological functor* if it is additive and the sequence

$$H(X) \xrightarrow{H(u)} H(Y) \xrightarrow{H(w)} H(Z)$$

in $\mathcal{A}$ is exact for any distinguished triangle

$$X \xrightarrow{u} Y \xrightarrow{v} Z \xrightarrow{w} X[1]$$

in $\mathcal{D}$ (compare III.6.14a)).

Axiom TR2 implies that if H is a cohomological functor, then the sequence

$$H(X[i]) \xrightarrow{H(u[i])} H(Y[i]) \xrightarrow{H(v[i])} H(Z[i]) \xrightarrow{H(w[i])} H(X[i+1])$$

is exact for any distinguished triangle in $\mathcal{D}$.

One of the main examples of a cohomological functor is the 0-cohomology functor $C^\bullet \to H^0(C^\bullet)$ considered as a functor from $K(\mathcal{A})$ or from $D(\mathcal{A})$ to $\mathcal{A}$ (the proof that $K(\mathcal{A})$ and $D(\mathcal{A})$ are triangulated will be given later, see Theorem IV.1.9 and Corollary IV.2.7); the fact that H^0 is a cohomological functor follows from the definition of distinguished triangles in $K(\mathcal{A})$ and in $D(\mathcal{A})$ and from the long exact cohomology sequence.

Another example of a cohomological functor is the functor $X \mapsto \mathrm{Hom}_{\mathcal{D}}(U, X)$ for any $U \in \mathrm{Ob}\,\mathcal{D}$ (see Proposition IV.1.3).

7. A Cone

By TR1c) and Corollary IV.1.4b), any morphism $u : X \to Y$ in a triangulated category determines an object $C(u)$ (uniquely up to an isomorphism) as the third object in a distinguished triangle $X \xrightarrow{u} Y \longrightarrow Z = C[u] \longrightarrow X[1]$. This object is called *a cone of the morphism u* (the indefinite article emphasizes the nonuniqueness of the choice). A cone is given together with mappings $Y \to C[u] \to X[1]$.

The explanation of this name is as follows. We will show below (Theorem IV.1.9) that for any abelian category $\mathcal{A}$ the corresponding category $K(\mathcal{A})$ with distinguished triangles defined in III.3.4 becomes a triangulated category. In III.3.4 distinguished triangles were represented by diagrams

$$K^\bullet \xrightarrow{\tilde{f}} \mathrm{Cyl}\, f \xrightarrow{\pi} C(f) \xrightarrow{\delta} K^\bullet[1]. \tag{IV.1}$$

On the other hand, the diagram $K^\bullet \xrightarrow{f} L^\bullet \xrightarrow{\pi} C(f) \xrightarrow{\delta} K^\bullet[1]$ is isomorphic to (IV.1) in $K(\mathcal{A})$. The corresponding isomorphism is represented by the following diagram (with the notation of Lemma III.3.3):

$$\begin{array}{ccccccc}
K^\bullet & \xrightarrow{f} & L^\bullet & \xrightarrow{\pi} & C(f) & \xrightarrow{\delta} & K^\bullet[1] \\
\| & & \downarrow \alpha & & \| & & \| \\
K^\bullet & \xrightarrow{\tilde{f}} & \mathrm{Cyl}\, f & \xrightarrow{\alpha} & C(f) & \xrightarrow{\delta} & K^\bullet[1]
\end{array}$$

The reader can easily verify that this diagram is commutative in $K(\mathcal{A})$ (although it is not commutative in $\mathrm{Kom}(\mathcal{A})$). The proof that α is an isomorphism in $K(\mathcal{A})$ is given in lemma III.3.3.

Now let us return to a general triangulated category $\mathcal{D}$. If we want to make C functorial we could try to proceed as follows:

a) define $C : \mathrm{Ob}(\mathrm{Mor}\ \mathcal{D}) \to \mathrm{Ob}\ (\mathcal{D})$ using the axiom of choice (here Mor $\mathcal{D}$ is the category of morphisms in $\mathcal{D}$, so that elements of $\mathrm{Ob}(\mathrm{Mor}\ \mathcal{D})$ are morphisms in $\mathcal{D}$);
b) define $C : \mathrm{Mor}(\mathrm{Mor}\ \mathcal{D}) \to \mathrm{Mor}\ \mathcal{D}$ using TR3 and the axiom of choice.

After that we would see no reason for the equality $C(u \circ v) = C(u) \circ C(v)$ to hold. This "nonfunctoriality of a cone" is the first symptom that something is going wrong in the axioms of a triangulated category. Unfortunately, at the moment, we don't have a more satisfactory version.

One can ask the following question. In various situations we get various categories $\mathcal{B}$ together with some construction called the "abstract cone" $C :$ Mor $\mathcal{B} \to \mathcal{B}$. Does there exist some set of useful general axioms for such a construction? These axioms, if they exist, should determine at least two structures:

a) functorial properties of C with respect to morphisms in Mor $\mathcal{B}$;
b) the behaviour of C under the composition of morphisms (with respect to morphisms in the category Mor $(\mathcal{B})$).

8. A Cone of the Composition and the Octahedron Axiom

Let $X \xrightarrow{u} Y, Y \xrightarrow{v} Z$ be two morphisms in a triangulated category $\mathcal{D}$. Let us construct from these morphisms two distinguished triangles with third objects $C(u)$, $C(v)$ respectively. Then we have a morphism $w : C(v) \to C[u](1)$, which is defined as the composition of the morphisms $C(v) \to Y[1]$ (from the second triangle) and $Y[1] \to C(u)[1]$ (from the first triangle). A part of the octahedron axioms can be represented by the following equality:

$$C(v \circ u) = C\left(C(v) \xrightarrow{w} C(u)[1]\right)[-1].$$

Indeed, let us consider the upper cap diagram constructed from the commutative triangle

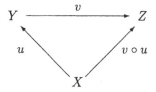

Using the notation of IV.1.1, we have

$$Z' = C(u), \quad X' = C(v),$$
$$Y' = C(v \circ u)$$

(the triangle on the right side of the lower cap).

On the other hand, from the left triangle of the lower cap, we get $Y' = C(w)$.

The rest of the octahedron axiom describes arrows in the distinguished triangle defining $C(v \circ u)$. For some additional remarks about the meaning of the octahedron axiom see IV.2.9.

In the remaining part of this section, we prove the following theorem:

9. Theorem. *Let $\mathcal{A}$ be an abelian category. The category $K(\mathcal{A})$ with the translation functor and with distinguished triangles as in III.3.4 is a triangulated category.*

The same is true for $K^{\pm}(\mathcal{A})$ and for $K^b(\mathcal{A})$.

To prove this theorem, we must verify the axioms.

10. Axiom TR1

Parts b) and c) follow from the definitions and from the remark in 6 that the triangle $X \xrightarrow{u} Y \longrightarrow C(u) \longrightarrow X[1]$ is distinguished. Part a) follows from the diagram

$$\begin{array}{ccccccc} X & \xrightarrow{\mathrm{id}} & X & \longrightarrow & 0 & \longrightarrow & X[1] \\ \| & & \| & & \downarrow & & \| \\ X & \xrightarrow{\mathrm{id}} & X & \longrightarrow & C(\mathrm{id}) & \longrightarrow & X[1] \end{array}$$

We must show that it is commutative and that it gives an isomorphism of triangles. The essential step is to check that $0 \to C(\mathrm{id})$ is homotopic to the identity morphism of $C(\mathrm{id}) = X[1] \oplus X$. The corresponding homotopy is

$$\mathrm{id}_{C(\mathrm{id})} = h \circ d + d \circ h, \quad h = \begin{pmatrix} 0 & 0 \\ \mathrm{id}_X & 0 \end{pmatrix}.$$

11. Axiom TR2

Let Δ_1 be a triangle $X \xrightarrow{u} Y \xrightarrow{v} Z \xrightarrow{w} X[1]$ and let Δ_2 be the corresponding triangle $Y \xrightarrow{v} Z \xrightarrow{w} X[1] \xrightarrow{-u[1]} Y[1]$. We show that if Δ_1 is distinguished, then Δ_2 is also distinguished. The converse statement might be proved similarly, or it can be deduced from the first one by applying it twice to Δ_2 (and using the appropriate translation).

To verify that Δ_2 is distinguished, we construct an isomorphism between it and the distinguished triangle

$$\Delta_3 : Y \xrightarrow{v} Z \xrightarrow{s} C(v) \xrightarrow{t} Y[1].$$

Since Δ_1 is distinguished, we can assume that $Z = C(u) = X[1] \oplus Y$ and that v, w are natural morphisms. Then

$$C(v) = Y[1] \oplus Z = Y[1] \oplus X[1] \oplus Z$$

and $d_{C(v)}$ is given by

$$d_{C(v)} = \begin{pmatrix} -d_Y & 0 & \mathrm{id}_Y \\ 0 & -d_X & u[1] \\ 0 & 0 & d_Y \end{pmatrix}.$$

Let us define a morphism of complexes $\theta : X[1] \to C(v)$ by

$$\theta^i \left(x^{i+1} \right) = \left(-u^{i+1} x^{i+1}, x^{i+1}, 0 \right)$$

for $x^{i+1} \in X[1]^i = X^{i+1}$. We show that the diagram

$$\begin{array}{ccccccc}
Y & \xrightarrow{v} & Z & \xrightarrow{w} & X[1] & \xrightarrow{-u[1]} & Y[1] \\
\mathrm{id}\downarrow & & \mathrm{id}\downarrow & & \theta\downarrow & & \mathrm{id}\downarrow \\
Y & \xrightarrow{v} & Z & \xrightarrow{s} & C(v) & \xrightarrow{t} & Y[1]
\end{array}$$

determines a morphism of triangles in $K(\mathcal{A})$. The only nonobvious part is that two morphisms s and $\theta \circ w$ from $Z = X[1] \oplus Y$ to $C(v) = Y[1] \oplus X[1] \oplus Y$ are homotopic. Using explicit formulas for s and $\theta \circ w$ one can easily see that $h^i : Z^i \to C(v)^{i-1}$ given by

$$h^i \left(x^{i+1}, y^i \right) = (y^i, 0, 0),$$

is the required homotopy.

Now let us show that this morphism of triangles is an isomorphism. To do this we must verify that $\theta : X[1] \to C(v)$ is an isomorphism in $K(\mathcal{A})$. Let $\psi : C(v) \to X[1]$ be the projection on the second summand. Then $\psi \circ \theta = \mathrm{id}_{X[1]}$ and $\theta \circ \psi$ is given by

$$(\theta \circ \psi)^i : C(v)^i \longrightarrow C(v)^i; \ (y^{i+1}, x^{i+1}, y^i) \mapsto (-f^{i+1} x^{i+1}, x^{i+1}, 0).$$

It is easy to see that the required homotopy between $\theta \circ \psi$ and $\mathrm{id}_{C(v)}$ is

$$h^i : C(v)^i \longrightarrow C(v)^{i-1}; \ (y^{i+1}, x^{i+1}, y^i) \mapsto (y^i, 0, 0).$$

12. Axiom TR3

It is sufficient to consider the diagram TR3 (see IV.1.1) in the case when $Z = C(u)$, $Z' = C(u')$, v, w, v', w' are standard morphisms. In this case h can be taken in the form

$$h = f[1] \oplus g.$$

Before proceeding to the octahedron axiom, we prove one auxiliary result. An exact sequence of complexes $0 \to X \xrightarrow{u} Y \xrightarrow{v} Z \to 0$ is said to be *semi-split* if for any i there exists a morphism $w^i : Z^i \to Y^i$ such that

248   IV. Triangulated Categories

$v^i w^i = \mathrm{id}_{Z^i}$ ($w = (w^i)$ need not be a morphism of complexes). For example, the exact sequence (IV.1)

$$0 \longrightarrow K \xrightarrow{\bar{f}} \mathrm{Cyl}(f) \xrightarrow{\pi} C(f) \longrightarrow 0$$

is semi-split. Therefore, any distinguished triangle in $K(\mathcal{A})$ is isomorphic to a triangle $X \to Y \to Z \to X[1]$ such that $0 \to X \to Y \to Z \to 0$ is a semi-split exact sequence.

The converse is also true.

13. Lemma. *Any semi-split exact sequence in $\mathrm{Kom}(\mathcal{A})$ can be completed to a distinguished triangle $X \xrightarrow{u} Y \xrightarrow{v} Z \longrightarrow X[1]$ in $K(\mathcal{A})$.*

Proof. Selecting a splitting, we may (and will) assume that $Y^i = X^i \oplus Z^i$, u is the injection, v is the projection, and

$$d_Y^i\left(x^i, z^i\right) = \left(d_X^i x^i - f^i(z^i), d_Z^i z^i\right).$$

It is easy to see that $d_Y^2 = 0$ if and only if $f^i : Z^i \to X^{i+1}$ form a morphism of complexes

$$f = (f^i) : Z \longrightarrow X[1]. \qquad (\mathrm{IV}.2)$$

Let us show that the triangle

$$X \xrightarrow{u} X \oplus Z \xrightarrow{v} Z \xrightarrow{f} X[1] \qquad (\mathrm{IV}.3)$$

is distinguished. To do this we prove that it is isomorphic to the distinguished triangle

$$\begin{array}{ccccccc}
X & \xrightarrow{u} & X \oplus Z & \xrightarrow{v} & Z & \xrightarrow{f} & X[1] \\
\downarrow{\mathrm{id}} \;\text{①} & & \downarrow{\mathrm{id}} \;\text{②} & & \downarrow{g} \;\text{③} & & \downarrow{\mathrm{id}} \\
X & \xrightarrow{u} & X \oplus Z & \xrightarrow{s} & X[1] \oplus X \oplus Z = C(u) & \xrightarrow{t} & X[1]
\end{array}$$

where $g = (f, 0, \mathrm{id}_Z)$.

Necessary steps:

a) *g is a morphism of complexes.* This is clear from the formula for $d_{C(u)}$:

$$d_{C(u)}\left(x^{i+1}, x^i, z^i\right) = \left(-d_X x^{i+1}, x^{i+1} + d_X x^i - f(z^i), d_Z z^i\right).$$

b) *Commutativity of the squares ① and ③.* Obvious.
c) *Commutativity of the square ② modulo homotopy.* We have

$$\begin{aligned}
s & : & (x^i, z^i) &\mapsto (0, x^i, z^i) \\
g \circ v & : & (x^i, z^i) &\mapsto (f^i(z^i), 0, z^i).
\end{aligned}$$

The required homotopy k is given by

$$k^i : (X \oplus Z)^i \longrightarrow (X[1] \oplus X \oplus Z)^{i-1} : (x^i, z^i) \mapsto (x^i, 0, 0).$$

d) *g is a homotopy equivalence.* Let $g' : X[1] \oplus X \oplus Z \to Z$ be the projection to the third summand. Then $g' \circ g = \mathrm{id}_Z$ and $g' \circ g$ is given by

$$(g \circ g')^i : (x^{i+1}, x^i, z^i) \longrightarrow (f^i(z^i), 0, z^i).$$

The homotopy ℓ between $g \circ g'$ and $\mathrm{id}_{C(u)}$ is

$$\ell^i : (x^{i+1}, x^i, z^i) \longrightarrow (x^i, 0, 0)$$

(cf. IV.1.12).

□

14. Octahedron Axiom TR4

Assume we are given an upper cap of the octahedron (in the notation of IV.1.1). By Lemma IV.1.13 we can assume that both distinguished triangles in this upper cap are semi-split, so that

$$Y^i = X^i \oplus Z'^i, \; Z^i = Y^i \oplus X'^i = X^i \oplus Z'^i \oplus X'^i,$$

the corresponding maps are natural injections and projections, and the arrows marked [1] are related to the splitting by (IV.2) and (IV.3).

Denote these arrows as follows:

$$f : Z' \longrightarrow X[1], \quad (g, h) : X' \longrightarrow Y[1] = X[1] \oplus Z'[1].$$

a) *Construction of the lower cap.* Denote

$$Y'^i = Z'^i \oplus X'^i, \quad d_{Y'}^i (z'^i, x'^i) = (d_{Z'}^i z'^i + h^i(x'^i), d_{X'}^i(x'^i)).$$

It is easy to see that $Y' = \{Y'^i, d_{Y'}^i\}$ is a complex. Two of the four diagonal morphisms of the lower cap are the injection $Z' \to Y'$ and the projection $Y' \to X'$. The morphism $Z \to Y'$ is the projection

$$Z = X \oplus Z' \oplus X' \longrightarrow Z' \oplus X' = Y'.$$

Finally, the morphism $Y' \longrightarrow X[1]$ is $f \oplus g : Y' = Z' \oplus X' \longrightarrow X[1]$.

We have to check that triangles marked ⌢ are commutative and those marked ∗ are distinguished. Commutativity immediately follows from the definition of the morphisms. Distinguishability can be proved by applying the constructions from the proof of Lemma IV.1.13 to the semi-split exact sequences

$$0 \longrightarrow Z' \longrightarrow Y' \longrightarrow X' \longrightarrow 0,$$
$$0 \longrightarrow X \longrightarrow Z \longrightarrow Y' \longrightarrow 0.$$

b) *Equality in $K(\mathcal{A})$ of the two morphisms $Y \to Y'$ and of the two morphisms $Y' \to Y[1]$.* First of all, both morphisms $Y \to Y'$ are given by the same formula

$$(x^i, z^i) \in Y^i \mapsto (z^i, 0) \in Y'^i.$$

Next, the composition
$$Y' \longrightarrow X[1] \longrightarrow Y[1]$$
is given by
$$(z'^i, x'^i) \mapsto \left(f^i(z'^i) + g^i(x'^i), 0\right)$$
and the composition
$$Y' \longrightarrow X' \longrightarrow Y[1]$$
is given by
$$(z'^i, x'^i) \mapsto \left(g^i(z'^i), h^i(x'^i)\right).$$
An easy computation shows that
$$k^i : (z'^i, x'^i) \in Y'^i = Z'^i \oplus X'^i \mapsto (0, z'^i) \in X^i \oplus Z'^i = Y^i$$
is the required homotopy $k = \{k^i : Y'^i \longrightarrow Y^i = Y[1]^{i-1}\}$.

Exercises

1. Triangulated Categories and Abelian Categories. Let us recall that an abelian category $\mathcal{A}$ is said to be semisimple if any exact triple in $\mathcal{A}$ splits (see III.2.3, III.5.8).

We sketch here the proof of the following result. Let a triangulated category $\mathcal{C}$ be abelian. Then $\mathcal{C}$ is semisimple and any distinguished triangle in $\mathcal{C}$ is isomorphic to a triangle of the form

$$X \xrightarrow{f} Y \xrightarrow{g} \ker f[1] \oplus \operatorname{Coker} f \xrightarrow{h} X[1]$$

(with natural morphisms g and h). Conversely, if $\mathcal{A}$ is a semisimple abelian category and $T : \mathcal{A} \to \mathcal{A}$ is an arbitrary automorphism, then T, together with the above triangles, defines the structure of a triangulated category on $\mathcal{A}$.

The converse is proved straightforwardly. To prove the direct statement it suffices to check the following. Let $\mathcal{C}$ be an arbitrary triangulated category and $f : X \to Y$ be a monomorphism in $\mathcal{C}$. Then f is an isomorphism of X with a direct summand of Y.

Indeed, let $Z[-1] \xrightarrow{\alpha} X \xrightarrow{f} Y \xrightarrow{\beta} Z$ be a distinguished triangle constructed from f. Then $f \circ \alpha = 0$, so that $\alpha = 0$ (f is a monomorphism). Hence, $\alpha[1] \circ \operatorname{id}_Z = 0$ and there exists $\gamma : Z \to Y$ with $\beta \circ \gamma = \operatorname{id}_Z$. To prove that $f \oplus \gamma : X \oplus Z \to Y$ is an isomorphism it suffices to prove that a cone of $f \oplus \gamma$ is isomorphic to 0. This follows, for example, from the octahedron whose upper cap contains the commutative triangle

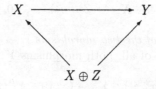

2. Examples of Triangulated Categories. Let $\mathcal{B}$ be an additive (but not necessarily abelian) category. Prove that, repeating definitions from III.4.1, we can introduce the category $K(\mathcal{B})$ of homotopic classes of complexes over $\mathcal{B}$. Show that the category $K(\mathcal{B})$ is triangulated (distinguished triangles and translation functor are defined as in the case of an abelian category).

In particular, let $\mathcal{M}$ be a family of objects of an abelian category $\mathcal{A}$, $\mathcal{M}^\oplus$ be a full subcategory of $\mathcal{A}$ consisting of finite direct sums of objects from $\mathcal{M}$. Then $\mathcal{M}^\oplus$ is an additive category and the corresponding triangulated category $K(\mathcal{M}^\oplus)$ is denoted by Tr $\mathcal{M}$.

Other examples of triangulated categories can be found in Beilinson et al. [1], Sect. I.1.

IV.2 Derived Categories Are Triangulated

1. Localizing Classes of Morphisms in Triangulated Categories

The derived category $\mathcal{D}^*(\mathcal{A})$ is obtained from $K^*(\mathcal{A})$ by localization with respect to some localizing class of morphisms (see III.2.6). We show that under certain conditions the localized category can be made triangulated in a natural way. Namely, the localizing class of morphisms should satisfy the following conditions of *compatibility with triangulation*:

a) $s \in S$ if and only if $T(s) \in S$.
b) If in the diagram TR3 (see IV.1.1) we have $f, g \in S$, then there exists a complementing morphism $h \in S$.

2. Theorem. *Let $\mathcal{D}$ be a triangulated category and S a localizing class of morphisms that is compatible with the triangulation. Define the translation functor T_S in $\mathcal{D}_S$ in the natural way ($T_S = T$ on $Ob\,\mathcal{D}_S = Ob\,\mathcal{D}$). A triangle in $\mathcal{D}_S$ is called distinguished if it is isomorphic to the image of a distinguished triangle in $\mathcal{D}$ under the localization $\mathcal{D} \to \mathcal{D}_S$.*

Then $\mathcal{D}_S$, with this translation functor and this class of distinguished triangles, is a triangulated category.

Proof. Recall first that a morphism $f : X \to Y$ in $\mathcal{D}_S$ is represented by a roof in $\mathcal{D}$, i.e., by a diagram in $\mathcal{D}$ of the form

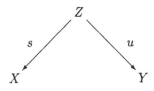

with $s \in S$. Two roofs are equivalent, i.e., represent the same morphism, if they can be included into one "common roof" (see III.2.8). Define T_S on morphisms by declaring that $T_S(f)$ for f as above is the equivalence class of the roof

252   IV. Triangulated Categories

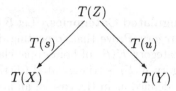

Here we use IV.2.1a): $T(s) \in S$. The reader can easily verify that this definition respects the equivalence of roofs.

Next we shall verify that these structures in $\mathcal{D}_S$ satisfy all the axioms TR1–TR4.

3. Axiom TR1

We must only check the property c). Let a morphism $X \xrightarrow{u} Y$ in $\mathcal{D}_S$ be represented by the roof $X \xleftarrow{s} Z \xrightarrow{u'} Y$, $s \in S$. Complete u' to a distinguished triangle
$$\Delta : Z \xrightarrow{u'} Y \xrightarrow{v} U \xrightarrow{w} Z[1]$$
in $\mathcal{D}$. Denote by Δ' the following triangle in $\mathcal{D}_S$:
$$\Delta' : X \xrightarrow{u} Y \xrightarrow{v} U \xrightarrow{s[1]w} X[1].$$
In $\mathcal{D}_S$ we have the morphism of triangles $(s, \mathrm{id}_Y, \mathrm{id}_U) : \Delta \to \Delta'$. It is in fact an isomorphism in $\mathcal{D}_S$, because s is invertible in $\mathcal{D}_S$. Therefore $X \xrightarrow{u} Y$ can be completed to a distinguished triangle.

4. Axiom TR2

This obviously follows from the definitions and from properties of T.

5. Axiom TR3

We can assume that the given distinguished triangles in $\mathcal{D}_S$ are represented by distinguished triangles in $\mathcal{D}$, and morphisms f, g in $\mathcal{D}_S$ are represented by roofs $(s, \tilde{f})$, $(t, \tilde{g})$. We must construct the dashed arrows r and $\tilde{h}$ in the following diagram:

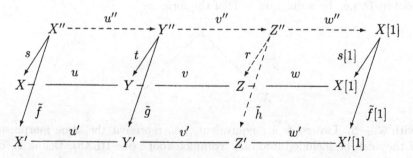

where the horizontal dashed arrows are some auxiliary morphisms.

IV.2 Derived Categories Are Triangulated   253

Step I. We claim that by changing, if necessary, the roof representing $f : X \to X'$ we can guarantee the existence of a morphism $u'' : X'' \to Y''$ in $\mathcal{D}$ such that both squares containing this morphism are commutative. Indeed, using the properties of a localizing class we can complete the following diagram to a commutative square in $\mathcal{D}$:

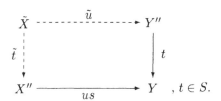

Replace X'' by $\tilde{X}$, s by $s\tilde{t}$, $\tilde{f}$ by $\tilde{f}\tilde{t}$. It is clear that $X \xleftarrow{s\tilde{t}} X'' \xrightarrow{\tilde{f}\tilde{t}} X'$ represents the same morphism $f : X \to X'$ in $\mathcal{D}_S$. Next, $\tilde{u} : \tilde{X} \to X''$ makes one of the two squares commutative (the back one): $t\tilde{u} = us\tilde{t}$. As for the second square, it commutes in $\mathcal{D}_S$ but not necessarily in $\mathcal{D}$: from $u'f = gu$ follows formally $u'\tilde{f}s^{-1} = gt^{-1}u = \tilde{g}\tilde{u}(s\tilde{t})^{-1}$, so $u'\tilde{f}\tilde{t} = \tilde{g}\tilde{u}$ in Mor $\mathcal{D}_S$.

To make the second square commutative in $\mathcal{D}$, we must change the representative of f once more. Let us consider two morphisms $\tilde{g}\tilde{u}, u\tilde{f}\tilde{t} : \tilde{X} \to Y'$ in $\mathcal{D}$. As they are equal in $\mathcal{D}_S$, we know they have a "left equalizer" $q : \tilde{\tilde{X}} \to \tilde{X}$, $q \in S$ (by one of the properties of a localizing system). Take $\tilde{\tilde{X}}$ as the new X''; the rest is clear.

Step II. Complete $u'' : X'' \to Y''$ to a distinguished triangle $X'' \xrightarrow{u''} Y'' \xrightarrow{v''} Z'' \xrightarrow{w''} X''[1]$ in $\mathcal{D}$. Using TR3 for $\mathcal{D}$, choose $\tilde{h}$ yielding a morphism of triangles. Construct r similarly, with the additional condition $r \in S$, using the property IV.2.1b). Denote by h the morphism $Z \to Z'$ in $\mathcal{D}$ represented by the roof $(r, \tilde{h})$. Obviously, (f, g, h) is the required morphism of triangles in $\mathcal{D}_S$.

6. Axiom TR4

Let

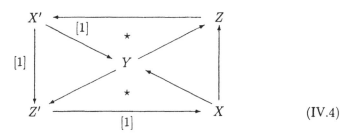

(IV.4)

254   IV. Triangulated Categories

be an upper cap in $\mathcal{D}_S$. Let f and g be represented by some roofs

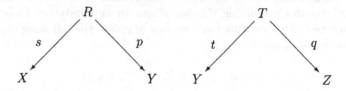

with $s, t \in S$. According to Lemma III.2.8, their composition is represented by the roof

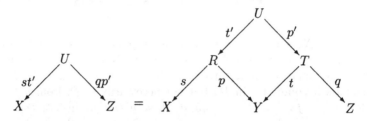

where $t' \in S$ and $pt' = tp'$. Since the roof

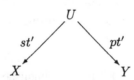

represents in $\mathcal{D}_S$ the same morphism f, we can assume, after relabelling, that morphisms $f, g, h = gf$ in $\mathcal{D}_S$ are represented by the roofs

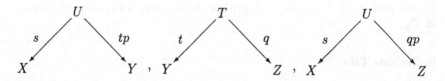

coming from the diagram

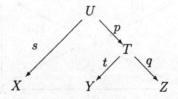

with $s, t \in S$.

Now let us consider the following cap in $\mathcal{D}$ constructed from morphisms $p: U \to T$, $q: T \to Z$:

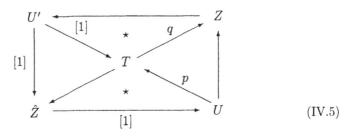

(IV.5)

We claim that the image of this upper cap in $\mathcal{D}_S$ is isomorphic to the original cap (IV.4). To prove this, consider the following diagram in $\mathcal{D}_S$ formed by distinguished triangles taken from (IV.4) and (IV.5):

$$\begin{array}{ccccccc} U & \xrightarrow{p} & T & \to & \hat{Z} & \to & U[1] \\ s\downarrow & & t\downarrow & & r\downarrow & & s[1]\downarrow \\ X & \xrightarrow{f} & Y & \to & Z' & \to & X[1] \end{array}$$

In this diagram, the left square is commutative and s, t are isomorphisms in $\mathcal{D}_S$. By the axiom TR3 for $\mathcal{D}_S$, which we have already verified, there exists a morphism $r : \hat{Z} \to Z'$ in $\mathcal{D}_S$ that makes the diagram commutative. By Corollary IV.1.4 (whose proof is based on axioms TR1–TR3 only) r is an isomorphism in $\mathcal{D}_S$, so that (s, t, r) is an isomorphism of triangles in $\mathcal{D}_S$.

Similar arguments applied to upper distinguished triangles in caps (IV.5) and (IV.4) show that there exists an isomorphism $r' : U' \to X'$ is $\mathcal{D}_S$ such that

$$\begin{array}{ccccccc} T & \xrightarrow{q} & Z & \to & U' & \to & T[1] \\ t\downarrow & & \text{id}\downarrow & & r'\downarrow & & t[1]\downarrow \\ Y & \xrightarrow{f} & Z & \to & X' & \to & Y[1] \end{array}$$

is an isomorphism of distinguished triangles. Therefore, (s, t, id, r, r') is an isomorphism of caps (IV.5) and (IV.4).

Now, using the axiom TR4 for $\mathcal{D}$, we complete the upper cap (IV.5) to an octahedron in $\mathcal{D}$ with the lower cap

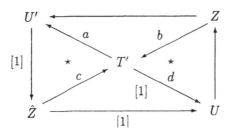

IV. Triangulated Categories

Then the lower cap

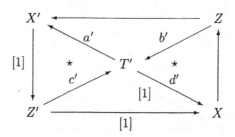

with $a' = r'a$, $b' = b$, $c' = cr^{-1}$, $d' = s[1]d$ is a lower cap in $\mathcal{D}_S$ which completes the upper cap (IV.4) to an octahedron in $\mathcal{D}_S$.

7. Corollary. *Derived categories $D^*(\mathcal{A})$ are triangulated.*

Proof. To apply Theorem IV.2.2 to the situation $D = K^*(\mathcal{A})$, $S = \{\text{quasi-isomorphisms}\}$ we must only verify that S is compatible with the structure of triangulated category in $K^*(\mathcal{A})$ given by Theorem IV.1.8.

But property IV.1.1a) for S is clear, and property IV.1.1b) follows from the five-lemma (Ex. 5.6) applied to the diagram consisting of two exact cohomology sequences (III.3.6) for two triangles in TR3. □

In $D^*(\mathcal{A})$ Lemma IV.1.13 from the previous section can be strengthened as follows.

8. Proposition. *Any exact triple of complexes $0 \longrightarrow X \xrightarrow{u} Y \xrightarrow{v} Z \longrightarrow 0$ in $\mathrm{Kom}(\mathcal{A})$ can be completed to a distinguished triangle in $D^*(\mathcal{A})$ by an appropriate morphism $Z \xrightarrow{w} X[1]$, and any distinguished triangle in $D^*(\mathcal{A})$ is isomorphic to one obtained in this way.*

Proof. Everything follows from the main diagram from Lemma III.3.3: the required distinguished triangle is

$$X \xrightarrow{\bar{u}} \mathrm{Cyl}\, u \longrightarrow C(u) \xrightarrow{\delta} X[1].$$

Proposition III.3.5 shows that the original exact triple is quasi-isomorphic to this triangle. The last assertion can be proved by arguments similar to those used for $K^*(\mathcal{A})$ in IV.1.12. □

9. Cones and the Octahedron Axiom for Morphisms Represented by Monomorphisms of Complexes

Using Proposition IV.2.8 we can give the following interpretation of a cone and of the octahedron axiom in $D^*(\mathcal{A})$.

If a morphism $u : X \to Y$ in $D^*(\mathcal{A})$ is represented by monomorphism of complexes then the cone $C(u)$ is the quotient $Y/u(X)$ and $Y \to C(u)$ is the factorization. Let us consider now the octahedron diagram (see 1), assuming the morphisms $X \to Y$ and $Y \to Z$ (and so $X \to Z$ also) are represented by monomorphisms of complexes. We claim that the octahedron axiom asserts essentially the existence of the natural isomorphism

$$Z/Y \simeq (Z/X)/(Y/X).$$

In fact, we have $X' = Z/Y$, $Z' = Y/X$. Next, the lower cap shows that $Y' = Z/X$ and that $Z' \to Y'$ is represented by the natural injection. Finally, the third vertex of the left distinguished triangle in the lower cap is represented by the quotient $Y'/Z' = (Z/X)/(Y/X)$, and the same element in the upper cap is represented by the quotient Z/Y.

10. Extensions

Proposition IV.2.8 justifies the following terminology. Let $\mathcal{D}$ be a triangulated category. An object Y in $\mathcal{D}$ is said to be an extension of Z by X if there exists a distinguished triangle $X \to Y \to Z \to X[1]$ in $\mathcal{D}$. A full subcategory $\mathcal{D}'$ in $\mathcal{D}$ is said to be stable under extensions if, for any distinguished triangle $X \to Y \to Z \to X[1]$, the condition $X, Z \in \mathrm{Ob}\, \mathcal{D}'$ implies $Y \in \mathrm{Ob}\, \mathcal{D}'$.

Exercises

1. Postnikov Systems and Convolutions. a) Let $\mathcal{D}$ be a triangulated category and $X^\bullet = \{X^0 \xrightarrow{d^0} X^1 \xrightarrow{d^1} \cdots \longrightarrow X^n\}$ be a finite complex over $\mathcal{D}$ (i.e., the composition of any two consequent d's is zero). By a *right Postnikov system* subordinated to $X^\bullet$ we mean a diagram of the form

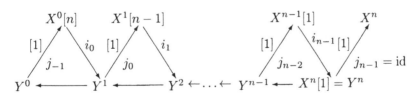

in which all triangles

$$X^\nu[n-\nu] \xrightarrow{i_\nu} Y^{\nu+1} \longrightarrow Y^\nu \xrightarrow{j_{\nu+1}} X^\nu[n-\nu+1]$$

are distinguished and $j_\nu \circ i_\nu : X^\nu[n-\nu] \longrightarrow X^{\nu+1}[n-\nu]$ coincides with $d^\nu[n-\nu]$ (so, in particular, $i_{n-1} = d^{n-1}[1]$).

By a *left Postnikov system* subordinated to $X^\bullet$ we mean a diagram of the form

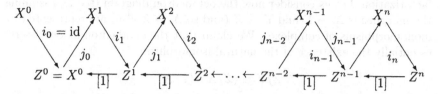

in which all triangles

$$Z^\nu \xrightarrow{j_\nu} X^{\nu+1} \xrightarrow{i_{\nu+1}} Z^{\nu+1} \longrightarrow Z^\nu[1]$$

are distinguished and $j_\nu \circ i_\nu$ coincides with d^ν (so, in particular, $j_0 = d^0$).

Right (resp. *left*) *convolution* of a complex $X^\bullet$ is an object $T \in \mathrm{Ob}\,\mathcal{D}$ such that there exists a right (resp. left) Postnikov system subordinated to $X^\bullet$ and satisfying $T = Y^0[n-1]$ (resp. $T = Z^n[n]$).

b) Prove that the class of all right convolutions of $X^\bullet$ coincides with the class of all left convolutions of $X^\bullet$.

Sketch of the Proof. The case $n = 1$ is clear. Let $n = 2$ and T be a right convolution of the complex $\left\{ X^0 \xrightarrow{d^0} X^1 \xrightarrow{d^1} X^2 \right\}$. Using the corresponding right Postnikov system we can construct the following lower cap:

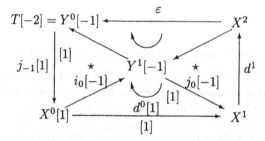

and complete it to an octahedron diagram with an upper cap

Then one can easily see that $T[-2][2] = T$ is the convolution of the following left Postnikov system:

IV.2 Derived Categories Are Triangulated   259

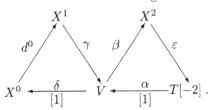

For $n > 2$ we use the induction. Let us consider the case $n = 3$. Let $T = Y^0[2]$ be the right convolution of the complex $X^\bullet = \{X^0 \longrightarrow X^1 \longrightarrow X^2 \longrightarrow X^3\}$ corresponding to the Postnikov system

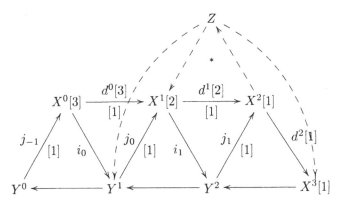

Let Z be the third vertex of a distinguished triangle of the form

$$X^2[1] \longrightarrow Z \longrightarrow X^1[2] \xrightarrow{d^1[2]} X^2[2].$$

Using the octahedron axiom we construct, similarly to the case $n = 2$, morphisms $r : Y^1[-1] \longrightarrow Z$, $q : Z \longrightarrow X^3[1]$ such that $d^2 = qu$, $j_0[-1] = vr$. Defining $p = r[-1]i_0[-2]$ let us consider the complex

$$X^0[1] \xrightarrow{p} Z[-1] \xrightarrow{q[-1]} X^3.$$

One can easily see that $Y^0[1]$ is its right convolution. Using the induction assumption, let

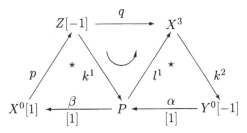

by the corresponding left Postnikov system. The reader can check that the diagram

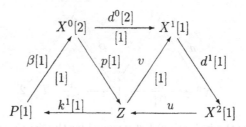

is a right Postnikov system subordinated to the complex $X^0 \longrightarrow X^1 \longrightarrow X^2$ so that $P[2]$ is its convolution. Using the induction once more one can construct the corresponding left Postnikov system for this complex. Joining it with the above left Postnikov system we get the required left Postnikov system subordinated to the initial complex $X^\bullet$, which shows that T is a left convolution of $X^\bullet$.

Denote the class of all convolutions of a complex $X^\bullet$ by Tot $X^\bullet$. Sometimes Tot $X^\bullet$ is empty, and sometimes it contains many nonisomorphic objects (for $n \geq 3$).

c) *Hypersimplices.* By a *distinguished $(n+1)$-dimensional hypersimplex* in a triangulated category $\mathcal{D}$ we mean a family G of objects $X^{[i,j]}, 0 \leq i \leq j \leq n$, and of morphisms

$$f_{ijk} : X^{[j+1,k]} \longrightarrow X^{[i,k]},$$
$$h_{ijk} : X^{[i,j]} \longrightarrow X^{[j+1,k]}[1],$$
$$g_{ijk} : X^{[i,k]} \longrightarrow X^{[i,j]},$$

for all triples $i \leq j \leq k$ such that all triangles

$$X^{[j+1,k]} \xrightarrow{f_{ijk}} X^{[i,k]} \xrightarrow{g_{ijk}} X^{[i,j]} \xrightarrow{h_{ijk}} X^{[j+1,k]}[1]$$

are distinguished in $\mathcal{D}$, and all triangles of the form

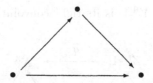

in G are commutative. For $n = 1$, G is a distinguished triangle, for $n = 2$ it is an octahedron diagram.

In general, a (k,n)-*hypersimplex* is a polyhedron $\triangle(k,n)$ in $\mathbb{R}^n$ whose vertices are centres of k-dimensional faces of the n-dimensional simplex $\triangle^n$. A distinguished $(n+1)$-dimensional hypersimplex in $\mathcal{D}$ is a diagram whose objects are placed in the vertices of the hypersimplex $\triangle(1, n+1)$ and morphisms correspond to edges of $\triangle(1, n+1)$ ($X^{[i,j]}$ for $0 \leq i \leq j \leq n$ is placed at the centre of the edge $(i, j+1)$ of $\triangle^{n+1}$). The hypersimplex $\triangle(1, n+1)$ had n-dimensional faces of two types: $(n+2)$ n-dimensional simplices $\triangle^n$ and $(n+2)$ hypersimplices $\triangle(1, n)$ (find them). The corresponding diagrams in a distinguished $(n+2)$-dimensional hypersimplex in $\mathcal{D}$ are commutative simplices and distinguished n-dimensional hypersimplices respectively (similarly to the division of faces of a distinguished octahedron into commutative and distinguished triangles). The *equator* of a distinguished $(n+1)$-dimensional hypersimplex is its subdiagram

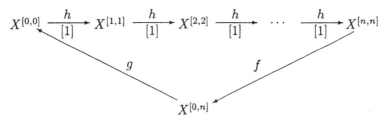

It can be defined as a unique oriented cycle E of length $n+1$ in G such that no two consecutive edges of E lie in the same n-dimensional face. The composition of any two consecutive morphisms of the equator is zero.

Prove that convolutions of a complex $X^0 \xrightarrow{d^0} X^1 \xrightarrow{d^1} \ldots \longrightarrow X^n$ can be defined as such objects T that for some $\alpha : T \longrightarrow T^0$, $\beta : X^n[-n] \longrightarrow T$ the diagram

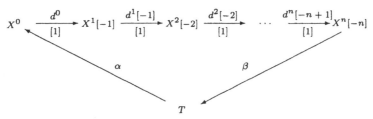

is the equator of some distinguished $(n+1)$-dimensional hypersimplex in $\mathcal{D}$. Deduce the existence of the following cyclic symmetry: if T is a convolution of $X^0 \longrightarrow X^1 \longrightarrow \ldots \longrightarrow X^n$, then X^0 is a convolution of $X^1 \longrightarrow X^2 \longrightarrow \ldots \longrightarrow X^n \longrightarrow T[n]$.

d) Let
$$X^\bullet = \{X^0 \longrightarrow X^1 \longrightarrow \ldots \longrightarrow X^n\},$$
$$Y^\bullet = \{Y^0 \longrightarrow Y^1 \longrightarrow \ldots \longrightarrow Y^m\}$$

be two complexes, $S \in \text{Tot } X^\bullet$, $T \in \text{Tot } Y^\bullet$ be their convolutions (so that we have morphisms $\alpha : X^n \longrightarrow S[n]$, $\beta : T[n] \longrightarrow Y^0[n]$) and $f : S \longrightarrow T$ be an arbitrary morphism contained in a distinguished triangle

$$S \xrightarrow{f} T \longrightarrow U \longrightarrow S[1].$$

Prove that $U \in \mathrm{Tot}\, Z^\bullet$, where $Z^\bullet$ is the complex
$$\left\{X^0 \longrightarrow X^1 \longrightarrow \ldots \longrightarrow X^n \xrightarrow{\varphi} Y^0[n] \longrightarrow Y^1[n] \longrightarrow \ldots \longrightarrow Y^m[n]\right\}$$
with $\varphi = \beta \circ f[n] \circ \alpha$.

Examples of Convolutions. e) Let $\mathcal{D} = D(\mathcal{A})$ be the derived category of an abelian category $\mathcal{A}$, $F^0 K^\bullet \subset F^1 K^\bullet \subset \ldots \subset F^n K^\bullet = K^\bullet$ be a complex over $\mathcal{A}$ with a finite increasing filtration. The diagram

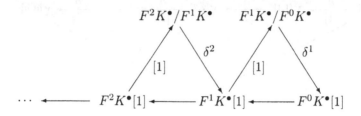

is the right Postnikov system subordinated to the complex
$$(K^\bullet / F^{n-1} K^\bullet)[-n] \longrightarrow \ldots$$
$$\ldots \longrightarrow (F^2 K^\bullet / F^1 K^\bullet)[-2] \longrightarrow (F^1 K^\bullet / F^0 K^\bullet)[-1] \longrightarrow F^0 K^\bullet$$
(here δ^i are morphisms in distinguished triangles corresponding to exact sequences of complexes $0 \longrightarrow F^{i-1} K^\bullet \longrightarrow F^i K^\bullet \longrightarrow F^i K^\bullet / F^{i-1} K^\bullet \longrightarrow 0$), whose convolution is $K^\bullet[n]$.

f) Let $\mathcal{A}$ be an additive category, $\mathcal{D} = K^b(\mathcal{A})$ the homotopy category of bounded complexes over $\mathcal{A}$. A *twisted complex* over $\mathcal{A}$ is a bigraded object $C = \{C^{ij}\}$, $C^{ij} = 0$ for all but a finite number of pairs (i, j), endowed with a family of endomorphisms $d_k : C \to C$ of bidegree $(k, 1-k)$, $k = 0, 1, \ldots$, such that $\left[\sum d_k\right]^2 = 0$.

In particular, $d_0^2 = 0$, which means that each $C^{i,\bullet}$ is a complex over $\mathcal{A}$, $d_0 d_1 + d_1 d_0 = 0$, which means that $d_1 : C^{i,\bullet} \to C^{i+1,\bullet}$ is a morphism of complexes, $d_1^2 = d_2 d_0 + d_0 d_2$, which means that d_1^2 is homotopic to 0. Hence, we have a complex over $K^b(\mathcal{A})$

$$\ldots \longrightarrow (C^{i,\bullet}, d_0) \xrightarrow{d_1} (C^{i+1,\bullet}, d_0) \longrightarrow \ldots .$$

Show that $(C^\bullet, d)$ where $C^n = \bigoplus_{i+j=n} C^{ij}$, $d = \sum d_k$ is a convolution of this complex in $K^b(\mathcal{A})$.

2. Spectral Sequences. Let $\mathcal{D}$ be a triangulated category, $H : \mathcal{D} \to \mathcal{A}$ be a cohomological functor from $\mathcal{D}$ to an abelian category $\mathcal{A}$, $H^p(X) = H(X[p])$ for $X \in \mathrm{Ob}\,\mathcal{D}$. Let $X^\bullet$ be a finite complex over $\mathcal{D}$, $T \in \mathrm{Tot}\, X^\bullet$ its convolution. Then there exists a spectral sequence with $E_1^{pq} = H^q(X^p)$, $E^n = H^n(T)$.

To construct it let us consider a subordinated to $X^\bullet$ right Postnikov system with $T = Y^0[n-1]$. Acting on various objects of this system by appropriate powers of the translation functor, we obtain a diagram of the following form (we assume that $X^j = 0$ for $j > n$):

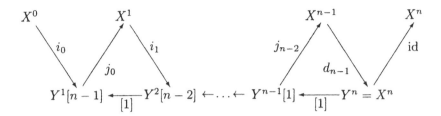

The functor $H^\bullet$ makes from this diagram a bigraded exact couple

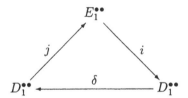

where $E_1^{pq} = H^q(X^p)$, $D_1^{pq} = H^q(Y^p[n-p])$, and i, j, δ have the bidegrees $(1,0)$, $(0,0)$, $(1,-1)$ respectively.

Prove that the spectral sequence corresponding to this exact couple converges to $H^n(T)$ (cf. Ex. III.7.3c)).

b) Let $\mathcal{A}$ be an abelian category with sufficiently many injective objects. Prove the following generalization of Proposition III.7.11. Let $K^\bullet \in \mathrm{Ob\,Kom}^+(\mathcal{A})$, $\{F^p K^\bullet\}$ a finite filtration of $K^\bullet$. Then all $F^p K^\bullet$ have matched Cartan–Eilenberg resolutions. More precisely, $K^\bullet$ has a Cartan–Eilenberg resolution $L^{\bullet\bullet} = \{L^{ij}\}$ with a finite filtration $\{F^p L^{\bullet\bullet}\}$ such that $F^p L^{\bullet\bullet}$ is a Cartan–Eilenberg resolution of $F^p K^\bullet$ (for the map $F^p K^\bullet \longrightarrow F^p L^{\bullet,0}$ induced by $K^\bullet \longrightarrow L^{\bullet,0}$).

Next, let $X^\bullet$, $Y^\bullet$ be two bounded complexes over $\mathcal{A}$. Then there exist two Eilenberg–Moore spectral sequences, both converging to $E_\infty^n = \mathrm{Hom}_{D^b(\mathcal{A})}(X^\bullet, Y^\bullet[n])$, with

$$^I E_1^{pq} = \bigoplus_{i-j=p} \mathrm{Ext}^q_\mathcal{A}(X^i, Y^j),$$

$$^{II} E_1^{pq} = \bigoplus_{i-j=p} \mathrm{Ext}^q_\mathcal{A}\left(H^i(X^\bullet), H^j(Y^\bullet)\right).$$

Sketch of the construction. Let us consider a somewhat more general situation where we have two bounded complexes $X^\bullet$, $Y^\bullet$ endowed with finite filtrations

$\{F^i X^\bullet\}$, $\{F^i Y^\bullet\}$. We construct a spectral sequence with the same limit E_∞^n as above, and with

$$E_1^{pq} = \bigoplus_{i-j=p} \mathrm{Hom}_{D^b(\mathcal{A})} \left(F^i X^\bullet / F^{i+1} X^\bullet, (F^j Y^\bullet / F^{j+1} Y^\bullet)[p+q] \right).$$

To do this we must extend the notion of a Cartan–Eilenberg resolution to complexes, to introduce on $I^\bullet(X^\bullet)$ a filtration $F^i I^\bullet(X^\bullet) = I^\bullet(F^i X^\bullet)$, and to consider the natural filtration in $\mathrm{Hom}^\bullet(I^\bullet(X^\bullet), Y^\bullet)$ induced by filtrations in $I^\bullet(X^\bullet)$ and in $Y^\bullet$. To obtain Eilenberg–Moore spectral sequences we have to take as $F^i X^\bullet$ and $F^i Y^\bullet$ stupid or canonical filtrations in $X^\bullet$ and $Y^\bullet$ (see III.7.5). The appearance of $^{II}E_2$ is due to the fact that in the grading related to the canonical filtration the differential acts by the chess springer move.

c) In the assumptions of b) we immediately obtain that if $\mathrm{Ext}_\mathcal{A}^q(X^i, Y^i) = 0$ for all $p > 0$ and all i, j, then $\mathrm{Hom}_{D^b(\mathcal{A})}(X^\bullet, Y^\bullet) = \mathrm{Hom}_{K^b(\mathcal{A})}(X^\bullet, Y^\bullet)$. Prove that this is true in any abelian category (not using Cartan–Eilenberg resolutions). This fact generalizes Ex. III.5.1a).

3. Massey Products. a) Let $\mathcal{D}$ be a triangulated category and $X^\bullet$ be a complex of length 4 over $\mathcal{D}$:

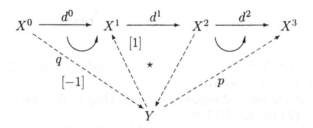

Let Y be the third vertex of a distinguished triangle containing d^1. Since $d^1 d^0 = d^2 d^1 = 0$, there exist morphisms $p : Y \longrightarrow X^3$, $q : X^0 \longrightarrow Y[-1]$ such that the first and the third triangles in this diagram are commutative (Proposition IV.1.3). The composition $p[-1] \circ q : X^0 \longrightarrow X^3[-1]$ is called the triple Massey product and is denoted by $\langle d^0, d^1, d^2 \rangle$. The nonuniqueness in the choice of p and q leads to the nonuniqueness of $\langle d^0, d^1, d^2 \rangle$. Prove that the class of $\langle d^0, d^1, d^2 \rangle$ modulo the subgroup

$$A = \mathrm{Hom}(X^1, X^3[-1]) \circ d^0 + d^2[-1] \circ \mathrm{Hom}(X^0, X^2[-1]) \subset \mathrm{Hom}(X^0, X^3[-1])$$

does not depend on the choices. Show that the complex $X^\bullet$ has at least one Postnikov system (so that Tot $X^\bullet$ is nonempty) if and only if the image of $\langle d^0, d^1, d^2 \rangle$ in $\mathrm{Hom}(X^0, X^3[-1])/A$ vanishes.

b) Let $X^\bullet$ be a complex of the length $n+1$:

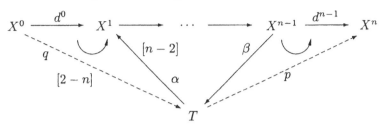

Denote by $\langle d^0, d^1, \ldots, d^{n-1}\rangle$ the subset of $\mathrm{Hom}(X^0, X^n[2-n])$ consisting of such θ that there exist a convolution T of the complex $X^1 \longrightarrow X^2 \longrightarrow \ldots \longrightarrow X^n$ and morphisms q and p (see the diagram) such that the first and the last triangles are commutative and $\theta = p[2-n] \circ q$ (here α and β come from the Postnikov system that represents T as a convolution). Prove that

(i) $\langle d^0, d^1, \ldots, d^{n-1}\rangle$ is nonempty if and only if $0 \in \langle d^1, d^2, \ldots, d^{n-1}\rangle$ and $0 \in \langle d^0, d^1, \ldots, d^{n-2}\rangle$.

(ii) $X^\bullet$ has at least one Postnikov system if and only if $0 \in \langle d^0, d^1, \ldots, d^{n-1}\rangle$.

c) Let $\mathcal{D} = K(\mathcal{A})$ for an abelian category $\mathcal{A}$. In this case one can give another definition of Massey products. For $U^\bullet, V^\bullet \in \mathrm{Ob}\,\mathrm{Kom}(\mathcal{A}) = \mathrm{Ob}\,K(\mathcal{A})$ denote by ∂ the differential in the complex $\mathrm{Hom}^\bullet(U^\bullet, V^\bullet)$, so that ∂-cocycles of degree i are morphisms of complexes $U^\bullet \longrightarrow V^\bullet[i]$, ∂-coboundaries are morphisms homotopic to 0, and $\mathrm{Hom}_{K(\mathcal{A})}(U^\bullet, V^\bullet[i]) = H^i(\mathrm{Hom}^\bullet(U^\bullet, V^\bullet))$.

Let $X^0 \xrightarrow{d^0} X^1 \xrightarrow{d^1} X^2 \xrightarrow{d^2} X^3$ be a complex of length 4 over $K(\mathcal{A})$, i.e., $\partial d^i = 0$, $d^1 d^0 = \partial u$, $d^2 d^1 = \partial v$ for some $u \in \mathrm{Hom}^{-1}(X^0, X^2)$, $v \in \mathrm{Hom}^{-1}(X^1, X^3)$. Prove that $\theta = d^2[-1]u - vd^0 \in \mathrm{Hom}^{-1}(X^0, X^3)$ is a ∂-cocycle, i.e., defines a morphism of complexes $X^0 \longrightarrow X^3[-1]$. Prove that its class in $\mathrm{Hom}_{K(\mathcal{A})}(X^0, X^3[-1])/A$, where A is the subgroup defined in a), does not depend on the choices and coincides with $\langle d^0, d^1, d^2\rangle$.

d) Let $X^0 \xrightarrow{d^0} X^1 \xrightarrow{d^1} X^2 \xrightarrow{d^2} X^3 \xrightarrow{d^3} X^4$ be a complex of length 5 in $K(\mathcal{A})$ and $\langle d^0, d^1, d^2\rangle = 0$, $\langle d^1, d^2, d^3\rangle = 0$ (modulo the corresponding subgroups). This means that there exist $u \in \mathrm{Hom}^{-1}(X^0, X^2)$, $v \in \mathrm{Hom}^{-1}(X^1, X^3)$, $w \in \mathrm{Hom}^{-1}(X^2, X^4)$, $\alpha \in \mathrm{Hom}^{-2}(X^0, X^3)$, $\beta \in \mathrm{Hom}^{-2}(X^1, X^4)$, $\tau \in \mathrm{Hom}^{-2}(X^1, X^3)$ such that $d^1 d^0 = \partial u$, $d^2 d^1 = \partial v$, $d^3 d^2 = \partial w$, $d^2 u - v d^0 = \partial \alpha + \tau d^0$, $d^3 v - w d^1 = \partial \beta - d^3 \tau$. Verify that $\theta = d^3 \alpha - wu + \beta d^0 \in \mathrm{Hom}^{-2}(X^0, X^4)$ is a ∂-cocycle, so that we get a morphism $X^0 \longrightarrow X^4[-1]$ in $\mathrm{Kom}(\mathcal{A})$. Prove that its class in $\mathrm{Hom}_{K(\mathcal{A})}(X^0, X^4[-2])$ is defined uniquely modulo elements from

$$B = d^3[-2] \circ \mathrm{Hom}(X^0, X^3[-2])$$
$$+ \mathrm{Hom}(X^2[-1], X^4[-2]) \circ \mathrm{Hom}(X^0, X^2[-1])$$
$$+ \mathrm{Hom}(X^1, X^4[-2]) \circ d^0.$$

Verify that $\langle d^0, d^1, d^2, d^3\rangle = \theta + B$.

e) Extend this to all n.

f) *Massey products and the tangent space.* Let X be an affine algebraic variety over $\mathbb{C}$, A its function algebra, $x \in X$ a (possibly singular) point, $\mathbb{C}_x = A/m_x$ the residue field in x, considered as an A-module. Prove that $\mathrm{Ext}^1_A(m_x, m_x)$ can be identified with the Zarisky tangent space $T_x X$ to X at x, and

$$\left\{ \xi \in \mathrm{Ext}^1_A(m_x, m_x), \quad \text{such that} \quad \overbrace{\langle \xi, \ldots, \xi \rangle}^{n \text{ times}} = 0 \quad \text{for all} \quad n \right\}$$

corresponds to the tangent cone $T\mathbb{C}_x X = \mathrm{Spec}\left(\oplus m_x^i / m_x^{i+1} \right)$ to X at x.

4. Thick Subcategories. There exists an alternative way to look at the localization in triangulated categories, when we start not from morphisms that become isomorphisms, as in Theorem IV.2.2, but from objects that become isomorphic to the zero object. A complete triangulated subcategory $\mathcal{C}$ of a triangulated category $\mathcal{D}$ is said to be *thick* (in French, *épaisse*) if it satisfies the following condition:

(T) Assume that a morphism $f: X \to Y$ in $\mathcal{D}$ can be factored through an object from $\mathcal{C}$ (i.e., f can be represented as a composition $X \longrightarrow V \longrightarrow Y$ with $V \in \mathrm{Ob}\, \mathcal{C}$) and enters a distinguished triangle $X \xrightarrow{f} Y \longrightarrow Z \longrightarrow X[1]$ with $Z \in \mathrm{Ob}\, \mathcal{C}$. Then $X, Y \in \mathrm{Ob}\, \mathcal{C}$.

A standard example of a thick subcategory is the category of all acyclic objects in $K(\mathcal{A})$. Indeed, the first condition in (T) means that $H^\bullet(f)$ is the zero morphism, and the second condition means that $H^\bullet(f)$ is an isomorphism, which implies that both X and Y are acyclic.

The relation between thick subcategories in $\mathcal{D}$ and localizing classes of morphisms is as follows. A localizing class S in $\mathrm{Mor}(\mathcal{D})$ is said to be *saturated* if $s \in S \Leftrightarrow \{\text{there exist morphisms } f, f' \text{ in } \mathcal{D} \text{ such that } f \circ s \in S \text{ and } s \circ f' \in S\}$.

Prove the following result. Let $\mathcal{D}$ be a triangulated category. Then

$$\mathcal{C} \mapsto \varphi(\mathcal{C}) = \left\{ \begin{aligned} & s \in \mathrm{Mor}\, \mathcal{C}, \quad s \text{ is contained in a distinguished} \\ & \text{triangle } X \xrightarrow{s} Y \longrightarrow Z \longrightarrow X[1] \text{ with } Z \in \mathrm{Ob}\, \mathcal{C}. \end{aligned} \right\}$$

determines a one-to-one correspondence between the set of thick subcategories in $\mathcal{D}$ and the set of saturated localizing classes in $\mathcal{D}$ compatible with triangulation.

The converse mapping associates with a class $S \subset \mathrm{Mor}\, \mathcal{D}$ the full subcategory $\psi(S)$ generated by such objects $Z \in \mathrm{Ob}\, \mathcal{D}$ that there exists a distinguished triangle $X \xrightarrow{s} Y \longrightarrow Z \longrightarrow X[1]$ with $s \in S$.

IV.3 An Example: The Triangulated Category of Λ-Modules

1. Grassmann Algebras

Let k be a field of characteristic $\neq 2$, and let E be a vector space over k of dimension $n+1$. We consider the $\mathbb{Z}$-graded exterior algebra

$$\Lambda = \Lambda(E) = \bigoplus_{i=0}^{n+1} \Lambda^i E.$$

Denote by $\mathcal{M}(\Lambda)$ the category of left unitary $\mathbb{Z}$-graded modules over Λ with degree preserving homomorphisms as morphisms. Let $\mathcal{M}^b(\Lambda)$ be the full subcategory in $\mathcal{M}(\Lambda)$ consisting of finitely generated (i.e., finite dimensional over k) Λ-modules.

2. Operations with Λ-Modules

a) Let V be a Λ-module. For $m \in \mathbb{Z}$ let

$$V(m)^i = V^{i-m}, \quad V(m) = \bigoplus_{i \in \mathbb{Z}} V(m)^i,$$

with the action of Λ on $V(m)$ induced by the action of Λ on V. It is clear that the degree shift $V \mapsto V(m)$ gives a functor $\mathcal{M}(\Lambda) \to \mathcal{M}(\Lambda)$ (which is the identity on morphisms: $f \mapsto f(m) = f$).

b) For two Λ-modules V, V' let

$$V \otimes V' = \bigoplus_\ell (V \otimes V')^\ell, \quad (V \otimes V')^\ell = \bigoplus_{i+j=\ell} V^i \otimes_k V'^j,$$

$$e(c \otimes v') = ev \otimes v' + (-1)^{\deg v} v \otimes ev', \quad e \in E \subset \Lambda, v \in V, v' \in V'.$$

So in this section we always take tensor products *over k, and not over Λ*.

c) Any left Λ-module has a canonical structure of a right Λ-module V_r given by

$$v\lambda = (-1)^{\deg v \deg \lambda} \lambda v.$$

It is clear that the map $V \mapsto V_r$ produces a functor that gives an isomorphism of the categories of left and right (graded) Λ-modules.

d) Let $V^* = \text{Hom}_k(V, k)$ be supplied with the Λ-action $(\lambda \varphi)(v) = (-1)^{\deg \lambda \deg v} \varphi(\lambda v)$. Together with the grading $(V^*)^i = \text{Hom}_k(V^{-i}, k)$, this defines a Λ-module structure on V^*.

3. Proposition. *For $V \in \mathcal{M}^b(\Lambda)$ the following conditions are equivalent:*

a) V is free (that is, $V = \bigoplus_i \Lambda(m_i)$ for some $m_i \in \mathbb{Z}$),

b) V is a projective object in $\mathcal{M}^b(\Lambda)$,

c) V is an injective object in $\mathcal{M}^b(\Lambda)$.

Proof. Obviously, a) ⇒ b). To prove b) ⇒ a), first note that $\Lambda(m)$ is indecomposable in $\mathcal{M}^b(\Lambda)$ (if $\Lambda(m) = V \oplus V'$, then $\dim V^m + \dim V'^m = 1$, and if, say, $\dim V^m = 1$, then $V = \Lambda(m)$ and $V'^k = \{0\}$). Now let V be a projective Λ-module, which can be assumed to be indecomposable. Then V is a direct summand of some free Λ-module $F = \oplus \Lambda(m_i)$. The Krull–Schmidt theorem says that the set of indecomposable summands of F is determined uniquely by F, so V is isomorphic to one of the $\Lambda(m_i)$s.

To prove the equivalence a) ⇔ c) let us choose a nonzero element $\omega \in \Lambda^{n+1}(E)$ and define a linear functional α on Λ by $\alpha(\omega) = 1$, $\alpha(\lambda) = 0$ for $\lambda \in \Lambda^j$, $j \neq n+1$. For any m define the pairing

$$\Lambda(m) \times \Lambda(-n-m-1) \longrightarrow k$$

by

$$\lambda \cdot 1(m) \times \lambda' 1 \cdot (-n-m-1) \longrightarrow \alpha(\lambda\lambda').$$

$(1(m), 1(-n-m-1)$ are the generators in $\Lambda(m), \Lambda(-n-m-1))$. We must check that this pairing is nondegenerate. To do this, one has to find, for any $0 \neq \lambda \in \Lambda$, a $\lambda' \in \Lambda$ such that $\lambda\lambda' = \omega$. Let $e_0 \ldots e_n$ be a basis in E such that $\omega = e_0 \ldots e_n$. Decompose λ into a sum of monomials in e_i's and let $ce_0^{\nu_0} \ldots e_n^{\nu_n}$, $c \neq 0$, $\nu_i = 0$ or 1, be one of the nonzero monomonials of the smallest degree in this decomposition. Then λ' can be taken to be $\lambda' = \pm c^{-1} e_0^{\mu_0} \ldots e_n^{\mu_n}$ with $\mu_i = 0, 1$ complementary to λ_i (that is, $\mu_i + \lambda_i = 1$ for all $i = 0, \ldots, n$).

The nondegeneracy of the above pairing implies that $\Lambda(m)^* = \Lambda(-n-m-1)$. Since $* : \mathcal{M}^b(\Lambda) \to \mathcal{M}^b(\Lambda)$ is a contravariant functor and $*^2 = \mathrm{Id}$, we see that $*$ interchanges projective and injective objects in $\mathcal{M}^b(\Lambda)$. So a) ⇔ c) follows from a) ⇔ b). □

4. The Category $\mathcal{M}^b(\Lambda)/\mathcal{F}$

Let $\mathcal{F} \subset \mathcal{M}^b(\Lambda)$ be the full subcategory of $\mathcal{M}^b(\Lambda)$ formed by all free graded Λ-modules. A morphism $f : V \to V'$ in $\mathcal{M}^b(\Lambda)$ is said to be equivalent to zero if it can be factored as $V \to F \to V'$ with $F \in \mathrm{Ob}\,\mathcal{F}$. It is clear that the set of all morphisms equivalent to zero is a two-sided ideal $I \subset \mathrm{Mor}\,\mathcal{M}^b(\Lambda)$. Define the category $\mathcal{M}^b(\Lambda)/\mathcal{F}$ by

$$\mathrm{Ob}\,\mathcal{M}^b(\Lambda)/\mathcal{F} = \mathrm{Ob}\,\mathcal{M}^b(\Lambda), \quad \mathrm{Mor}\,\mathcal{M}^b(\Lambda)/\mathcal{F} = \left(\mathrm{Mor}\,\mathcal{M}^b(\Lambda)\right)/I.$$

The main result of this section is the following theorem.

5. Theorem. *$\mathcal{M}^b(\Lambda)/\mathcal{F}$ has the natural structure of a triangulated category.*

We shall describe this structure below. Its most unusual feature is the form taken by the translation functor T, namely

$$T(V) = \left(\Lambda(-n) \underset{k}{\otimes} V\right)/i(V)(-n), \quad n+1 = \dim E, \qquad (\mathrm{IV}.6)$$

where $i(V) = \Lambda^{n+1}(E) \underset{k}{\otimes} V \subset \Lambda \underset{k}{\otimes} V$ (see IV.3.7 below).

6. The Plan of the Proof

Any Λ-module $V = \oplus V^j$ determines a family of complexes of linear spaces indexed by elements $e \in E$:

$$L_e(V) = \ldots \longrightarrow V^{j-1} \xrightarrow{d^{j-1}(e)} V^j \xrightarrow{d^j(e)} V^{j+1} \longrightarrow \ldots,$$

where $d^j(e) : V^j \to V^{j+1}$ is the multiplication by e. It is clear that $L_e(V)$ and $L_{ce}(V)$ for $c \in k^*$ are isomorphic, so the $L_e(V)$ are indexed essentially by points of the projectivization $\mathcal{P}(E)$ of E. Using the language of algebraic geometry, we say that a complex of quasi-coherent $\mathcal{O}$-modules on $\mathcal{P}(E)$ is *rigid* if it is isomorphic to a complex

$$L : \ldots \longrightarrow V^j \otimes \mathcal{O}(j) \longrightarrow V^{j+1} \otimes \mathcal{O}(j+1) \longrightarrow \ldots,$$

where the V^j are linear spaces. The rigid complex L is said to be *finite* if it is bounded and all V^j are finite dimensional.

Given a rigid complex L we construct the graded Λ-module $V(L)$ by setting

$$V(L) = \bigoplus_j V(L)^j = \bigoplus_j \Gamma\left(L^j(-j)\right).$$

To define the Λ-action on $V(L)$ let us consider the mapping

$$a = \Gamma\left(d^j(-j)\right) : V(L)^j \longrightarrow V(L)^{j+1} \otimes E^*$$

where we use the canonical isomorphism $\Gamma(\mathcal{O}(1)) \simeq E^*$. For $e \in E$, $v \in V(L)^j$ define

$$ev = (-1)^j (\mathrm{id} \otimes s_e) a(v),$$

where $s_e : E^* \to k$ is the convolution with e. The equality $d^{j+1} d^j = 0$ implies that $e^2 v = 0$, so that $V(L)$ is a graded Λ-module.

It is easy to check that the above construction gives a functor from the category *Rig* of rigid complexes to $\mathcal{M}(\Lambda)$. This functor is an equivalence of categories. The (quasi)-inverse functor has, in fact, already been constructed: a family of mappings $d^j(e) : V^j \to V^{j+1}$ which is linear in e is the same as a linear mapping $V^j \to V^{j+1} \otimes E^*$, and this mapping determines a morphism of sheaves $V^j \otimes \mathcal{O} \to V^{j+1} \otimes \mathcal{O}(1)$ on $\mathcal{P}(E)$. Tensoring it with $\mathrm{id}_{\mathcal{O}(j)}$, we get the differential $d^j : V^j \otimes \mathcal{O}(j) \to V^{j+1} \otimes \mathcal{O}(j+1)$. The objects of $\mathcal{M}^b(\Lambda)$ correspond under this equivalence to finite rigid complexes.

Now let us consider the composition of functors

$$\Phi : \mathcal{M}^b(\Lambda) \longrightarrow Rig^b \longrightarrow \mathcal{D}^b,$$

where $\mathcal{D}^b$ is the derived category of bounded complexes of quasi-coherent sheaves on $\mathcal{P}(E)$.

Theorem IV.3.5 is a consequence of the following facts, to be proved below:

a) The essential image of Φ (i.e., the full subcategory of $\mathcal{D}^b$ consisting of all objects that are isomorphic to objects of the form $\Phi(V)$) is closed under the translation functor T.

b) There is a natural isomorphism
$$\operatorname{Hom}_{\mathcal{M}^b(\Lambda)}(V, W) \bmod I \xrightarrow{\sim} \operatorname{Hom}_{\mathcal{D}^b}(\Phi(V), \Phi(W)).$$

c) The construction used to verify axioms TR1c and TR4 (see IV.1.1), when applied to diagrams in the essential images of Φ, stays within this essential image.

Indeed, using a)–c), we see that the essential image of Φ has the structure of a triangulated category induced from the one on $\mathcal{D}^b$; the structure passes to the equivalent category $\mathcal{M}^b(\Lambda)/\mathcal{F}$. (The essential image of Φ can be proved to coincide with $\mathcal{D}^b$; see IV.3.1).

7. The Translation Functor

Define $T(V)$ for $V \in \mathcal{M}^b(\Lambda)$ as in (IV.6). We must prove that $\Phi(T(V))$ is isomorphic to $T(\Phi(V))$ in $\mathcal{D}^b$.

Let $W \in \operatorname{Rig}^b$ be a complex $\{W^j = V^{j+1} \otimes \mathcal{O}(j), d_W = -d_V\}$. Let $\mathcal{O}(1)$ be a complex with a single nonzero term in degree 0, equal to $\mathcal{O}(1)$. It is clear that $T(\Phi(V)) = W \otimes \mathcal{O}(1)$ (tensor product of complexes of $\mathcal{O}$-modules). The complex $\mathcal{O}(1)$ is quasi-isomorphic to the rigid complex
$$\tilde{\mathcal{O}}(1) : \ldots \longrightarrow 0 \longrightarrow \mathcal{O}(-n) \longrightarrow E \otimes \mathcal{O}(-n+1) \longrightarrow \ldots \longrightarrow \Lambda^n(E) \otimes \mathcal{O} \longrightarrow 0$$
corresponding to the Λ-module $\Phi\bigl(\Lambda/\Lambda^{n+1}(E)(-n)\bigr)$. Let us choose a quasi-isomorphism $q : \tilde{\mathcal{O}}(1) \to \mathcal{O}(1)$. (This choice is equivalent to the choice of a nonzero vector in the one-dimensional space $\Lambda^{n+1}(E)$). It is easy to see that $\operatorname{id} \otimes q : W \otimes \tilde{\mathcal{O}}(1) \to W \otimes \mathcal{O}(1)$ is also a quasi-isomorphism. But looking at (IV.6) one can infer that $W \otimes \tilde{\mathcal{O}}(1)$ is isomorphic to $\Phi(T(V))$.

To compare morphisms in $\mathcal{M}^b$ and $\mathcal{D}^b$ we need some auxiliary results.

8. Lemma. *In the notation of III.5.4 we have, for finite-dimensional linear spaces V, W,*
$$\operatorname{Hom}_{\mathcal{D}^b}(V \otimes \mathcal{O}(i)[a], W \otimes \mathcal{O}(j)[b]) = \operatorname{Hom}(V, W) \otimes \operatorname{Ext}^{b-a}(\mathcal{O}(i), \mathcal{O}(j)),$$
$$\operatorname{Ext}^{\ell}(\mathcal{O}(i), \mathcal{O}(j)) = \begin{cases} S^{j-i}(E^*) & \text{for } \ell = 0, j \geq i, \\ S^{-n-1+i-j}(E) \otimes \Lambda^{n+1}(E) & \text{for } \ell = n, j \leq i - n - 1, \\ 0 & \text{otherwise.} \end{cases}$$

Proof. The first equality follows from the additivity of Hom and from the definition of Ext^k (see remark III.5.5b)). The second equality follows from the classical results about cohomology of sheaves $\mathcal{O}(i)$ on projective spaces (see Hartshorne [2] or Okonek et al. [1]). $\square$

IV.3 An Example: The Triangulated Category of Λ-Modules 271

9. Surjectivity of Φ on Morphisms

Let $\varphi : \Phi(V) \to \Phi(W)$ be a morphism of rigid complexes in the derived category. We show that it is induced by the usual morphism $\tilde{\varphi}$ of complexes.

We denote by V (resp. W) both the Λ-module and the corresponding rigid complex; the components of the first one are V^i (resp. W^i), and those of the second one are $V^i(i) = V^i \otimes \mathcal{O}(i)$ (resp. $W^i(i) = W^i \otimes \mathcal{O}(i)$).

We construct $\tilde{\varphi} : V \to W$ in Rig^b by induction on the length of the complex $V \oplus W$. The case of length 0 (both complexes are zero) is trivial. Now let i be the maximum number such that $V^i \oplus W^i \ne \{0\}$. Let us consider two distinguished triangles

$$\begin{array}{ccccc} V^i(i)[-i] & \longrightarrow & V & \longrightarrow & V' \\ & & \downarrow \varphi & & \\ W^i(i)[-i] & \longrightarrow & W & \longrightarrow & W' \end{array} \quad (\text{IV.7})$$

where $V' = \sigma_{<i}V$, $W' = \sigma_{<i}W$. (Recall that $(\sigma_{<k}C)^j = C^j$ for $j < k$, $(\sigma_{<k}C)^j = 0$ for $j \ge k$). Let us check that we can apply Corollary IV.1.5. To do this we will prove that $\text{Hom}_{\mathcal{D}^b}\left(V^i(i)[-i], W'\right) = 0$ using induction on the length of W'. Let ℓ be the maximum number with $W'^\ell \ne 0$ (it is essential that $\ell < i$). From the triangle

$$W'^\ell(\ell)[-\ell] \longrightarrow W' \longrightarrow W''$$

we obtain the inclusion $\text{Hom}_{\mathcal{D}^b}\left(V^i(i)[-i], W'\right) \to \text{Hom}_{\mathcal{D}^b}\left(V^i(i)[-i], W''\right)$ (note that $\text{Hom}_{\mathcal{D}^b}\left(V^i(i)[-i], W'^\ell(\ell)[-\ell]\right) = 0$ because $\ell < i$ and because of Lemma IV.3.8). The latter group is zero by the induction hypothesis.

Now Corollary IV.1.5 shows that there exist morphisms $\tilde{\varphi}^i : V^i(i)[-i] \to W^i(i)[-i]$ and $\varphi' : V' \to W'$ in $\mathcal{D}^b$ such that $(\tilde{\varphi}^i, \varphi, \varphi')$ is a morphism of triangles. By Lemma IV.3.8, $\tilde{\varphi}^i$ is a morphism of complexes, since it is induced by the unique morphism of vector spaces $V^i \to W^i$. By the induction assumption φ' is also induced by the morphism of rigid complexes $\tilde{\varphi}^{<i} : V' \to W'$. Let us show that the concatenation $(\tilde{\varphi}^{<i}, \tilde{\varphi}^i) = \tilde{\varphi}$ is a morphism of complexes. To do this we consider the commutative diagram

$$\begin{array}{ccccc} V^{i-1}(i-1)[-i+1] & \longrightarrow & V' & \longrightarrow & V^i(i)[-i+1] \\ \downarrow \tilde{\varphi}^{i-1} & & \downarrow \tilde{\varphi}^{<i} & & \downarrow \tilde{\varphi}^i \\ W^{i-1}(i-1)[-i+1] & \longrightarrow & W' & \longrightarrow & W^i(i)[-i+1] \end{array}$$

Here the right square is the continuation of (IV.7) one step to the right, and the left square is the part of the diagram (IV.7) for $(V', W', \varphi^{<i})$ instead of (V, W, φ). The composition of morphisms in the upper (resp. lower) row of this diagram is $d_V^{i-1} : V^{i-1}(i-1) \to V^i(i)$ (resp. $d_W^{i-1} : W^{i-1}(i-1) \to W^i(i)$). The commutativity of the outer square means that $(\tilde{\varphi}^{<i}, \tilde{\varphi}^i)$ is a morphism of complexes.

10. The Kernel of Φ on Morphisms

a) I is in the kernel because Φ transforms free Λ-modules into acyclic rigid complexes, and those are isomorphic to zero in $\mathcal{D}^b$.

b) Let $\varphi = \Phi(\tilde{\varphi}) = 0$. We show that this implies $\tilde{\varphi} = 0$ if we impose an additional condition on the Λ-module W, namely $\Lambda^{n+1}(E)W = 0$. Such W's will be called *reduced* modules.

Let us show that if W is reduced, then $\operatorname{Hom}_{\mathcal{D}^b}\left(V^i(i)[-i], W'[-1]\right)$ acts trivially on the set of extensions $\tilde{\varphi}^i$ of morphisms of triangles in (IV.7) (see Lemma IV.1.5). Indeed, $\tilde{\varphi}^i$ is determined uniquely up to a morphism $V^i(i)[-i] \to W^i(i)[-i]$ that can be factored through $W'[-1] \to W^i(i)[-i]$. Splitting off the components of $W'[-1]$ one by one (starting from the right) as above, and using Lemma IV.3.9 each time to calculate $\operatorname{Hom}_{\mathcal{D}^b}$, we see that $\operatorname{Hom}\left(V^i(i)[-i], W'[-1]\right)$ is a quotient of $\operatorname{Hom}\left(V^i, W^{i-n-1} \otimes \Lambda^{n+1} E\right)$. An element of this latter Hom yields the morphism $V^i \to W^i$ equal to its composition with the Λ-multiplication $W^{i-n-1} \otimes \Lambda^{n+1} E \to W^i$. So if W is reduced, $\tilde{\varphi}^i$ is determined uniquely. In particular, $\tilde{\varphi}^i = 0$ whenever $\varphi = 0$.

But if W is reduced, then W' is also reduced. So, by induction, $\tilde{\varphi} = 0$.

c) A Λ-module W is isomorphic to a direct sum of a free module and a reduced module. Indeed, let $L = \Lambda^{n+1}(E)W \subset W$ and let $\ell_1, \ldots, \ell_r$ be a homogeneous basis in L. Let $w_1, \ldots, w_r$ be homogeneous elements in W such that $\ell_i = \omega w_i$ (ω is a fixed nonzero element in $\Lambda^{n+1}E$; see IV.3.3). It follows from the proof of nondegeneracy of the pairing in IV.3.3 that $F = \Lambda(w_1, \ldots, w_r) \subset W$ is a free submodule of W, $F = \oplus \Lambda(m_i)$, $m_i = \deg w_i$. By Proposition IV.3.3, F is an injective module, so $W = F \oplus W_0$ for some W_0. It is clear that W_0 is reduced.

Let us note also that, by the Krull–Schmidt theorem, both F and W_0 are unique up to an isomorphism. However, the decomposition $W = F \oplus W_0$ is not unique.

d) Now we can complete the proof. Let $\tilde{\varphi} : V \to W$, $W = F \oplus W_0$ with F free, W_0 reduced, and let $\Phi(\tilde{\varphi}) = 0$. By b) the composition $V \xrightarrow{\tilde{\varphi}} W \xrightarrow{\text{pr}} W_0$ (pr is the projection) is zero. Therefore, $\tilde{\varphi}$ is of the form $V \longrightarrow F \xrightarrow{(\text{id},0)} W$, so that $\tilde{\varphi} \in I$.

11. Axiom TR1c

We must prove that any morphism u in $\mathcal{M}^b(\Lambda)/\mathcal{F}$ can be included into a distinguished triangle. Working instead with the essential image $\Phi(\operatorname{Rig}^b)$, we can find the required triangle using a cone of u, and then construct the quasi-isomorphic object in Rig^b as in IV.3.6.

Another proof will be useful in IV.3.12. First, any morphism $V \to W$ in $\mathcal{M}^b(\Lambda)/\mathcal{F}$ is the composition of a monomorphism in $\mathcal{M}^b(\Lambda)$ and an isomorphism mod $\mathcal{F}$. Indeed, we can embed V into a free module $V \otimes \Lambda \otimes \Lambda^{n+1}(E^*)(-n-1)$, add this free module to W and then project onto W.

So we may assume that u is an embedding of a submodule. Then the corresponding morphism of rigid complexes is also an embedding and we can take the third object in the distinguished triangle to be the corresponding quotient complex, which is the image of the quotient module in $\mathcal{M}^b(\Lambda)$.

12. Axiom TR4

It is sufficient to show that for any upper cap consisting of rigid complexes one can construct a lower cap also consisting of rigid complexes. By IV.3.11, we can assume that $X \to Y$ and $Y \to Z$ in the upper cap are embeddings. But then IV.2.9 shows that the object Y' in the lower cap is Z/X.

Exercises

1. S-Modules and Λ-Modules. Let, as in IV.3.1, $\Lambda = \Lambda(E)$ be the exterior algebra of a finite-dimensional vector space E, $S = S(E^*)$ the symmetric algebra of the dual space E^*. The Serre theorem (see Hartshorne, [2], Chap. II, Sect. 5) describes coherent sheaves on the projective space $\mathsf{P}(E)$ in terms of S-modules. Our Theorem IV.3.5 can be obtained from the Serre theorem and the results below on the relation between categories of S-modules and Λ-modules.

a) Let $\mathcal{M}(\Lambda)$ be the category from IV.3.1 and $\mathcal{C}(\Lambda)$ the category of bounded complexes
$$\cdots \longrightarrow V_j \xrightarrow{\partial_j} V_{j+1} \longrightarrow \cdots$$
in which V_j are objects of $\mathcal{M}(\Lambda)$ and ∂_j are linear grading preserving mappings that anticommute with Λ (i.e., $\partial_j ev = -e\partial_j v$ for $v \in V_j$, $e \in E \subset \Lambda$). Morphisms in $\mathcal{C}(\Lambda)$ are morphisms of complexes (commuting with Λ). Let $\mathcal{D}(\Lambda)$ be the localization of $\mathcal{C}(\Lambda)$ by quasi-isomorphisms.

On the other hand, let $\mathcal{M}(S)$ be the category of graded S-modules (we take $\deg x = 1$ for $x \in E^* \subset S$), $\mathcal{C}(S)$ the category of complexes over $\mathcal{M}(S)$ and $\mathcal{D}(S)$ the corresponding derived category (localization of $\mathcal{M}(S)$ by quasi-isomorphisms).

Let also $\mathcal{C}^b(\Lambda)$, $\mathcal{C}^b(S)$ be the full subcategories of $\mathcal{C}(\Lambda)$, $\mathcal{C}(S)$ formed by bounded complexes of finitely generated complexes and $\mathcal{D}^b(\Lambda)$, and let $\mathcal{D}^b(S)$ be their localizations by quasi-isomorphisms.

Prove that (with the natural definition of the translation functors and of distinguished triangles) $\mathcal{D}(\Lambda)$, $\mathcal{D}(S)$ are triangulated categories and $\mathcal{D}^b(\Lambda)$, $\mathcal{D}^b(S)$ are their full triangulated subcategories.

b) *The functor $F : \mathcal{C}(\Lambda) \to \mathcal{C}(S)$.* Let $V^\bullet = (V_j, \partial_j) \in \mathrm{Ob}\, \mathcal{C}(\Lambda)$, and let $V_j = \underset{i}{\oplus} V_j^i$ be the decomposition into homogeneous components. Define
$$W_k^l = \underset{m}{\oplus} \left(S^m \otimes V_{k+l-m}^{m-l} \right), \quad W_k = \underset{l}{\oplus} W_k^l$$

(where S^m is the homogeneous component of degree m in S). The formula $s(s_1 \otimes v) = ss_1 \otimes v$ makes each W_k an object of $\mathcal{M}(S)$. Let $\{e_p\}$, $\{x_p\}$ be dual bases in E, E^*. Defining differentials $d_k : W_k \longrightarrow W_{k+1}$ by $d_k(s \otimes v) = \sum x_p s \otimes e_p v + s \otimes \partial v$ we obtain an object $W^\bullet$ of $\mathcal{C}(S)$. Verify that the map $V^\bullet \mapsto W^\bullet$ can be extended to a functor $F : \mathcal{C}(\Lambda) \longrightarrow \mathcal{C}(S)$ such that $F\left(\mathcal{C}^b(\Lambda)\right) \subset \mathcal{C}^b(S)$.

c) Define the functor $G : \mathcal{C}(S) \longrightarrow \mathcal{C}(\Lambda)$ as follows: for $W^\bullet = (W_k, d_k) \in$ Ob $\mathcal{C}(S)$, $W_k = \bigoplus_l W_k^l$, let

$$V_j^i = \bigoplus_m \mathrm{Hom}\left(\Lambda^m, W_{i+j+m}^{-i-m}\right), \quad V_j = \oplus V_j^i,$$

$$e\varphi(\lambda) = \varphi(\lambda e), \quad \partial\varphi(\lambda) = -\sum x_p \varphi(\lambda e_p) + d(\varphi(\lambda)).$$

Verify that these formulas define the functor $G : \mathcal{C}(S) \longrightarrow \mathcal{C}(\Lambda)$, which is right adjoint to F.

d) Prove that F can be extended to a functor $F_D : \mathcal{D}^b(\Lambda) \longrightarrow \mathcal{D}^b(S)$ which is an equivalence of categories (and G defines a quasi-inverse to F_D). To do this use the properties of the corresponding Koszul complex (see Ex. I.7.5).

e) Algebras $S(E^*)$, $\Lambda(E)$ are mutually dual quadratic algebras, and some of the results above remain valid in a more general situation (see Happel [1]).

f) Let $\mathcal{P}$ be the full subcategory of $\mathcal{D}^b(S)$ formed by complexes isomorphic (in $\mathcal{D}^b(S)$) to bounded complexes of finite-dimensional modules. Use the Serre theorem to verify that $\mathcal{D}^b(S)/\mathcal{P} = \mathcal{D}^b$ (see IV.3.6).

g) Let $\mathcal{I}$ be the full subcategory of $\mathcal{D}^b(\Lambda)$ formed by complexes isomorphic (in $\mathcal{D}^b(\Lambda)$) to bounded complexes of free Λ-modules. Prove that $\mathcal{D}^b(\Lambda)/\mathcal{I} = \mathcal{M}^b(\Lambda)$.

h) Verify that F_D establishes an equivalence between $\mathcal{P}$ and $\mathcal{I}$ and, therefore, between $\mathcal{M}^b(\Lambda)/\mathcal{F}$ and the bounded derived category of the category $\mathrm{Coh}_{\mathsf{P}(E)}$ of coherent algebraic sheaves on $\mathsf{P}(E)$.

2. Other Descriptions of $\mathcal{D}^b$. Two other descriptions of the derived category $\mathcal{D}^b = \mathcal{D}^b(\mathrm{Coh}_{\mathsf{P}(F)})$ were obtained by A.A. Beilinson [1].

a) Let $n = \dim E$ and let $\mathcal{M}_{[0,n]}(\Lambda)$ be the full subcategory of $\mathcal{M}^b(\Lambda)$ formed by finite direct sums of free modules $\Lambda[i]$ with $0 \leq i \leq n$. Denote by $\mathcal{C}_{[0,n]}(\Lambda)$ the corresponding full subcategory of $\mathcal{C}^b(\Lambda)$ and by $K_{[0,n]}(\Lambda)$ the corresponding homotopy category.

Similarly, let $S[i]$ be the free graded S-module with one generator of degree i, $\mathcal{M}_{[0,n]}(S)$ the full subcategory of $\mathcal{M}^b(S)$ formed by finite direct sums of modules $S[i]$, $0 \leq i \leq n$, $\mathcal{C}_{[0,n]}(S)$ the corresponding full subcategory of $\mathcal{C}^b(S)$ and $K_{[0,n]}(S)$ the corresponding homotopy category.

Beilinson's theorem says that each of triangulated categories $K_{[0,n]}(\Lambda)$, $K_{[0,n]}(S)$ is equivalent to $D^b(\mathrm{Coh}_{\mathsf{P}(E)})$. To establish the equivalence with $K_{[0,n]}(\Lambda)$ we replace a complex of sheaves by a quasi-isomorphic complex whose terms are finite direct sums of sheaves $\mathcal{O}(i)$, $0 \leq i \leq n$, and to establish the equivalence with $K_{[0,n]}(S)$ we use complexes whose terms are finite direct sums of sheaves $\Omega^i(i)$, $0 \leq i \leq n$.

b) Generalizing the approach adopted in this section one can describe some other categories of coherent algebraic sheaves (on flag manifolds, on quadrics and their complete intersections, etc.) as $\text{Tr}\mathcal{C}$ for appropriate additive categories $\mathcal{C}$ (see Kapranov [1–3]).

The next batch of exercises gives yet another method to construct triangulated categories; this method can be considered as a generalization of Theorem IV.3.5 in this section.

3. Exact Categories. a) Let $\mathcal{A}$ be an abelian category, $\mathcal{B}$ its full additive subcategory. Assume $\mathcal{B}$ is closed under extensions. This means, by definition, that in each exact triple $0 \to X' \to X \to X'' \to 0$ in $\mathcal{A}$ with $X', X'' \in \text{Ob}\mathcal{B}$ the object X is isomorphic to an object from $\mathcal{B}$. The pair $(\mathcal{B}, \mathcal{E})$, where $\mathcal{E}$ is the class of triples in $\mathcal{B}$ that are exact in $\mathcal{A}$, is called an *exact category*. In particular, each abelian category $\mathcal{A}$ is exact ($\mathcal{E}$ is the class of all exact triples in $\mathcal{A}$).

One can give the definition of an exact category $(\mathcal{B}, \mathcal{E})$ that does not involve a larger abelian category $\mathcal{A}$ (see, e.g., Quillen [4]). There exists a canonical way to represent $\mathcal{B}$ as a full subcategory of an abelian category, namely, of the category of additive functors $F : \mathcal{B}^0 \to \text{Ab}$ such that the sequence, $0 \to F(X'') \to F(X) \to F(X')$, of abelian groups is exact for any triple $(X' \to X \to X'') \in \mathcal{E}$.

Each additive category can be made an exact category in at least one way (for example, taking for $\mathcal{E}$ the class of all split triples $X \to X \oplus Y \to Y$).

Using the fact that $\mathcal{B}$ is closed under extensions prove that if $X, Y, Z \in \text{Ob}\mathcal{B}$ and

is a cartesian square in $\mathcal{A}$, then it is cartesian in $\mathcal{B}$ as well. Formulate and prove the corresponding property of cocartesian squares.

b) *$\mathcal{E}$-injective and $\mathcal{E}$-projective objects.* Let $(\mathcal{B}, \mathcal{E})$ be an exact category. An object $I \in \text{Ob } \mathcal{E}$ is said to be *$\mathcal{E}$-injective* if any triple $(I \to Y \to Z) \in \mathcal{E}$ splits. The class of all injective objects will be denoted by $\mathcal{I}_\mathcal{E}$. Similarly, $P \in \text{Ob } \mathcal{E}$ is said to be *$\mathcal{E}$-projective* if any triple $(X \to Y \to P) \in \mathcal{E}$ splits. The class of all projective objects will be denoted by $\mathcal{P}_\mathcal{E}$.

Verify the following property of injective objects. If

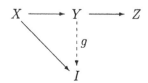

is a diagram in $\mathcal{B}$ with $(X \to Y \to Z) \in \mathcal{E}$ and $I \in \mathcal{I}_\mathcal{E}$ then there exists a morphism $g : Y \to I$ which makes it commutative.

Formulate and prove the corresponding property of $\mathcal{E}$-projective objects.

4. Frobenius Categories. a) An exact category $(\mathcal{B}, \mathcal{E})$ is called a *Frobenius category* if $\mathcal{I}_\mathcal{E} = \mathcal{P}_\mathcal{E}$ and for any $X \in \operatorname{Ob} \mathcal{B}$ there exist triples $Y \to I \to X$ and $X \to I' \to Y'$ in $\mathcal{E}$ with $I, I' \in \mathcal{I}_\mathcal{E}$. In other words, projective and injective objects coincide and there are sufficiently many of them in $\mathcal{B}$.

In b)–d) examples of Frobenius categories are given.

b) An abelian category $\mathcal{M}^b(\Lambda)$ of finite-dimensional graded Λ-modules is a Frobenius category (Proposition IV.3.3).

c) An abelian category of finite-dimensional modules over the group algebra $k[G]$ of a finite group G is a Frobenius category. More generally, the category of finite-dimensional modules over any Frobenius k-algebra (for the definition see Curtis and Reiner [1]) is a Frobenius category.

d) Let $\mathcal{B}'$ be an additive category with split idempotents (i.e., any morphism $\alpha : X \to X$ in $\mathcal{B}'$ with $\alpha^2 = \alpha$ is the projection onto a direct summand). Let $\mathcal{B} = \operatorname{Kom}^b(\mathcal{B}')$; define $\mathcal{E}$ as the class of all triples $X^\bullet \to Y^\bullet \to Z^\bullet$ such that for any i the sequence $X^i \to Y^i \to Z^i$ splits. Then $\mathcal{E}$ is a Frobenius category and $\mathcal{E}$-projective ($= \mathcal{E}$-injective) complexes are finite direct sums of complexes of the form $\dots 0 \longrightarrow X \xrightarrow{\operatorname{id}} X \longrightarrow 0 \dots$ with $X \in \operatorname{Ob} \mathcal{B}'$.

5. Stable Category. a) Let $\mathcal{B}$ be a Frobenius category. For $X, Y \in \operatorname{Ob} \mathcal{B}$ denote by $I(X, Y)$ the set of all morphisms $f : X \to Y$ in $\mathcal{B}$ that can be factored through an object from $\mathcal{I}_\mathcal{E}$. Define the corresponding *stable category* $\mathcal{B}_0$ by setting $\operatorname{Ob} \mathcal{B}_0 = \operatorname{Ob} \mathcal{B}$, $\operatorname{Hom}_{\mathcal{B}_0}(X, Y) = \operatorname{Hom}_\mathcal{B}(X, Y)/I(X, Y)$. Verify that the composition in $\mathcal{B}_0$ is well defined and that $\mathcal{B}_0$ is an additive category.

6. Suspension. a) Let $\mathcal{B}$ be a Frobenius category and

$$\begin{array}{ccccc} X & \longrightarrow & I & \longrightarrow & Y \\ \| & & \downarrow u & & \downarrow v \\ X & \longrightarrow & I' & \longrightarrow & Y' \end{array}$$

be a diagram with rows from $\mathcal{E}$ and with $I, I' \in \mathcal{I}_\mathcal{E}$. Verify that there exist morphisms $u : I \to I'$, $v : Y \to Y'$ making this diagram commutative. Verify further that if (u, v), $(\tilde{u}, \tilde{v})$ are two such pairs of morphisms, then the images of v and $\tilde{v}$ in $\operatorname{Hom}_{\mathcal{B}_0}(Y, Y')$ coincide. This implies that for any such pair (u, v) the image of v in $\operatorname{Hom}_{\mathcal{B}_0}(Y, Y')$ is an isomorphism. Show that the existence of a canonical extension $v \in \operatorname{Hom}_{\mathcal{B}_0}(Y, Y')$ enables us to define the *suspension functor* $T : \mathcal{B}_0 \to \mathcal{B}_0$ such that for any $X \in \operatorname{Ob} \mathcal{B}_0 = \operatorname{Ob} \mathcal{B}$ there exists a triple $(X \to I \to TX) \in \mathcal{E}$ with $I \in \mathcal{I}_\mathcal{E}$.

b) Using the equality $\mathcal{I}_\mathcal{E} = \mathcal{P}_\mathcal{E}$ prove that T is an autoequivalence of the category $\mathcal{B}_0$. In order to do this, construct a quasi-inverse to T using arguments dual to those in a).

c) Show that, replacing possibly the category $\mathcal{B}_0$ by an equivalent one, we may assume T to be an automorphism.

IV.3 An Example: The Triangulated Category of Λ-Modules 277

7. Distinguished Triangles. Now let $X, Y \in \mathrm{Ob}\,\mathcal{B}$, $u : X \to Y$ an arbitrary morphism in $\mathcal{B}$ and $X \xrightarrow{i} I \xrightarrow{p} TX$ a triple from $\mathcal{E}$ with $I \in \mathcal{I}_\mathcal{E}$. Check that in the diagram

$$\begin{array}{ccccc} X & \xrightarrow{i} & I & \xrightarrow{p} & TX \\ \downarrow u & & \downarrow t & & \| \mathrm{id} \\ Y & \xrightarrow{v} & C & \dashrightarrow{w} & TX \end{array}$$

in which the left square is cocartesian there exists a unique morphism w that makes it commutative.

Triangles $X \xrightarrow{u} Y \xrightarrow{v} C \xrightarrow{w} TX$ in $\mathcal{B}$ that can be embedded into such a diagram, as well as their images in $\mathcal{B}_0$, are called *standard triangles*. Any triangle isomorphic to a standard one is called a *distinguished triangle*.

8. Stable Category Is Triangulated. Let $\mathcal{B}$ be a Frobenius category, $\mathcal{B}_0$ the corresponding stable category. Let us assume that the suspension functor is an automorphism of $\mathcal{B}_0$. The main result of the present batch of exercises is that *the category $\mathcal{B}_0$ with T as the translation functor and distinguished triangles defined as in the previous exercises is triangulated* (this gives, of course, another proof of Theorem IV.3.5).

As an example of arguments that are used to prove this result we verify the axiom TR3.

First of all, it is clear that both distinguished triangles in this axiom can be assumed to be standard. Let

$$\begin{array}{ccccc} X & \xrightarrow{i} & I & \xrightarrow{p} & TX \\ \downarrow u & & \downarrow t & & \| \mathrm{id} \\ Y & \xrightarrow{v} & C_u & \xrightarrow{w} & TX \end{array} \qquad \begin{array}{ccccc} X' & \xrightarrow{i'} & I' & \xrightarrow{p'} & TX' \\ \downarrow u' & & \downarrow t' & & \| \mathrm{id} \\ Y' & \xrightarrow{v'} & C_{u'} & \xrightarrow{w'} & TX' \end{array}$$

be the corresponding commutative diagrams. We have two morphisms $f : X \to X'$, $g : Y \to Y'$ in $\mathcal{B}$ such that $ft' \equiv tg \bmod I(X, Y')$. Ex. 3b) implies that there exists $\alpha : I \to Y'$ such that $ft' - tg = i\alpha$. Next, one can easily verify that there exists $s : I \to I'$ such that the diagram

$$\begin{array}{ccc} X & \longrightarrow & I & \longrightarrow & TX \\ \downarrow f & & \downarrow s & & \downarrow Tf \\ X' & \longrightarrow & I' & \longrightarrow & TX' \end{array}$$

is commutative. Morphisms $gv' : Y \to C_{u'}$ and $st' + \alpha v' : I \to C_{u'}$ have the property that the two composite morphisms $X \to Y \to C_{u'}$ and $X \to I \to C_{u'}$ coincide. Since C_u is the cocartesian product, there exists $h : C_u \to C_{u'}$ such that $vh = gv'$, $th = st' + \alpha v'$. After that one can easily verify that the image of h in $\mathcal{B}_0$ completes f, g to a morphism of distinguished triangles.

IV.4 Cores

1. What Is the Problem?

An important discovery in the homological algebra in the last few years was the fact that the derived categories of two absolutely different abelian categories can be equivalent as triangulated categories. In this section we describe an axiomatic approach to a technique that allows us to see various abelian subcategories inside a given triangulated category. This technique is called the formalism of t-structures.

The axioms of a t-structure formalize the following situation. Let $\mathcal{A}$ be an abelian category and $\mathcal{D} = D^*(\mathcal{A})$ be its derived category. Denote by $\mathcal{D}^{\geq n}$ (resp. $\mathcal{D}^{\leq n}$) the full subcategory of $\mathcal{D}$ formed by complexes $K^\bullet$ with $H^i(K^\bullet) = 0$ for $i < n$ (resp. for $i > n$).

By Proposition III.5.2, the full subcategory $\mathcal{D}^{\geq 0} \cap \mathcal{D}^{\leq 0}$ coincides with $\mathcal{A}$; more explicitly, the functor $\mathcal{A} \longrightarrow \{$the category of H^0-complexes$\} = \mathcal{D}^{\geq 0} \cap \mathcal{D}^{\leq 0}$ is the equivalence of categories.

It turns out that to prove that the intersection $\mathcal{D}^{\geq 0} \cap \mathcal{D}^{\leq 0}$ is abelian we need only the following formal properties.

2. Definition. *A t-structure on a triangulated category $\mathcal{D}$ is a pair of strictly full subcategories $(\mathcal{D}^{\leq 0}, \mathcal{D}^{\geq 0})$ satisfying the conditions a)–c) below. Let $\mathcal{D}^{\leq n} = \mathcal{D}^{\leq 0}[-n]$, $\mathcal{D}^{\geq n} = \mathcal{D}^{\geq 0}[-n]$.*

a) $\mathcal{D}^{\leq 0} \subset \mathcal{D}^{\leq 1}$ and $\mathcal{D}^{\geq 1} \subset \mathcal{D}^{\geq 0}$.
b) $\mathrm{Hom}(X,Y) = 0$ for $X \in \mathrm{Ob}\,\mathcal{D}^{\leq 0}$, $Y \in \mathrm{Ob}\,\mathcal{D}^{\geq 1}$.
c) For any $X \in \mathrm{Ob}\,\mathcal{D}$ there exists a distinguished triangle $A \to X \to B \to A[1]$ with $A \in \mathrm{Ob}\,\mathcal{D}^{\leq 0}$, $B \in \mathrm{Ob}\,\mathcal{D}^{\geq 1}$.

The core of the t-structure is the full subcategory $\mathcal{A} = \mathcal{D}^{\geq 0} \cap \mathcal{D}^{\leq 0}$.

3. Proposition. *If $\mathcal{D} = D^*(\mathcal{A})$ is the derived category of an abelian category $\mathcal{A}$, then the pair $(\mathcal{D}^{\leq 0}, \mathcal{D}^{\geq 0})$ described in IV.4.1 is a t-structure with the core $\mathcal{A}$.*

Proof. We must verify conditions b) and c). To prove b) we repeat, with slight modifications, the arguments from III.5.6. Let a morphism $\varphi : X \to Y$ in $D^*(\mathcal{A})$ with $X \in \mathrm{Ob}\,\mathcal{D}^{\leq 0}$, $Y \in \mathrm{Ob}\,\mathcal{D}^{\geq 1}$ be represented by a roof $X \xleftarrow{s} K \xrightarrow{f} Y$, where s is a quasi-isomorphism. First of all, as $Y \in \mathrm{Ob}\,\mathcal{D}^{\geq 1}$, the complex $Y/\tau_{\leq 0} Y$ (see III.7.5 for the definition of $\tau_{\leq 0}$) is quasi-isomorphic to Y. Hence we can assume that $Y^i = 0$ for $i < 0$ and $d_Y^0 : Y^0 \to Y^1$ is a monomorphism. Next, as $X \in \mathrm{Ob}\,\mathcal{D}^{\leq 0}$ and s is a quasi-isomorphism, we have $K \in \mathrm{Ob}\,\mathcal{D}^{\leq 0}$, so that the natural morphism $r : \tau_{\leq 0} K \longrightarrow K$ is a quasi-isomorphism and the roof $X \xleftarrow{sr} \tau_{\leq 0} K \xrightarrow{fr} Y$ also represents the morphism φ. Let us prove that $fr = 0$. Indeed, for all $i \neq 0$ we have either $X^i = 0$ or $(\tau_{\leq 0} K)^i = 0$ and for $i = 0$ we have $d_Y^0 (fr)^0 = (fr)^1 d_{\tau_{\leq 0} K}^0 = 0$ so that $(fr)^0 = 0$ because d_Y^0 is a monomorphism.

Finally, c) follows from the exact sequence of complexes

$$0 \longrightarrow A = \tau_{\leq 0}X \longrightarrow X \longrightarrow X/\tau_{\leq 0}X = B \longrightarrow 0. \qquad \square$$

4. Theorem. *The core* $\mathcal{A} = \mathcal{D}^{\geq 0} \cap \mathcal{D}^{\leq 0}$ *of any t-structure in* $\mathcal{D}$ *is an abelian category.*

The proof occupies IV.4.5–IV.4.9.
We begin with the construction of the truncation functors τ corresponding to a given t-structure.

5. Lemma. *a) There exist functors* $\tau_{\leq n} : \mathcal{D} \to \mathcal{D}^{\leq n}$ *(resp.* $\tau_{\geq n} : \mathcal{D} \to \mathcal{D}^{\geq n}$*) that are right (resp. left) adjoint to the corresponding embedding functors.*
b) For any $X \in \mathrm{Ob}\,\mathcal{D}$ *there exists a distinguished triangle of the form*

$$\tau_{\leq 0}X \longrightarrow X \longrightarrow \tau_{\geq 1}X \xrightarrow{d.} \tau_{\leq 0}X[1] \qquad (\mathrm{IV}.8)$$

and any two distinguished triangles $A \to X \to B \to A[1]$ *with* $A \in \mathrm{Ob}\,\mathcal{D}^{\leq 0}$, $B \in \mathrm{Ob}\,\mathcal{D}^{\geq 1}$ *are canonically isomorphic.*

Proof. Let us verify the existence of $\tau_{\leq 0}$ and $\tau_{\geq 1}$. To deal with other cases one has to apply the translation functor in $\mathcal{D}$.

For any X let us choose a distinguished triangle $A \to X \to B \to A[1]$ with $A \in \mathrm{Ob}\,\mathcal{D}^{\leq 0}$, $B \in \mathrm{Ob}\,\mathcal{D}^{\geq 1}$ and define $\tau_{\leq 0}$ and $\tau_{\geq 1}$ on objects by the formulas $\tau_{\leq 0}X = A$ and $\tau_{\geq 1}X = B$. Let $f : X \to Y$ be a morphism in $\mathcal{D}$ and $A' \to Y \to B' \to A'[1]$ be a triangle corresponding to the object Y. Let us show that the composition $A \longrightarrow X \xrightarrow{f} Y$ can be uniquely factored through A'. Indeed, we have the exact sequence

$$\mathrm{Hom}(A, B'[1]) \longrightarrow \mathrm{Hom}(A, A') \longrightarrow \mathrm{Hom}(A, Y) \longrightarrow \mathrm{Hom}(A, B').$$

By IV.4.2a) and IV.4.2b) the left and the right groups in this sequence vanish. Hence $f : X \to Y$ yields a morphism $\tau_{\leq 0}(f) : A \to A'$ and the family of these morphisms for all f's complete $\tau_{\leq 0}$ to a functor. Similarly, one establishes the functoriality of $\tau_{\geq 1}$ and the uniqueness of triangles $A \to X \to B \to A[1]$ (cf. Corollary IV.1.5).

To prove that $\tau_{\leq 0}$ is adjoint to the embedding $\mathcal{D}^{\leq 0} \to \mathcal{D}$ we use the just-constructed isomorphism of functors (in Y)

$$\mathrm{Hom}_{\mathcal{D}^{\leq 0}}(A, \tau_{\leq 0}Y) \cong \mathrm{Hom}_{\mathcal{D}}(A, Y), \quad A \in \mathrm{Ob}\,\mathcal{D}^{\leq 0}.$$

Similarly one considers the functor $\tau_{\geq 1}$. $\qquad \square$

6. Relations Among Truncation Functors

Next we shall show that the functors τ possess properties that are obvious in the case $\mathcal{D} = D^*(\mathcal{A})$.

a) $\tau_{\leq n} X = 0$ iff $X \to \tau_{\geq n+1} X$ is an isomorphism.

It suffices to consider the case $n = 0$ when the result follows from Lemma IV.4.5b).

b) For $m \leq n$ there exist natural isomorphisms $\tau_{\leq m} X \to \tau_{\leq m} \tau_{\leq n} X$ and $\tau_{\geq n} X \to \tau_{\geq n} \tau_{\geq m} X$.

As $\mathcal{D}^{\leq m} \subset \mathcal{D}^{\leq n}$ there exists a canonical morphism $\tau_{\leq m} \to \tau_{\leq n}$ of functors that are adjoint to embeddings of these subcategories, and after one more application of $\tau_{\leq m}$ this morphism becomes an isomorphism. The second assertion is proved similarly.

c) For $m \leq n$ there exists a natural isomorphism $\tau_{\geq m} \tau_{\leq n} X \to \tau_{\leq n} \tau_{\geq m} X \left(\stackrel{\text{def}}{=} \tau_{[m,n]} X \right)$.

Let us verify first that *all functors τ map each of the categories $\mathcal{D}^{\geq p}$, $\mathcal{D}^{\leq p}$ into itself*. Indeed, the embeddings $\tau_{\leq q} \mathcal{D}^{\leq p} \subset \mathcal{D}^{\leq p}$ and $\tau_{\geq q} \mathcal{D}^{\geq p} \subset \mathcal{D}^{\leq p}$ follow from b). To verify, say, that $\tau_{\leq q} \mathcal{D}^{\leq p} \subset \mathcal{D}^{\leq p}$, it suffices to verify that $\mathcal{D}^{\leq p}$ is stable under extensions; indeed the triangle

$$Y \longrightarrow \tau_{\geq q} Y \longrightarrow (\tau_{\leq q-1} Y)[1] = \tau_{\leq q}(Y[1]) \longrightarrow Y[1]$$

shows that $\tau_{\geq q} Y$ for any $Y \in \text{Ob } \mathcal{D}^{\leq p}$ is an extension of two objects from $\mathcal{D}^{\leq p}$. But, by a), $Z \in \text{Ob } \mathcal{D}$ belongs to $\mathcal{D}^{\leq p}$ if and only if $\text{Hom}_{\mathcal{D}}(Z, U) = 0$ for all $U \in \text{Ob } \mathcal{D}^{\geq p+1}$. The exact sequence of Homs corresponding to a distinguished triangle shows that this property is stable under extensions.

Now we can construct the required morphism of functors. Let us consider the diagram

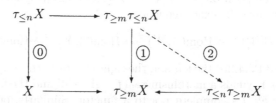

in which the solid arrows come from the definition of functors τ. Here the arrow ① is the action of $\tau_{\geq m}$ on ⓪. By the above, $\tau_{\geq m} \tau_{\leq n} X$ lies in $\mathcal{D}^{\leq n}$. Hence there exists a unique factorization of ① through $\tau_{\leq n} \tau_{\geq m} X$, which gives ②.

It remains to prove that ② is an isomorphism. It is rather instructive to compare the arguments below with the discussion about the meaning of the octahedron axiom in IV.2.8. Our task here is quite similar: in $D^*(\mathcal{A})$ the complex X has a filtration $\tau_{\leq m-1} X \subset \tau_{\leq n} X \subset X$ and we want to establish that $\tau_{\leq n} X / \tau_{\leq m-1} X$ is isomorphic to a subobject of $X / \tau_{\leq m-1} X = \tau_{\geq m} X$.

Formally, let us consider the upper cap of the octahedron starting from the right commutative triangle

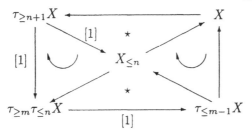

and complete it with a lower cap

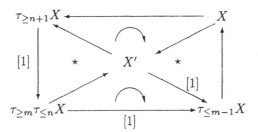

By Lemma IV.4.6b), its right triangle can be canonically identified with the triangle $\tau_{\leq m-1}X \to X \to \tau_{\geq m}X \to (\tau_{\leq m-1}X)[1]$, so that $X' = \tau_{\geq m}X$. Similarly, the left triangle can be canonically identified with the triangle that extends the morphism $X' \to \tau_{\geq n+1}X$:

$$\begin{array}{ccccccc}
\tau_{\leq n}\tau_{\geq m}X & \longrightarrow & \tau_{\geq m}X = X' & \longrightarrow & \tau_{\geq n+1}X & \longrightarrow & (\tau_{\leq n}\tau_{\geq m}X)[1] \\
\downarrow & & \| & & \| & & \downarrow \\
\tau_{\geq m}\tau_{\leq n}X & \longrightarrow & X' & \longrightarrow & \tau_{\geq n+1}X & \longrightarrow & (\tau_{\geq m}\tau_{\leq n}X)[1]
\end{array}$$

Comparing this diagram with the one used to define the morphism $\tau_{\geq m}\tau_{\leq n}X \to \tau_{\leq n}\tau_{\geq m}X$ we see that this morphism is just the isomorphism f.

7. Kernels and Cokernels in $\mathcal{A} = \mathcal{D}^{\geq 0} \cap \mathcal{D}^{\leq 0}$

Now let $f : X \to Y$ be a morphism in $\mathcal{A}$. Denote by Z a cone of f and set

$$K = \tau_{\leq -1}Z, \quad C = \tau_{\geq 0}Z.$$

Define k and c as compositions $k : \tau_{\leq -1}Z \to Z \to X[1]$, $c : Y \to Z \to \tau_{\geq 0}Z$. We claim that $(K[-1], k[-1])$ and (C, c) are respectively the kernel and the cokernel of f.

Let us prove, say, the statement about the cokernel (the one about the kernel is proved similarly). First of all, $C \in \mathrm{Ob}\, \mathcal{D}^{\geq 0}$ because $C = \tau_{\geq 0}Z$. Next, from the triangle $Y \to Z \to X[1] \to Y[1]$ and the stability of $\mathcal{D}^{\leq 0}$ under extensions we see that $Z \in \mathrm{Ob}\, \mathcal{D}^{\leq 0}$. Applying τ, we get $C \in \mathrm{Ob}\, \mathcal{D}^{\leq 0}$. Thus $C \in \mathrm{Ob}\, \mathcal{A}$.

282   IV. Triangulated Categories

Now let $T \in \mathrm{Ob}\,\mathcal{A}$ be an arbitrary object. Then $\mathrm{Hom}(K,T) = \mathrm{Hom}(K[1],T) = 0$ because $K \in \mathrm{Ob}\,\mathcal{D}^{\leq -1}$, $K[1] \in \mathrm{Ob}\,\mathcal{D}^{\leq -2}$. From the exact sequence associated with the triangle $K \to Z \to C \to K[1]$ we see that $\mathrm{Hom}(C,T) = \mathrm{Hom}(Z,T)$. Next, the exact sequence associated with the triangle $X \to Y \to Z \to X[1]$ has the form

$$\mathrm{Hom}(C,T)$$
$$\|$$
$$0 = \mathrm{Hom}(X[1],T) \longrightarrow \mathrm{Hom}(Z,T) \longrightarrow \mathrm{Hom}(Y,T) \longrightarrow \mathrm{Hom}(X,T)$$

(The left equality follows from $X[1] \in \mathrm{Ob}\,\mathcal{D}^{\leq -1}$.) But the exactness of this sequence means that (C,c) is the cokernel of f.

The above construction of the kernel and of the cokernel can be represented by the following diagram of the form "lower cap":

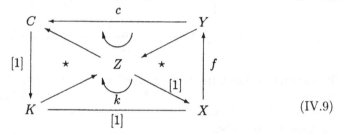

(IV.9)

In special cases when f is either a monomorphism or an epimorphism in $\mathcal{A}$ the picture becomes somewhat simpler. If, say, f is a monomorphism, then $K = 0$, so that $Z \to C$ is an isomorphism. Hence, $Z \in \mathrm{Ob}\,\mathcal{A}$ and $C = \mathrm{Coker}\,f$ is included into the distinguished triangle $X \xrightarrow{f} Y \xrightarrow{c} C \to X[1]$.

Similarly, if f is an epimorphism in $\mathcal{A}$, then $K[-1] = \ker f$ is included into a distinguished triangle $K[-1] \xrightarrow{k[-1]} X \xrightarrow{f} Y \to K$.

8. Canonical Decomposition of a Morphism in $\mathcal{A}$

Now we have to prove that for any $f : X \to Y$ the cokernel of $k[-1] : K[-1] \to X$ is isomorphic to the kernel of $c : Y \to C$. This common object I lies in the center of the upper cap

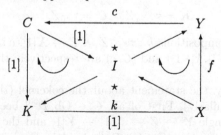

which corresponds to the lower cap (IV.9).

It is clear that in $\mathcal{A}$ the morphism $c : Y \to C$ is an epimorphism (as the cokernel of f) and $k : K \to X[1]$ is a monomorphism (as the kernel of $f[1]$). Thus the remark at the end of IV.4.7 shows that I is the cokernel of the kernel of f and the kernel of the cokernel of f.

9. Direct Sums and Products in $\mathcal{A}$

In $\mathcal{D}$ the direct sum of two objects exists and is their extension. Since $\mathcal{D}^{\geq m}$ and $\mathcal{D}^{\leq n}$ are stable under extensions (see the proof of Lemma IV.4.5c)), direct sums of objects from $\mathcal{A}$ lie in $\mathcal{A}$.

So the proof of Theorem 4 is completed. $\square$

10. Cohomology Functors

Let $\mathcal{D}$ be a triangulated category, $\mathcal{A} = \mathcal{D}^{\geq 0} \cap \mathcal{D}^{\leq 0}$ the core of some t-structure in $\mathcal{D}$. Let

$$H^0 = \tau_{[0,0]} : \mathcal{D} \to \mathcal{A}, \quad H^i(X) = H^0(X[i]).$$

In case $\mathcal{D} = D^*(\mathcal{A})$ with the t-structure from IV.4.1, H^i is the usual cohomology of a complex.

11. Theorem. *a)* H^0 *is a cohomology functor (see IV.1.6).*

Let, in addition, $\cap_n \mathrm{Ob}\,\mathcal{D}^{\leq n} = \cap_n \mathrm{Ob}\,\mathcal{D}^{\geq n} = \{0\}$ (the zero object in $\mathcal{D}$). Then

b) A morphism $f : X \to Y$ in $\mathcal{D}$ is an isomorphism if and only if all $H^i(f)$ are isomorphisms in $\mathcal{A}$.

c) $\mathrm{Ob}\,\mathcal{D}^{\leq n} = \{X \in \mathrm{Ob}\,\mathcal{D} \mid H^i(X) = 0 \text{ for all } i > n\}$. *Similarly,* $\mathrm{Ob}\,\mathcal{D}^{\geq n} = \{X \in \mathrm{Ob}\,\mathcal{D} \mid H^i(X) = 0 \text{ for all } i < n\}$.

Proof (of Part a)). We prove that for any distinguished triangle $X \to Y \to Z \to X[1]$ the sequence $H^0(X) \to H^0(Y) \to H^0(Z)$ is exact in $\mathcal{A}$.

a) If all objects X, Y, Z belong to $\mathcal{D}^{\leq 0}$, then the sequence

$$H^0(X) \longrightarrow H^0(Y) \longrightarrow H^0(Z) \longrightarrow 0 \qquad (\text{IV.10})$$

is exact. If $U \in \mathrm{Ob}\,\mathcal{D}^{\leq 0}$, $V \in \mathrm{Ob}\,\mathcal{D}^{\geq 0}$, then $H^0(U) = \tau_{\geq 0} U$ and $H^0(V) = \tau_{\leq 0} V$. Since the functors $\tau_{\geq 0}$ and $\tau_{\leq 0}$ are adjoint to embeddings of $\mathcal{D}^{\leq 0}$ and $\mathcal{D}^{\geq 0}$ into $\mathcal{D}$ (from the corresponding sides), we have

$$\mathrm{Hom}\,(H^0(U), H^0(V)) = \mathrm{Hom}\,(U, H^0(V)) = \mathrm{Hom}(U, V).$$

For any $W \in \mathrm{Ob}\,\mathcal{A} = \mathrm{Ob}\,\mathcal{D}^{\leq 0} \cap \mathrm{Ob}\,\mathcal{D}^{\geq 0}$ we have $\mathrm{Hom}(X[-1], W) = 0$ so that the sequence

$$0 \longrightarrow \mathrm{Hom}(Z, W) \longrightarrow \mathrm{Hom}(Y, W) \longrightarrow \mathrm{Hom}(X, W)$$

is exact. As $H^0(W) = W$, this sequence coincides with

$$0 \longrightarrow \text{Hom}\,(H^0(Z), W) \longrightarrow \text{Hom}\,(H^0(Y), W) \longrightarrow \text{Hom}\,(H^0(X), W).$$

Since W is an arbitrary object of $\mathcal{A}$, the sequence (IV.10) is exact.

b) *If $X \in \text{Ob}\,\mathcal{D}^{\leq 0}$, then the sequence (IV.10) is exact.* Let us show that $\tau_{\geq 1} Y \to \tau_{\geq 1} Z$ is an isomorphism. Indeed, for an arbitrary $U \in \text{Ob}\,\mathcal{D}^{\geq 1}$ we have $\text{Hom}(X[1], U) = \text{Hom}(X, U) = 0$ so that the exact sequence of Homs (Proposition IV.1.3) shows that the morphism $\text{Hom}(Z, U) \to \text{Hom}(Y, U)$ is an isomorphism. The required statement follows from the fact that the functor $\tau_{\geq 1}$ is adjoint to the embedding of $\mathcal{D}^{\geq 1}$ into $\mathcal{D}$.

Now we complete the lower cap in the right diagram below by an upper cap shown in the left diagram:

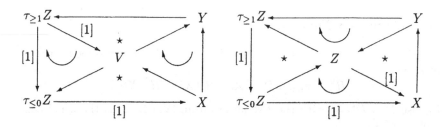

Since $\tau_{\geq 1} Z$ is isomorphic to $\tau_{\geq 1} Y$, the upper distinguished triangle of the left cap is isomorphic to $\tau_{\leq 0} Y \to Y \to \tau_{\geq 1} Z \to (\tau_{\leq 0} Y)[1]$, so that $V = \tau_{\leq 0} Y$. Hence the lower distinguished triangle in the left cap is of the form $X \to \tau_{\leq 0} Y \to \tau_{\leq 0} Z \to X[1]$ and we can apply to it the case a). It remains to note that $H^0(Y) = H^0(\tau_{\leq 0} Y)$ and similarly for Z.

c) The dual arguments show that *if $Z \in \text{Ob}\,\mathcal{D}^{\geq 0}$, then the sequence*

$$0 \longrightarrow H^0(X) \longrightarrow H^0(Y) \longrightarrow H^0(Z)$$

is exact.

d) *The general case.* Let us consider the octahedron

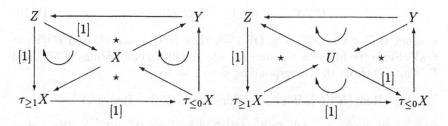

Applying b) to the distinguished triangle $(\tau_{\leq 0} X, Y, U)$ we get an exact sequence $H^0(X) \to H^0(Y) \to H^0(U) \to 0$. Applying c) to the distinguished triangle $(U, Z, \tau_{\geq 1} X[1])$ we get an exact sequence $0 \to H^0(U) \to H^0(Z)$, which means that $H^0(X) \to H^0(Y) \to H^0(Z)$ is also exact. □

Proof (of Part b)). Let us take first an object $X \in \mathrm{Ob}\,\mathcal{D}$ such that $H^i(X) = 0$ for all i and show that $X = 0$. If $X \in \mathrm{Ob}\,\mathcal{D}^{\geq 0}$, then the condition $H^0(X) = 0$ means that $\tau_{\leq 0} X = 0$, so that from the distinguished triangle (IV.8) we get $X = \tau_{\geq 1} X \in \mathrm{Ob}\,\mathcal{D}^{\geq 1}$. Repeating these arguments we get $X \in \cap_n \mathrm{Ob}\,\mathcal{D}^{\geq n} = \{0\}$. Dual arguments show that $X \in \mathrm{Ob}\,\mathcal{D}^{\leq 0}$ also implies $X = 0$. The general case follows from the fact that $H^i(\tau_{\geq 1} X) = H^i(\tau_{\leq 0} X) = 0$ for all i so that $\tau_{\geq 1} X = \tau_{\leq 0} X = 0$ and the distinguished triangle (IV.8) shows that $X = 0$.

Assume further that $f : X \to Y$ belongs to a distinguished triangle $X \to Y \to Z \to X[1]$ and that all $H^i(f)$ are isomorphisms. Since H^0 is a cohomological functor, the exact sequence corresponding to the above triangle shows that $H^i(Z) = 0$ for all i. Hence $Z = 0$ and f is an isomorphism. The converse is clear. □

Proof (of Part c)). If $H^i(X) = 0$ for all $i > 0$, then $H^i(\tau_{\geq 1} X) = 0$ for all i, so that $\tau_{\geq 1} X = 0$ by b) and $X = \tau_{\leq 0} X \in \mathrm{Ob}\,\mathcal{D}^{\leq 0}$ by (IV.8). Conversely, if $X \in \mathrm{Ob}\,\mathcal{D}^{\leq 0}$, then $X = \tau_{\leq 0} X$, so that $\tau_{\geq 1} X = 0$ and $H^i(X) = H^i(\tau_{\geq 1} X) = 0$ for $i > 0$. Similarly, if $X \in \mathrm{Ob}\,\mathcal{D}^{\geq 0}$, then $H^i(X) = 0$ for $i < 0$. Taking $X[n]$ for X we get c). □

12. *t*-Exact Functors

Let $\mathcal{D}, \tilde{\mathcal{D}}$ be two triangulated categories, each equipped with a *t*-structure, and let $F : \mathcal{D} \to \tilde{\mathcal{D}}$ be an exact functor (so that F commutes with translations and maps distinguished triangles into distinguished triangles). The functor F is said to be *left t-exact* if $F(\mathcal{D}^{\geq 0}) \subset \tilde{\mathcal{D}}^{\geq 0}$, *right t-exact* if $F(\mathcal{D}^{\leq 0}) \subset \tilde{\mathcal{D}}^{\leq 0}$, and *t-exact* if it is both left *t*- and right *t*-exact.

This definition models, of course, the situation when $\mathcal{D} = D(\mathcal{A})$, $\tilde{\mathcal{D}} = D(\tilde{\mathcal{A}})$ for two abelian categories $\mathcal{A}, \tilde{\mathcal{A}}$, and F is the derived functor (left or right) of a functor $\varphi : \mathcal{A} \to \tilde{\mathcal{A}}$ (which is respectively left exact, right exact, or exact).

Let us remark that φ can be reconstructed from F by the formula

$$\varphi(X) = H^0(F(X)), \quad X \in \mathrm{Ob}\,\mathcal{A} \subset \mathrm{Ob}\,\mathcal{D},$$

where H^0 is the cohomology functor in $\tilde{\mathcal{D}}$.

13. Remarks

The existence of a *t*-structure in a triangulated category $\mathcal{D}$ does not automatically imply that $\mathcal{D}$ is the derived category $D(\mathcal{A})$ of the core $\mathcal{A}$ of this *t*-structure. Moreover, in the general case there is no obvious relation of $\mathcal{D}$ with the category of complexes over $\mathcal{A}$. This is caused by the nonuniqueness of a cone $C(f)$ of a morphism f. Namely, to construct a functor $\mathrm{Kom}(\mathcal{A}) \to \mathcal{D}$ we have to be able to associate an object of $\mathcal{D}$ with a complex in $\mathcal{A}$ of the form, say, $\ldots 0 \longrightarrow A \stackrel{f}{\longrightarrow} B \longrightarrow 0 \ldots$. A natural candidate for the role of such an object is the third vertex $C(f)$ of the triangle $A \xrightarrow{f} B \longrightarrow C \longrightarrow A[1]$ (recall that $\mathcal{A}$ is a full subcategory of $\mathcal{D}$). However, such an association is not functorial because C is determined only up to a noncanonical isomorphism (see the discussion in IV.1.7). This gives one more evidence that the situation with derived categories is not quite satisfactory (see, however, Exs. IV.2.1–3).

Exercises

1. Let $\mathcal{D}$ be a triangulated category. A t-structure $(\mathcal{D}^{\leq 0}, \mathcal{D}^{\geq 0})$ is said to be *bounded* if it satisfies the condition of Theorem IV.4.11 (that is, $\cap_n \operatorname{Ob} \mathcal{D}^{\geq n} = \{0\}$), and, moreover, for any $X \in \operatorname{Ob} \mathcal{D}$ only a finite number of objects $H^i(X) \in \operatorname{Ob} \mathcal{A}$ is nonzero.

   a) Prove that the standard t-structure in $D^b(\mathcal{A})$ is bounded, while the standard t-structure in $D(\mathcal{A})$ is not.

   b) Let $\mathcal{A}$ be the core of a bounded t-structure $(\mathcal{D}^{\leq 0}, \mathcal{D}^{\geq 0})$. For $X, Y \in \operatorname{Ob} \mathcal{A}$ define
   $$\operatorname{Ext}^i_{\mathcal{D}}(X, Y) = \operatorname{Hom}_{\mathcal{D}}(X, Y[n]).$$

   For $\mathcal{D} = D^b(\mathcal{A})$ the groups $\operatorname{Ext}^i_{\mathcal{D}}(X, Y)$ coincide, clearly, with the groups $\operatorname{Ext}^i_{\mathcal{A}}(X, Y)$ (see III.5.3). Moreover, one can define the composition
   $$\operatorname{Ext}^i_{\mathcal{D}}(X, Y) \times \operatorname{Ext}^j_{\mathcal{D}}(Y, Z) \longrightarrow \operatorname{Ext}^{i+j}_{\mathcal{D}}(X, Z)$$

   (similarly to III.5.4b)). The difference between $\mathcal{D}$ and $D^b(\mathcal{A})$ is controlled by the difference between $\operatorname{Ext}^i_{\mathcal{D}}(X, Y)$ and $\operatorname{Ext}^i_{\mathcal{A}}(X, Y)$. Namely, let $F : D^b(\mathcal{A}) \to \mathcal{D}$ be a t-exact functor (see IV.4.12). Then it is an equivalence of categories if and only if $\operatorname{Ext}^*_{\mathcal{D}}$ is generated by $\operatorname{Ext}^1_{\mathcal{D}}$ (that is, any element $\alpha \in \operatorname{Ext}^i_{\mathcal{D}}(X, Y)$, $X, Y \in \operatorname{Ob} \mathcal{A}$, is a linear combination of monomials $\beta_1 \beta_2 \ldots \beta_i$, $\beta_j \in \operatorname{Ext}^1_{\mathcal{D}}(X_j, X_{j+1})$ with $X_1 = X$, $X_{i+1} = Y$).

   Let us remark that the interpretation of $\operatorname{Ext}^i_{\mathcal{A}}(X, Y)$ by Yoneda (see III.3.5d)) shows that $\operatorname{Ext}^*_{\mathcal{A}}$ is generated by $\operatorname{Ext}^1_{\mathcal{A}}$, so that the above condition is clearly necessary.

2. **Gluing t-Structures.** Let
$$\mathcal{C} \xrightarrow{P} \mathcal{D} \xrightarrow{Q} \mathcal{E} \qquad \qquad \text{(IV.11)}$$

be an exact triple of triangulated categories (see Ex. IV.2.4). t-structures on $\mathcal{C}, \mathcal{D}, \mathcal{E}$ are said to be *compatible* (or (IV.11) is said to be t-exact) if P and Q are t-exact functors.

   a) Prove that a t-structure on $\mathcal{D}$ determines unique compatible t-structures on $\mathcal{C}$ and $\mathcal{E}$ (namely $\mathcal{C}^{\leq 0} = \mathcal{C} \cap \mathcal{D}^{\leq 0}$, $\mathcal{E}^{\leq 0} = Q\mathcal{D}^{\leq 0}$ and similarly for $\mathcal{C}^{\geq 0}$, $\mathcal{E}^{\geq 0}$).

   b) Conversely, for two t-structures on $\mathcal{C}$ and $\mathcal{E}$ there exists at least one compatible t-structure on $\mathcal{D}$. More precisely, let $\mathcal{C} \xrightarrow{P} \mathcal{D} \xrightarrow{Q} \mathcal{E}$ be a t-exact triple of triangulated categories. Define

$$^\perp(P\mathcal{C}^{>0}) = \{X \in \mathrm{Ob}\,\mathcal{D} \mid \mathrm{Hom}(X,Y) = 0 \quad \text{for all} \quad Y \in P\mathcal{C}^{>0}\},$$
$$(P\mathcal{C}^{<0})^\perp = \{X \in \mathrm{Ob}\,\mathcal{D} \mid \mathrm{Hom}(Y,X) = 0 \quad \text{for all} \quad Y \in P\mathcal{C}^{<0}\}.$$

Prove that the t-structure on $\mathcal{D}$ is determined by t-structures on $\mathcal{C}$ and $\mathcal{E}$ as follows:

$$\mathcal{D}^{\leq 0} = \{X \in {}^\perp(P\mathcal{C}^{>0}) \mid Q(X) \in \mathrm{Ob}\,\mathcal{E}^{\leq 0}\},$$
$$\mathcal{D}^{\geq 0} = \{X \in (P\mathcal{C}^{<0})^\perp \mid Q(X) \in \mathrm{Ob}\,\mathcal{E}^{\geq 0}\}.$$

c) It might happen that for given t-structures on $\mathcal{C}$ and $\mathcal{E}$ there exists no compatible t-structure on $\mathcal{D}$. However, there is a case where the existence of a compatible t-structure on $\mathcal{D}$ can be guaranteed. Namely, let $\mathcal{C} \xrightarrow{P} \mathcal{D} \xrightarrow{Q} \mathcal{E}$ be a t-exact triple of triangulated categories, and assume that P has left and right adjoint functors (this is equivalent to the existence of left and right adjoint functors for Q). Then for any t-structures on $\mathcal{C}$ and $\mathcal{E}$ there exists a compatible t-structure on $\mathcal{D}$.

3. Categories of Sheaves on a Topological Space and on Its Subspaces.

Let X be a topological space, $U \subset X$ an open subspace, $Y = X \setminus U$ the complement, $i : Y \to X$, $j : U \to X$ the inclusions. Denote by $\mathcal{A}_X, \mathcal{A}_Y, \mathcal{A}_U$ the categories of sheaves of abelian groups on X, Y, U and by $\mathcal{D}_X, \mathcal{D}_Y, \mathcal{D}_U$ the corresponding bounded derived categories.

Consider the following six functors:

$$Rj_\bullet, j_! : \mathcal{D}_U \longrightarrow \mathcal{D}_X, \quad j^\bullet : \mathcal{D}_X \longrightarrow \mathcal{D}_U,$$
$$i_\bullet : \mathcal{D}_Y \longrightarrow \mathcal{D}_X, \quad i^\bullet, i^! : \mathcal{D}_X \longrightarrow \mathcal{D}_Y$$

(see Sect. III.8, the functor $i^!$ can be defined as the right derived functor of the left exact functor "sections supported on Y", see III.8.25). Prove the following properties of these functors.

a) All these functors are exact functors between the corresponding triangulated categories.

b) $i^\bullet$ and $i^!$ are respectively left and right adjoint to $i_\bullet$.

c) $j_!$ and $Rj_\bullet$ are respectively left and right adjoint to $j^\bullet$.

d) $j^\bullet i_\bullet = 0$. By adjointness, we have also $i^\bullet j_! = 0$, $i^! Rj_\bullet = 0$ and

$$\mathrm{Hom}_{\mathcal{D}_X}(j_!\mathcal{H}^\bullet, i_\bullet\mathcal{G}^\bullet) = 0, \quad \mathrm{Hom}_{\mathcal{D}_X}(i_\bullet\mathcal{G}^\bullet, Rj_\bullet\mathcal{H}^\bullet) = 0,$$

for $\mathcal{G}^\bullet \in \mathrm{Ob}\,\mathcal{D}_Y$, $\mathcal{H}^\bullet \in \mathrm{Ob}\,\mathcal{D}_U$.

e) There exist (functorial in $\mathcal{F}^\bullet \in \mathrm{Ob}\,\mathcal{D}_X$) morphisms

$$w : i_\bullet i^\bullet \mathcal{F}^\bullet \longrightarrow j_! j^\bullet \mathcal{F}^\bullet[1], \quad w' : Rj_\bullet j^\bullet \mathcal{F}^\bullet \longrightarrow i_\bullet i^! \mathcal{F}^\bullet[1]$$

such that the triangles

$$j_!j^{\bullet}\mathcal{F}^{\bullet} \xrightarrow{u} \mathcal{F}^{\bullet} \xrightarrow{v} i_{\bullet}i^{\bullet}\mathcal{F}^{\bullet} \xrightarrow{w} j_!j^{\bullet}\mathcal{F}^{\bullet}[1],$$
$$i_{\bullet}i^!\mathcal{F}^{\bullet} \xrightarrow{u'} \mathcal{F}^{\bullet} \xrightarrow{v'} Rj_{\bullet}j^{\bullet}\mathcal{F}^{\bullet} \xrightarrow{w'} i_{\bullet}i^!\mathcal{F}^{\bullet}[1]$$

are distinguished (here u, u', v, v' are the adjunction morphisms corresponding to the functors from b) and c)). By d) and Corollary IV.1.5, w and w' are determined uniquely.

f) The adjunction morphisms,

$$i^{\bullet}i_{\bullet}\mathcal{G}^{\bullet} \longrightarrow \mathcal{G}^{\bullet} \longrightarrow i^!i_{\bullet}\mathcal{G}^{\bullet}, \quad \mathcal{G}^{\bullet} \in \mathrm{Ob}\, \mathcal{D}_Y,$$
$$j^{\bullet}Rj_{\bullet}\mathcal{H}^{\bullet} \longrightarrow \mathcal{H}^{\bullet} \longrightarrow j^{\bullet}j_!\mathcal{H}^{\bullet}, \quad \mathcal{H}^{\bullet} \in \mathrm{Ob}\, \mathcal{D}_U,$$

are isomorphisms.

4. Gluing. A collection consisting of three triangulated categories (not necessarily related to sheaf categories) and six exact functors satisfying the conditions a)–f) of the previous exercise is called the *gluing data*. One of examples was described above. One can also verify that gluing data can be obtained if one defines appropriate functors in categories of coherent sheaves (X, U, Y are algebraic varieties or schemes) or in categories of sheaves in étale topologies.

5. Let us assume that we have gluing data as in Ex. 3. Prove that $\mathcal{D}_Y \xrightarrow{i_{\bullet}} \mathcal{D}_X \xrightarrow{j^{\bullet}} \mathcal{D}_U$ is an exact triple of triangulated categories. Since $i_{\bullet}$ has both left and right adjoint functors, Ex. 2c) shows that any pair of t-structures $(\mathcal{D}_Y^{\leq 0}, \mathcal{D}_Y^{\geq 0})$ on $\mathcal{D}_Y$ and $(\mathcal{D}_U^{\leq 0}, \mathcal{D}_U^{\geq 0})$ on $\mathcal{D}_U$ determines a unique compatible t-structure on $\mathcal{D}_X$. Moreover, $\mathcal{D}_X^{\leq 0}$ and $\mathcal{D}_X^{\geq 0}$ for this structure are given by

$$\mathcal{D}_X^{\leq 0} = \left\{\mathcal{F} \in \mathrm{Ob}\, \mathcal{D}_X \mid j^{\bullet}\mathcal{F} \in \mathrm{Ob}\, \mathcal{D}_U^{\leq 0},\ i^{\bullet}\mathcal{F} \in \mathrm{Ob}\, \mathcal{D}_Y^{\leq 0}\right\},$$
$$\mathcal{D}_X^{\geq 0} = \left\{\mathcal{F} \in \mathrm{Ob}\, \mathcal{D}_X \mid j^{\bullet}\mathcal{F} \in \mathrm{Ob}\, \mathcal{D}_U^{\geq 0},\ i^!\mathcal{F} \in \mathrm{Ob}\, \mathcal{D}_Y^{\geq 0}\right\}.$$

This result about the gluing of t-structures was proved by Beilinson et al. [1, n. 1.4]]. They applied it for the construction of the so called perverse sheaves. Using such gluing data one can also construct some nonstandard t-structures on $\mathcal{D}_X$ gluing together translated t-structures on $\mathcal{D}_U$ and on $\mathcal{D}_Y$.

6. Representations of Quivers. a) By a *quiver* we mean a finite oriented graph $\Gamma = (X(\Gamma), E(\Gamma))$ without multiple edges and oriented cycles ($X(\Gamma)$ is the set of vertices, $E(\Gamma)$ is the set of edges). A (finite-dimensional) *representation of a quiver* is a family of finite-dimensional linear spaces V_α, $\alpha \in X(\Gamma)$ and morphisms $\varphi_{\alpha\beta} : V_\alpha \to V_\beta$, $(\alpha\beta) \in E(\Gamma)$.

Verify that the category Rep_Γ of all finite-dimensional representations of a quiver Γ (with obvious definition of morphisms) is an abelian category.

b) Prove that any simple object V of Rep_Γ has the following form: one of the spaces V_α is one-dimensional, and all others are zero spaces. Therefore there exists a one-to-one correspondence between simple objects of Rep_Γ and vertices of Γ. Prove that for any two simple objects $V^{(\alpha)}$, $V^{(\beta)}$ ($\alpha, \beta \in X(\Gamma)$)

$$\dim \text{Ext}^1_{\text{Rep}_\Gamma}\left(V^{(\alpha)}, V^{(\beta)}\right) = \begin{cases} 1 & \text{if } (\alpha\beta) \in E(\Gamma), \\ 0 & \text{if } (\alpha\beta) \notin E(\Gamma). \end{cases}$$

Therefore the quiver Γ is determined by Rep_Γ uniquely up to an isomorphism.

c) For a given quiver Γ denote by $A(\Gamma)$ an algebra with generators e_α, $\alpha \in X(\Gamma)$, $f_{\alpha\beta}$, $(\alpha\beta) \in E(\Gamma)$ and relations $e_\alpha^2 = e_\alpha$, $e_\alpha e_\beta = 0$ for $\alpha \neq \beta$, $e_\beta f_{\alpha\beta} = f_{\alpha\beta} e_\alpha = f_{\alpha\beta}$. Prove that the category Rep_Γ is equivalent to the category of finite-dimensional left $A(\Gamma)$-modules.

d) For any vertex α of a quiver Γ denote by $\sigma_\alpha \Gamma$ the graph that is obtained from Γ by the reversion of the directions of all edges adjacent to α. Prove that $\sigma_\alpha \Gamma$ is a quiver.

e) Let Γ be a quiver and $\mathcal{D}(\Gamma)$ the bounded derived category of Rep_Γ. Let α be a source in Γ, i.e., such vertex that no edge of Γ ends in α (any quiver contains at least one source). We construct the functor $R_\alpha^+ : \mathcal{D}(\Gamma) \to \mathcal{D}(\sigma_\alpha \Gamma)$ as follows. Let

$$V^\bullet = \{\ldots \longrightarrow V^i \longrightarrow V^{i+1} \longrightarrow \ldots\}$$

be an object of $\mathcal{D}(\Gamma)$ so that each V^i is an object of Rep_Γ, $V^i = \{V^i_\alpha, \varphi^i_{\alpha\beta}\}$. Define

$$R_\alpha^+ V^\bullet = W^\bullet = \{\ldots \longrightarrow W^i \longrightarrow W^{i+1} \longrightarrow \ldots\} \in \text{Ob}\,\mathcal{D}(\sigma_\alpha \Gamma)$$

$$W^i = (W^i_\alpha, \psi^i_{\alpha\beta}) \in \text{Rep}_{\sigma_\alpha \Gamma},$$

by the following formulas:

$$W^i_\beta = V^i_\beta \quad \text{for} \quad \beta \neq \alpha,$$

$$W^i_\alpha = \left(\bigoplus_{(\alpha\beta) \in E(\Gamma)} V^i_\beta\right) \oplus V^{i-1}_\alpha;$$

$$\psi^i_{\beta\gamma} = \varphi^i_{\beta\gamma} \quad \text{if } (\beta\gamma) \in E(\sigma_\alpha \Gamma), \beta \neq \alpha, \gamma \neq \alpha,$$

$\psi^i_{\beta\alpha} : W^i_\beta \longrightarrow W^i_\alpha$ is the embedding of a direct summand if $(\beta\alpha) \in E(\sigma_\alpha \Gamma)$, so that $(\alpha\beta) \in E(\Gamma)$.

Introduce differentials $d^i_W : W^i \to W^{i+1}$ in such a way that R_α^+ becomes a functor from $\mathcal{D}(\Gamma)$ to $\mathcal{D}(\sigma_\alpha \Gamma)$.

f) Similarly, for a sink in $E(\Gamma)$ (i.e., for a vertex α such that Γ does not contain edges starting at α) construct a functor $R_\alpha^- : \mathcal{D}(\Gamma) \to \mathcal{D}(\sigma_\alpha \Gamma)$.

g) Clearly σ_α makes a source α into a sink and vice versa. Prove that if $\alpha \in X(\Gamma)$ is a source, then the functor $R_\alpha^- \circ R_\alpha^+ : \mathcal{D}(\Gamma) \to \mathcal{D}(\Gamma)$ is isomorphic to the identify functor. Similarly, if α is a sink, then $R_\alpha^+ \circ R_\alpha^- : \mathcal{D}(\Gamma) \to \mathcal{D}(\Gamma)$ is isomorphic to the identity functor. Therefore R_α^- and R_α^+ determine an equivalence of triangulated categories $\mathcal{D}(\Gamma)$ and $\mathcal{D}(\sigma_\alpha \Gamma)$ (compare with b)).

7. Module Categories. Results stated in g) in the previous exercise can be obtained from the following more general results about derived categories of categories of modules over finite-dimensional algebras.

a) Let $\mathcal{A} = A$-mod for an algebra A with dim $A < \infty$. Let $M \in \mathrm{Ob}\,\mathcal{A}$ satisfies the following conditions:

(i) $\mathrm{Ext}^i_A(M, M) = 0$ for $i > 0$.
(ii) A has a finite projective dimension $n = \mathrm{dhp}\,A$ (i.e., any A-module N has a projective resolution of the length $\leq n$).
(iii) A as an object of $\mathcal{A}$ has a finite M-codimension, i.e., there exists an exact sequence
$$0 \longrightarrow A \longrightarrow M^0 \longrightarrow \ldots \longrightarrow M^r \longrightarrow 0$$
with $M^i \in \mathrm{Ob}(\mathrm{add}\,M)$ (see Ex. 4).

Prove that the functor $\varphi_M : K^b(\mathrm{add}\,M) \to D^b(\mathcal{A})$ is an equivalence of categories (use Ex. III.5.4b)).

b) Let a ring A and a module M satisfy the conditions (i)–(iii) from a). Define $B = \mathrm{End}_A M$, $\mathcal{B} = \mathrm{mod}\text{-}B$. Considering M as a right B-module define a functor $F : \mathcal{A} \to \mathcal{B}$ by $F(X) = \mathrm{Hom}_A(M, X)$.

Prove that F induces an equivalence of triangulated categories (preserving the triangulation) $\tilde{F} : D^b(\mathcal{A}) \to D^b(\mathcal{B})$.

Hint. We have $F(M) = B$, so that F induces an equivalence of add M with $\mathcal{P}_\mathcal{B}$ (the class of projective B-modules) and hence the equivalence $\tilde{F} : K^b(\mathrm{add}\,M) \to K^b(\mathcal{P}_\mathcal{B})$. By a), $K^b(\mathrm{add}\,M)$ is equivalent to $D^b(\mathcal{A})$. Finally, the condition (ii) in a) easily implies that B has a finite projective dimension and by III.5.20 (more precisely, by the corresponding result for bounded complexes), $K^b(\mathcal{P}_\mathcal{B})$ is equivalent to $D^b(\mathcal{B})$.

V. Introduction to Homotopic Algebra

V.1 Closed Model Categories

1. Definition. *Let $\mathcal{C}$ be an arbitrary category, L and R two classes of morphisms in $\mathcal{C}$. The class L is said to be* left complementary *to R (and R is said to be* right complementary *to L) if the following condition is satisfied: for any solid arrow commutative square*

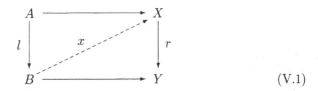
$$(V.1)$$

with $l \in L$, $r \in R$, there exists a diagonal morphism x making both triangles commutative.

In the paper by Quillen [1], where this notion was first introduced, morphisms from L (resp. from R) are said to have the left (resp. the right) lifting property with respect to morphisms from R (resp. from L).

2. Properties of Complementary Classes

a) The class of all isomorphisms in $\mathcal{C}$ is both right and left complementary to any other class.

b) For any class L there exists a maximum class right complementary to L; we denote it by $\rho(L)$. Similarly, for any class R there exists a maximum class $\lambda(R)$ left complementary to R.

Indeed, say, $\rho(L)$ is the union of all classes right complementary to L.

c) Operations λ and ρ satisfy the following properties:

$$\rho\lambda\rho = \rho, \quad \lambda\rho\lambda = \lambda. \qquad (V.2)$$

Indeed, both ρ and λ reverse inclusions: if, say, $L' \subset L$, then $\rho(L) \subset \rho(L')$. Next, $\lambda\rho(L) \supset L$ and $\rho\lambda(R) \supset R$ by definition. Applying ρ to the first inclusion we get $\rho\lambda\rho(L) \subset \rho(L)$. Substituting $R = \rho(L)$ into the second

292   V. Introduction to Homotopic Algebra

inclusion we get the converse inclusion. The second equality in (V.2) is proved in the same way.

d) The classes $\rho(L)$ and $\lambda(R)$ are closed under the composition of morphisms.

Indeed, let, say, $r', r'' \in \rho(L)$. To prove that $r'r'' \in \rho(L)$ we construct consecutively morphisms in the following diagram:

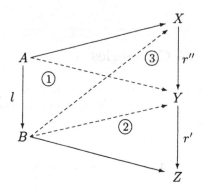

Here ① is $r''a$, ② is constructed by applying the diagram (V.1) to the square $ABYZ$, and ③ is constructed by applying (V.1) to the square $ABXY$. The existence of ③ shows that $r'r'' \in \rho(L)$.

e) In passing from the category $\mathcal{C}$ to the dual category $\mathcal{C}^\circ$ the complementary classes remain complementary classes with the words "right" and "left" interchanged.

f) If, in the diagram (V.1), either l is an epimorphism or r is a monomorphism, then x, whenever it exists, is unique.

g) The classes $\rho(L)$ and $\lambda(R)$ are closed under retractions. More precisely, a morphism $f : X \to Y$ is said to be a *retraction* of a morphism $g : X' \to Y'$ if f and g can be included into a commutative diagram,

$$\begin{array}{ccccc} X & \xrightarrow{u} & X' & \xrightarrow{u'} & X \\ \downarrow f & & \downarrow g & & \downarrow f \\ Y & \xrightarrow{v} & Y' & \xrightarrow{v'} & Y, \end{array}$$

with $u' \cdot u = \mathrm{id}_X$, $v' \cdot v = \mathrm{id}_Y$. We claim that $g \in \lambda(R)$ implies $f \in \lambda(R)$. To prove the claim let us consider a square,

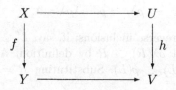

with $h \in R$, and the associated solid arrow diagram,

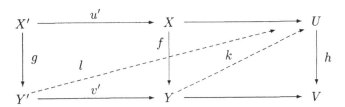

Morphism l exists because $g \in \lambda(R)$; morphism $k : Y \to U$ is $k = lv$. Similarly, $g \in \rho(L)$ implies $f \in \rho(L)$.

3. Complementary Classes and Products

Consider a cocartesian square $ABCD$ with $g : A \to B$, $f : C \to D$. Then $g \in \lambda(R)$ implies $f \in \lambda(R)$. Indeed, adjoin to $ABCD$ the square $CDXY$ with $h \in R$ as shown below:

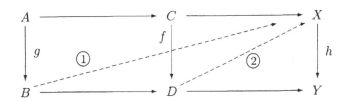

Morphism ① exists because $g \in \lambda(R)$. Morphism ② exists by the universal property of the cocartesian square $ABCD$.

Similarly (or using V.1.2d)) one can prove the following property:

b) If $ABCD$ is a cartesian square, then $f \in \rho(L)$ implies $g \in \rho(L)$.

4. Closed Model Categories

A category $\mathcal{C}$ is said to be a closed model category if three classes of morphisms in $\mathcal{C}$ are specified: a class F of *fibrations*, a class G of *cofibrations*, and a class W of *weak equivalences*. These data must satisfy the following axioms.

CM0. $\mathcal{C}$ is closed under finite inductive and projective limits.
CM1. Any two of the classes F, G, W determine the third class as follows:
  a) $W = \rho(F) \cdot \lambda(G)$, where for any two classes of morphisms S and T we define $S \cdot T = \{st \mid s \in S, t \in T\}$.
  b) $G = \lambda(F \cap W)$.
  c) $F = \rho(G \cap W)$.
CM2. Mor $\mathcal{C} = F \cdot (G \cap W) = (F \cap W) \cdot G$.

CM3. W contains all isomorphisms. If any two of the three morphisms f, g, fg belong to W, then the third one also belongs to W.

If $\mathcal{C}$ is a closed model category then $\mathcal{C}^\circ$ has the dual structure of a closed model category with F° as cofibrations, G° as fibrations and W° as weak equivalences.

5. Remarks

Axioms CM0–CM3 were suggested by D. Quillen [1]. If $\mathcal{C}$ satisfies these axioms, then the localization with respect to W (see III.2.2) enables us to introduce the corresponding homotopy category Ho$\mathcal{C}$ and the universal functor $\mathcal{C} \to \text{Ho}\,\mathcal{C}$ which makes all weak equivalences isomorphisms.

Let Top be the category of topological spaces and continuous mappings. Define F as the class of all Serre fibrations, and W as the class of all weak homotopy equivalences. Define G by the construction from the axiom CM1b). Then Top becomes a closed model category and Ho Top is the natural category for the homotopy theory (see Exercises to Sect. V.2).

Let us assume now that a topological space is the geometric realization $|X|$ of a simplicial set X. How can one infer the homotopy properties of $|X|$ from X? This problem was solved by Kan. Following Quillen, Kan's result can be formulated as a theorem saying that $\Delta^\circ Set$ can be endowed with the structure of a closed model category in such a way that Ho$\Delta^\circ Set$ and Ho Top become (modulo some minor details) equivalent. The description of this structure on $\Delta^\circ Set$ is the main theme of the first two sections of this chapter.

The situation when quite dissimilar model categories $\mathcal{C}$ give equivalent (or close to being equivalent) homotopy categories Ho$\mathcal{C}$ is not at all unusual, and the choice of a convenient model category for treating a given homotopy problem may lead to a substantial simplification. The category of differential graded algebras (see Sect. IV.3) is an example of a closed model category with remarkably simple algebraic properties.

6. Fibrations, Cofibrations, and Weak Equivalences of Simplicial Sets

By a *horn* we mean the simplicial set $V(n,k)$ with

$$V(n,k)_m = \{f : [m] \to [n] \mid f \text{ is nondecreasing, } \text{Im } f \neq [n], [n]\setminus\{k\}\}.$$

Evidently, $V(n,k) \subset \Delta[n]$; the canonical inclusion $V(n,k) \hookrightarrow \Delta[n]$ will be called the horn morphism. Faces and degeneration morphisms are induced by the corresponding morphisms in $\Delta[n]$. The geometric realization $|V(n,k)|$ is the boundary of the n-simplex with the k-th face (which is the $(n-1)$-dimensional simplex opposite to the k-th vertex) removed.

V.1 Closed Model Categories

Let us recall also that the simplicial sphere $\dot{\Delta}[n]$ is the following simplicial set:
$$\dot{\Delta}[n]_m = \{f : [m] \to [n] \mid f \text{ is nondecreasing, Im } f \neq [n]\}.$$
The natural inclusion morphism $\dot{\Delta}[n] \hookrightarrow \Delta[n]$ is called the sphere morphism. Let *hor* and *sph* be the classes of horn and sphere morphisms. Define
$$F = \rho(hor), \quad G = \lambda\rho(sph), \quad W = \rho(sph) \cdot \lambda\rho(hor). \tag{V.3}$$

7. Theorem. *The category $\Delta^\circ Set$ with fibrations, cofibrations, and weak equivalences as in (V.3) is a closed model category.*

The verification of all the axioms occupies the remaining part of Sect. V.1 (axioms CM0–CM2) and the entire Sect. V.2 (axiom CM3).

8. Axiom CM0

An easy verification using II.3.21.

9. CM1a) and CM1b)

The equality $W = \rho(F) \cdot \lambda(G)$ follows from (V.3) and (V.2). The equality $G = \lambda(F \cap W)$ follows similarly from the equality $\rho(G) = F \cap W$ which we verify next.

We have $\rho(G) = \rho\lambda\rho(sph) = \rho(sph)$. Therefore we have only to prove that $\rho(sph) = F \cap W$.

a) $\rho(sph) \subset W$ because $\lambda\rho(hor)$ contains all identity morphisms (see (V.3)).

b) To prove the inclusion $\rho(sph) \subset F = \rho(hor)$ let us consider the following commutative diagram:

$$\begin{array}{ccc} \dot{\Delta}[n-1] & \xrightarrow{\text{rim}} & V(n,k) \\ \downarrow & & \downarrow \\ \Delta[n-1] & \xrightarrow{D_n^k} & \dot{\Delta}[n] \end{array} \tag{V.4}$$

Here the two vertical arrows are evident (the right one is completely determined by the requirement that $V(n,k) \to \dot{\Delta}[n-1] \hookrightarrow \Delta[n]$ is the horn morphism). Next,
$$\text{rim}_m : \dot{\Delta}[n-1]_m \longrightarrow V(n,k)_m$$
maps $f : [m] \to [n]$ into $\partial_k^n \cdot f$. Finally,
$$(D_k^n)_m : \Delta[n-1]_m \longrightarrow \dot{\Delta}[n-1]_m$$
is given by the same formula.

V. Introduction to Homotopic Algebra

Geometrically, the above diagram describes the simplicial sphere $\dot\Delta[n]$ as the sealing of the horn $V(n,k)$ by the lead $\Delta[n-1]$ along the rim $\dot\Delta[n-1]$. Therefore this diagram must be a cocartesian square, and this fact has an easy formal verification.

Let us consider now a morphism $f \in \rho(sph)$. To prove that $f \in \rho(hor)$ we have to construct a dashed arrow ③ in the right lower square of the following diagram:

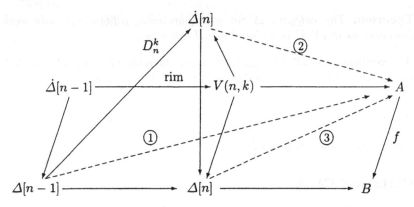

To do this we first complete our square by the auxiliary morphisms (solid arrows at the diagram) as follows. The nonplane "square" $(\Delta[n-1], \dot\Delta[n-1], V(n,k), \dot\Delta[n])$ is (V.4). The morphism $\Delta[n-1] \to \Delta[n]$ is the "k-th face" inclusion of simplicial sets. The commutativity is clear.

The dashed diagonal ①: $\Delta[n-1] \to A$ exists as $f \in \rho(sph)$. The dashed morphism ② is uniquely determined by morphisms ① and $V(n,k) \to A$ because (V.4) is a cocartesian square. Finally, to construct the diagonal ③ in the square $(\dot\Delta[n], \Delta[n], A, B)$ we again use the fact that $f \in \rho(sph)$. The required commutativity of two triangles can be checked without any difficulties.

The geometrical meaning of this proof is quite transparent. Namely, $(f : A \to B) \in \rho(sph)$ means that if we are given a simplex in the base B and its boundary is lifted to A, then we can lift to A the whole simplex. Next, $f \in \rho(hor)$ means that if we are given a simplex in B with the part of its boundary, the horn, lifted to A, then the whole simplex can be lifted to A. To construct this lifting we first consider the rim of the horn as the boundary of its lead and lift the whole lead to A. Thus we get the lifting of the boundary of our simplex, and we can lift the simplex itself.

c) $F \cap W \subset \rho(sph)$. For the proof let us look at the definition (V.3): $f \in F \cap W$ means that $f \in \lambda(hor)$ and that $f = gh$, where $g \in \rho(sph)$ and $h \in \lambda\rho(hor)$. We prove that f is a retraction of g; by V.1.2g) this implies the required property.

Let us consider the following commutative diagram,

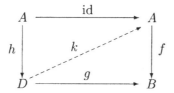

in which k exists because $f \in \rho(hor)$ and $h \in \lambda\rho(hor)$. The required retraction diagram is

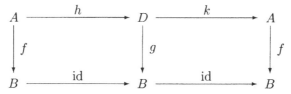

10. Axiom CM1c)

We shall prove that $\lambda(F) = G \cap W$; applying ρ to both parts we get the required equality.

a) $\lambda(F) \subset W$. Indeed, $\lambda(F) \subset \lambda\rho(hor) = \rho(sph) \cdot \lambda\rho(hor) = W$.

b) $\lambda(F) \subset G$. Indeed, we have already proven that $\rho(sph) \subset \rho(hor)$. It remains to apply λ.

c) $G \cap W \subset \lambda(F)$. Rewrite the definition (V.3): $f \in G \cap W$ means that $f \in \lambda\rho(sph)$ and that $f = gh$ with $g \in \rho(sph)$ and $h \in \lambda\rho(hor)$. We shall prove that f is a retraction of h; by V.1.2g) this will imply the required property.

To do this we consider the following commutative diagram,

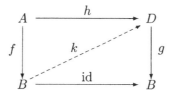

in which k exists because $g \in \rho(sph)$ and $f \in \lambda q \rho(sph)$. The required retraction diagram is

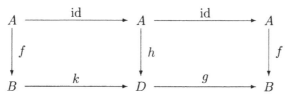

11. Axiom CM2 (First Part)

We want to represent an arbitrary morphism $f : A \to B$ as the composition $f = pi$, where $p \in F = \rho(hor)$, $i \in G \cap W = \lambda(F) = \lambda\rho(hor)$ (the last equality was proven in V.1.10). In order to do this we construct p and i explicitly and then prove that they belong to the corresponding classes.

Denote by $\mathcal{H}(f)$ the set of all commutative squares of the form

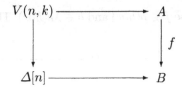

(the left arrow is a horn). Define the simplicial set $\mathrm{Ex}^1(f)$ and the morphism $i^1(f) = A \to \mathrm{Ex}^1(f)$ by the following cocartesian square,

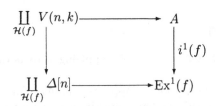

Let us note that the left arrow in this diagram, being a direct sum of horns, belongs to $\lambda\rho(hor)$. Therefore, by V.1.3a), $i^1(f) \in \lambda\rho(hor)$ for any f.

By the universality of a cocartesian square there exists a canonical morphism $p^1(f) : \mathrm{Ex}^1(f) \to B$. Define inductively $\mathrm{Ex}^{k+1}(f) = \mathrm{Ex}^1\left(p^k(f)\right)$, $i^{k+1}(f) = i^1\left(p^k(f)\right) \circ i^k(f)$, $p^{k+1}(f) = p^1\left(p^k(f)\right)$.

Finally, let $\mathrm{Ex}^\infty(f) = \varinjlim \mathrm{Ex}^k(f)$ (with respect to $i^k(f)$), $i = \varinjlim i^k(f)$, $p = \varprojlim p^k(f)$. One can easily see that all $i^k(f)$ are inclusions (because horn morphisms are inclusions), so that $\mathrm{Ex}^\infty(f)$ is, in fact, the union of all $\mathrm{Ex}^k(f)$. All $p^k(f)$ being compatible, $p : \mathrm{Ex}^\infty(f) \to B$ clearly exists, and the composite morphism $p \cdot i : A \to B$ coincides with f. Next, by the above remark all the morphisms

$$i^1\left(p^k(f)\right) : \mathrm{Ex}^k(f) \longrightarrow \mathrm{Ex}^{k+1}(f)$$

belong to $\lambda\rho(hor)$. Therefore, given a commutative diagram

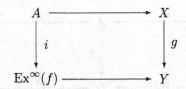

with $g \in \rho(hor)$ we can use the induction on k to construct a compatible system of liftings $\text{Ex}^k(f) \to X$. Therefore $i \in \lambda\rho(hor)$.

It remains to verify that $p \in \rho(hor)$. Let us consider a square of the form

The image of $V(n,k)$ lies, evidently, in $\text{Ex}^k(f)$ for some k. By the construction of $\text{Ex}^{k+1}(f)$ there exists a lifting $\Delta[n] \to \text{Ex}^{k+1}(f)$ and the verification is completed.

12. Axiom CM2 (Second Part)

We want to represent an arbitrary morphism $f : A \to B$ as the composition $f = jq$, where $j \in F \cap W = \rho(sph)$ (see V.1.9) and $q \in G = \lambda\rho(sph)$.

Denote by $\text{Sph}(f)$ the set of all commutative squares of the form

The arguments similar to those in the previous subsection (with $\mathcal{H}(f)$ replaced by $\text{Sph}(f)$) immediately give the required result.

V.2 Homotopic Characterization of Weak Equivalences

1. Definition. *A simplicial set E is called a* Kan set *if $p_E \in F$, where $p_E : E \to \Delta[0]$ is the projection to the point.*

Thus any morphism of a horn $V(n,k)$ to a Kan set can be extended to a morphism of the simplex $\Delta[n]$ filling $V(n,k)$. This property is an imitation of the "homotopy extension" property. In the next subsection we shall give the precise definitions, while now we attempt to seize the generality of the notion.

a) In the situation of V.1.11, let us consider the decomposition $X \to \text{Ex}^\infty(p_X) \to \Delta[0]$ of the morphism $p_X : X \to \Delta[0]$. Evidently, $\text{Ex}^\infty(p_X)$ is a Kan set. Hence any simplicial set can be embedded into a Kan set in such a way that the embedding is a weak equivalence (i.e., belongs to W).

b) *Any simplicial group is a Kan set.* Indeed, let $G = \{G_n\}$ be a simplicial group. We must prove that any morphism $\varphi : V(n,k) \to G$ can be extended to a morphism $\psi : \Delta[n] \to G$. As φ is completely determined by the elements $\varphi\left(\partial_n^i\right) \in G_{n-1}$, $i \neq k$, and ψ is completely determined by the element $\psi\left(\mathrm{id}_{[n]}\right) \in G_n$, we have to prove the following assertion.

Suppose that we are given $x_0, \ldots, x_{k-1}, x_{k+1}, \ldots, x_n \in G_{n-1}$ such that
$$d_{n-1}^i x_j = d_{n-1}^{j-1} x_i, \quad 0 \leq i < j \leq n, \quad i, j \neq k$$
(we use the notations $d_l^i = G\left(\partial_l^i\right)$, $s_l^i = G\left(\sigma_l^i\right)$). The claim is that there exists $x \in G_n$ such that
$$d_n^i x = x_i, \quad 0 \leq i \leq n, \quad i \neq k.$$
To find such an x let us construct first $u \in G_n$ such that
$$d_n^i u = x_i \quad \text{for} \quad 0 \leq i < k.$$
If $k = 0$ there is no condition on u, and we set
$$u = e_n \quad \text{(the unit element of the group } G_n).$$
For $k > 0$ let
$$u_0 = s_{n-1}^0 x_0,$$
and define $u_r \in G_n$ for $0 < r \leq k-1$ inductively as follows:
$$y_{r-1} = s_{n-1}^r \left(\left(d_n^r u_{r-1}\right)^{-1} x_r\right), \quad u_r = u_{r-1} y_{r-1}.$$
An easy induction by r shows that
$$d_n^i u_r = x_i \quad \text{for} \quad 0 \leq i \leq r$$
(we have to use that $d_n^i y_{r-1} = e_{n-1}$ for $0 \leq i \leq r-1$). Now we can set $u = u_{k-1}$.

Next, let
$$v_0 = u$$
and, for $0 < r \leq n - k$, let
$$z_{r-1} = s_{n-1}^{n-r}\left(\left(d_n^{m-r+1} v_{r-1}\right)^{-1} x_{n-r+1}\right), \quad v_r = v_{r-1} z_{r-1}.$$
Using once again the induction by r one can easily see that
$$d_n^i v_r = x_i \quad \text{for} \quad 0 \leq i < k \quad \text{and for} \quad i > n-1$$
(we have to use that $d_n^i z_{r-1} = e_{n-1}$ for $0 \leq i < k$ and for $i > n-r+1$). Now $x = v_{n-k}$ is the required element. □

c) For $n \geq 1$ the simplex $\Delta[n]$ is *not* a Kan set (contrary to a naive expectation).

The reason is that morphisms of simplicial sets preserve the internal order of vertices. So, for example, the morphism of the zero-dimensional horn $V(1,0)$ into $\Delta[1]$ which maps "the end of the one-dimensional simplex" into "the beginning" does not extend to a morphism of the entire $\Delta[1]$.

The Kan set $\text{Ext}^\infty(p_{\Delta[1]})$ which might be thought of as a homotopy equivalent of the segment, can be understood as an infinite-dimensional cell obtained by gluing consecutive simplices with various orientations together. In general the role played by Kan sets in the simplicial category is similar to the role played by function spaces in the homotopic topology.

2. Homotopies of Simplicial Sets

a) Let X be a simplicial set. Vertices $x, y \in X_0$ are said to be *strongly connected* if $d_1^0 z = x$, $d_1^1 z = y$ for some $z \in X_1$ (again we define $d_l^i = X(\partial_l^i)$). Vertices $x, y \in X_0$ are said to be *connected* if there exists a chain $x = x_0, x_1, x_2, \ldots, x_p = y$ such that either (x_i, x_{i+1}) or (x_{i+1}, x_i) are strongly connected for all $0 \le i \le p-1$. Let us remark that if E is a Kan set, then any two connected vertices are strongly connected. To prove this it suffices to verify that the relation of being strongly connected is symmetric and transitive. Both properties are proved using liftings of appropriate horns $V(2, k)$.

For any X denote by $\pi_0(X)$ the set of the equivalence classes of X_0 with respect to the connectedness relation. Evidently, this is the same as $\pi_0(|X|)$.

b) For two simplicial sets X, Y let

$$[X, Y] = \pi_0(\mathbf{Hom}(X, Y)),$$

where the inner **Hom** in $\Delta^\circ Set$ is defined in II.4.24.

c) Let $f, g : X \to Y$ be two morphisms in $\Delta^\circ Set$. Being elements from $\mathbf{Hom}(X, Y)_0$, they define classes $[f], [g] \in [X, Y]$. Morphisms f, g are called *homotopic* if $[f] = [g]$. The mapping $p : X \to Y$ is called a *homotopy equivalence* if there exists $q : Y \to X$ such that pq is homotopic to id_Y and qp is homotopic to id_X.

One can give a direct definition of homotopy between two morphisms as in a). Namely, $f, g : X \to Y$ are said to be strongly homotopic if there exists a morphism $k : X \times \Delta[1] \to Y$ (homotopy between f and g) such that $k^0 = f$, $k^1 = g$, where

$$k^i = k \circ \left(\text{id}_X \times \varepsilon\left(\partial_1^i\right)\right) : X = X \times \Delta[0] \longrightarrow X \times \Delta[1] \longrightarrow Y.$$

The equivalence relation generated by the strong homotopy is the homotopy. If E is a Kan set, then homotopic morphisms $f : X \to E$ are strongly homotopic.

The main goal of the present section is to prove the following theorem.

3. Theorem. *A morphism $f : X \to Y$ in $\Delta^\circ Set$ is a weak equivalence if and only if for any Kan set E the induced map*

$$[f, E] : [Y, E] \longrightarrow [X, E]$$

is one-to-one.

Axiom CM3 for $\Delta^\circ Set$ obviously follows from this theorem.

4. Proof of the Theorem: The Plan

The implication $f \in W \Rightarrow \{[f, E]$ is one-to-one$\}$ is proved by a direct construction of all the necessary homotopies in V.2.5–V.2.7 below.

To prove the converse implication we start with a reduction. We want to prove that bijectivity of $[f, E]$ for all Kan sets E implies that $f \in \rho(G)\lambda(F)$. Let $f = jq$, $j \in F \cap W = \rho(G)$, $q \in G$ (see V.1.9 and V.1.12). Then $[f, E] = [q, E] \circ [j, E]$. Now $[j, E]$ is one-to-one because $j \in F \cap W \subset W$ (we use the direct implication). Hence $[q, E]$ is also one-to-one.

Thus it remains to prove that if $q \in G$ and $[q, E]$ is one-to-one for all Kan sets E, then $q \in \lambda(F)$. This is proved in two steps.

Let us introduce the following class of morphisms:

$$F_K = \{\text{fibrations of Kan sets}\}.$$

We prove first that $q \in \lambda(F_K)$ (see V.2.8–V.2.15), and then that $G \cap \lambda(F_K) = G \cap \lambda(F)$ (see V.2.16–V.2.37). For the proof we have to develop some techniques which might be of independent interest.

We begin by giving a very useful characterization of cofibrations.

5. Lemma. *A morphism $f : X \to Y$ is a cofibration if and only if it is an embedding (i.e., all $f_n : X_n \to Y_n$ are embeddings).*

Proof. a) $\Leftarrow$. Let f be an embedding. Considering X as a simplicial subset of Y, we define

$$Y^{(i)} = \mathrm{Sk}^i Y \cup X \subset Y, \qquad (Y^{(-1)} = X),$$

and denote by $l_i : Y^{(i-1)} \to Y^{(i)}$ the natural embedding. Clearly, $Y = \cup Y^{(i)}$. By V.1.2d), it suffices to prove that $l_i \in G$ for any i.

Any element $y \in \left(Y^{(i)}\right)_i - \left(Y^{(i-1)}\right)_i$ determines a morphism $j_y : \Delta[i] \to Y^{(i)}$ with $j_y(\dot\Delta[i]) \subset Y^{(i-1)}$. The family of all morphisms j_y, $y \in \left(Y^{(i)}\right)_i - \left(Y^{(i-1)}\right)_i$ determines the morphisms $j : \coprod_y \Delta[i] \to Y^{(i)}$ and $j' : \coprod_y \dot\Delta[i] \to Y^{(i-1)}$. One can easily see that these morphisms enter the following cocartesian square

$$\begin{array}{ccc} \coprod_y \dot\Delta[i] & \xrightarrow{j'} & Y^{(i-1)} \\ {\scriptstyle s}\downarrow & & \downarrow{\scriptstyle l} \\ \coprod_y \Delta[i] & \xrightarrow{j} & Y^{(i)} \end{array}$$

(i.e., $Y^{(i)}$ is obtained by gluing to $Y^{(i-1)}$ nondegenerate i-simplices along their boundaries). It is clear that in this square we have $s \in sph \subset \lambda\rho(sph) = G$. By V.1.3, $l_i \in G$.

b) ⇒. Let $f \in G$. Applying to f the arguments from V.1.12 we get a decomposition $f = jq$, where the morphism $q \in G$ constructed in V.1.12 is an embedding. Let us consider the following commutative square,

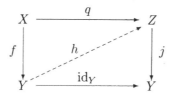

Since $f \in G$ and $j \in F \cap W = \rho(G)$ we see that there exists a morphism $h : Y \to Z$ making the upper triangle commutative. Since q is an embedding, f is also an embedding. □

Let us continue now with the proof that $f \in W \Rightarrow \{[f, E]$ is one-to-one$\}$. Decompose f as $f = pi$, where $p \in \rho(G)$, $i \in \lambda(F)$. Since $[f, E] = [i, E][p, E]$, it remains to verify that $[i, E]$ and $[p, E]$ are one-to-one.

6. Lemma. *If $p \in \rho(G)$, then $[p, E]$ is one-to-one.*

Proof. We prove that any $p : X \to Y$ from $\rho(G)$ is a homotopy equivalence. This immediately implies that $[p, Z]$ is one-to-one for any Z.

To construct the homotopy inverse $q : Y \to X$ to p we consider the following square,

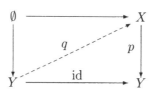

in which the left morphism belongs to G by Lemma V.2.5. A homotopy between qp and id_X is given by the diagonal in the following square,

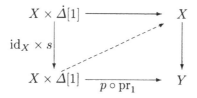

in which the left morphism again belongs to G by Lemma V.2.5. □

7. Lemma. *If $i \in \lambda(F)$, then $[i, E]$ is one-to-one.*

Proof. a) $[i, E]$ is surjective because any morphism $g : X \to E$ can be represented as $g = hi$, where h is the diagonal in the following square,

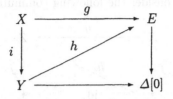

The right morphism in this square belongs to F by the definition of a Kan set.

b) Let us prove now that $[i, E]$ is injective. Let $\alpha, \beta : Y \to E$ be morphisms such that $\alpha \cdot i$ is strongly homotopic to $\beta \cdot i$, and let $k : X \times \Delta[1] \to E$ be the corresponding homotopy. We want to prove that α and β are homotopic. In order to do this we will construct in a moment a simplicial set Z and a morphism $(g : E \to Z) \in \lambda(F)$ such that $g \cdot \alpha$ is homotopic to $g \cdot \beta$. After that the first part of the proof would imply that $[g, E] : [Z, E] \to [E, E]$ is surjective. Any preimage of id_E under this mapping gives a morphism $h : Z \to E$ such that $h \cdot g$ is homotopic to id_E, and then $h \cdot g \cdot \alpha$, $h \cdot g \cdot \beta$ will be homotopic to each other and to α and β respectively.

First we construct the set U using the following cocartesian square.

$$\begin{array}{ccc} X \times \dot\Delta[1] & \xrightarrow{\mathrm{id} \times s} & X \times \Delta[1] \\ {\scriptstyle i \times \mathrm{id}} \downarrow & & \downarrow \\ Y \times \Delta[1] & \longrightarrow & U \end{array}$$

The morphisms $(\alpha, \beta) : Y \times \dot\Delta[1] \to E$ and $k : X \times \Delta[1] \to E$ yield, by the universality property, a morphism $\gamma : U \to E$. Similarly, the morphisms $\mathrm{id} \times s : Y \times \dot\Delta[1] \to Y \times \Delta[1]$ and $i \times \mathrm{id} : X \times \Delta[1] \to Y \times \Delta[1]$ yield a morphism $\delta : U \to Y \times \Delta[1]$. Now complete (γ, δ) to the following cocartesian square,

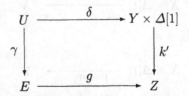

One can easily see that k' is a homotopy between $g \cdot \alpha$, $g \cdot \beta$, and that $g \in \lambda(F)$. □

Thus, the first part of Theorem V.2.3 is proved.

The next lemma delivers a large class of morphisms from $G \cap W$.

8. Lemma. *Let $f : X \to Y$ be an embedding (i.e., $f \in G$). Then the natural embedding*
$$g : (X \times \Delta[n]) \cup (Y \times V(n,k)) \longrightarrow Y \times \Delta[n]$$
induced by embeddings $f \times \mathrm{id}_{\Delta[n]}$ and $\mathrm{id}_Y \times (V(n,k) \longrightarrow \Delta[n])$, belongs to $G \cap W = \lambda\rho(hor)$.

The proof of this lemma is similar to the proof of the first part of Lemma V.2.5. Namely, we represent $Y \times \Delta[n]$ as the union of the increasing sequence of simplicial sets
$$Y(i) = \left(Y^{(i)} \times \Delta[n]\right) \cup (Y \times V(n,k)),$$
where $Y^{(i)}$ is as in the proof of Lemma V.2.5. One can easily verify that the natural embedding $Y(i-1) \to Y(i)$ can be completed to a cocartesian square,

$$\begin{array}{ccc}
\coprod_y \left(\dot\Delta[i] \times \Delta[n]\right) \cup (\Delta[i] \times V(n,k)) & \longrightarrow & Y(i-1) \\
\downarrow{\scriptstyle s} & & \downarrow \\
\coprod_y (\Delta[i] \times \Delta[n]) & \longrightarrow & Y(i)
\end{array}$$

($\coprod_y$ is taken over all $y \in \left(Y^{(i)}\right)_i - \left(Y^{(i-1)}\right)_i$). By 3, it suffices to verify that for any i, k, n the embedding
$$\left(\dot\Delta[i] \times \Delta[n]\right) \cup (\Delta[i] \times V(n,k)) \longrightarrow \Delta[i] \times \Delta[n] \qquad (V.5)$$
belongs to $G \cap W = \lambda\rho(hor)$.

To do this let us define the increasing family of simplicial subsets $D_0, \ldots, D_N$ in $\Delta[i] \times \Delta[n]$ satisfying the following conditions:

a) $D_0 = \left(\dot\Delta[i] \times \Delta[n]\right) \cup (\Delta[i] \times V(n,k))$.

b) The embedding $D_j \to D_{j+1}$ can be completed to a cocartesian square

$$\begin{array}{ccc}
V(n+i, k_j) & \longrightarrow & D_j \\
\downarrow & & \downarrow \\
\Delta[n+i] & \longrightarrow & D_{j+1}
\end{array} \qquad (V.6)$$

c) $D_N = \Delta[i] \times \Delta[n]$.

Speaking somewhat loosely, in passing from D_j to D_{j+1} we add to D_j a nondegenerate $(n+i)$-dimensional simplex in $\Delta[i] \times \Delta[n]$ whose "horn" belongs to D_j. A rigorous construction of D_j's can be easily obtained from the description of the triangulation of $\Delta[i] \times \Delta[n]$ given in I.1.5.

The square (V.6) being cocartesian, the embedding $D_j \to D_{j+1}$ belongs to $\lambda\rho(hor)$. Hence the composition of all these embeddings, which is the embedding (V.5), also belongs to $\lambda\rho(hor)$. □

9. Morphisms of Hom's and of Products

Let $f : A \to B$, $g : X \to Y$ be morphisms of simplicial sets. Natural morphisms of simplicial sets

$$\mathbf{Hom}(A,X) \xrightarrow{g_*} \mathbf{Hom}(A,Y) \xleftarrow{f^*} \mathbf{Hom}(B,Y)$$

determine the cartesian product

$$\mathbf{Hom}(f,g) = \mathbf{Hom}(A,X) \underset{\mathbf{Hom}(A,Y)}{\times} \mathbf{Hom}(B,Y) \qquad (V.7)$$

and a morphism

$$f/g : \mathbf{Hom}(B,X) \longrightarrow \mathbf{Hom}(f,g).$$

One can interpret $\mathbf{Hom}(f,g)$ and f/g as follows. An element $z \in \mathbf{Hom}(f,g)_n$ is a commutative square

$$z = \begin{array}{c} A \times \Delta[n] \longrightarrow X \\ {\scriptstyle f \times \mathrm{id}} \downarrow \quad \nearrow_{k} \quad \downarrow {\scriptstyle g} \\ B \times \Delta[n] \longrightarrow Y \end{array}$$

and an element $k \in \mathbf{Hom}(f,g)_n = \mathrm{Hom}_{\Delta\circ\mathbf{Set}}(B \times \Delta[n], X)$ with $(f/g)(k) = z$ is a diagonal in this square making the whole diagram commutative.

A similar construction can be applied to products. The natural morphisms

$$A \times Y \xrightarrow{\mathrm{id} \times g} A \times X \xleftarrow{f \times \mathrm{id}} B \times X$$

determine the cocartesian product

$$\mathbf{Prod}(f,g) \overset{\mathrm{def}}{=} (A \times Y) \coprod_{A \times X} (B \times X), \qquad (V.8)$$

and the morphism

$$f \square g : \mathbf{Prod}(f,g) \longrightarrow B \times Y.$$

The next relation between f/g and $f \square g$ generalizes Theorem II.4.24a).

10. Proposition. Let $f : A \to B$, $g : X \to Y$, $h : U \to V$ be three morphisms of simplicial sets. There exists a natural isomorphism of simplicial sets $\varepsilon : \mathbf{Hom}(f \square h, g) \to \mathbf{Hom}(h, f/g)$ which is functorial in f, g, h and can be included into the following commutative diagram:

$$\begin{array}{ccc}
\mathbf{Hom}(B \times V, X) & \xrightarrow{\delta} & \mathbf{Hom}(V, \mathbf{Hom}(B, X)) \\
{\scriptstyle (f \square h)/g} \downarrow & & \downarrow {\scriptstyle h/(f/g)} \\
\mathbf{Hom}(f \square h, g) & \xrightarrow{\varepsilon} & \mathbf{Hom}(h, f/g)
\end{array} \qquad (V.9)$$

(Here δ is the isomorphism that establishes the adjunction of functors $X \times \bullet$ and $\mathbf{Hom}(\bullet, B)$ from Theorem II.4.24a).)

Proof. To construct ε let us note that an element of $\mathbf{Hom}(f \square h, g)_n$ is a commutative square

$$\begin{array}{ccc}
\mathbf{Prod}(f, h) \times \Delta[n] & \longrightarrow & X \\
{\scriptstyle (f \square h) \times \mathrm{id}} \downarrow & & \downarrow {\scriptstyle g} \\
B \times V \times \Delta[n] & \longrightarrow & Y
\end{array} \qquad (V.10)$$

By the universality of the cocartesian product (V.8) the upper morphism in this square is a pair of morphisms

$$A \times V \times \Delta[n] \longrightarrow X \longleftarrow B \times U \times \Delta[n]$$

such that the associated two composite morphisms

$$A \times U \times \Delta[n] \longrightarrow X$$

coincide.

The commutativity of (V.10) means that the two composite morphisms

$$A \times V \times \Delta[n] \longrightarrow Y$$

(one via $B \times V \times \Delta[n]$ and one via X) coincide, as well as the analogous two composite morphisms

$$B \times U \times \Delta[n] \longrightarrow Y.$$

On the other hand, an element of $\mathbf{Hom}(h, f/g)_n$ is a commutative square

$$\begin{array}{ccc}
U \times \Delta[n] & \longrightarrow & \mathbf{Hom}(B, X) \\
{\scriptstyle h \times \mathrm{id}} \downarrow & & \downarrow {\scriptstyle f/g} \\
V \times \Delta[n] & \longrightarrow & \mathbf{Hom}(f, g)
\end{array} \qquad (V.11)$$

308   V. Introduction to Homotopic Algebra

By the universality of the cartesian product (V.7) the lower morphism in this square is a pair of morphisms
$$\mathbf{Hom}(B,Y) \longleftarrow V \times \Delta[n] \longrightarrow \mathbf{Hom}(A,X),$$
such that the associated two composite morphisms
$$V \times \Delta[n] \longrightarrow \mathbf{Hom}(A,Y)$$
coincide.

The commutativity of (V.10) means that the two composite morphisms
$$U \times \Delta[n] \longrightarrow \mathbf{Hom}(B,Y)$$
(one through $V \times \Delta[n]$ and one through $\mathbf{Hom}(B,X)$) coincide, and the two composite morphisms
$$U \times \Delta[n] \longrightarrow \mathbf{Hom}(A,X)$$
also coincide.

Now it is clear that the natural isomorphisms of the form
$$\mathrm{Hom}_{\Delta \circ \mathrm{Set}}(A \times V \times \Delta[n], X) \longrightarrow \mathrm{Hom}_{\Delta \circ \mathrm{Set}}(V \times \Delta[n], \mathbf{Hom}(A,X))$$
give bijections between the sets of squares of the form (V.10) and of the form (V.11). Easy verifications that these bijections commute with simplicial operations in $\mathbf{Hom}(f\,\Box\,h, g)$ and in $\mathbf{Hom}(h, f/g)$ and that (V.9) is commutative are left to the reader. □

We shall often use this proposition in the situations where $f : A \to B$ and $h : U \to V$ are embeddings (cofibrations); in particular, h will usually be the embedding of the horn $V(n,k)$ or of the sphere $\dot{\Delta}[n]$ into $\Delta[n]$. One can easily see that in this case $\mathbf{Prod}(f,g)$ is $(B \times U) \cup (A \times V)$ (the union inside $B \times V$) and Proposition 10 can be reformulated as follows.

11. Corollary. *Commutative squares*

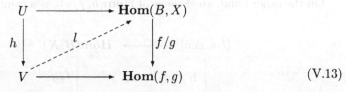

(V.12)

are in one-to-one correspondence with commutative squares

(V.13)

and the liftings k in (V.12) are in one-to-one correspondence with the liftings l in (V.13).

V.2 Homotopic Characterization of Weak Equivalences

The next proposition describes the properties of the morphism $f/g :$ $\mathbf{Hom}(B, X) \to \mathbf{Hom}(f, g)$ in the case when f is a cofibration and g is a fibration.

12. Proposition. *Let us assume that* $(f : A \to B) \in G$, $(g : X \to Y) \in F$. *Then* $f/g \in F$. *If, moreover, either* $f \in W$ *or* $g \in W$, *then* $f/g \in F \cap W$.

Proof. We must verify that $f/g \in \rho(hor)$, i.e., that any square formed by solid arrows in the following diagram:

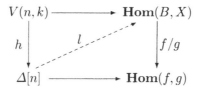

admits a lifting l. By Corollary V.2.11 this is equivalent to the existence of a lifting k in the corresponding square

$$\begin{array}{ccc}
(B \times V(n,k)) \cup (A \times \Delta[n]) & \longrightarrow & X \\
{\scriptstyle f \square g} \downarrow \quad \nearrow^{k} & & \downarrow {\scriptstyle g} \\
B \times \Delta[n] & \longrightarrow & Y
\end{array}$$

The existence of k follows from the fact that $g \in F = \rho(hor)$ and $f \square g \in \lambda \rho(hor)$ by Lemma V.2.8.

Let us prove the second assertion. As $F \cap W = \rho(G) = \rho(sph)$, we have to find a lifting l in the square

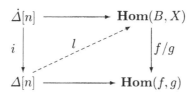

i.e., a morphism k in

$$\begin{array}{ccc}
(B \times \dot{\Delta}[n]) \cup (A \times \Delta[n]) & \longrightarrow & X \\
{\scriptstyle f \square i} \downarrow \quad \nearrow^{k} & & \downarrow {\scriptstyle g} \\
B \times \Delta[n] & \longrightarrow & Y
\end{array}$$

310   V. Introduction to Homotopic Algebra

If $g \in F \cap W = \rho(G)$, then k exists because $f \square i$, being an embedding, belongs to G (Lemma V.2.5). If $f \in F \cap W$, we can apply Corollary V.2.11 once again and conclude that the existence of k is equivalent to the existence of a morphism m in the following square:

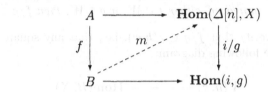

Since i, being an embedding, belongs to G, $i/g \in F$ and such an m exists. $\square$

13. Corollary. *a) If E is a Kan set then, for any simplicial set X, the set $\mathrm{Hom}(X, E)$ is a Kan set.*
   b) If $(i : A \to B) \in G$ and E is a Kan set then
$$(i^* : \mathrm{Hom}(B, E) \longrightarrow \mathrm{Hom}(A, E)) \in G.$$

The next result will be needed later.

14. Lemma. *Let $g : X \to Y$ be a fibration, $j : Y' \to Y$ an arbitrary embedding, and $q : Y \to Y'$ a retraction of j (i.e., $qj = \mathrm{id}_{Y'}$). Let $g' : X' \to Y'$ be the induced fibration and $j : X' \to X$ the corresponding embedding. Then for any homotopy $h : \Delta[1] \times Y \to Y$ between id_Y and jq, there exists a morphism $r : X \to X'$ with $rj' = \mathrm{id}_{X'}$ and a homotopy $k : \Delta[1] \times X \to X$ between id_X and $j'r$ such that the diagram*

$$\begin{array}{ccc} \Delta[1] \times X & \xrightarrow{k} & X \\ {\scriptstyle \mathrm{id} \times g} \downarrow & & \downarrow {\scriptstyle g} \\ \Delta[1] \times X & \xrightarrow{h} & Y \end{array}$$

commutes (i.e., k covers h).
   A similar result holds for homotopies between jq and id_Y.

Proof. One can easily see that the required homotopy should be a one-dimensional simplex of $\mathrm{Hom}(X, X)$ which starts at the point id_X and which is mapped by $j'/g : \mathrm{Hom}(X, X) \to \mathrm{Hom}(j', g)$ into the one-dimensional simplex $x \in \mathrm{Hom}(j', g)_1$ defined by the following pair of composite mappings

$$a : \Delta[1] \times X' \xrightarrow{\mathrm{pr}} X' \xrightarrow{j'} X,$$
$$b : \Delta[1] \times X \xrightarrow{\mathrm{id} \times g} \Delta[1] \times Y \xrightarrow{h} Y.$$

Since $j'X' \to X$ is a cofibration (being an embedding) and g is a fibration, j'/g is also a fibration. Hence the required simplex $k \in \mathbf{Hom}(X,X)_1$ exists. □

15. Theorem (Covering Homotopy Theorem). *Let $i : A \to B$ be a cofibration and $p : X \to Y$ a fibration. Let also a commutative diagram*

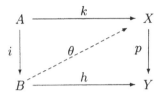

and its extension

$$\begin{array}{ccc} A \times \Delta[1] & \xrightarrow{K} & X \\ {\scriptstyle i \times \mathrm{id}_{\Delta[1]}} \downarrow & \overset{\Theta}{\nearrow} & \downarrow p \\ B \times \Delta[1] & \xrightarrow{H} & Y \end{array} \qquad \qquad (V.14)$$

(in the sense that $K \circ (\mathrm{id}_A \times \varepsilon(\partial_1^0)) : A = A \times \Delta[0] \to A \times \Delta[1] \to X$ coincides with k and similarly for H) be given. Then there exists a morphism $\Theta : B \times \Delta[1] \to X$ extending θ.

Proof. It follows from Proposition V.2.12 applied to the square

$$\begin{array}{ccc} \Delta[0] = V(1,0) & \longrightarrow & \mathbf{Hom}(B,X) \\ \downarrow & & \downarrow {\scriptstyle i/p} \\ \Delta[1] & \longrightarrow & \mathbf{Hom}(i,p) \end{array}$$

obtained from (V.14) by using Corollary V.2.11. □

16. Proposition. *Let $q \in G$ be such that $[q, E]$ is one-to-one for all Kan sets E. Then $q \in \lambda(F_K)$, where F_K is the class of all fibrations of Kan sets.*

Proof. Let us prove first that

$$F_K \cap W = F_K \cap \{\text{homotopy equivalences}\}.$$

We have $F_K \cap W \subset F \cap W = \rho(G)$. In proving Lemma V.2.6 we have seen that any morphism in $\rho(G)$ is a homotopy equivalence, so that the right-hand side contains the left-hand side.

Conversely, let $p : E \to E'$ be a fibration of Kan sets and $q : E' \to E$ its homotopy inverse. Replacing q by a homotopic mapping $\tilde{q}$ we can assume that $p\tilde{q} = \mathrm{id}_{E'}$. Indeed, since pq and $\mathrm{id}_{E'}$ are homotopic and E' is a Kan set, there exists a commutative square

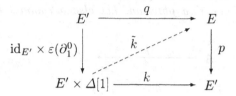

To construct $\tilde{k}$ we note that $p \in F$ and $\mathrm{id}_{E'} \times \varepsilon(\partial_1^0) \in \lambda(F)$. Define $\tilde{q} = \tilde{k}^1 = \tilde{k} \circ (\mathrm{id}_{E'} \times \varepsilon(\partial_1^0)) : E' \to E$. Then $\tilde{q}$ is homotopic to q and $p\tilde{q} = \mathrm{id}_{E'}$.

Now let us consider a strong homotopy $h : E \times \Delta[1] \to E$ between $\tilde{q}p$ and id_E and show that it can be replaced by a "homotopy relative to p", i.e., by such a homotopy $\tilde{h}$ that the following square is commutative,

$$\begin{array}{ccc} E \times \Delta[1] & \xrightarrow{\tilde{h}} & E \\ p \times \mathrm{id} \downarrow & & \downarrow p \\ E' \times \Delta[1] & \xrightarrow{\mathrm{pr}_1} & E' \end{array}$$

Before presenting a formal construction let us look at the geometrical picture. The condition $p\tilde{q} = \mathrm{id}_{E'}$ means that $\tilde{q}(E')$ is a section of the fibration p. The homotopy h joins each point of the total space E by a path with the corresponding point in the section. We want to replace h by a homotopy $\tilde{h}$ such that the whole joining path belongs to the corresponding fibre. Let us consider a horn in E formed by the initial path and its projection to the section. Projection of this horn to the base consists of two identical 1-simplices. Hence it can be completed to a degenerate 2-simplex in the base with three coinciding vertices. The third edge of this 2-simplex is also degenerate. As p is a fibration, this 2-simplex can be lifted to E and its degenerate edge becomes a path in the fibre which joins the point in E with the corresponding point in the section.

Formally, let us consider the diagram

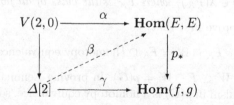

Here α is given by two elements from $\mathbf{Hom}(E, E)_1$ with the common beginning h, and $\tilde{q}ph \in \mathrm{Hom}_{\Delta^\circ\mathbf{Set}}(E \times \Delta[1], E)$. Next, γ corresponds to the composition

$$E \times \Delta[2] \xrightarrow{\mathrm{id}_E \times \varepsilon(\sigma_1^1)} E \times \Delta[1] \xrightarrow{h} E \xrightarrow{p} E.$$

Finally, p_* is induced by p. Commutativity is immediate. A lift β exists because $p_* \in F = \rho(hor)$.

Let us consider this β as a morphism $\tilde{\beta} : E \times \Delta[2] \to E$. One can easily see that the composition $\tilde{h} = \tilde{\beta} \circ (\mathrm{id}_E \times \varepsilon(\partial_2^0)) : E \times \Delta[1] \to E$ is the required morphism.

Recall now that our aim is to prove that $p \in F_K \cap W$, or, since $p \in F_K$, that $p \in F \cap W = \rho(G)$. To do this let us consider an arbitrary commutative square

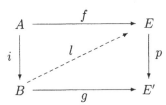

with $i \in G$, and prove that it admits an extension l. Indeed, let

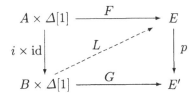

be a square with $F = \tilde{h} \circ (f \times \mathrm{id}_{\Delta[1]})$, $G = \mathrm{pr}_{E'} \circ (g \times \mathrm{id}_{\Delta[1]})$. It is commutative because $\tilde{h}$ is a homotopy relative to p. Considering the subsets $A \times \varepsilon(\partial_1^a)(\Delta[0])$ and $B \times \varepsilon(\partial_1^a)(\Delta[0])$ respectively, we obtain the initial square (for $a = 1$) and the square

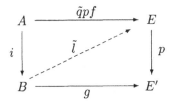

(for $a = 0$). In the last square we set $\tilde{l} = \tilde{q}g$; commutativity follows from $p\tilde{q} = \mathrm{id}_{E'}$. By Theorem V.2.15 (about covering homotopy) $\tilde{l}$ can be extended to L, and for l we can take the restriction of L. □

314   V. Introduction to Homotopic Algebra

The proof of the next proposition occupies the rest of this section.

17. Proposition. $G \cap \lambda(F_K) = G \cap \lambda(F)$.

Plan of the Proof

Clearly, $\lambda(F) \subset \lambda(F_K)$, so that we have to prove that $G \cap \lambda(F_K) \subset \lambda(F)$. Fix once for all a morphism $f : A \to B$ from $G \cap \lambda(F_K)$. We want to construct dashed arrows in various squares of the form

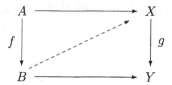

with $g \in F$. For some g this is easy. Namely, if $g \in F_K$, then we just use the definition of f. If g is induced by a fibration from F_K, then the problem can be solved without any difficulties. Namely, we have the diagram

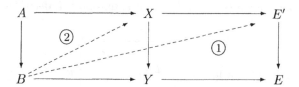

in which $h \in F_K$ and $XYEE'$ is a cocartesian square. The arrow ① exists because $f \in \lambda(F_K)$ and the arrow ② exists by universality.

Unfortunately, we cannot guarantee that any fibration is induced by a fibration from F_K. We will introduce a special class of fibrations – the so-called minimal fibrations – and prove that any fibration is in a sense equivalent to a minimal one. On the other hand, for any Kan set Z we will construct a universal fibration $p : E \to B$ with the fibre Z, prove that B (and, therefore, also E) is a Kan set, and prove that any minimal fibration with the fibre Z is induced by p.

18. Definition. *Let $g : X \to Y$ be a fibration, and $i : X' \hookrightarrow X$ an embedding of a simplicial set. The restriction $g' : X' \to Y$ of g to X' is said to be a fibrewise strong deformation retract of g if there exists a morphism $s : X \to X'$ with $g's = g$, $si = \mathrm{id}_{X'}$, and a homotopy $h : \Delta[1] \times X \to X$ between id_X and is over Y (i.e., such that $gh = g \cdot \mathrm{pr}_2 : \Delta[1] \times X \to Y$).*

19. Proposition. *A strong fibrewise deformation retract is a fibration.*

V.2 Homotopic Characterization of Weak Equivalences

Proof. Let us consider the diagram

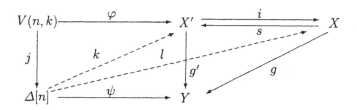

Since $g : X \to Y$ is a fibration, there exists a morphism $l : \Delta[n] \to X$ such that $lj = i\varphi$, $gl = \psi$. Let $k = sl$. Using the equalities $si = \text{id}_{X'}$, $g's = g$, $gi = g'$, we get $kj = slj = si\varphi = \varphi$, $g'k = g'sl = gl = \psi$. □

Let us remark that in proving this proposition we did not use the existence of h. In the next proposition the existence of h is already used.

20. Proposition. *Let $g' : X' \to Y$ be a strong fibrewise deformation retract of $g : X \to Y$, and let $f : A \to B$ be an arbitrary cofibration. Then $f \in \lambda(g')$ implies $f \in \lambda(g)$.*

Proof. Let us consider the diagram

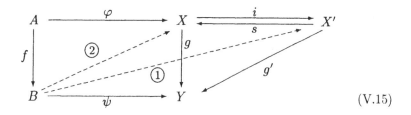

(V.15)

As $f \in \lambda(g')$, there exists a diagonal ① in the square $ABX'Y$. Consider the square

$$\begin{array}{ccc} V(1,0) = \Delta[0] & \xrightarrow{\Phi} & \text{Hom}(B, X) \\ {\scriptstyle j}\downarrow & & \downarrow{\scriptstyle f/g} \\ \Delta[1] & \xrightarrow{\Psi} & \text{Hom}(f, g) \end{array}$$

in which j is a natural embedding, Φ corresponds to the 0-simplex $B \xrightarrow{①} X' \xrightarrow{s} X$ from $\text{Hom}(B, X)_0 = \text{Hom}_{\Delta° \text{Set}}(B, X)$, and ψ corresponds to the 1-simplex $\theta \in \text{Hom}(f, g)_1$ which is defined as follows: we have

$$\text{Hom}(f, g)_1 = \text{Hom}(A \times \Delta[1], X) \underset{\text{Hom}(A \times \Delta[1], Y)}{\times} \text{Hom}(B \times \Delta[1], Y),$$

and θ corresponds to the pair (θ_1, θ_2), where

$$\begin{aligned}\theta_1 &= h \circ (\varphi \times \mathrm{id}_{\Delta[1]}) : A \times \Delta[1] \longrightarrow X, \\ \theta_2 &= \psi \times \mathrm{pr}_1 : B \times \Delta[1] \longrightarrow Y\end{aligned}$$

(One can easily see that θ_1 and θ_2 induce equal morphisms $A \times \Delta[1] \longrightarrow Y$). Since h is a homotopy between id_X and is, the above square is commutative.

By Proposition V.2.12, f/g is a fibration so that there exists $\Theta : \Delta[1] \to \mathbf{Hom}(B, X)$ such that the diagram remains commutative. Define $\Delta[0] \to \mathbf{Hom}(B, X)$ as the composition $\Delta[0] = V(1,1) \to \Delta[1] \xrightarrow{\Theta} \mathbf{Hom}(B, X)$. The image of this morphism determines a 0-simplex in $\mathbf{Hom}(B, X)$, i.e., a morphism of simplicial sets ②: $B \to X$. We leave to the reader an easy verification of the commutativity of (V.15).

21. Connected Simplices

Let $g : X \to Y$ be a fibration. Denote by $i_n : \dot\Delta[n] \to \Delta[n]$ the inclusion of the boundary of the standard simplex and consider the following fibration (see Proposition V.2.12):

$$i_n/g : \mathbf{Hom}(\Delta[n], X) \longrightarrow \mathbf{Hom}(i_n, g).$$

Two simplices $x_1, x_2 \in X_n$ are said to be *g-equivalent* if the corresponding morphisms $\tilde{x}_1, \tilde{x}_2 \in \mathbf{Hom}(\Delta[n], X)_0 = \mathrm{Hom}_{\Delta^\circ \mathbf{Set}}(\Delta[n], X)$ lie in one connected component of some fibre of the fibration i_n/g.

Since i_n/g is a fibration, all its fibres are Kan sets. Hence two elements of some fibre in i_n/g are g-equivalent if and only if they are vertices of a 1-simplex in this fibre, and the definition of g-equivalence can be reformulated as follows:

$x_1, x_2 \in X_n$ are g-equivalent if and only if there exists a diagram

$$\Delta[1] \times \dot\Delta[n] \coprod_{\{e\} \times \dot\Delta[n]} \{e\} \times \Delta[n] \xrightarrow{a} X$$

with maps h_1, h_2 and $\Delta[1] \times \Delta[n] \xrightarrow{b} Y$, g (V.16)

in which $e = 0$ or 1, $\{e\} = \varepsilon(\partial_1^0)(\Delta[0]) \subset \Delta[1]$, j is a natural embedding, $h_1 j = h_2 j = a$, $gh_1 = gh_2 = b$, $h_1(\{1-e\} \times \delta_n) = x_1$, $h_2(\{1-e\} \times \delta_n) = x_2$; recall that $\delta_n \in \Delta[n]_n$ denotes the unique nondegenerate n-simplex in $\Delta[n]$.

The reader can easily verify that two degenerate simplices are g-equivalent iff they coincide.

V.2 Homotopic Characterization of Weak Equivalences

22. Definition. *A fibration $g : X \to Y$ is said to be* minimal *if any two g-equivalent simplices coincide.*

Roughly speaking, the minimality of g means that there are no redundant relative homotopies in X: any two homotopic morphisms are equal. In particular, the reader can easily verify that if $g : X \to Y$ is minimal, then any connected component of any fibre of g contains exactly one 0-simplex.

The proof of following lemma is easy.

23. Lemma. *Suppose that $g : X \to Y$ is a minimal fibration and $\varphi : Y' \to Y$ is an arbitrary morphism. Then the induced fibration $g' : X \underset{Y}{\times} Y' = X' \to Y'$ is also minimal.*

24. Proposition. *For any fibration $g : X \to Y$ there exists a simplicial subset $X' \subset X$ such that $g' : X' \to Y$ is a minimal fibration and a fibrewise deformation retract of g.*

Proof. a) Evidently, g-equivalence is an equivalence relation. Mark one representative in each equivalence class of g-equivalent simplices in X in such a way that the following condition is satisfied: if there is a degenerate simplex in a given class, it is to be marked (by the remark just before Definition V.2.22, a degenerate simplex in a given class is unique). Then all degenerate simplices will be marked. Denote by X' the simplicial set which consists of marked simplices and is maximal among subsets with this property. Let us show that X' satisfies all the conditions of the theorem.

b) First, if $x \in X_n$ is a marked simplex and $d_n^i x \in X'$ for all i, $0 \leq i \leq n$, then $x \in X'$. Indeed, assume the contrary. Add to X' the simplex x and all its degenerations, i.e., all simplices of the form $X(\sigma)x$ for all surjective $\sigma : [m] \to [n]$. The set of simplices obtained in such a way is clearly invariant under all operators $X(f)$, so that we get a simplicial subset $X'' \subset X$. On the other hand, each simplex $\tilde{x} \in X''$ is either degenerate, or equal to x, or belongs to X'. In each case $\tilde{x}$ is marked, a contradiction with the maximality of X'.

c) Let now $\mathcal{N}$ be the set of all pairs (Z, h) consisting of a simplicial subset $Z \subset X$ containing X' and of a homotopy $\Delta[1] \times Z \to X$ of the form $h = l\tilde{h}$, where $l : Z \to X$ is an embedding and $\tilde{h}$ is a retraction $\Delta[1] \times Z \to Z$ over Y (i.e., $\tilde{h}$ is a homotopy between id_Z and iq over Y, where $i : X' \to Z$ is the embedding and $q : Z \to X'$ is the left inverse to i, $qi = \mathrm{id}_{X'}$).

The set $\mathcal{N}$ is nonempty (it contains the pair $(X', i_{X' \to X} \circ \mathrm{pr}_2)$) and partially ordered $((Z, h) < (Z', h')$ iff $Z \subset Z'$ and the restriction of h' to Z is h). Let (Z, h) be a maximum element of $\mathcal{N}$ with respect to this ordering. We shall show that $Z = X$. This would imply that X' is a strong relative deformation retract of X, so that by Proposition V.2.19 $g' : X' \to X$ is a fibration, and, by the choice of X', this fibration is minimal.

So, let (Z, h) be a maximum pair such that $Z \neq X$, and let $x \in X_n$ be a simplex of minimum dimension among those not belonging to Z. Let

318   V. Introduction to Homotopic Algebra

Z' be the smallest simplicial subset of X containing Z and x. We extend $h : \Delta[1] \times Z \to X$ to $h' : \Delta[1] \times Z' \to Z$ using the construction described below.

First, it is clear that $d_n^i x \in Z$ for all i. Let $\tilde{j} : \dot{\Delta}[n] \to Z$ be the corresponding morphism. Define $\varphi : \Delta[1] \times \dot{\Delta}[n] \coprod_{\{0\} \times \dot{\Delta}[n]} \{0\} \times \Delta[n] \to X$ as the morphism coinciding with $h \circ (\mathrm{id}_{\Delta[1]} \times \tilde{j})$ on the first factor and with the morphism $\tilde{x} : \Delta[n] = \{0\} \times \Delta[n] \to X$ determined by the simplex x on the second factor. Such a morphism φ gives the commutative square (see Corollary V.2.11)

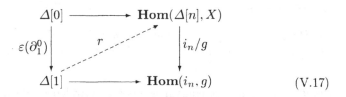

(V.17)

where $i_n : \dot{\Delta}[n] \to \Delta[n]$ is the natural embedding. Since i_n is a cofibration and g is a fibration, i_n/g is also a fibration (Proposition 12). Hence the square (V.17) admits a lifting $r : \Delta[1] \to \mathbf{Hom}(\Delta[n], X)$. Moreover, since i_n/g is a fibration, r can be chosen in such a way that $r(\{1\}) \in \mathbf{Hom}(\Delta[n], X)_0 = \mathrm{Hom}_{\Delta°\mathbf{Set}}(\Delta[n], X)$ corresponds to a marked simplex $x' \in X_n$. (This is because any connected component of any fibre of i_n/g contains a marked point. The reader can easily reconstruct the corresponding arguments using liftings of horns $V(i, 2) \to \Delta[2]$.)

Thus, we have constructed a marked n-simplex x' whose boundary lies in $h(\{1\} \times Z)$, i.e., in X'. By part b) of the proof, $x' \in X'$. Let us extend $q : Z \to X'$ to $q' : Z' \to X'$ by setting $q'(x) = x'$. Finally, using $r : \Delta[1] \to \mathbf{Hom}(\Delta[n], X)$ we extend the homotopy $h : \Delta[1] \times Z \to Z$ to a homotopy $\tilde{h}' : \Delta[1] \times Z' \to Z'$ between $\mathrm{id}_{Z'}$ and $i'q' : Z' \to Z'$. The pair $(Z', h' = f'\tilde{h}')$ where f' is the embedding $f' : Z' \to X$ is clearly greater (with respect to the ordering in $\mathcal{N}$) then the pair (Z, h), in contradiction with the maximality of (Z, h). Hence $Z = X$.   □

As the next step we prove that a minimal fibration over $\Delta[n]$ is trivial. To do this we need the following important property of minimal fibrations.

25. Lemma. *Let $g : X \to Y$ be a minimal fibration and $u : X \to X$ a morphism which is homotopic over Y to the identity morphism id_X and satisfies $gu = g$. Then u is an isomorphism in the category $\Delta°\mathbf{Set}$.*

Proof. Let $g_* : \mathbf{Hom}(X, X) \to \mathbf{Hom}(X, Y)$ be the mapping induced by the composition with g. The condition $gu = g$ means that the zero-dimensional simplices u and id_X belong to the same fibre of g_*, and the existence of a homotopy between u and id_X over Y means that these simplices belong to the same connected component of this fibre. Since g_* is a fibration (by Corollary V.2.13b)), there exists a strong homotopy between u and id_X in this connected component, i.e., there exists a morphism $h : \Delta[1] \times X \to Y$ connecting u and id_X and satisfying $gh = g \cdot \mathrm{pr}_2$. We want to prove that u_n is one-to-one for any n. Letting $X_{-1} = \emptyset$ we have the required property for $n = -1$. Hence we can use the induction in n.

a) Assume that u_r is injective for all $r < n$. Let $x, x' \in X_n$ be two simplices such that $u_n(x) = u_n(x')$ and $\tilde{x}, \tilde{x}' : \Delta[n] \to X$ the corresponding morphisms. Then $g \circ \tilde{x} = h \circ \tilde{x}'$ and, by the induction assumption, $\tilde{x} \mid_{\dot\Delta[n]} = \tilde{x}' \mid_{\dot\Delta[n]}$. Hence we get a commutative diagram:

$$(\Delta[1] \times \dot\Delta[n]) \cup (\{1\} \times \Delta[n]) \longrightarrow X$$
$$\downarrow \qquad \qquad \qquad \qquad \downarrow g$$
$$\Delta[1] \times \Delta[n] \xrightarrow{g \circ \tilde{x} \circ \mathrm{pr}_2} Y$$

in which the left vertical arrow is a natural embedding, the two diagonals are $h_1 = h \circ (\mathrm{id}_{\Delta[1]} \times \tilde{x})$ and $h_2 = h \circ (\mathrm{id}_{\Delta[1]} \times \tilde{x}')$ and the upper horizontal arrow is the restriction of the latter two mappings to $(\Delta[1] \times \dot\Delta[n]) \cup (\{1\} \times \Delta[n])$ (the two restrictions coincide). This diagram means that the simplices $h_1(\{0\} \times \delta_n) = x$ and $h_2(\{0\} \times \delta_n) = x'$ are g-equivalent (see diagram (V.16)). Since g is a minimum fibration, we get $x = x'$.

b) Assume, in addition, that u_r is surjective for all $r < n$. Let x be an arbitrary n-simplex in X. By the induction assumption, for any i, $0 \le i \le n$, there exists $x'_i \in X_{n-1}$ with $u_{n-1}(x'_i) = d^i_n x$ and, by a), these x'_i are unique. Hence the family $\{x'_i\}$ determines a morphism $\varphi : \dot\Delta[n] \to X$.

Evidently, $g \circ \varphi = g \circ \tilde{x} \mid_{\dot\Delta[n]}$. The restriction to $\{1\} \times \dot\Delta[n]$ of the composition

$$\Delta[1] \times \dot\Delta[n] \xrightarrow{\mathrm{id}_{\Delta[1]} \times \varphi} X \xrightarrow{h} X$$

equals

$$\{1\} \times \dot\Delta[n] = \dot\Delta[n] \xrightarrow{\tilde{x} \mid_{\dot\Delta[n]}} X.$$

Thus $h \circ (\mathrm{id}_{\Delta[1]} \times \varphi)$ and $\tilde{x}$ determine a morphism ψ such that the square

$$(\Delta[1] \times \dot\Delta[n]) \cup (\{1\} \times \Delta[n]) \xrightarrow{\psi} X$$
$$j \downarrow \qquad \overset{\Psi}{\nearrow} \qquad \downarrow g$$
$$\Delta[1] \times \Delta[n] \longrightarrow Y$$

is commutative. Since $g \in F$ and $j \in G \cap W = \lambda(F)$ (see Lemma V.2.8), there exists a diagonal $\Psi : \Delta[1] \times \Delta[n] \to X$ such that the whole diagram is commutative. Define $\tilde{z}: \Delta[n] \to X$ as the restriction of Ψ to $\{0\} \times \Delta[n] = \Delta[n]$ and let $z \in X_n$ be the corresponding n-simplex. Then we have the following commutative diagram:

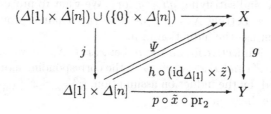

in which Ψ and $h \circ (\mathrm{id}_{\Delta[1]} \times \tilde{z})$ coincide on $\{1\} \times \Delta[n]$ by the minimality of g. This means, evidently, that $x = u_n(z)$, and the surjectivity of u_n is proved. □

26. Corollary. *Two minimal fibrations $g: X \to Y$, $g': X' \to Y$ with the same homotopic type over Y are isomorphic over Y.*

Proof. The condition of the corollary means that there exist morphisms $u: X \to X'$ and $v: X' \to X$ over Y such that uv and vu are homotopic over Y to the identity morphisms $\mathrm{id}_{X'}$ and id_X respectively. By Lemma V.2.25, uv and vu are isomorphisms over Y. □

27. Proposition. *Any minimal fibration over $\Delta[n]$ is isomorphic over $\Delta[n]$ to a direct product.*

Proof. a) Let $g: X \to Y$ be a minimal fibration and $\varphi, \psi: Y' \to Y$ be two homotopic morphisms. We prove first that the induced fibrations $g_\varphi: X_\varphi \to Y'$ and $g_\psi: X_\psi \to Y'$ defined by the cartesian squares,

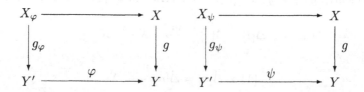

are isomorphic. Without loss of generality we can assume that φ and ψ are connected by a simple homotopy $\chi: \Delta[1] \times Y' \to Y$. Let us consider the fibration $g_\chi: X_\chi \to \Delta[1] \times Y'$ induced by χ and form the commutative diagram

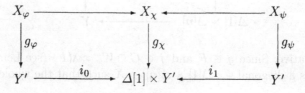

in which $i_0 = \varepsilon(\partial_1^1) \times \mathrm{id}_{Y'}$ is induced by the embedding of the initial point into $\Delta[1]$, $i_1 = \varepsilon(\partial_1^0) \times \mathrm{id}_{Y'}$ is induced by the embedding of the end point into $\Delta[1]$, and embeddings j_0, j_1 are determined by i_0, i_1. Let h_0 be a homotopy between $\varepsilon(\partial_1^1 \circ \sigma_0^0) : \Delta[1] \to \Delta[1]$ and $\mathrm{id}_{\Delta[1]}$ (pulling out the initial point of the segment) and h_1 a homotopy between $\mathrm{id}_{\Delta[1]}$ and $\varepsilon(\partial_1^0 \circ \sigma_0^0)$ (contracting to the end point of the segment). By Lemma V.2.14, there exist $r_0 : X_\chi \to X_\varphi$ with $r_0 j_0 = \mathrm{id}_{X_\varphi}$ and a homotopy $k_0 : \Delta[1] \times X_\chi \to X_\chi$ between $j_0 r_0$ and id_{X_χ} which fits into the commutative diagram

$$\begin{array}{ccc} \Delta[1] \times X_\chi & \xrightarrow{k_0} & X_\chi \\ \mathrm{id}_{\Delta[1]} \times g_\chi \downarrow & & \downarrow g_\chi \\ \Delta[1] \times (\Delta[1] \times Y') & \xrightarrow{h_0} & \Delta[1] \times Y' \end{array}$$

Similarly, there exist $r_1 : X_\chi \to X_\psi$ such that $r_1 j_1 = \mathrm{id}_{X_\psi}$ and a homotopy $k_1 : \Delta[1] \times X_\chi \to X_\chi$ between id_{X_χ} and $j_1 r_1$ which fits into the commutative diagram

$$\begin{array}{ccc} \Delta[1] \times X_\chi & \xrightarrow{k_1} & X_\chi \\ \mathrm{id}_{\Delta[1]} \times g_\chi \downarrow & & \downarrow g_\chi \\ \Delta[1] \times (\Delta[1] \times Y') & \xrightarrow{h_1} & \Delta[1] \times Y' \end{array}$$

This obviously implies that $r_1 j_0 : X_\varphi \to X_\psi$ and $r_0 j_1 : X_\psi \to X_\varphi$ are mutually inverse modulo homotopies (i.e., their compositions are homotopic to the corresponding identity morphisms). By Lemma V.2.23, g_φ and g_ψ are minimal fibrations. By Corollary V.2.26, they are isomorphic over Y'.

b) Now let $X \to \Delta[n]$ be a minimal fibration over $\Delta[n]$, $\varphi : \Delta[n] \to \Delta[n]$ the identity morphism and $\psi : \Delta[n] \to \Delta[n]$ the morphism induced by the mapping $[n] \to [n]$ which sends the whole $\Delta[n]$ to the point $0 \in [n]$. Then g_φ equals g and g_ψ is isomorphic to the direct product $\Delta[n] \times Z \to Z$ where Z is the fibre of g over the 0-th vertex (corresponding to the embedding $[0] \to [n]$, $0 \to 0$) of $\Delta[n]$. On the other hand, φ and ψ are homotopic: the homotopy $\Delta[1] \times \Delta[n] \to \Delta[n]$ is given by the mapping $v : [1] \times [n] \to [n]$ such that

$$v(0, i) = 0, \quad v(1, i) = i.$$

By a), $g : X \to \Delta[n]$ is isomorphic over $\Delta[n]$ to the direct product $\Delta[n] \times Z \to Z$. □

28. Definition. *A fibration $g : X \to Y$ is said to be locally trivial if for any simplex $y \in Y_n$ the fibration $g_y : X_y \to \Delta[n]$ induced from g by the* *corresponding morphism* $\tilde{y} : \Delta[n] \to Y$, *is isomorphic over* $\Delta[n]$ *to a direct product.*

In other words, a fibration g is locally trivial if for any $y \in Y_n$ there exists a commutative diagram

$$\begin{array}{ccccc} \Delta[n] \times Z & \xrightarrow{\alpha(y)} & Y_y & \longrightarrow & X \\ {\scriptstyle \mathrm{pr}_1} \downarrow & & \downarrow {\scriptstyle g_y} & & \downarrow {\scriptstyle g} \\ \Delta[n] & \xrightarrow{\mathrm{id}} & \Delta[n] & \xrightarrow{\tilde{y}} & Y \end{array}$$

in which the right square is cartesian and $\alpha(y)$ an isomorphism.

The family $\{\alpha(y)\}, y \in Y_n, n = 0, 1, 2, \ldots$, is called an *atlas* of the given locally trivial fibration.

Later we shall always assume that the base Y of g is connected. Then it is clear that the fibres Z over all simplices y of Y are isomorphic. Denote by $A(Z) = \{A_n(Z)\}$ the simplicial group of automorphisms of the simplicial set Z (see V.2.34 and Ex. 1).

The atlas $\{\alpha(y)\}$ of g is nonunique. Namely, associate with any y an element $\gamma(y) \in A_n(Z) \subset \mathrm{Hom}_{\Delta \circ \mathrm{Set}}(\Delta[n] \times Z, Z)$. Define $\gamma_I(y) : \Delta[n] \times Z \to \Delta[n] \times Z$ by $\gamma_I(y)(t,z) = (t, \gamma(y)z)$. Clearly, $\bar{\alpha}(y) = \alpha(y)\gamma_I(y)$ is again an atlas of g. Conversely, if $\bar{\alpha}(y), \alpha(y)$ are two atlases then $\alpha(y)^{-1}\bar{\alpha}(y) = \gamma_I(y)$ for some $\gamma(y) \in A_n(Z)$. The families $\{\alpha(y)\}$ and $\{\bar{\alpha}(y)\}$ related to one another as above are said to be equivalent.

29. Gluing Functions

Let us study now the relation between isomorphisms $\alpha(y)$ and simplicial operations in Y. Let $y \in Y_n$ and $f : [m] \to [n]$ be a nonincreasing map. Then $\widetilde{Y(f)y} = \tilde{y} \circ \varepsilon_f : \Delta[m] \to Y$ ($\varepsilon_f : \Delta[m] \to \Delta[n]$ is defined in II.4.24). Hence the transitivity of forming the induced fibration means that there exists a cartesian square of the form

$$\begin{array}{ccc} X_{Y(f)y} & \longrightarrow & X_y \\ {\scriptstyle g_{Y(f)y}} \downarrow & & \downarrow {\scriptstyle g_y} \\ \Delta[m] & \xrightarrow{\varepsilon_f} & \Delta[n] \end{array}$$

Let us include this square (as the bottom face) into the following commutative diagram:

V.2 Homotopic Characterization of Weak Equivalences

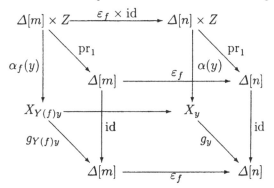

By the universality property of cartesian squares there exists a unique morphism $\alpha_f(y) : \Delta[m] \times Z \to X_{Y(f)y}$ which is an isomorphism (because three other vertical arrows are isomorphisms and the upper square is cartesian).

On the other hand, with the simplex $Y(f)y \in Y_m$ we have associated an isomorphism $\alpha(Y(f)y) : \Delta[m] \times Z \to X_{Y(f)y}$. The ratio of these two isomorphisms $\alpha(Y(f)y)^{-1} \circ \alpha_f(y) : \Delta[m] \times Z \to \Delta[m] \times Z$ defines, clearly, an element $\mu(f, y) \in A_m(Z)$ by the formula

$$\alpha(Y(f)y)^{-1} \circ \alpha_f(y) = \mu_I(f, y).$$

The family $\{\mu(g, y)\}$ for all y and f is called the *gluing functions* of the given locally trivial fibration. Their properties are described in the following proposition.

30. Proposition. *a) The cocycle condition: for* $[k] \xrightarrow{f_1} [m] \xrightarrow{f_2} [n]$ *and* $y \in Y_n$ *we have*

$$\mu(f_2 f_1, y) = \mu(f_1, Y(f_2)y) \cdot (A(Z)(f_1))(\mu(f_2, y)) \qquad (V.18)$$

(*·* *is the product in* $A_k(Z)$).

b) If an atlas $\{\alpha(y)\}$ *is replaced by an equivalent one,* $\bar{\alpha}(y) = \alpha(y)\gamma_I(y)$, *the gluing functions are replaced by*

$$\bar{\mu}(f, y) = \gamma(Y(f)y)^{-1} \cdot \mu(f, y) \cdot (A(Z)(f))(\gamma(y)). \qquad (V.19)$$

c) A locally trivial fibration can be reconstructed from its gluing functions: two locally trivial fibrations are isomorphic (over Y) if and only if the corresponding gluing functions satisfy (V.19) for some $\{\gamma(y)\}$.

A rather easy and purely formal verification of this proposition is left to the reader.

31. Regular Gluing Functions

Among various families of gluing functions corresponding to all choices of atlases of a given locally trivial fibration there exist some particularly simple gluing functions. Namely, a family of gluing functions $\{\mu(f,y)\}$ is said to be *regular* if

$$\mu(f,y) = e_n \quad \text{(the unit element of the group } A_n(Z))$$

for all degeneracy operators $f = \sigma_n^i$, $i = 0, 1, \ldots, n$ and for all face operators $f = \partial_n^i$, $i = 1, 2, \ldots, n$, except ∂_n^0.

32. Proposition. *Any locally trivial fibration possesses a regular family of gluing functions.*

Proof. Let us ensure first the conditions $\mu(\partial_n^i, y) = e_n$, $i = 1, 2, \ldots, n$, for all nondegenerate $y \in Y_n$. To do this we will consecutively change $\alpha(y)$, $y \in Y_n$, as follows.

Leave all $\alpha(y)$, $y \in Y_0$, unchanged. Let $n \geq 1$ and $\alpha(y')$ for $y' \in Y_r$, $r < n$, satisfy $\mu(\partial_r^i, y') = e_r$ for $r < n$, $1 \leq i \leq r$, $y' \in Y_r$. For a nondegenerate simplex $y \in Y_n$ define $\xi^i = \mu(\partial_n^i, y) \in A_{n-1}(Z)$. By (V.18), $d_{n-1}^i \xi^j = d_{n-1}^{j-1} \xi^i$ for $1 \leq i < j \leq n-1$. Since $A(Z)$ is a Kan set (see V.1.1b)) there exists $\xi \in A_n(Z)$ such that $\xi^i = d_n^i \xi$. One can easily see that replacing $\alpha(y)$ by $\bar{\alpha}(y) = \alpha(y)\xi_I^{-1}$ we get the required equality $\mu(\partial_n^i, y) = e_n$ and do not change $\mu(\partial_n^i, \bar{y})$ for other nondegenerate $\bar{y} \in Y_r$, $r \leq n$.

Let us define now $\alpha(y)$ for degenerate y by the formula

$$\alpha(s_n^i y) = \alpha_{\sigma_n^i}(y).$$

One can easily verify that the family of gluing functions corresponding to the constructed atlas is a regular one. $\square$

Now we describe the gluing functions of an induced fibration. Let $g : X \to Y$ be a locally trivial fibration with the fibre Z, and $\mu(f,y)$ a family of regular gluing functions of g. Let $\varphi : Y' \to Y$ be a morphism of simplicial sets and $g' : X' \to Y'$ the induced fibration defined by the cartesian square

$$\begin{array}{ccc} X' & \xrightarrow{\varphi'} & X \\ g' \downarrow & & \downarrow g \\ Y' & \xrightarrow{\varphi} & Y \end{array}$$

The fibre of g' coincides, obviously, with Z. Definition V.2.28 shows that g' is also a locally trivial fibration.

33. Lemma. *A regular family of gluing functions $\mu'(f, y')$ for g' is given by*

$$\mu'(f, y') = \mu(f, \varphi_n(y')).$$

An easy proof of this lemma is left to the reader. $\square$

V.2 Homotopic Characterization of Weak Equivalences   325

To complete the proof of Proposition V.2.16 we construct a universal fibration $p : E \to B$ with the fibre Z, prove that its base B is a Kan set and that any minimal fibration $f : X \to Y$ with the fibre Z is induced by some morphism $\varphi : Y \to B$.

34. Universal Fibration

Suppose that we are given a simplicial set Z, and let $A(Z) = \{A_n(Z)\}$ be the simplicial group of automorphisms of Z. Let us recall the definition of $A(Z)$. An element $\gamma \in A_n(Z)$ is a morphism of simplicial sets $\gamma : \Delta[n] \times Z \to Z$ such that $\gamma_I = (\mathrm{pr}_1, \gamma) : \Delta[n] \times Z \to \Delta[n] \times Z$ is an isomorphism. The formula $z \to \gamma(\delta_n, z)$, where δ_n is the only nondegenerate n-simplex in $\Delta[n]$, determines an invertible map (which will also be denoted by γ) $Z_n \to Z_n$. One can easily verify that in such a way we obtain an action of the group $A_n(Z)$ on Z_n and all these actions are glued together to the action of the simplicial group $A(Z)$ on the simplicial set Z.

Denote by e_n the identity element of the group $A_n(Z)$.

a) *Definition of B.* Let

$$B_0 = [\bullet] \quad \text{(one-element set)},$$

$$B_n = A_{n-1}(Z) \times \ldots \times A_0(Z) \quad \text{for} \quad n \geq 1.$$

Thus, an element of B_n is a family $(\gamma_{n-1}, \ldots, \gamma_0)$, $\gamma_i \in A_i(Z)$. Define $d_n^i : B_n \to B_{n-1}$, $s_n^i : B_n \to B_{n+1}$ by

$$d_0^i(\gamma_0) = [\bullet] \quad \text{for} \quad i = 0, 1, \quad \gamma_0 \in B_1 = A_0(Z),$$

$$d_n^0(\gamma_{n-1}, \ldots, \gamma_0) = (\gamma_{n-2}, \ldots, \gamma_0),$$

$$d_n^i(\gamma_{n-1}, \ldots, \gamma_0) = (d_{n-1}^{i-1}\gamma_{n-1}, \ldots, d_{n-i+1}^1 \gamma_{n-i+1},$$
$$\gamma_{n-i-1} \cdot d_{n-i}^0 \gamma_{n-i}, \gamma_{n-i-2}, \ldots, \gamma_0),$$

$$s_n^i(\gamma_{n-1}, \ldots, \gamma_0) = (s_{n-1}^{i-1}\gamma_{n-1}, \ldots, s_{n-i}^0 \gamma_{n-i}, e_{n-i}, \gamma_{n-i-1}, \ldots, \gamma_0).$$

One can easily verify that these mappings d and s satisfy the required commutation conditions, so that we get a simplicial set $B = \{B_n\}$.

b) *Definition of E.* Let

$$E_0 = Z_0,$$

$$E_n = Z_n \times B_n = Z_n \times A_{n-1}(Z) \times \ldots \times A_0(Z) \quad \text{for} \quad n \geq 1.$$

$$d_n^0(z_n, \gamma_{n-1}, \ldots, \gamma_0) = (\gamma_{n-1} \cdot d_n^0 z_n, \gamma_{n-2}, \ldots, \gamma_0),$$

$$d_n^i(z_n, b_n) = (d_n^i z_n, d_n^i b_n) \quad \text{for} \quad i \geq 1,$$

$$s_n^i(z_n, b_n) = (s_n^i z_n, s_n^i b_n).$$

In the last two formulas $d_n^i b_n$ and $s_n^i b_n$ are as in a).

Again one easily sees that $E = \{E_n\}$ is a simplicial set.

c) *Definition of $p : E \to B$*. Let
$$p_0(z_0) = [\bullet],$$
$$p_n(z_n, b_n) = b_n, \quad \text{for} \quad n \geq 1.$$
$p = \{p_n\}$ is clearly a morphism of simplicial sets.

35. Proposition. *a) $p : E \to B$ is a fibration.*
b) B is a Kan set.

Proof. We have to prove that $p \in F = \rho(hor)$, i.e., that there exists a morphism h in the diagram

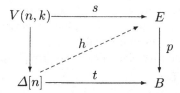

The morphism s is determined by a family of $(n - 1)$-simplices $x_0, \ldots, x_{k-1}, x_{k+1}, \ldots, x_n \in E_{n-1}$ such that
$$d^i_{n-1} x_j = d^{j-1}_{n-1} x_i, \quad i < j, \quad i, j \neq k. \tag{V.20}$$

The morphism t is determined by an n-simplex $y \in B_n$, and the commutativity of the square means that
$$p(x_i) = d^i_n y, \quad i \neq k.$$

The existence of h is equivalent to the existence of an n-simplex $x \in E_n$ such that
$$p(x) = y, \quad d^i_n x = x_i, \quad i \neq k.$$

Let $y = (\gamma_{n-1}, \ldots, \gamma_0)$, $x_i = (z_{n-1,i}, \gamma_{n-2,i}, \ldots, \gamma_{0,i})$, $i \neq k$, where $\gamma_l, \gamma_{l,i} \in A_l(Z)$, $z_{n-1,i} \in Z_{n-1}$. The condition $p(x) = y$ means that the required element x is of the form
$$x = (z_n, \gamma_{n-1}, \ldots, \gamma_0)$$
for some $z_n \in Z_n$. The conditions $d^i_n x = x_i$ imply that for z_n we can take any element in Z_n such that
$$d^i_n z_n = z_{n-1,i}, \quad i \neq k.$$

By (V.20),
$$d^i_{n-1} z_{n-1,j} = d^{j-1}_{n-1} z_{n-1,i}, \quad i < j, \quad i, j \neq k.$$

V.2 Homotopic Characterization of Weak Equivalences

Since Z is a Kan set (being a fibre of the fibration $f : X \to Y$), an element z_n with the required properties always exists.

b) Let us assume that $y_0, \ldots, y_{k-1}, y_{k+1}, \ldots, y_n \in B_{n-1}$ such that

$$d^i_{n-1} y_j = d^{j-1}_{n-1} y_i, \qquad i < j, \quad i, j \neq k. \tag{V.21}$$

We must find $y \in B_n$ such that

$$d^i_n y = y_i, \qquad i \neq k.$$

Let

$$y_i = (y_{n-2,i}, \ldots, \gamma_{0,i}).$$

(i) $k > 0$. The condition $d^0_n y = y_0$ implies that y must be of the form

$$y = (\gamma_{n-1}, \gamma_{n-2,0}, \ldots, \gamma_{0,0}).$$

for some $\gamma_{n-1} \in A_{n-1}(Z)$. Next, the conditions

$$d^i_n y = y_i, \qquad i \neq 0, k$$

imply that γ_{n-1} might be any element for which

$$d^i_{n-1} \gamma_{n-1} = \gamma_{n-2, i+1}, \qquad i \neq k-1.$$

By (V.21),

$$d^i_{n-2} \gamma_{n-2, j+1} = d^{j-1}_{n-2} \gamma_{n-2, i+1}, \qquad i < j, \quad i, j \neq k-1.$$

Since $A(Z)$ is a Kan set (being a simplicial group, see V.2.1b)), the required element γ_{n-1} always exists.

(ii) $k = 0$. Let us remark, first of all, that, by (V.21),

$$d^i_{n-2} \gamma_{n-2, j+1} = d^{j-1}_{n-2} \gamma_{n-2, i+1}, \qquad 0 < i < j.$$

Similarly to the above arguments we can prove that there exists $\gamma_{n-1} \in A_{n-1}(Z)$ such that

$$d^i_{n-1} \gamma_{n-1} = \gamma_{n-2, i+1}, \qquad i > 0.$$

Let us set now

$$y = \left(\gamma_{n-1}, \gamma_{n-2,1} \cdot \left(d^0_{n-1} \gamma_{n-1} \right)^{-1}, \gamma_{n-3,1}, \ldots, \gamma_{0,1} \right).$$

To verify the equality $d^i_n y = y_i$ we argue slightly differently for $i = 1$ and for $i > 1$.

By the definition of simplicial operations in B (see V.2.34a)) we have

$$\begin{aligned} d^1_n y &= \left(\gamma_{n-2,1} \cdot \left(d^0_{n-1} \gamma_{n-1} \right)^{-1} \cdot d^0_{n-1} \gamma_{n-1}, \gamma_{n-3,1}, \ldots, \gamma_{0,1} \right) \\ &= (\gamma_{n-2,1}, \gamma_{n-3,1}, \ldots, \gamma_{0,1}) = y_1. \end{aligned}$$

Next,
$$d_n^2 y = \left(d_{n-1}^1 \gamma_{n-1}, \gamma_{n-3,1} \cdot d_{n-2}^0 \gamma_{n-2,1} \cdot d_{n-2}^0 \left(d_{n-1}^0 \gamma_{n-1}\right)^{-1}, \gamma_{n-4,1}, \ldots, \gamma_{0,1}\right).$$

Now
$$d_{n-1}^1 \gamma_{n-1} = \gamma_{n-2,1}, \tag{V.22}$$

$$\begin{aligned}
d_{n-2}^0 \left(d_{n-1}^0 \gamma_{n-1}\right)^{-1} &= d_{n-2}^0 d_{n-1}^0 \gamma_{n-1}^{-1} = d_{n-2}^0 d_{n-1}^1 \gamma_{n-1}^{-1} \\
&= d_{n-2}^0 \left(d_{n-1}^1 \gamma_{n-1}\right)^{-1} = d_{n-2}^0 \gamma_{n-2,1}.
\end{aligned} \tag{V.23}$$

where the first and the third equalities in (V.23) follow from the fact that simplicial operations in $A(Z)$ commute with the taking of the inverse in the group.

Finally, rewriting the equality
$$d_{n-1}^1 y_2 = d_{n-1}^1 y_1$$

componentwise we get
$$\gamma_{n-3,1} \cdot d_{n-2}^0 \gamma_{n-2,1} = \gamma_{n-3,2} \cdot d_{n-2}^0 \gamma_{n-2,2}, \tag{V.24}$$
$$\gamma_{j,1} = \gamma_{j,2}, \quad 0 \le j \le n-4. \tag{V.25}$$

Now (V.22)–(V.25) imply
$$d_n^2 y = y_2.$$
Similarly (using $d_{n-1}^1 y_i = d_{n-1}^{i-1} y_1$) we get $d_n^i y = y_i$. $\square$

Properties of the universal fibration $p : E \to B$ are described by the following:

36. Proposition. *The fibration $p : E \to B$ is locally trivial; a regular family of gluing functions can be chosen in such a way that*

$$\mu_p\left(\partial_n^0, (\gamma_{n-1}, \ldots, \gamma_0)\right) = \gamma_{n-1}. \tag{V.26}$$

Proof. Let us show how to choose an atlas yielding a regular family of gluing functions. Let $b = (\gamma_{n-1}, \ldots, \gamma_0) \in B_n$. Then the fibre E_b of p over b consists of all families $(z, \tilde{b}) \in E$ such that $z \in Z_m$, $\tilde{b} = B(f)b$ for some $f : [m] \to [n]$. Define $\alpha(b) : \Delta[n] \times Z \to E_b$ by $\alpha(b)(f, z) = (T(f, b)z, B(f)b)$ for $f : [m] \to [n]$, $z \in Z_m$, where $T(f, b) \in A_m(Z)$ is defined as follows:

a) If $f(0) = 0$ then $T(f, b) = e_m$.
b) Let $f(0) = r > 0$. Then f can be represented in the form
$$f = f_{\text{red}} \circ \partial_{m+r}^0 \circ \partial_{m+1}^0, \tag{V.27}$$

where $f_{\text{red}} : [m+r] \to [n]$, $f_{\text{red}}(0) = 0$.
Let $\tilde{b} = B(f_{\text{red}})b = (\tilde{\gamma}_{m+r-1}, \ldots, \tilde{\gamma}_0) \in B_{m+r}$. Then

$$T(f,b) \stackrel{\text{def}}{=\!=} \tilde{\gamma}_m \cdot d_{m+1}^0 \tilde{\gamma}_{m+1} \cdot d_{m+1}^0 d_{m+2}^0 \tilde{\gamma}_{m+2} \cdot \ldots \cdot d_{m+1}^0 \ldots d_{m+r-1}^0 \tilde{\gamma}_{m+r-1}.$$

V.2 Homotopic Characterization of Weak Equivalences   329

To complete the proof of the proposition we must verify that:

a) $\alpha(y)$ does not depend on the decomposition (V.27) (which, in general, is nonunique).
b) $\alpha(y)$ is an isomorphism.
c) The gluing functions corresponding to $\{\alpha(y)\}$ form a regular family and satisfy (V.26).

A straightforward (although rather cumbersome) verification is left to the reader. □

37. Proposition. *Let $g : X \to Y$ be a locally trivial fibration with a connected base Y and with a fibre Z. Let also $p : E \to B$ be the universal fibration with the fibre Z. Then there exists a morphism of simplicial sets $\varphi : Y \to B$ such that the induced fibration $p_\varphi : Y \times_B E \to Y$ is isomorphic to $g : X \to Y$.*

Proof. To construct φ let $\mu(f,y)$ be a regular family of gluing functions for the fibration g (Proposition V.2.32). Define

$$\varphi(y) = \left(\mu\left(\partial_n^0, y\right), \mu\left(\partial_{n-1}^0, d_n^0 y\right), \ldots, \mu\left(\partial_1^0, d_2^0 \ldots d_n^0 y\right)\right).$$

Let us verify that φ is a morphism of simplicial sets, i.e., that $\varphi(Y(f)y) = B(f)\varphi(y)$ for $y \in Y_n$, $f : [m] \to [n]$. It suffices to consider $f = \sigma_n^i$ and $f = \partial_n^i$.

a) $f = \sigma_n^i$. We need the following property of regular gluing functions:

$$\mu\left(\partial_{k+1}^0, s_k^j \tilde{y}\right) = \begin{cases} e_{k+1} & \text{for } j = 0, \\ s_{k-1}^{j-1} \mu(\partial_k^0, \tilde{y}) & \text{for } j > 0, \end{cases} \qquad \text{(V.28)}$$

for any $\tilde{y} \in Y_k$. Indeed,

$$\mu\left(\partial_{k+1}^0, s_k^j \tilde{y}\right) = \mu\left(\partial_{k+1}^0, s_k^j \tilde{y}\right) \left[d_k^0 \mu\left(\sigma_k^j, \tilde{y}\right)\right] = \mu\left(\sigma_k^i \partial_{k+1}^0, \tilde{y}\right).$$

Here the first equality follows from $\mu\left(\sigma_k^i, \tilde{y}\right) = e_k$ and the second one from the cocycle condition (V.18). Now for $j = 0$ we have $\sigma_k^0 \partial_{k+1}^0 = \text{id}_{\Delta[k]}$ and $\mu\left(\text{id}_{\Delta[k]}, \tilde{y}\right) = e_k$. For $j > 0$ we have $\sigma_k^j \partial_{k+1}^0 = \partial_k^0 \sigma_{k-1}^{j-1}$ so that again from (V.18) and from the regularity we have

$$\mu\left(\partial_{k+1}^0, s_k^j \tilde{y}\right) = \mu\left(\sigma_{k-1}^{j-1}, d_k^0 \tilde{y}\right) \cdot \left[s_{k-1}^{j-1} \mu\left(\partial_k^0, \tilde{y}\right)\right] = s_{k-1}^{j-1} \mu\left(\partial_k^0, \tilde{y}\right)$$

and (V.28) is proved.
    Now

$$\varphi(s_n^i y) = \left(\mu\left(\partial_{n+1}^0, s_n^i y\right), \ldots, \mu\left(\partial_1^0, d_2^0 \ldots d_{n+1}^0 s_n^i y\right)\right).$$

Using the equality

$$d_j^0 \ldots d_{n+1}^0 s_n^i y = \begin{cases} s_{j-2}^{i+j-n-2} d_{j-1}^0 \ldots d_n^0 y & \text{for } j \geq n-i+2, \\ d_j^0 \ldots d_n^0 y & \text{for } j < n-i+2, \end{cases}$$

together with (V.28) and the exact formulas for the action of s_n^i on B we get
$$\varphi(s_n^i y) = s_n^i \varphi(y).$$

b) $f = \partial_n^i$. The equality
$$\varphi(d_n^i y) = d_n^i \varphi(y)$$
is clear for $i = 0$. If $i > 0$ then it is proved using the regularity of μ and the cocycle condition similarly to part a). Details are left to the reader.

It remains to prove that the induced fibration $p_\varphi : Y \times_B E \to Y$ is isomorphic to $g : X \to Y$. But Lemma V.2.33 and Proposition V.2.36 imply that
$$\mu_\varphi(\partial_n^0, y) = \mu_p(\partial_n^0, \varphi(y)) = \mu(\partial_n^0, y)$$
is a regular family of gluing function of the fibration p_φ. By Proposition V.2.30b), p_φ is isomorphic to g over Y. □

38. Remark. The map φ inducing the given fibration g from the universal one p, is essentially unique. Namely, there exists the following classification theorem. Let Z be a simplicial set with an effective action of a simplicial group G. Let $p : E \to B$ be the universal fibration constructed from G as in V.2.34. For any morphism $\varphi : Y \to B$ let $p_\varphi : Y \times_B E \to Y$ be the induced fibration. Then $\varphi \to p_\varphi$ is an one-to-one correspondence between homotopy classes of morphisms $\varphi : Y \to B$ and isomorphism classes of locally trivial G-fibrations $g : X \to Y$ with the fibre Z. This theorem, which belongs, essentially, to Eilenberg and Maclane, is proved, for example, in May [1].

Exercises

1. Simplicial Group of Automorphisms of a Simplicial Set. a) Prove that the product $Y * Z$ of two simplicial sets (see II.4.23) satisfies the following property:
$$\mathrm{Hom}_{\Delta \circ Set}(X, Y * Z) = \mathrm{Hom}_{\Delta \circ Set}(X, Y) \times \mathrm{Hom}_{\Delta \circ Set}(X, Z)$$
for any $X \in \mathrm{Ob}\, \Delta° Set$.

b) Let X, Y, Z, U be four simplicial sets, $f = (f_n) \in \mathrm{Hom}_{\Delta \circ Set}(X * Y, Z)$, $g = (g_n) \in \mathrm{Hom}_{\Delta \circ Set}(X * Z, U)$. Prove that the family of maps $h_n : X_n \times Y_n \to Z_n$, $n = 1, 2, \ldots$, where $h_n(x, y) = g_n(x, f_n(x, y))$, $x \in X_n$, $y \in Y_n$, defines a morphism $h = f * g \in \mathrm{Hom}_{\Delta \circ Set}(X * Y, U)$ so that we get a map
$$\mathrm{Hom}_{\Delta \circ Set}(X * Y, Z) \times \mathrm{Hom}_{\Delta \circ Set}(X * Z, U) \longrightarrow \mathrm{Hom}_{\Delta \circ Set}(X * Y, U).$$

c) Using a) and b) together with the first part of Theorem II.4.24 construct a simplicial map
$$\mathbf{Hom}(Y, Z) * \mathbf{Hom}(Z, U) \longrightarrow \mathbf{Hom}(Y, U).$$
Prove that this morphism makes $\mathbf{Hom}(Y, Y)$ into a simplicial semigroup $E(Y)$. Prove that invertible elements of this semigroup (i.e., invertible elements of all semigroups $E_n(Y)$) form the simplicial group $A(Y)$ from V.2.34.

2. Complexes as a Closed Model Category. Let $\mathcal{A}$ be an abelian category with sufficiently many injective objects, $\mathcal{C} = \text{Kom}^-(\mathcal{A})$ the category of left-bounded complexes over $\mathcal{A}$. By definition, the class F of *fibrations* in $\mathcal{C}$ consists of all epimorphisms in $\mathcal{C}$, the class G of *cofibrations* consists of monomorphisms in $\mathcal{C}$ whose cokernels are complexes of projective objects in $\mathcal{A}$, and the class W of *weak equivalences* consists of quasi-isomorphisms in $\mathcal{C}$. Prove that the axioms CM0–CM3 are satisfied.

3. Hausdorff Topological Spaces as a Closed Model Category. The structure of a closed model category is determined by the following data.

A *fibration* is a continuous map $f: X \to Y$ satisfying the Serre condition: $f \in \rho\{\text{faces of a cube}\}$, where a *face of a cube* is a map $[0,1]^{r-1} \to [0,1]^r$ of the form $(x_1, \ldots, x_{r-1}) \mapsto (x_1, \ldots, x_{r-1}, 0)$.

A *weak equivalence* is a map $f: X \to Y$ inducing isomorphisms of homotopy groups $\pi_q(X, x) \to \pi_q(Y, f(x))$ for all $q \geq 0$, $x \in X$ (i.e., a weak homotopy equivalence in the topological sense).

Finally, *cofibrations* are defined by the formula $G = \lambda(F \cap W)$.

The geometric realization functor maps a weak equivalence f of simplicial sets into a weak equivalence $|f|$ of topological spaces. Conversely, if $|f|$ is a weak equivalence, f also is.

4. Homological Classification of Weak Equivalences of Simplicial Sets. Prove that a morphism $f: X \to Y$ of simplicial sets is a weak equivalence if and only if it induces isomorphisms

$$\pi_0(X) \xrightarrow{\sim} \pi_0(Y), \quad \pi_1(X, x) \xrightarrow{\sim} \pi_1(Y, f(x)),$$

$$H^q(Y, \mathcal{A}) \xrightarrow{\sim} H^q(X, f^*(\mathcal{A})), \quad q \geq 0,$$

where $\mathcal{A}$ is any homological coefficient system of abelian groups.

A sketch of the proof. a) Let f be a weak equivalence. Then $|f|$ is also a weak equivalence. It suffices to consider the case when both $|X|$ and $|Y|$ are connected. If $|X|$ and $|Y|$ are also simply connected, then any homological coefficient system is trivial and f induces an isomorphism on cohomology. Finally, the general case reduces to the above using the Leray spectral sequences of the universal coverings $\tilde{X} \to X$ and $\tilde{Y} \to Y$.

b) Let f satisfies the above conditions. To prove that f and $|f|$ are weak equivalences it suffices, by the Whitehead theorem, to prove that f induces isomorphisms $\pi_q(|X|, x) \to \pi_q(|Y|, f(x))$ for all $q \geq 2$. If $|X|$ and $|Y|$ are simply connected, this directly follows from the isomorphisms $H^q(Y, \mathcal{A}) \xrightarrow{\sim} H^q(X, f^*(\mathcal{A}))$, and in the general case one must use the Leray spectral sequences as above.

5. Simplicial Groups as a Closed Model Category. A morphism of simplicial groups is said to be a *fibration* if it is a fibration as a morphism of simplicial sets.

To define a weak equivalence we need the notion of Moore homotopy groups. Let $G = \{G_n\}$ be a simplicial group. Define

332   V. Introduction to Homotopic Algebra

$$N_0(G) = G,$$
$$N_q(G) = \cap_{i>0} \ker G(\partial_q^i) \quad \text{for} \quad q \geq 1;$$
$$\partial_q : N_q(G) \longrightarrow N_{q-1}(G), \quad \text{the map induced by} \quad G(\partial_q^0).$$

One can verify that $\operatorname{Im} \partial_{q+1}$ is a normal subgroup of $\ker \partial_q$. Let

$$\pi_q(G) = \ker \partial_q / \operatorname{Im} \partial_{q+1}.$$

Then $\pi_0(G) = \pi_0(|G|)$ and $\pi_q(G)$ is an abelian group for $q \geq 1$.

A morphism of simplicial groups is said to be a *weak equivalence* if it induces an isomorphism of groups π_q for all q.

Finally, *cofibrations* are again defined by the formula $G = \lambda(F \cap W)$.

6. Fibrant and Cofibrant Objects. An object X of a closed model category $\mathcal{C}$ is said to be a *fibrant* if $X \to \omega$ (where ω is a final object of $\mathcal{C}$) is a fibration. Similarly, an object X is said to be *cofibrant* if $\alpha \to X$ (where α is an initial object of $\mathcal{C}$) is a cofibration.

In $\Delta^\circ \mathit{Set}$ any object is cofibrant and Kan sets are fibrant. In $\mathrm{Kom}^-(\mathcal{A})$ any object is fibrant and complexes formed by projective objects are cofibrant. In *Top* any object is fibrant and CW-complexes are cofibrant.

7. Homotopy Categories. a) Let $\mathcal{C}$ be a closed model category. For $X, Y \in \mathrm{Ob}\,\mathcal{C}$ denote by $X \coprod Y$ their coproduct in $\mathcal{C}$. For two morphisms $f : X \to Z$, $g : Y \to Z$ denote by $f + g : X \coprod Y \to Z$ the morphism constructed by the universality property of $X \coprod Y$. Similarly one can define $(h, k) : Z \to X \times Y$ for a pair of morphisms $h : Z \to X$, $k : Z \to Y$.

b) *Cylinder for an object.* Let $X \in \mathrm{Ob}\,\mathcal{C}$. By a cylinder for the object X we mean a family $(\mathrm{Cyl}\,X, \delta_0, \delta_1, \sigma)$ consisting of an object $\mathrm{Cyl}\,X \in \mathrm{Ob}\,\mathcal{C}$ and three morphisms $\delta_0, \delta_1 : X \to \mathrm{Cyl}\,X$, $\sigma : \mathrm{Cyl}\,X \to X$ such that $\sigma \delta_0 = \sigma \delta_1 = \mathrm{id}_X$, σ is a weak equivalence, $\delta_0 + \delta_1 : X \coprod X \to \mathrm{Cyl}\,X$ is a cofibration (whence δ_0, δ_1 are also cofibrations).

c) *Path space for an object* X is defined dually as a collection (X^I, d_0, d_1, τ) where $X^I \in \mathrm{Ob}\,\mathcal{C}$, $d_0, d_1 : X^I \to X$, $\tau : X \to X^I$ with the properties $d_0 \tau = d_1 \tau = \mathrm{id}_X$, τ is a weak equivalence, $(d_0, d_1) : X^I \to X \times X$ is a fibration (whence d_0, d_1 are also fibrations).

Both a cylinder and a path space need not exist, and if they do exist, they may be nonunique. In particular, $X \mapsto \mathrm{Cyl}\,X$ is not necessarily a functor from $\mathcal{C}$ into itself.

In $\Delta^\circ \mathit{Set}$ functorial cylinders and path spaces always exist: $\mathrm{Cyl}\,X = X * \Delta[1]$, X^I is constructed from $\mathrm{Hom}(\Delta[1], X)$ as in V.2.11. In *Top* they also exist but are not necessarily given by the standard construction. For example, the product of a space X and the unit interval is a cylinder in the sense of b) if and only if X is a cofibrant space (i.e., a CW-complex). In $\mathrm{Kom}^-(\mathcal{A})$ a cylinder of a complex X is the cylinder of the identity morphism id_X in the sense of III.3.2. Verify all these assertions.

d) *Homotopies.* Let $f, g : X \to Y$ be two morphisms in $\mathcal{C}$. A *left homotopy* between f and g is a morphism $h : \text{Cyl}\, X \to Y$ (where $(\text{Cyl}\, X, \delta_0, \delta_1, \sigma)$ is a cylinder for X) such that $h\delta_0 = f$, $h\delta_1 = g$. A *right homotopy* between f and g is a morphism $k : X \to Y^I$ (where (Y^I, d_0, d_1, τ) is a path space of Y) such that $d_0 k = f$, $d_1 k = g$. The corresponding morphisms f and g are said to be *homotopic from the left* (resp. *from the right*).

e) *Left and right homotopies.* Let $f, g : X \to Y$ be two morphisms in $\mathcal{C}$ with X cofibrant and Y fibrant. Using axioms of a closed model category Quillen [1] proved that if f, g are homotopic from the right (resp. from the left), then a right homotopy exists for any cylinder for X (resp. a left homotopy exists for any path space for Y) and that the notions of right and of left homotopy between f and g are equivalent.

f) *The category* ho $\mathcal{C}$. Objects of ho $\mathcal{C}$ are objects of $\mathcal{C}$ that are both fibrant and cofibrant. Morphisms in ho $\mathcal{C}$ are classes of homotopic (does not matter, from the left or from the right) morphisms in $\mathcal{C}$. Verify that the composition in ho $\mathcal{C}$ is well defined.

g) *The category* Ho $\mathcal{C}$ *and the equivalence.* The category Ho $\mathcal{C}$ is defined as the localization of $\mathcal{C}$ with respect to the class W of weak equivalences, Ho $\mathcal{C} = \mathcal{C}[W^{-1}]$. One of the main theorems in Quillen [1] says that the localization functor $Q : \mathcal{C} \to \text{Ho}\,\mathcal{C}$ induces an equivalence of categories ho $\mathcal{C} \to$ Ho $\mathcal{C}$. Specialize this theorem to $\mathcal{C} = \text{Kom}^-(\mathcal{A})$ and to $\mathcal{C} = \text{Top}$.

V.3 DG-Algebras as a Closed Model Category

1. DG-Algebras

By a *DG-algebra* (short for *D*ifferential *G*raded algebra) over a field k of characteristic 0 we mean the following data:

a) A complex of linear spaces over k: $A = (A^i, d)$, $A^i = 0$ for $i < 0$, $d : A^i \to A^{i+1}$.

b) A multiplication $A^i \underset{k}{\otimes} A^j \to A^{i+j}$, $a \otimes b \mapsto ab$, which makes $\oplus_i A^i$ into an associative algebra with unit $1 \in A^0$; this algebra is denoted by the same symbol A. The multiplication must be graded- (or super-) commutative,

$$ab = (-1)^{ij} ba, \qquad a \in A^i, \quad b \in A^j,$$

and d must be a derivation of degree 1,

$$d(ab) = da \cdot b + (-1)^i a \cdot db, \qquad a \in A^i.$$

A morphism of *DG*-algebras is a morphism of complexes which is also a morphism of algebras with unit.

Let $\mathcal{DGA}$ denote the category of *DG*-algebras (the field k is assumed to be fixed).

2. Examples

a) *The de Rham Algebra.* Let X be a C^∞-manifold, $\Omega^\bullet(X)$ the algebra of smooth differential forms on X (over $\mathbb{R}$ or $\mathbb{C}$). It is a DG-algebra and $\Omega^\bullet$ determines, evidently, a functor from the category of C^∞-manifolds to the category $\mathcal{DGA}$.

b) *The Polynomial de Rham Algebra.* For $X = \mathbb{R}^n$ let us consider the subalgebra of $\Omega^\bullet(\mathbb{R}^n)$ consisting of forms with polynomial coefficients. The restriction of d on this subalgebra defines on it the structure of a DG-algebra. We use this algebra to define the de Rham functor on the category of simplicial sets. To do this we have to modify slightly the definitions.

Define

$$\nabla(n) = \oplus \nabla(n)^q, \quad \text{the } DG\text{-algebra of polynomial differential forms in variables } x_0(n), \ldots, x_n(n) \text{ subject to the relation } x_0(n) + \ldots + x_n(n) = 1.$$

For any nondecreasing map $f : [m] \to [n]$ define

$$\nabla(f)(x_i(n)) = \sum_{j: f(j) = i} x_j(m),$$

and extend $\nabla(f)$ to maps $\nabla(f)^q : \nabla(n)^q \to \nabla(n)^q$,

$$d \circ \nabla(f)^{q-1} = \nabla(f)^q \circ d.$$

We get, for any $q \geq 0$, a simplicial set (even a simplicial linear space) $\nabla^q = (\nabla(n)^q)$. The collection of all these simplicial sets, $\nabla = (\nabla(n), n \geq 0)$, has the structure of a simplicial DG-algebra (i.e., of a simplicial object in the category $\mathcal{DGA}$. For any simplicial set X define

$$\Omega^q(X) = \text{Hom}_{\Delta^\circ \text{Set}}(X, \nabla^q).$$

In other words, one element of $\Omega^q(X)$ is a family of polynomial q-forms, one on each simplex of X, which are compatible with face and degeneration operators. One can easily see that $\Omega^\bullet(X) = \underset{q \geq 0}{\oplus} \Omega^q(X)$ has a natural structure of a DG-algebra over k. In particular, $\Omega^\bullet(\Delta(n)) = \nabla(n)$.

The map $X \mapsto \Omega^\bullet(X)$ determines, evidently, a contravariant functor $\Delta^\circ \text{Set} \to \mathcal{DGA}$.

c) Let A be an arbitrary DG-algebra. Its cohomology $H^\bullet(A)$ inherits the multiplication from the multiplication in A: (class of a) × (class of b) = (class of ab). Define $d = 0$ on $H^\bullet(A)$. Then $H^\bullet(A)$ becomes a DG-algebra and $H^\bullet$ yields a covariant functor $\mathcal{DGA} \to \mathcal{DGA}$.

3. Definition. *a) A morphism $f : A \to B$ of DG-algebras is said to be a weak equivalence if $H^\bullet(f) : H^\bullet(A) \to H^\bullet(B)$ is an isomorphism (i.e., f is a quasi-isomorphism).*

b) A morphism of DG-algebras is said to be a fibration if it is a surjection.

c) A morphism of DG-algebras is said to be a *cofibration* if it belongs to $\lambda(F \cap W)$, where F is the class of fibrations and W is the class of weak equivalences (see V.1.1).

4. Theorem. *The category $\mathcal{DGA}$ with the above classes of weak equivalences, fibrations, and cofibrations is a closed model category.*

This theorem will be proved below in V.3.5–V.3.11. It shows that the Sullivan–de Rham functor $\Omega^\bullet : \Delta^\circ Set \to \mathcal{DGA}$ is a contravariant functor between two closed model categories. However, the functor does not map one list of structures into the dual of another one. In particular, the image of a fibration in $\Delta^\circ Set$ might fail to be a cofibration in $\mathcal{DAG}$.

Nevertheless, the functor $\Omega^\bullet$ induces an equivalence of appropriate localized categories.

5. Verification of CM0 (Sect. V.1.4)

The algebra k is both the initial and the final object in $\mathcal{DGA}$. Any diagram $B \xleftarrow{f} A \xrightarrow{g} C$ can be completed to a cocartesian square as follows:

$$\begin{array}{ccc} A & \xrightarrow{f} & B \\ {\scriptstyle g}\downarrow & & \downarrow{\scriptstyle g'} \\ C & \xrightarrow{f'} & B \underset{A}{\otimes} C \end{array} \qquad f'(c) = 1 \otimes c, \quad g'(b) = b \otimes 1,$$

where $B \underset{A}{\otimes} C$ is the usual relative tensor product of graded k-algebras, and $d(b \otimes c) = db \otimes c + (-1)^i b \otimes dc$ with $i = \deg b$. Finally, any diagram $B \xrightarrow{f} A \xleftarrow{g} C$ can be completed to a cartesian square as follows:

$$\begin{array}{ccc} B \underset{A}{\times} C & \longrightarrow & B \\ \downarrow & & \downarrow{\scriptstyle f} \\ C & \xrightarrow{g} & A \end{array} \qquad B \underset{A}{\times} C = \{(b,c), f(b) = g(c)\} \subset B \times C.$$

6. Polynomial DG-Algebras

To verify the axioms CM1–CM3 (Sect. V.1.4) we introduce some useful classes of DG-algebras. For a graded linear space $V = \underset{n \geq 0}{\oplus} V_n$ over k let $V[-1] = \underset{n \geq 0}{\oplus} V[-1]_n$, where $V[-1]_n = V_{n-1}$. Define DG-algebras $S(V)$ and $T(V)$ as follows:

$S(V)$ is the free supercommutative DG-algebra generated by V and equipped with the zero differential d;

$T(V) = S(V \oplus V[-1])$, $d|_{V_n} = \text{id}: V_n \longrightarrow V[-1]_{n+1} = V_n$, $d|_{V[-1]} = 0$.

One can easily see that for any DG-algebra A and for any graded vector space V morphisms $\alpha : T(V) \to A$ are in one-to-one correspondence with degree-preserving linear maps $\alpha' : V \to A$; $\alpha'(v) = \alpha(v)$ for $v \in V \subset T(V)$. Similarly, morphisms $\beta : S(V) \to A$ are in one-to-one correspondence with degree-preserving linear maps $\beta' : V \to A$ satisfying the condition $\beta'(V) \subset Z_A$, where Z_A is the space of cycles in A, $\beta'(v) = \beta(v)$ for $v \in V \subset S(V)$. In particular, as elements of $V[-1]$ are cycles in $T(V)$, there exists a canonical morphism $\gamma : S(V[-1]) \to T(V)$.

7. Lemma. *a) For any DG-algebra A and for any V, V' the morphism $i : A \to A \underset{k}{\otimes} S(V) \underset{k}{\otimes} T(V')$, $a \mapsto a \otimes 1 \otimes 1$, is a cofibration.*

b) Similarly, for any A and V as above, the morphism $j : A \to A \underset{k}{\otimes} T(V)$, $a \mapsto a \otimes 1$, is a weak equivalence.

c) For any V the morphism $\gamma : S(V[-1]) \to T(V)$ is a cofibration.

Proof. a) We must prove that any diagram of the form

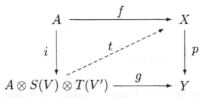

with $p \in F \cap W$, admits a lifting t. By the remark at the end of V.3.6, to construct such a morphism t we must define a morphism of DG-algebras $t_0 : A \otimes 1 \otimes 1 \to X$ and two degree-preserving linear maps $t_1 : V \to X$, $t_2 : V' \to X$ such that $t_1(V) \subset Z_X$ (cycles in X) and the following compatibility conditions hold: $t_0 i = f$, $pt_1 = g_1$, $pt_2 = g_2$; here $g_1 : V \to Y$, $g_2 : V' \to Y$ are the restrictions of g to $1 \otimes V \otimes 1$ and to $1 \otimes 1 \otimes V'$ respectively. First of all, let $t_0 = f$. Next, since p belongs to F, it is an epimorphism and t_2 can be any (degree-preserving) lifting of $g_2 : V' \to Y$. Finally, one can easily check that the conditions $p \in F$ (i.e., p is an epimorphism) and $p \in W$ (i.e., p is a quasi-isomorphism) imply that the restriction $p|_{Z_X}: Z_X \to Z_Y$ is also an epimorphism. Hence t_1 can be any degree-preserving lifting of $g_1 : V \to Z_Y$.

b) Define a morphism $l : A \otimes T(V) \to A$ of DG-algebras by the formula $l(a \otimes 1) = a$, $l(a \otimes \omega) = 0$ for $\omega \in (V \oplus V[-1])T(V) \overset{\text{def}}{=} T^+(V)$. Evidently, $lj = \text{id}_A$. We prove that jl is homotopic to $\text{id}_{A \otimes T(V)}$ (as morphisms of complexes), i.e., that there exists $h : A \otimes T(V) \to A \otimes T(V)$ such that $(hd+dh)(a \otimes 1) = 0$, $(hd + dh)(a \otimes \omega) = a \otimes \omega$ for $\omega \in T^+(V)$. To construct h, let us fix a homogeneous basis $v_1, \ldots, v_m$ in V and denote by $V_i \subset V$ the one-dimensional space generated by v_i, so that $T(V) = T(V_1) \otimes \ldots \otimes T(V_m)$. First, let us construct the map $h_i : T(V_i) \to T(V_i)$ such that

$$(h_i d + d h_i)(1) = 0, \quad (h_i d + d h_i)(\omega) = \omega \quad \text{for} \quad \omega \in T^+(V_i). \quad (V.29)$$

If $\deg v_i$ is even then any element $\omega \in T(V_i)$ can be uniquely written in the form $\omega = p(v_i) + q(v_i)dv_i$ for some polynomials p and q, and we set $h_i(\omega) = \tilde{q}(v_i)$ where $\tilde{q}(s) = \int_0^s q(\sigma)d\sigma$. If $\deg v_i$ is odd then ω can be uniquely written in the form $\omega = p(dv_i) + q(dv_i)v_i$ where p and q are again polynomials, and we define $h_i(\omega)$ by the following formulas: $h_i(1) = 0$, $h_i(q(dv_i)v_i) = 0$, $h_i(p(dv_i)) = \{p(dv_i)/dv_i\} v_i$ if $p(0) = 0$. One can easily see that in both cases the condition (V.29) is satisfied.

Next we define $h : A \otimes T(V) \to A \otimes T(V)$. It is completely determined by the elements $h(a \otimes \omega_1 \otimes \ldots \otimes \omega_m)$, where a is homogeneous and each $\omega_i \in T(V_i)$ is homogeneous and either equals 1 or belongs to $T^+(V_i)$. Let l be the number of those i that $\omega_i \in T^+(V_i)$. Define

$$h(a \otimes \omega_1 \otimes \ldots \otimes \omega_m) = \frac{1}{2^l} \sum_{i=1}^m (-1)^{P_i} a \otimes \omega_1 \otimes \ldots \otimes h_i(\omega_i) \otimes \ldots \otimes \omega_m$$

where $P_i = \deg a + \deg \omega_1 + \ldots + \deg \omega_m$. Easy computations show that h satisfies the required conditions, so that j is a homotopy equivalence, hence a quasi-isomorphism.

c) Similarly to a) we must construct a lifting t in the diagram

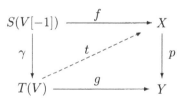

with $p \in F \cap W$. By the remark at the end of V.3.6, f corresponds to a linear map $f_1 : V[-1] \to X$ and g corresponds to a linear map $g_1 : V \to Y$ (both degree-preserving) such that $f_1(V[-1]) \subset Z_X$ and $dg_1(v) = p(f_1(v))$ for any $v \in V[-1]_n = V_{n-1}$. To construct t it suffices to define $t_1 : V \to X$ (also degree preserving) such that $pt_1 = g_1$, $dt_1 = f_1$.

Let us remark first that p is a quasi-isomorphism and $p(f_1(v)) \subset B_Y$ (boundaries in Y) so that $f_1(v) \in B_X$ for any $v \in V[-1]$. Hence our linear map t_1 should make the following diagram of linear spaces commutative:

338   V. Introduction to Homotopic Algebra

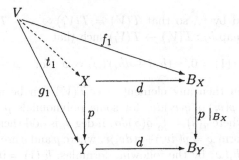

The existence of such t_1 will be proved if we show that the square in this diagram is cocartesian. One can easily see that this property follows from the surjectivity of $p\,|_{B_X}\colon B_X \to B_Y$ (which, in turn, follows from the surjectivity of p) and from the formula $\ker(p \circ d) = \ker p + Z_X$. To prove this formula let us note first that the left-hand side contains the right-hand side. Conversely, let $pdx = dpx = 0$ for $x \in X$, i.e., $px \in Z_Y$. As we have already remarked in the proof of a), $p\,|_{Z_X}\colon Z_X \to Z_Y$ is an epimorphism. So there exists $z \in Z_X$ with $pz = px$ and $x = (x - z) + z$ is the required decomposition. $\square$

8. Verification of CM2 (First Part)

a) Let $f : A \to B$ be a morphism in $\mathcal{DGA}$. We can represent it as the composition
$$A \xrightarrow{i} A \otimes T(B) \xrightarrow{p} B,$$
where $i(a) = a \otimes 1$, $p(a \otimes b) = f(a)\beta(b)$ for $a \in A$, $b \in T(B)$ and $\beta : T(B) \to B$ is the morphism corresponding to the identity morphism $B \to B$ (see V.3.6).

Now Lemma V.3.7 implies that $i \in G \cap W$ and p, being surjective, belongs to F.

9. Verification of CM2 (Second Part)

Again, we are given $f : A \to B$ which we want to represent as a composition
$$A \xrightarrow{j} C \xrightarrow{q} B$$
with $j \in G$, $q \in F \cap W$. As the 0-th approximation we take
$$A \xrightarrow{j_0} C_0 = A \otimes S(Z_B) \otimes T(B) \xrightarrow{q_0} B,$$
where Z_B are cycles in B, $j_0(a) = a \otimes 1 \otimes 1$, $q_0(a \otimes z \otimes b) = f(a)\alpha(z)\beta(b)$, and $\alpha : S(Z_B) \to Z_B$ corresponds to $\mathrm{id} : Z_B \to Z_B$. We have $j_0 \in G$, $q_0 \in F$, but instead of $q_0 \in W$ we can only guarantee that q_0 is surjective but not necessarily injective on cohomology.

To fix up this defect we add to C_0 new elements whose boundaries are cycles that become boundaries after being mapped to B. Since this construction generates new elements of the same type, this procedure has to be iterated an infinite number of times.

Formally, assume that we have already constructed the diagram

$$A \xrightarrow{j_n} C_n \xrightarrow{q_n} B, \quad n \geq 0.$$

Define the graded linear space $V_n \subset C_n \oplus B$ as follows:

$$V_n = \{(c, b) \mid dc = 0, q_n(c) = db, \deg b = \deg c - 1\};$$

$$\deg(c, b) = \deg c.$$

Let C_{n+1} be the vertex of the following cocartesian square:

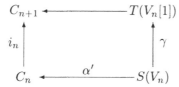

Here γ is the morphism defined in V.3.6, α' is the composition $S(V_n) \to S(V_n') \to C_n$, where $V_n' \subset C_n$ is the image of V_n under the projection $V_n \to C_n$. Let also $j_{n+1} = i_n \circ j_n : A \to C_{n+1}$.

Finally, define $q_{n+1} : C_{n+1} \to B$ by the universality of the above cocartesian square using the morphism $q_n : C_n \to B$ and the morphism $\beta' : T(V_n[-1]) \to B$, which is constructed similarly to α'.

The morphism γ is a cofibration by Lemma V.3.7c), so that $i_n \in G = \lambda(F \cap W)$ by V.1.3. The morphism $j_{n+1} \in G$ being the composition of cofibrations, is itself a cofibration. One can easily see that $j = \varinjlim j_n : A \to C = \varinjlim C_n$ is also a cofibration.

It remains to verify that $q = \varprojlim q_n : C \to B$ belongs to W (the surjectivity of q is clear because all q_n are surjective). But $H^\bullet(q)$ is clearly surjective because all $H^\bullet(q_n)$ are surjective. On the other hand, if a nonzero cohomology class in C_n is in the kernel of $H^\bullet(q_n)$, then the cycle representing this class becomes a boundary in C_{n+1} and, therefore, in C.

10. Verification of CM3

Clearly follows from the definition of a quasi-isomorphism.

11. Verification of CM1

CM1b) is the definition of G. We verify c) and then deduce a).

Let us show first that $F \subset \rho(G \cap W)$. Consider a morphism $j: A \to B$ from $G \cap W$ and an arbitrary morphism $f \in F$. Represent j as the composition $A \xrightarrow{i} A \otimes T(B) \xrightarrow{p} B$, where $i(a) = a \otimes 1$, $p(a \otimes b) = f(a)\beta(b)$ for $a \in A$, $b \in T(B)$. We have to prove the existence of the morphism ③ in the diagram

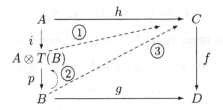

We construct ① and ② and then define ③=①·②.

Since $f \in F$, f is a surjection, so that there exists a linear degree-preserving map $l: B \to C$ such that $g = lf$. Since $T(B) = S(B \oplus B[-1])$, the morphism ① is completely determined if we require that it coincides with h on A and with l on B. The morphism ② is defined by the diagram

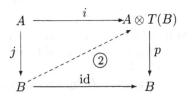

in which $j \in G = \lambda(F \cap W)$ and $p \in F \cap W$ because p is a surjection and $i, j \in W$.

Let us prove now the converse inclusion $\rho(G \cap W) \subset F$. Let $f \in \rho(G \cap W)$, $f = pi$, $i \in G \cap W$, $p \in F$. Then there exists a dashed arrow in the diagram

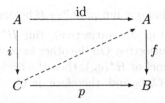

Since p is surjective, f is also surjective, i.e., $f \in F$. So, the axiom CM1c) is verified.

Finally, let us prove that $W = \rho(G)\lambda(F)$. Let $f \in W$, $f = pi$, $i \in G \cap W$, $p \in F$. By CM3, $p \in F \cap W$. Next, by CM1b), $F \cap W \subset \rho\lambda(F \cap W) = \rho(G)$ and similarly, by CM1c), $G \cap W \subset \lambda\rho(G \cap W) = \lambda(F)$. Hence $p \in \rho(G)$, $i \in \lambda(F)$, and $W \subset \rho(G)\lambda(F)$.

To verify the converse inclusion we prove that $\rho(G) \subset W$ and $\lambda(F) \subset W$. Let $g \in \rho(G)$, $g = pi$, $i \in G$, $p \in F \cap W$. Let us construct a lifting h in the following diagram:

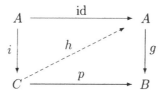

Since $hi = \text{id}$, $gh = p$, and p is a quasi-isomorphism, g is also a quasi-isomorphism. Hence $\rho(G) \subset W$ (and even $\rho(G) \subset F \cap W$).

Similarly, let $g \in \lambda(F)$, $g = pi$, $i \in G \cap W$, $p \in F$. From the diagram

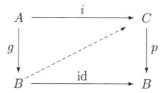

we get, using similar arguments, that $\lambda(F) \subset G \cap W$.

Exercises

Exercises below describe the relation between closed model categories $\mathcal{DGA}$ and $\Delta^\circ \text{Set}$, which leads to the equivalence of corresponding homotopy categories (see Sect. V.5).

1. Simplicial Morphisms of DG-Algebras. For $A, B \in \text{Ob } \mathcal{DGA}$ let

$$S(A, B)_n = \text{Hom}_{\mathcal{DGA}}(A, \nabla(n) \otimes B),$$

where $\nabla(n)$ is the algebra of polynomial differential forms on the standard n-simplex (see V.3.2). With any nondecreasing map $f : [m] \to [n]$ we associate the morphism $S(A, B)_n \to S(A, B)_m$ induced by the homomorphism $\nabla(f) \otimes \text{id}_B : \nabla(n) \otimes B \to \nabla(m) \otimes B$. Prove that these morphisms determine on $\{S(A, B)_n\}$ the structure of a simplicial set, and the composition

$$S(B, C)_n \times S(A, B)_n \longrightarrow S(A, C)_n,$$

$(\varphi, \psi) \mapsto \varphi \circ \psi =$
$$\left\{ A \xrightarrow{\psi} \nabla(n) \otimes B \xrightarrow{\text{id}_{\nabla(n)} \otimes \varphi} \nabla(n) \otimes \nabla(n) \otimes C \xrightarrow{\mu \otimes \text{id}_C} \nabla(n) \otimes C \right\},$$

where $\mu : \nabla(n) \otimes \nabla(n) \to \nabla(n)$ is the multiplication morphism, determines a morphism of simplicial sets $S(B, C) * S(A, B) \to S(A, C)$.

Verify that $S(A,B)$ is a functor
$$\mathcal{DGA}^\circ \times \mathcal{DGA} \longrightarrow \Delta^\circ Set.$$

2. Adjoint Functors. For $A, B \in \text{Ob}\,\mathcal{DGA}$, $X \in \text{Ob}\,\Delta^\circ Set$ define the map
$$\Phi: \text{Hom}_{\mathcal{DGA}}(A, \Omega^\bullet(X) \otimes B) \longrightarrow \text{Hom}_{\Delta^\circ Set}(X, S(A,B))$$
functorially in A, B, X (for $\varphi: A \to \Omega^\bullet(X) \otimes B$, $x \in X_n$, take as $\Phi(\varphi)(x) \in S(A,B)_n$ the composition
$$A \xrightarrow{\varphi} \Omega^\bullet(X) \otimes B \xrightarrow{\Omega^\bullet(\tilde{x}) \otimes \text{id}_B} \Omega^\bullet(\Delta[n]) \otimes B = \nabla[n] \otimes B$$
where $\tilde{x}: \Delta[n] \to X$ is the morphism of simplicial sets that maps the only nondegenerate n-simplex in $\Delta[n]$ into x.)

Verify that Φ is an isomorphism in each of the following two cases:

a) X is finite (i.e., the number of nondegenerate simplexes in X is finite);
b) $\dim B^n < \infty$ for any n.

3. Fibrations in $\mathcal{DGA}$ and in $\Delta^\circ Set$. Let $i: A \to B$, $p: C \to D$ be two morphisms of DG-algebras. Construct the canonical map of simplicial sets
$$S(i,p): S(B,C) \longrightarrow S(A,C) \underset{S(A,D)}{\times} S(B,D),$$
where the fibre product in $\Delta^\circ Set$ is taken with respect to the maps $\varphi \mapsto p \circ \varphi$, $\phi \mapsto \phi \circ i$.

The role (and the proof) of the following result is similar to the one of Proposition V.2.12. Let $i \in F$, $p \in G$ in $\mathcal{DGA}$. Then

a) $S(i,p) \in F$ in $\Delta^\circ Set$.
b) If either $i \in W$ or $p \in W$, then $S(i,p) \in W$.

V.4 Minimal Algebras

1. Notation

Let $(A = \underset{i \geq 0}{\oplus} A^i, d)$ be a DG-algebra over a field k of characteristic 0. Define $A^+ = \underset{i \geq 1}{\oplus} A^i$ and for any $n \geq 0$ let $A(n)$ be the subalgebra generated by $\underset{i=0}{\overset{n}{\oplus}} A^i$ and dA^n. Evidently, $dA(n) \subset A(n)$ and $A = \cup_{n=0}^{\infty} A(n)$. Let also $A(-1) = k$.

We have $A(-1) \subset A(0) \subset A(1) \subset \ldots$. Introduce intermediate subalgebras $A(n,q)$,
$$A(n-1) = A(n,0) \subset A(n,1) \subset \ldots \subset A(n),$$
inductively as follows: $A(n,q)$ is generated by $A(n,q-1)$ and by $x \in A^n$ such that $dx \in A(n,q-1)$.

V.4 Minimal Algebras

2. Definition. *An algebra A is said to be* minimal *if $A^0 = k$, A is free (as a commutative graded $\mathbb{Z}$-algebra) and the differential d in A is such that $A(n) = \cup_{q \geq 0} A(n, q)$ for all $n \geq 0$.*

This definition is useful for the following three reasons:

a) Minimal algebras admit a rather explicit inductive description (see V.4.6–V.4.10) and are well adapted to computation of homotopy groups (see V.5.12, V.5.13, V.5.14).
b) Any simply connected DG-algebra is weakly equivalent to a minimal algebra (V.4.11) and this correspondence has rather nice functorial properties (V.5.5).
c) Cofibrant DG-algebras can be completely described in terms of minimal ones (V.5.4).

The first simple result about minimal algebras is given by the following lemma.

3. Lemma. *If A is a minimal algebra, then its differential d is decomposable; by definition, this means that $dA^+ \subset A^+ \cdot A^+$.*

Proof. Denote by $a(n, q)$ the union of the following two statements:

$$a(n,q) = \begin{cases} A(n,q)^i \subset [A(n,q)^+]^2 & \text{for all } i > n, \\ dA(n,q) \subset [A(n,q)^+]^2. \end{cases}$$

a) $a(0, 0)$ is true because $A(0, 0) = k$.

b) $a(n, q-1)$ implies $a(n, q)$. Indeed, any element of $A(n, q)$ is a sum of elements of the following three types: elements from $A(n, q-1)$, elements $x \in A^n$ such that $dx \in A(n, q-1)$, and products of elements from A^n that belong to $[A(n,q)^+]^2$. For elements of the first type both inclusions in $a(n, q)$ follow from $a(n, q-1)$. For elements of the third type they are evident. For elements of the second type only the second inclusion should be verified, and it follows from the first inclusion in $a(n, q-1)$.

c) If $a(n, q)$ is true for all q, then $a(n+1, q)$ is true. This follows from the minimality of A.

By the minimality, any element of A belongs to one of the subalgebras $A(n, q)$ and the second inclusion in $a(n, q)$ shows that its boundary is decomposable. □

4. Example (Cochains of a Lie Algebra). Let $\mathfrak{G}$ be a finite-dimensional Lie algebra over k. Its cochain complex $C^\bullet(\mathfrak{G})$ with coefficients in the trivial $\mathfrak{G}$-module k (see I.7.8) is a DG-algebra. Let us recall that $C^n(\mathfrak{G})$ consists of skew-symmetric n-linear functionals on $\mathfrak{G}$ and

$$dc(x_1, \ldots, x_{n+1}) = \sum_{1 \leq j < l \leq n+1} (-1)^{j+l-1} c([x_j, x_l], x_1, \ldots, \hat{x}_j, \ldots, \hat{x}_l, \ldots, x_{n+1}).$$

344   V. Introduction to Homotopic Algebra

Introduce in $C^\bullet(\mathfrak{G}) = \Lambda(\mathfrak{G}^*)$ the natural exterior multiplication. One can easily verify that d satisfies the Leibniz formula.

The algebra $C^\bullet(\mathfrak{G})$ is obviously supercommutative and free. Moreover, the differential is decomposable because $C^\bullet(\mathfrak{G})$ is generated by $C^1(\mathfrak{G})$. However, $C^\bullet(\mathfrak{G})$ is not necessarily minimal.

5. Proposition. *$C^\bullet(\mathfrak{G})$ is minimal if and only if $\mathfrak{G}$ is a nilpotent Lie algebra.*

Proof. Let $\mathfrak{G}_0 = \mathfrak{G}$, $\mathfrak{G}_i = [\mathfrak{G}_{i-1}, \mathfrak{G}]$. By definition, Lie algebra $\mathfrak{G}$ is nilpotent iff $\mathfrak{G}_i = \{0\}$ for sufficiently large i. Let $\mathfrak{G}_i^\perp = \{c \in \mathfrak{G}^* \mid \mathfrak{G}_i \in \ker c\}$. Below we will show that for $A = C^\bullet(\mathfrak{G})$ we have $A(1,q)^1 = \mathfrak{G}_q^\perp$. This equality implies that $C^\bullet(\mathfrak{G})$ is not minimal for a non-nilpotent $\mathfrak{G}$ because in this case $A(1) \neq \cup_{q \geq 0} A(1,q)$. On the other hand, if $\mathfrak{G}$ is nilpotent then $A(1) = \cup_{q \geq 0} A(1,q)$ and $A(1) = A(2) = \ldots$ because A is generated by A^1. Hence A is a minimal algebra.

The formula $A(1,q)^1 = \mathfrak{G}_q^\perp$ will be proved by induction on q. For $q = 0$ we have $A(1,0) = k$ so that $A(1,0)^1 = \{0\} = \mathfrak{G}^\perp$.

Assume now that the required formula is true for $q-1$. We have $A(1,q)^1 = \{c \in A^1 \mid dc$ belongs to the algebra generated by k and by $A(1, q-1)^1 = \mathfrak{G}_{q-1}^\perp\}$. Identify $\Lambda(\mathfrak{G}_{q-1}^\perp)$ with a subalgebra in $\Lambda(\mathfrak{G}^*) = \Lambda(\mathfrak{G})^*$. Then the annihilator of $\Lambda(\mathfrak{G}_{q-1}^\perp)$ in $\Lambda(\mathfrak{G})$ is $\mathfrak{G}_{q-1}\Lambda(\mathfrak{G})$. In particular,

$$A(1,q)^1 = \{c \in A^1 \mid c([x_1, x_2]) = 0 \quad \text{for any} \quad x_1 \in \mathfrak{G}_{q-1}\}.$$

But this just means that $A(1,q)^1 = \mathfrak{G}_q^\perp$.   □

Now we show that an appropriate generalization of Proposition V.4.5 yields a rather detailed characterization of all minimal algebras.

6. Definition. *A Lie superalgebra over a field k is a $\mathbb{Z}_2$-graded vector space $\mathfrak{G} = \mathfrak{G}_{(0)} \oplus \mathfrak{G}_{(1)}$ endowed with a bilinear composition law $[\cdot, \cdot]$ (supercommutator). These data should satisfy the following conditions (where we define $\tilde{x} = i \in \mathbb{Z}_2$ for $x \in \mathfrak{G}_{(i)}$):*

a) $[x, y] = -(-1)^{\tilde{x}\tilde{y}}[y, x]$.
b) $(-1)^{\tilde{x}\tilde{z}}[x, [y, z]] + (-1)^{\tilde{y}\tilde{z}}[z, [x, y]] + (-1)^{\tilde{x}\tilde{y}}[y, [z, x]] = 0$.

We shall consider only $\mathbb{Z}$-graded Lie superalgebras $\bigoplus_{i \geq 0} \mathfrak{G}^i$ satisfying the following additional conditions:

c) $[\mathfrak{G}^i, \mathfrak{G}^j] = \mathfrak{G}^{i+j}$.
d) $\mathfrak{G}_{(i)} = \bigoplus_{k = i \bmod 2} \mathfrak{G}^k$.

An important example of superalgebra is the following one. For a $\mathbb{Z}_2$-graded associative ring $A = A_{(0)} \oplus A_{(1)}$ let $A_L = (A, [\cdot, \cdot])$ where $[a, b] = ab - (-1)^{\tilde{a}\tilde{b}} ba$.

7. From a $\mathbb{Z}$-Graded Lie Superalgebra to a DG-Algebra

a) For a $\mathbb{Z}$-graded Lie superalgebra $\mathfrak{G}$ denote by $C^{(n)}(\mathfrak{G})$ the space of n-linear forms on $\mathfrak{G}$ satisfying the following condition:

$$c(x_1, \ldots, x_i, x_{i+1}, \ldots, x_n) = (-1)^{(\tilde{x}_{i+1}+1)(\tilde{x}_i+1)} c(x_1, \ldots, x_{i+1}, x_i, \ldots, x_n)$$

(in this formula, as well as in similar formulas below, notations of the type $\tilde{x}_i, \tilde{x}_{i+1}$ assume that x_i, x_{i+1} are homogeneous; extension to the general case is by linearity).

b) Define multiplication in $\bigoplus_{n\geq 0} C^{(n)}(\mathfrak{G})$ as follows. Let $\varepsilon : S_n \times \mathbb{Z}_2^n \to \{\pm 1\}$ (S_n is the symmetric group of degree n) be the function which is uniquely determined by the conditions

$$\varepsilon(\sigma_{i,i+1}; \tilde{x}_1, \ldots, \tilde{x}_n) = (-1)^{(\tilde{x}_{i+1}+1)(\tilde{x}_i+1)},$$

$$\varepsilon\left(\sigma; \tau\tilde{X}\right) \varepsilon\left(\tau; \tilde{X}\right) = \varepsilon\left(\sigma\tau, \tilde{X}\right);$$

here $\sigma_{i,j}$ is the permutation of elements i and j. For $c_1 \in C^{(m)}(\mathfrak{G})$, $c_2 \in C^{(n)}(\mathfrak{G})$ let

$$c_1 c_2(x_1, \ldots, x_{m+n}) = \frac{1}{(m+n)!} \sum_{\sigma \in S_{m+n}} \varepsilon(\sigma; \tilde{x}_1, \ldots, \tilde{x}_{m+n})$$
$$\times c_1\left(x_{\sigma(1)}, \ldots, x_{\sigma(m)}\right) c_2\left(x_{\sigma(m+1)}, \ldots, x_{\sigma(m+n)}\right).$$

One can easily see that the algebra $\bigoplus_{n\geq 0} C^{(n)}(\mathfrak{G})$ is generated by $C^{(1)}(\mathfrak{G})$.

c) Introduce a $\mathbb{Z}$-grading on $\bigoplus_{n\geq 0} C^{(n)}(\mathfrak{G})$ as follows: an element $c \in C^{(1)}(\mathfrak{G})$ which equals zero on all homogeneous components other than $\mathfrak{G}^i$ has the degree $\deg c = i + 1$. The condition for compatibility with the multiplication extends the grading to the whole algebra $\bigoplus_{n\geq 0} C^{(n)}(\mathfrak{G})$. With this $\mathbb{Z}$-grading (more precisely, with the induced $\mathbb{Z}_2$-grading) the multiplication is supercommutative: $c_1 c_2 = (-1)^{\deg c_1 \cdot \deg c_2} c_2 c_1$. This should be verified only for linear forms $c(x)$, and for such forms we have

$$c_1 c_2(x, y) = \frac{1}{2}\left(c_1(x) c_2(y) + (-1)^{(\tilde{x}+1)(\tilde{y}+1)} c_1(y) c_2(x)\right).$$

Let $\deg c_1 = i$, $\deg c_2 = j$. Then we can assume $x \in \mathfrak{G}^{i-1}$, $y \in \mathfrak{G}^{j-1}$ and

$$(-1)^{ij} c_2 c_1(x, y) = (-1)^{ij} \frac{1}{2}\left(c_2(x) c_1(y) + (-1)^{ij} c_1(x) c_2(y)\right) = c_1 c_2(x, y).$$

Denote by $C^n(\mathfrak{G})$ the space of elements of degree n (should not be confused with $C^{(n)}(\mathfrak{G})$!).

d) Define a linear mapping d of the supercommutative algebra $C^\bullet(\mathfrak{G}) = \underset{n \geq 0}{\oplus}\, C^{(n)}(\mathfrak{G})$ into itself by the formula

$$dc(x_1, \ldots, x_{n+1}) = \sum_{i \leq j < l \leq n+1} \varepsilon\left(\sigma_{j,l}; \tilde{x}_1, \ldots, \tilde{x}_n\right)(-1)^{\tilde{x}_j}$$

$$c\left([x_j, x_l], x_1, \ldots, \hat{x}_j, \ldots, \hat{x}_l, \ldots, x_{n+1}\right).$$

Let us verify that d makes $C^\bullet(\mathfrak{G})$ into a DG-algebra. First of all, d increases the degree by 1 and satisfies the Leibniz formula. Next, $C^\bullet(\mathfrak{G})$ is generated $C^{(1)}(\mathfrak{G})$, so that it suffices to verify that $d^2 c = 0$ for $c \in C^{(1)}(\mathfrak{G})$. But for such elements it easily follows from the Jacobi identity V.4.6b) on $\mathfrak{G}$. Hence $C^\bullet(\mathfrak{G})$ is a DG-algebra. Evidently, $C^{(0)}(\mathfrak{G}) = k$, $C^\bullet(\mathfrak{G})$ is free (as supercommutative algebra), and the differential d is decomposable.

8. From a DG-Algebra to a $\mathbb{Z}$-Graded Lie Superalgebra

Now let A be a free DG-algebra such that $A^0 = k$ and the differential is decomposable. Starting from it we construct a $\mathbb{Z}$-graded Lie superalgebra $L^\bullet(A)$.

a) Let $I = A^+/(A^+)^2$, and let I^p be the image of A^p in I. Clearly, I is a graded linear space. Define $L^p(A)$ as $(I^{p+1})^*$ and $L^\bullet(A) = \underset{p \geq 0}{\oplus}\, L^p(A)$.

b) There exists an obvious map $S^2(I) \to (A^+)^2/(A^+)^3$, where $S^2(I)$ is again defined in the superalgebra sense as the quotient of $I^{\otimes 2}$ by the relation $a \otimes b = (-1)^{\deg a \cdot \deg b} b \otimes a$. The algebra A being free, this map is an isomorphism. Hence d induces the map

$$\delta : I = A^+/(A^+)^2 \longrightarrow (A^+)^2/(A^+)^3 = S^2(I).$$

Let us consider the dual map

$$\delta^* : S^2(I^*) \longrightarrow I^*$$

and use it to define the supercommutator on $L^\bullet(A) = I^*$:

$$[x, y] = (-1)^{\tilde{x}} \delta^*(xy),$$

where the $\mathbb{Z}$-degree of $x \in L^p$ is, by definition, $\tilde{x} = p \bmod 2 = (\deg x + 1) \bmod 2$. This formula immediately gives $[x, y] = -(-1)^{\tilde{x}\tilde{y}}[y, x]$ and the Jacobi identity follows from $d^2 = 0$ (the verification is practically converse to the proof that $d^2 c = 0$ for $x \in C^{(1)}(\mathfrak{G})$ in V.4.7d)).

9. Theorem (Comparison Theorem). *a) Let $\mathfrak{G}$ be a $\mathbb{Z}$-graded Lie superalgebra, $C^\bullet(\mathfrak{G})$ the DG-algebra constructed in V.4.7. Then $L^\bullet(C^\bullet(\mathfrak{G}))$ is canonically isomorphic to $\mathfrak{G}$.*

b) Let A be a free DG-algebra such that $A^0 = k$ and the differential d in A is decomposable, and let $L^\bullet(A)$ be the $\mathbb{Z}$-graded Lie superalgebra constructed in V.4.8. Then $C^\bullet(L^\bullet(A))$ is (noncanonically) isomorphic to A as a graded k-algebra.

Sketch of the Proof. a) The natural linear map $\mathfrak{G} \to \mathfrak{L}^\bullet(\mathfrak{C}^\bullet(\mathfrak{G}))$ is obtained by considering $\mathfrak{G}$ as the space of linear forms on $I(C^\bullet(\mathfrak{G})) = \mathfrak{C}^{(1)}(\mathfrak{G}) = \mathfrak{G}^*$. We leave to the reader the verification that this gives an isomorphism of $\mathbb{Z}$-graded Lie superalgebras.

b) Let us choose first a linear section, $s : I \to A^+$, which is compatible with the gradings. It determines an isomorphism $S^\bullet(I) \xrightarrow{\sim} A$. On the other hand, the algebra $C^\bullet(L^\bullet(A))$ is freely generated by $C^{(1)}(L^\bullet(A)) = (L^\bullet(A))^* = I$. Hence s determines an isomorphism $C^\bullet(L^\bullet(A)) = A$. However, the differentials in $C^\bullet(L^\bullet(A))$ and in A coincide only modulo $(A^+)^2$. $\square$

10. Theorem. *Let A be a free DG-algebra such that $A^0 = k$, the differential d in A is decomposable, and $\dim A^i < \infty$ for all i. Then A is a minimal algebra iff $L^\bullet(A)$ is nilpotent in the following sense: for any $p \geq 0$ there exists $i_0(p)$ such that*

$$L^\bullet(A)_i \cap L^p(A) = \{0\} \quad \text{for} \quad i \geq i_0(p)$$

(here $L^\bullet(A)_i$ are the elements of the lower central series of the Lie superalgebra $L^\bullet(A)$).

Sketch of the Proof. First of all, repeating with small modifications (and taking into account $\mathbb{Z}$- and $\mathbb{Z}_2$-gradings) the proof of Proposition V.4.5 we get that $L^\bullet(A)$ is nilpotent iff $C^\bullet(L^\bullet(A))$ is minimal.

Now we prove that A and $C^\bullet(L^\bullet(A))$ are either both minimal or both nonminimal. Indeed, let (A, d_1), (A, d_2) be two DG-algebras with $A^0 = k$ whose differentials differ from each other in such a way that $(d_1 - d_2)A \subset (A^+)^3$. Then these two algebras are either both minimal or both nonminimal. First of all, $d_1 = d_2$ on A^1. Hence the filtrations $A(n)$ defined by d_1 and by d_2 (see 1) coincide. By induction on q we get that the filtrations $A(n,q)$ defined by d_1 and by d_2 also coincide. $\square$

11. Theorem. *Let a DG-algebra A over k be cohomologically connected, i.e., $H^0(A) = k$. Then there exists a weak equivalence $f : M \to A$ with a minimal algebra M.*

Such an algebra M is called the minimal model for A.

Proof. We construct M consecutively adding elements to $M(0) = k$ and extending the natural morphism $f(0) : k \to A$. We shall make all $H^n(f)$ isomorphisms by iterating two constructions described below: the first of these constructions kills $\operatorname{Coker} H^n(\cdot)$ and the second one (countably repeated) kills $\ker H^n(\cdot)$.

Killing the Cokernel. Let $h : N \to A$ be a morphism of DG-algebras. Fix some n and choose a subspace $C \subset Z^n(A)$ that maps isomorphically onto $\operatorname{Coker} H^n(f)$ under the composite map $Z^n(A) \to H^n(A) \to H^n(A)/\operatorname{Im} H^n(h)$. Define $N' = N \otimes S(C)$, where $S(C)$, as before, is the symmetric algebra of the graded space $C = C^n$ with the zero differential. Define the extension $h' : N' \to A$ of the morphism h by the formula $h'(1 \otimes c) = c$ for $c \in C$.

For $m \leq n-1$ we have $(N')^m = N^m$ and $H^m(h') = H^m(h)$. Next,

$$(N')^n = N^n \oplus C, \quad dC = 0, \quad H^n(h') = (H^n(h), a),$$

where a is the composition $C \hookrightarrow Z^n(A) \to H^n(A)$, so that $H^n(h')$ is surjective.

Killing the Kernel. Again we fix a morphism $h : N \to A$ of DG-algebras and an integer $n \geq 1$. Let $K \subset Z^n(N)$ be a subspace that maps isomorphically onto $\ker H^n(h)$. Then $h(K) \subset B^n(A) = d(A^{n-1})$. Let us choose also a subspace $\tilde{K} \subset A^{n-1}$ such that the restriction $d : \tilde{K} \to h(K)$ is an isomorphism. We shall denote by $h^{-1} \circ d$ the resulting isomorphism $\tilde{K} \to K$.

Let $N' = N \otimes S(\tilde{K})$ as a graded k-algebra (but not as a DG-algebra!). Introduce the differential in N' defining it on generators by the following formula:

$$d(x \otimes y) = dx \otimes y + (-1)^{\deg x} x \cdot (h^{-1} \circ d)(y) \otimes 1, \quad x \in N, \quad y \in \tilde{K}.$$

Finally, define a morphism $h' : N' \to A$ by the formula

$$h'(x \otimes y) = h(x) \cdot y, \quad x \in N, \quad y \in \tilde{K}.$$

Similarly to the situation with the cokernel we have $(N')^m = N^m$ and $H^m(h') = H^m(h)$ for $m \leq n-2$. Next, $(N')^{n-1} = N^{n-1} \oplus \tilde{K}$, $Z^{n-1}(N') = Z^{n-1}(N)$ (because $\tilde{K} \cap B^n(N) = \{0\}$) and $B^{n-1}(N') = B^{n-1}(N)$. Hence $H^{n-1}(h') = (H^{n-1}(h), 0)$.

To understand what is going on in dimension n let us consider the following diagram:

$$\begin{array}{ccccc} & & H^n(N') & & \\ & \nearrow H^n(i) & \uparrow & \searrow H^n(h') & \\ K \bmod B^n(N) & \longrightarrow & H^n(N) & \xrightarrow{H^n(h)} & H^n(A) \end{array}$$

$$i : N \hookrightarrow N'$$

By construction, $K \bmod B^n(N)$ coincides with the kernel of $H^n(h)$.

The main property of this first step of "killing the kernel" procedure is that $K \bmod B^n(N) = \ker H^n(i)$. Indeed, for any $y \in K$ we have $i(y) = y \otimes 1 = d(1 \otimes \tilde{y})$, where $\tilde{y} = (d^{-1} \circ h)(y)$.

However, we cannot just replace N by N' because $\ker H^n(h')$ might very well be nonzero. Hence we must iterate this first step a countable number of times by setting $N^{(1)} = N'$, $h^{(1)} = h'$, $N^{(i+1)} = N^{(i)'}$, $h^{(i+1)} = h^{(i)'}$ and, finally, $N^{(\infty)} = \varinjlim N^{(i)}$, $h^{(\infty)} = \varinjlim h^{(i)}$. We claim that $H^m(h^{(\infty)})$ is an isomorphism for $m \leq n-1$ and a monomorphism for $m = n$. Indeed, the first statement is true for all $H^m\left(h^{(i)}\right)$. Next, let γ be an element in $\ker H^n\left(h^{(\infty)}\right)$. Then it has a representative $\gamma^{(i)} \in H^n\left(h^{(i)}\right)$ for some i and the image of $\gamma^{(i)}$ in $H^n\left(N^{(i+1)}\right)$ is zero. Hence $\gamma = 0$.

Let us remark in conclusion that in the case when $H^1(N) = 0$ the first step suffices: $H^n(h')$ will be a monomorphism. In the construction of the minimal model this will be the case if $H^1(A) = 0$.

Let us begin to construct the minimal model.

Induction step. Let us assume that we are given a minimal DG-algebra $M(n)$ and a morphism $f(n) : M(n) \to A$ with the following properties:

a$_n$) $M(n)$ as a DG-algebra is generated by elements of degree $\leq n$.

b$_n$) $H^m(f(n))$ is an isomorphism for $m \leq n$ and a monomorphism for $m = n+1$.

The algebra $M(0) = k$ satisfies the conditions a$_0$) and b$_0$) because $H^0(A) = k$.

Let us define $M(n+1)$ and $f(n+1)$ as the result of the consecutive application to $f(n)$ of two operations: killing of the cokernel in the dimension $n+1$ and killing of the cokernel in the dimension $n+2$. It is evident that $M(n) \subset M(n+1)$, the morphism $f(n+1)$ is an extension of $f(n)$ and satisfies a$_{n+1}$) and b$_{n+1}$).

Let us prove that the algebra $M(n+1)$ is minimal. To simplify the notations define $M(n+1) = B$. Let $B(n)$ and $B(n,q)$ be the subalgebras of B defined as in V.4.1. It is clear that $B(m) = M(m)$ for $m \leq n+1$ and $B(m) = B$ for $m > n+1$. Using the minimality of $M(m)$ for $m \leq n$ we see that it suffices to prove that $M(n+1) = \cup_{q \geq 0} B(n+1,q)$.

First of all, $B(n+1,1) \supset M(n)'$, where $M(n)'$ is the result of killing the cokernel of $H^{n+1}(f(n))$. Indeed, $M(n)'$ is generated by $B(n+1,0) = M(n)$ and by cycles c of degree $n+1$.

Next, each step in the "killing of the kernel" construction increases the degree q by at most 1, because we add generators of degree $n+1$ whose differentials belong to the previous algebra.

So we see that $M = \varinjlim M(n)$, $f = \varinjlim f(n)$ will be the minimal model for A. □

Let us remark that the above construction is not functorial due to arbitrary choices of spaces C, K, $\tilde{K}$ used for kernel and cokernel killing. But, similarly to the case of a derived category, the minimal model becomes functorial after localizing with respect to the class of homotopy equivalences. In the category $\mathcal{DGA}$ the notion of homotopy equivalence should be slightly modified because the multiplication should be taken into account.

In the next section we shall formulate the main facts about the homotopy theory of DG-algebras and about its relation to the homotopy theory of simplicial sets.

Exercises

1. Graded Differential Lie Superalgebras as a Closed Model Category. a) A graded differential Lie superalgebra is a $\mathbb{Z}$-graded Lie algebra $\mathfrak{G}$ (satisfying the conditions a)–d) of V.4.6) together with a degree 1 linear map $\partial : \mathfrak{G} \to \mathfrak{G}$ such that $\partial^2 = 0$ and $\partial([x,y]) = [\partial x, y] + (-1)^{\bar{x}}[x, \partial y]$. Denote by $\mathcal{DGL}$ the corresponding category (with naturally defined morphisms).

Let us define *fibrations* in $\mathcal{DGL}$ as morphisms $\varphi : \mathfrak{G} \to \mathfrak{G}'$ that are epimorphisms in all positive degrees, *weak equivalences* as quasi-isomorphisms, and *cofibrations* by the axiom CM1b) from Sect. V.1.4. Following the proof of Theorem V.3.4 prove that $\mathcal{DGL}$ is a closed model category.

b) Define $\mathcal{DGL}_1$ as a full subcategory of $\mathcal{DGL}$ consisting of algebras $\mathfrak{G} = \oplus \mathfrak{G}^i$ with $\mathfrak{G}^0 = \{0\}$. Introduce *fibrations* in $\mathcal{DGL}_1$ as morphisms that are epimorphisms in all degrees $i \geq 2$, *weak equivalences* as quasi-isomorphisms, and *cofibrations* by the axiom CM1b) from V.4.1. Verify that $\mathcal{DGL}_1$ is a closed model category.

c) *Free Lie Algebras.* Let $V = \underset{i \geq 0}{\oplus} V^i$ be a graded vector space. The free Lie algebra $L(V)$ generated by V is the representing object in the category of $\mathbb{Z}$-graded Lie algebras (without a differential) for the functor $\mathfrak{G} \mapsto \mathrm{Hom}_{\mathrm{Gr\,Vect}}(V, \mathfrak{G})$ where $\mathrm{Gr\,Vect}$ is the category of graded vector spaces. Give an explicit construction for $L(V)$.

d) *Cofibrations in $\mathcal{DGL}_1$.* Prove that a morphism $f : (\mathfrak{G}, \partial) \to (\tilde{\mathfrak{G}}, \tilde{\partial})$ is a cofibration in $\mathcal{DGL}_1$ if and only if it can be represented as the composition $f = \rho \circ i$, where $i : (\mathfrak{G}, \partial) \to (\mathfrak{G}', \partial')$, $\rho : (\mathfrak{G}', \partial') \to (\tilde{\mathfrak{G}}, \tilde{\partial})$ are morphisms in $\mathcal{DGL}_1$ that can be embedded into a commutative diagram of the form

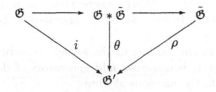

in which $\mathfrak{G} * \tilde{\mathfrak{G}}$ is the direct sum (free product) in the category $\mathcal{DGL}_1$ and θ is an isomorphism of Lie algebras (not necessarily differential).

e) Prove that in $\mathcal{DGL}_1$ each object is fibrant and that cofibrant objects are just algebras of the form $(L(V), \partial)$ for some graded vector space V (with arbitrary ∂).

2. Minimal Lie DG-Algebras. There exists a theory of minimal Lie DG-algebras that is parallel to theory of minimal DG-algebras described in this section. An object $\mathfrak{G}$ of the category $\mathcal{DGL}_1$ is said to be *minimal* if $\mathfrak{G}$ is a free Lie algebra (i.e., $\mathfrak{G} = L(V)$ for some graded vector space V) and ∂ is decomposable (i.e., $\partial(V) \subset [L(V), L(V)]$).

Modifying the proof of Theorem V.4.11 prove the following minimal model theorem: for $(\mathfrak{G}, \partial) \in \mathrm{Ob}\ \mathcal{DGL}_1$ there exists a minimal Lie DG-algebra $(L(V), \tilde{\partial})$ and a weak equivalence $(L(V), \tilde{\partial}) \to (\mathfrak{G}, \partial)$.

3. Homotopies in $\mathcal{DGL}_1$. Let us show how the homotopy theory from Ex. 2.7 is specialized to $\mathcal{DGL}_1$.

a) Let $(\mathfrak{G}, \partial) \in \mathrm{Ob}\ \mathcal{DGL}_1$ with $\mathfrak{G} = L(V)$ for some V. Define $(\tilde{G}, D) \in \mathrm{Ob}\ \mathcal{DGL}_1$ as follows: $\tilde{\mathfrak{G}} = L(W)$, where $W = V \oplus V' \oplus V''$ with $V' = V$, $V'' = V[-1]$, D equals ∂ on V, zero on V', and the identity map $V''_n = V_{n-1} = V'_{n-1} \to V'_{n-1}$ on V''.

Next, let $\delta_0 : (\mathfrak{G}, \partial) \to (\tilde{\mathfrak{G}}, D)$ be the canonical embedding and $\delta_1 : (\mathfrak{G}, \partial) \to (\tilde{\mathfrak{G}}, D)$ be defined by the formula

$$\delta_1(v) = v + sv + \sum_{n \geq 1} \frac{(i \circ D)^n}{n!} v, \qquad (*)$$

where $s : V \to V'$ is the identity mapping and $i : \tilde{\mathfrak{G}} \to \tilde{\mathfrak{G}}$ is the derivation of degree $+1$ defined on $W = V \oplus V' \oplus V''$ by the linear map $W \to W$ with the matrix

$$\begin{pmatrix} 0 & 0 & \mathrm{id}_V \\ 0 & 0 & 0 \\ 0 & 0 & 0 \end{pmatrix}$$

(the sum in $(*)$ is finite for each $v \in V$). Let us define also $\sigma : (\tilde{\mathfrak{G}}, D) \to (\mathfrak{G}, \partial)$ as the map induced by the projection $W = V \oplus V' \oplus V'' \to V$.

Verify that $\left((\tilde{\mathfrak{G}}, D), \delta_0, \delta_1, \sigma\right)$ is a cylinder for $(\mathfrak{G}, \partial)$ in the sense of Ex. 2.7b).

b) Define a right homotopy between morphisms $f_1, f_2 : (\mathfrak{G}, \partial) \to (\mathfrak{G}', \partial')$ in $\mathcal{DGL}_1$ with a free $\mathfrak{G} = L(V)$ as in Ex. 2.7d) using the cylinder from a). Prove that the existence of right homotopy does not depend on the choice of V with $\mathfrak{G} = L(V)$.

c) Prove that the right homotopy $f_1 \sim f_2$ is an equivalence relation in $\mathrm{Hom}_{\mathcal{DGL}_1}((\mathfrak{G}, \partial), (\mathfrak{G}', \partial'))$ with free $\mathfrak{G}$.

d) Let a diagram

$$(L(W), \partial_W) \xrightarrow{g} (L(V), \partial_V) \underset{f_2}{\overset{f_1}{\rightrightarrows}} (\mathfrak{G}', \partial') \xrightarrow{h} (\mathfrak{G}'', \partial'')$$

in $\mathcal{DGL}_1$ be given. Prove that $f_1 \sim f_2$ implies $h \circ f_1 \circ g \sim h \circ f_2 \circ g$. Prove that if h is a weak equivalence in $\mathcal{DGL}_1$, then $h \circ f_1 \sim h \circ f_2 \iff f_1 \sim f_2$. For this use the covering extension lemma: let $p : (L(V), \partial_V) \to (\mathfrak{G}', \partial')$ and $q : (\mathfrak{G}, \partial) \to (\mathfrak{G}', \partial')$ be morphisms in $\mathcal{DGL}_1$ with q being a quasi-isomorphism. Then there exists $r : (L(V), \partial_V) \to (\mathfrak{G}, \partial)$ such that $p = qr$.

e) Deduce from the covering extension lemma that a minimal model of a Lie DG-algebra is unique up to a quasi-isomorphism.

f) Let $f : (\mathfrak{G}, \partial) \to (\mathfrak{G}', \partial')$ be a morphism in $\mathcal{DGL}_1$ and $\varphi : (L(V), \partial_V) \to (\mathfrak{G}, \partial)$, $\varphi' : (L(V'), \partial_{V'}) \to (\mathfrak{G}', \partial')$ be minimal models of $(\mathfrak{G}, \partial)$ and $(\mathfrak{G}', \partial')$ respectively. Prove that there exists a morphism $\hat{f} : (L(V), \partial_V) \to (L(V'), \partial_{V'})$ such that $\varphi' \circ \hat{f} = f \circ \varphi$ (a minimal model for f). Prove that any two such morphisms are homotopic. Prove that if $\mathfrak{G}$ is free and $f, g : (\mathfrak{G}, \partial) \to (\mathfrak{G}', \partial')$ are two morphisms with minimal models $\hat{f}, \hat{g}$ respectively, then $f \sim g \Leftrightarrow \hat{f} \sim \hat{g}$.

4. Minimal Models of de Rham Algebras. Examples. The following method enables us in some cases to compute a minimal model M of the de Rham algebra $\Omega^\bullet(X)$ of a simplicial set X using only the cohomology algebra $A = H^\bullet(X, k)$ (we work in the category $\mathcal{DGA}$, a field assumed to be of zero characteristic).

Let us assume that A can be represented in the form $A = S(V)/I$, where V is a graded vector space and I is an ideal in $S(V)$ generated by the sequence $f_1, \ldots, f_n \in S(V)$ of elements in $S(V)$ such that:

a) all f_i are homogeneous and decomposable in $S(V)$;
b) f_i is not a zero divisor in $S(V)/(f_1, \ldots, f_{i-1})$.

Let $M = S(V \oplus kg_1 \oplus \ldots \oplus kg_n)$, where $\deg g_i = \deg f_i - 1$; define the differential in M by the formula $dv = 0$ for $v \in V$, $dg_i = f_i$ (essentially, M is the Koszul resolution of the $S(V)$-module A, see Ex. I.7.5).

One can verify that M is a minimal algebra and the natural map $\varepsilon : M \to \Omega^\bullet(X)$ is a weak equivalence, so that M is a minimal model of $\Omega^\bullet(X)$ in $\mathcal{DGA}$.

For example, let X be the two-dimensional sphere S^2 (more precisely, a simplicial set whose geometric realization is homeomorphic to the sphere). Then M is freely generated by $x \in M^2$, $y \in M^3$ with $dx = 0$, $dy = x^2$. For $X = \mathbb{CP}^2$ the algebra M is freely generated by $x \in M^2$, $y \in M^5$ with $dx = 0$, $dy = x^3$.

V.5 Equivalence of Homotopy Categories

This section contains no proofs. We present here the main results about the equivalence of homotopic categories following the paper by Bousfield and Gugenheim [1].

1. Definition. *a) An elementary homotopy between two morphisms $f_0, f_1 : B \to A$ of DG-algebras is a morphism $F : B \to A \underset{k}{\otimes} T(V)$, where $V = V^0 = kv$, satisfying the following property. Let morphisms $\delta_i : A \underset{k}{\otimes} T(V) \to A$, $i = 0, 1$, be defined by the formulas*

$$\delta_0(a \otimes p(v)) = p(0)a, \quad \delta_0(a \otimes p(v)dv) = 0,$$
$$\delta_1(a \otimes p(v)) = p(1)a, \quad \delta_1(a \otimes p(v)dv) = 0.$$

Then we require that $\delta_0 \circ F = f_0$, $\delta_1 \circ F = f_1$.

b) Two morphisms of DG-algebras are said to be homotopic *if they are connected by a chain of elementary homotopies.*

2. Lemma. *Homotopy is an equivalence relation on morphisms of DG-algebras which is compatible with the composition of morphisms. Homotopic morphisms of DG-algebras induce homotopic morphisms of underlying complexes, and, therefore, equal maps in cohomology.*

For two DG-algebras A, B denote by $[B, A]$ the set of equivalence classes of morphisms $B \to A$. For generic A, B the elementary homotopy is a non-transitive relation. But for cofibrant algebras the following is true.

3. Proposition. *a) If B is a cofibrant algebra (i.e., $k \to B$ is a cofibration), then for any DG-algebra A the elementary homotopy is a transitive relation on $[B, A]$.*
  b) If B is a cofibrant algebra, then for any weak equivalence $f : A \to A'$ the induced mapping $f_* : [B, A] \to [B, A']$ is one-to-one.

Connected cofibrant algebras admit the following description.

4. Proposition. *The class of connected cofibrant algebras coincides with the class of algebras of the form $M \otimes T(V)$, where M is a minimal algebra and $T(V)$ is the tensor algebra of a graded linear space V.*

5. Theorem (Uniqueness Theorem). *Let $f : M \to A$, $g : N \to A$ be two weak equivalences, where M and N are minimal, $H^0(A) = k$. Then there exists an isomorphism $h : M \xrightarrow{\sim} N$ such that morphisms f and $g \cdot h$ are (elementary) homotopic.*

6. Homotopy Groups of DG-Algebras

To define homotopy groups we must work in the category $\mathcal{DGA}_0$ whose objects are augmented DG-algebras (i.e., equipped with morphisms $\varepsilon : A \to k$) and morphisms are the morphisms in $\mathcal{DGA}$ commuting with augmentations. The classes W, F, G in $\mathcal{DGA}_0$ are defined as intersections of corresponding classes in $\mathcal{DGA}$ with morphisms in $\mathcal{DGA}_0$. Looking through the proofs in Sect. V.3 one can see that $\mathcal{DGA}_0$ is a closed model category.

For $A \in \mathrm{Ob}\,\mathcal{DGA}_0$ define $A_+ = \ker \varepsilon$.

7. Definition. *Homotopy groups of A are defined as $\pi^n(A) = H^n(A_+/A_+ \cdot A_+)$ where $A_+/A_+ \cdot A_+$ is considered as the complex with the grading induced by A^i and with the differential induced by $d : A \to A$.*

Let us remark that if $A^0 = k$ and d is decomposable (for example, if A is minimal), then $\pi^n(A) = A_+^n \bmod (A_+ \cdot A_+)$. According to V.4.8–V.4.9, in this case there exists a canonical structure of Lie superalgebra on $\bigoplus_{n \geq 0} \mathrm{Hom}_k(\pi^n(A), k)$. In algebraic topology the corresponding composition law is called the Whitehead multiplication. To give the description of $\pi^n(A)$ that looks like the definition of homotopy groups in topology, let us define the algebra $S_n \in \mathrm{Ob}\, \mathcal{DGA}_0$, $n \geq 0$. as follows:

$$S_0 = k,$$

$$S_n^i = \left\{ \begin{array}{ll} k & \text{for}\quad i = 0 \quad \text{or} \quad i = n, \\ 0 & \text{otherwise.} \end{array} \right\}.$$

$\varepsilon = 0$ on S_n^n (for $n > 0$); the multiplication on S_n^n is zero.

8. Proposition. *For any algebra $A \in \mathrm{Ob}\, \mathcal{DGA}_0$ there exists a natural one-to-one mapping*

$$[A, S_n] \xrightarrow{\sim} \mathrm{Hom}_k(\pi^n(A), k)$$

(with homotopy classes of morphisms in $\mathcal{DGA}_0$ at the left).

9. Proposition. *a) Homotopic morphisms of augmented DG-algebras $A \to B$ induce equal mappings $\pi^\bullet(A) \to \pi^\bullet(B)$.*

b) A weak equivalence of cofibrant algebras $A \to B$ induces an isomorphism $\pi^\bullet(A) \xrightarrow{\sim} \pi^\bullet(B)$.

This proposition, together with the theorem about the existence of a minimal model, implies that it suffices to compute homotopic groups for minimal algebras only.

Let us formulate the main results about the relation between homotopy theories of simplicial sets and of DG-algebras.

10. Two Homotopy Categories

Denote by $h\mathcal{S}$ the following category:

$$\mathrm{Ob}\, h\mathcal{S} = \text{Kan sets in} \quad \Delta^\circ \mathit{Set},$$

$$\mathrm{Hom}_{h\mathcal{S}}(X, Y) = [X, Y] \qquad (\text{see V.2.2}).$$

Similarly define the category $h\mathcal{A}$ as follows:

$$\mathrm{Ob}\, h\mathcal{A} = \text{cofibrant algebras in} \quad \mathcal{DGA}.$$

$$\mathrm{Hom}_{h\mathcal{A}}(A, B) = [A, B] \qquad (\text{see V.5.2})$$

(the composition of homotopy classes of morphisms is defined via the composition of the corresponding representatives).

There are the following functors between these two categories:

a) *The functor $S : h\mathcal{A} \to h\mathcal{S}^\circ$.* For $A \in \mathrm{Ob}\, \mathcal{DGA}$ let

$$S(A)_n = \mathrm{Hom}_{\mathcal{DGA}}(A, \nabla(n)),$$

where $\nabla(n)$ is the algebra of polynomial differential forms of the n-simplex (see V.3.2). On $S(A) = (S(A)_n)$ there exists a natural structure of a simplicial set which is functorial in A under morphisms in $\mathcal{DGA}$. For a cofibrant algebra A the set $S(A)$ is a Kan set. If morphisms f,g of two cofibrant algebras are homotopic, then $S(f)$ and $S(g)$ are homotopic in $\Delta^\circ Set$. Hence S determines a functor $h\mathcal{A} \to h\mathcal{S}^\circ$, which will be denoted by the same letter.

b) *The functor* $M : h\mathcal{S}^\circ \to h\mathcal{A}$. This functor associates with a Kan set the class of its de Rham algebra. Unfortunately, the de Rham algebra of a Kan set is not, in general, a cofibrant algebra. Hence to define M one must use the choice axiom. Let us choose, for each Kan set X, a weak equivalence $M(X) \to \Omega^\bullet(X)$, where $M(X)$ is a cofibrant algebra (which is minimal for a connected X). After that one can prove that M can be extended to homotopy classes of morphisms and is unique up to a homotopy. This guarantees the existence of the required functor.

11. Nilpotent Rational Simplicial Sets

a) Let G be a group. Let us recall that its *lower central series* $G = Z_1(G) \supset Z_2(G) \supset \ldots$ is defined inductively as follows:

$Z_{i+1}(G)$ is generated by commutators $xyx^{-1}y^{-1}$, $x \in Z_i(G)$, $y \in G$. The group G is called *nilpotent* if $Z_i(G) = \{e\}$ for $i \geq i_0$.

b) Let N be a G-module. Define $Z_1(N) = N$ and let $Z_{i+1}(N)$ be generated by $gn-n$ for $g \in G, n \in Z_i(N)$. A module N is called nilpotent if $Z_i(N) = \{0\}$ for $i \geq i_0$.

c) A group G is called a uniquely divisible group if for any $g \in G$, $k \geq 1$, the equation $x^k = g$ in G has exactly one solution.

d) A connected simplicial set or topological space X is called *nilpotent* if its fundamental group $\pi_1(X)$ is nilpotent (for one and hence for any choice of a base point), and all $\pi_1(X)$-modules $\pi_n(X)$ are also nilpotent.

e) A nilpotent simplicial set or topological space X is called *rational* if all the groups $\pi_n(X)$, $n \geq 1$, are uniquely divisible groups.

f) A nilpotent simplicial set or topological space X is said to have a *finite $\mathbb{Q}$-type* if all the homology spaces $H_n(X, \mathbb{Q})$, $n \geq 1$, are finite dimensional. The equivalent requirement is that $H_1(X, \mathbb{Q})$ and all $\pi_n(X) \otimes \mathbb{Q}$, $n \geq 2$, are finite dimensional.

g) Let A be a cofibrant DG-algebra over $\mathbb{Q}$ with $H^0(A) = \mathbb{Q}$ and let $M \to A$ be a weak equivalence with a minimal algebra M. The algebra A is said to have a *finite $\mathbb{Q}$-type* if M^i are finite dimensional for $i \geq 0$.

Denote by $h_f \mathcal{S}$ (resp. by $h_f \mathcal{A}$) the full subcategories of $h\mathcal{S}$ (resp. of $h\mathcal{A}$) formed by rational nilpotent Kan sets of finite $\mathbb{Q}$-type (resp. by cofibrant DG-algebras of finite $\mathbb{Q}$-type).

The categories $h_f \mathcal{S}_0$ (resp. $h_f \mathcal{A}_0$) are defined similarly using sets with a base point (resp. augmented algebras).

12. Theorem. *The restrictions of functors S and M to $h_f \mathcal{S}$ and $h_f \mathcal{A}$ respectively induce adjoint equivalences of these categories. The same is true for $h_f \mathcal{S}_0$ and $h_f \mathcal{A}_0$.*

13. Homotopy Groups

Theorem V.5.12 gives a rather simple procedure to compute homotopy groups of simplicial sets (up to torsion). Let X be a simplicial set whose geometric realization $|X|$ is a connected rational nilpotent space of finite $\mathbb{Q}$-type. Then the following results are true.

  a) A simplicial set X is weakly equivalent to a Kan set $\mathrm{Sing}\,|X|$, which is connected, nilpotent and has finite $\mathbb{Q}$-type.

  b) Let $\pi^j(M(X))$ be the homotopy groups of a minimal model $M(X)$ of the de Rham algebra of $\mathrm{Sing}\,|X|$ over $\mathbb{Q}$. Then the following fact holds.

14. Theorem. *For $j \geq 2$ there exist functorial isomorphisms $\pi^j(M(X)) = \mathrm{Hom}\,(\pi_j(X), \mathbb{Q})$. If $\pi_1(X)$ is abelian, then the same is true for $j = 1$.*

(For $\pi_1(X) \neq \{1\}$ one has to work with based spaces and augmented algebras).

The proof of this result and of its various generalizations is based on the following localization technique which enables one to give a geometric meaning to the replacement of $\pi_j(X)$ by $\pi_j(X) \otimes \mathbb{Q}$ (or by $\mathrm{Hom}(\pi_j(X), \mathbb{Q})$).

15. Theorem. *Let X be a connected nilpotent simplicial set of finite $\mathbb{Q}$-type. Then the natural morphism $X \to S \circ M(X)$ is universal in the class of morphisms $X \to Y$ in hS_0 such that Y is a rational simplicial set of finite $\mathbb{Q}$-type.*

References

A. Collections

[AAT] Algebra, algebraic topology and their interaction / Ed.: J.-E. Roos, Lect. Notes Math. v. 1183, Berlin – Heidelberg: Springer, 1986.

[AN] Algebraic number theory / Eds. J.W.S. Cassels, A. Fröhlich, London – New York: Academic Press, 1967.

[ES] Analyse et topologie sur les éspaces singuliers, II–III / Eds.: B. Teissier, J.-L. Verdier, Astérique, 1983, v. 101–102.

[HG] Homological group theory (Ed.: C.T.C. Wall, London Math. Soc. Lect. Notes, v. 36, Cambridge: Cambridge Uni.-Press, 1979.

IH] Seminar on intersection homology / Ed.: A. Borel, Boston: Birkhäuser, 1984.

[KT1] Algebraic K-theory. I, II, III / Lect. Notes Math., v. 341, 342, 343, Berlin – Heidelberg: Springer, 1973.

[KT2] Applications of algebraic K-theory to algebraic geometry and number theory. Parts 1, 2 / Contemporary Math., v. 55, 1986.

[SD] Systèmes différentiels et singularités / Eds.: A. Galligo, M. Granger, Ph. Maisonobe, Astérisque, 1985, v. 130.

B. Books and Papers

André M.
1. Categories of functors and adjoint functors / Am. J. Math., 1966, v. 88, no. 3, p. 529–547.
2. Méthodes simpliciales en algèbre homologique et algèbre commutative / Lect. Notes Math. v. 34, Berlin – Heidelberg: Springer, 1973.

Artin M.
1. Grothendieck topologies / Lect. Notes Math. Dept. Harvard Univ., 1962
2. Algebraic spaces / Preprint Yale Univ., 1969.

Artin M., Tate J.
1. Class field theory / Lect. Notes Math. Dept. Harvard Univ., 1961.

Avramov L., Halperin S.
1. Through the looking glass: a dictionary between rational homotopy theory and local algebra, in: [AAT], p. 1–27.

Bass H.
1. Algebraic K-theory / New York – Amsterdam: Benjamin Inc., 1968.

Beilinson A.
1. Coherent sheaves in P^n and problems of linear algebra / Funkts. Anal. Prilozh. 1978, v. 12, no. 3, p. 68–69 [Russian].
2. Higher regulators and values of the L-function / Itogi Nauki Tekh.. Ser. Sovrem. Probl. Mat., Novejshie Dostizh. 1984, v. 24, p. 181–238 [Russian].

Beilinson A., Bernstein J., Deligne P.
1. Faisceaux pervers / Astérisque, 1982, v. 100.

Bernstein J.N., Gelfand I.M., Gelfand S.I.
1. Algebraic bundles on P^n and problems of linear algebra / Funkts. Anal. Prilozh. 1978, v. 12, no. 3, p. 66–67 [Russian].

Bernstein J.N., Gelfand I.M., Ponomarev V.A.
1. Coxeter functors and Gabriel theorem / Usp. Mat. Nauk., 1973, v. 28, no. 1, p. 19–38 [Russian].

Berthelot P.
1. Cohomologie crystalline des schémas de charactéristique $p > 0$ / Lect. Notes Math. v. 407, Heidelberg – Berlin: Springer, 1974.

Boardman J., Vogt R.
1. Homotopy invariant algebraic structures on topological spaces / Lect. Notes Math. v. 347, Heidelberg – Berlin: Springer, 1973.

Borel A., Wallach N.
1. Continuous cohomology, discrete subgroups, and representations of reductive groups / Princeton: Princeton Univ. Press, 1980.

Bott R., Tu I.
1. Differential forms in algebraic topology / New York: Springer, 1982.

Bourbaki N.
1. Algèbre, Ch. 10. Algèbre homologique / Paris: Masson, 1980.

Bousfield A.K., Guggenheim V.K.A.M.
1. On PL de Rham theory and rational homotopy type / Mem. Am. Math. Soc., 1976, v. 8, no. 179.

Bredon G.
1. Sheaf theory / New York: McGraw-Hill, 1967.

Brenner Sh., Butler M.C.P.
1. Generalization of the Bernstein – Gelfand – Ponomarev reflection functors / Lect. Notes Math., Berlin – Heidelberg: Springer, 1980, v. 832, p. 103–170.

Brown K.S.
1. Cohomology of groups / New York: Springer, 1982.

Cartan H.
1. Variétés algébriques complexes et cohomologie / Colloque sur les fonctions de plusieurs variables, Bruxelles, Mars 11–14, 1953, Liège – Paris, 1953.

Cartan H., Chevalley C.
1. Géométrie algébrique, Sem. Cartan – Chevalley / Paris: Secrétariat Math. 1955/56.

Cartan H., Eilenberg S.
1. Homological algebra / Princeton, New Jersey: Princeton Univ. Press, 1956.

Cartier P.
1. Homologie cyclique: Rapport sur les travaux récents de Connes, Karoubi, Loday, Quillen, ... Sém. Bourbaki, no. 621 / Astérisque, 1985, v. 121–122., p. 123–146.

Connes A.
1. Non-commutative differential geometry / Publ. Math. IHES., 1986, v. 62, p. 41–144.

Curtis C.W., Reiner I.
1. Representation theory of finite groups and associative algebras. New York – London: Interscience Publ., 1982.

Dedeker P.
1. Sur la cohomologie non-abélienne I, II / Can. J. Math., 1960, v. 12, p. 231–251; 1963, v. 15, p. 84–93.

Deligne P.
1. Théorie de Hodge, II, III / Publ. Math. IHES, 1971. v. 40, p. 5–58; 1974, v. 44, p. 5–77.
2. La conjecture de Weil, I, II / Publ. Math. IHES, 1974, v. 43, p. 273–307; 1980, v. 52, p. 137–252.

Deligne P., Griffiths Ph., Morhan J., Sullivan D.
1. Real homotopy theory of Kähler manifolds / Invent. Math., 1975, v. 29, no. 3, p. 245–274.

Demazure M., Gabriel P.
1. Groupes algébriques, v. 1 / Paris: Masson, 1970.

Dieudonné J.
1. Panorama des mathématiques pures. La choix bourbachique / Paris: Gauthier-Villars, 1977.

Dold A.
1. Lectures on algebraic topology / Berlin – Heidelberg: Springer, 1972.

Dold A., Puppe D.
1. Homologie nicht-additiver Funktoren: Anwendungen / Ann. Inst. Fourier, 1961, v. 11, p. 201–312.

Dubrovin B.A., Novikov S.P., Fomenko A.T.
1. Contemporary geometry: methods of homology theory / Moscow: Nauka, 1984 [Russian].

Duskin J.
1. Simplicial methods and the interpretation of triples cohomology / Mem. Am. Math. Soc., 1975, no. 163.
2. Higher dimension torsors and the cohomology of topoi: the abelian theory / Lect. Notes Math., Heidelberg – Berlin: Springer, 1977, v. 753, p. 255–279.

Eckmann B., Hilton P.J.
1. Exact couples in an Abelian category / J. Algebra, 1966, v. 3, no. 1, p. 38–87.
2. Composition functors and spectral sequences / Comment. Math. Helv., 1966, v. 41, no. 3, p. 187–221.

Faith C.
1. Algebra: modules, rings and categories / Berlin – Heidelberg: Springer, 1973.

Feigin B.L., Tsygan B.L.
1. Cohomology of the Lie algebra of generalized jacobian matrices / Funkts. Anal. Prilozh., 1983, v. 17, no. 2, p. 86–87 [Russian].
2. Additive K-theory / Lect. Notes Math., Berlin – Heidelberg: Springer, 1970, v. 1289, p. 67–209.

Fuks D.B.
1. Homotopic topology / Moscow: Idz. MGU, 1969 [Russian].
2. Cohomology of infinite-dimensional Lie algebras / Moscow: Nauka, 1984 [Russian].

Fulton W., MacPherson R.
1. Categorical framework for the study of singular spaces / Mem. Am. Math. Soc., 1981, v. 31, no. 243.

Gabriel P., Zisman M.
1. Calculus of fractions and homotopy theory / Berlin – Heidelberg: Springer, 1967.

Gelfand I.M., Shilov B.E.
1. Commutative normed rings / Moscow: Fizmatgiz, 1960 [Russian].

Giraud J.
1. Cohomologie non Abéliennes / Lect. Notes Math., Berlin – Heidelberg: Springer, 1971.

Godement R.
1. Topologie algébrique et théorie des faisceaux / Paris: Hermann, 1958.

Goldblatt R.
1. Topoi. The categorial analysis of logic / Amsterdam – New York – Oxford: North-Holland, 1979.

Golovin V.D.
1. Homology of analytic sheaves and duality theorems / Moscow: Nauka, 1986 [Russian].

Goresky M., MacPherson R.
1. Intersection homology. II / Invent. Math., 1983, v. 72, no. 1, p. 77–130.

Gorodentsev A.L., Rudakov A.N.
1. Exceptional bundles on projective spaces / Duke Math. J., 1987, v. 54, p. 115–130.

Govorov V.E.
1. On flat modules / Sib. Math. J., 1965, v. 6, no. 2, p. 300–304.

Grey J.W.
1. Fragments of the history of sheaf theory / Lect. Notes Math., Heidelberg – Berlin: Springer, 1979, v. 753, p. 1–79.

Grothendieck A.
1. Sur quelques points d'algèbre homologique / Tôhoku Math. J., 1957, v. 9, no. 2, p. 119–183; no. 3, p. 185–221.
2. Fondements de la géométrie algébrique, Sém. Bourbaki, 1957–1962 / Paris: Secrétariat Math., 1962.
3. Local cohomology / Lect. Notes Math., Heidelberg – Berlin: Springer, 1967, v. 41.
4. Dix exposés sur la cohomologie de schémas / Amsterdam, North Holland, 1968.
5. Récoltes et semailles. Réflexions et témoignage sur un passé de mathématicien / Prepubl. Univ. des Sciences et Techniques de Languedoc, Montpellier, et CNRS, 1985.

Grothendieck A., Dieudonné J.
1. Eléments de Géométrie Algébrique (EGA), I, II, III, IV / Publ. Math. IHES, 1960, v. 4; 1961, v. 8; 1963, v. 17; 1964, v. 20; 1965, v. 24; 1966, v. 28; 1967, v. 32.
2. Eléments de Géométrie Algébrique (new version) / Berlin – Heidelberg: Springer, 1971.

Grothendieck A. et al.
1. Séminaire de géométrie algébrique (SGA)
   [SGA 2] Cohomologie locale des faisceaux cohérents et théorèmes de Lefschetz locaux et globaux / Amsterdam: North-Holland, 1968.
   [SGA 4] (with M. Artin, J.-L. Verdier) Théorie des topos et cohomologie étale de schémas / Lect. Notes Math., Berlin – Heidelberg: Springer 1972, v. 269, v. 270; 1973, v. 305.
   [SGA 4 1/2] (by P. Deligne, with J.F. Boutot, L. Illusie, J.-L. Verdier) Cohomologie étale / Lect. Notes Math., Berlin – Heidelberg: Springer. 1977, v. 569.
   [SGA 6] (with P. Berthelot, L. Illusie) Théorie de l'intersection et théorème de Riemann – Roch/Lect. Notes Math., Berlin – Heidelberg: Springer. 1971, v. 225.

Guichardet A.
1. Cohomologie des groupes topologiques et des algèbres de Lie / Paris: Cedic/Fernand Nathan, 1980.

Gulliksen T., Levin S.
1. Homology of local rings / Queen's Pap. Pure Appl. Math., 1969, no. 20.

Halperin S.
1. Lectures on minimal models / Mém. Soc. Math. Fr., 1983, v. 9/10.

Happel D.
1. Triangulated categories in the representation theory of finite-dimensional algebras / Lect. Notes London Math. Soc., 1988, v. 119.

Happel D., Ringel K.M.
1. Tilted algebras / Trans. Am. Math. Soc., 1982, v. 274, no. 2, p. 399–444.

Hartshorne R.
1. Residues and duality / Lect. Notes Math., Berlin – Heidelberg: Springer, 1966, v. 20.
2. Algebraic geometry / Berlin – Heidelberg: Springer, 1977.

Helemsky A.Ja.
1. Homology in Banach and topological algebras / Moscow, Izd. MGU, 1986.

Hilton P.
1. Homotopy theory and duality / New York: Gordon & Breach, 1965.

Hilton P., Stammbach U.
1. A course in homological algebra / Berlin – Heidelberg: Springer, 1970.

Hilton P., Wiley S.
1. Homology theory / Cambridge: Cambridge Univ. Press, 1960.

Hinich V.A., Shechtman V.B.
1. Geometry of a category of complexes and algebraic K-theory / Duke Math. J., 1985, v. 52, no. 2, p. 399–430.

Hochschild G.
1. On the cohomology groups of an associative algebra / Ann. Math., 1945, v. 46, p. 58–67.

Illusie L.
1. Complexe cotangent et déformations. I / Lect. Notes Math., Berlin – Heidelberg: Springer, 1971, v. 239.

Iversen B.
1. Cohomology of sheaves. Berlin – Heidelberg: Springer, 1986.

Johnson B.E.
1. Cohomology in Banach algebras / Mem. Am. Math. Soc., 1973, no. 127.

Johnstone P.T.
1. Topos theory / London – New York: Academic Press, 1977.

Kan D.M.
1. Adjoint functors / Trans. Am. Math. Soc., 1958, v. 87, p. 294–329.

Kapranov M.M.
1. On the derived category of coherent sheaves on Grassmann manifolds / Izv. Akad. Nauk SSSR, Ser. Math., 1984, v. 48, no. 1, p. 192–202 [Russian].
2. Derived category of coherent sheaves on a quadric / Funkts. Anal. Prilozh., 1986, v. 20, no. 2, p. 67 [Russian].
3. On the derived category of coherent sheaves on some homogeneous spaces / Invent. Math., 1988, v. 92, no. 3, p. 479–508.

Karoubi M.
1. Homologie cyclique des groupes et des algèbres / C. R. Acad. Sci., Paris Sér. 1, 1983, v. 297, p. 381–384.
2. Homologie cyclique et K-théorie algébrique, I, II / C. R. Acad. Sci., Paris Sér. 1, 1983, v. 297, p. 447–450; p. 513–516.

Kashiwara M.
1. Systems of microdifferential equations / Boston: Birkhäuser, 1983.
2. Riemann – Hilbert problem for holonomic systems / Publ. Res. Inst. Math. Sci., 1984, v. 20, no. 1, p. 319–365.

Knudsen F., Mumford D.
1. The projectivity of the moduli spaces of stable curves / Math. Scand., 1976, v. 39, no. 1, p. 19–35.

Knutson D.
1. Algebraic spaces / Lect. Notes Math., Berlin – Heidelberg: Springer, 1971, v. 203.

Lazard D.
1. Autour de la platitude / Bull. Soc. Math. Fr., 1969, v. 97, no. 1, p. 81–128.

Leites D.A.
1. Introduction to the theory of supermanifolds / Usp. Mat. Nauk., 1980, v. 35, no. 1, p. 3–57 [Russian].
2. Theory of supermanifolds / Petrozavodsk: Karelia Branch of AN SSSR, 1983 [Russian].

Lemann D.
1. Théorie homotopique des formes différentielles / Astérisque, 1977, v. 45.

Leray J.
1. L'anneau d'une représentation / C. R. Acad. Sci., Paris 1946, v. 222, p. 1366–1368.

Loday J.L., Quillen D.
1. Cyclic homology and the Lie algebra homology of matrices / Comment. Math. Helv., 1984, v. 59, no. 4, p. 565–591.

Lofwall C.
1. On the subalgebra generated by one-dimensional elements in the Yoneda Ext-algebra / in: [AAT], p. 291–338.

MacLane S.
1. Homology / Berlin – Heidelberg: Springer, 1963.
2. Categories for the working mathematician / New York: Springer, 1971.

Manin Y.I.
1. Some remarks on Koszul algebras and quantum groups / Ann. Inst. Fourier, 1987, v. 37, no. 4, p. 191–205.

Massey W.S.
1. Exact couples in algebraic topology / Ann. Math., 1952, v. 56, p. 363–396.
2. Homology and cohomology theory / New York – Basel: Marcel Dekker, 1978.

May J.P.
1. Simplicial objects in algebraic topology / Princeton: Van Nostrand, 1967.

Mazur B.
1. Notes on etale cohomology of number fields / Ann. Sc. Norm. Super Pisa, Cl. Sci., IV. Ser., 1973, v. 6, p. 521–556.

McCleary J.
1. User's guide to spectral sequences / Publish or Perish: Wilmington, Delaware (USA), 1985, 423 pp.

Meltzer H.
1. Tilting bundles, repetitive algebras and derived categories of coherent sheaves / Humboldt-Universität zu Berlin, Sek. Math., 1988, prepr. 193.

Milne J.S.
1. Etale cohomology / Princeton: Princeton Univ. Press, 1980.

Morgan J.
1. The algebraic topology of smooth algebraic varieties / Publ. Math. IHES, 1978, v. 48, p. 137–204.

Morita K.
1. Duality for modules and its applications to the theory of rings / Sci. Rep. Tokyo, Kyoiku Daigaku, Sec. A, 1958, v. 6, p. 83–142.

Morozov A.Yu.
1. Anomalies in gauge theories / Usp. Mat. Nauk., 1986, v. 150, no. 3, p. 337–416.

Okonek Ch., Schneider M., Spindler H.
1. Vector bundles on complex projective spaces / Basel – Boston – Stuttgart: Birkhäuser, 1980.

Priddy S.B.
1. Koszul resolutions and the Steenrod algebra / Bull. Am. Math. Soc., 1970, v. 76, no. 4, p. 834–839.
2. Koszul resolutions / Trans. Am. Math. Soc., 1970, v. 152, no. 1, p. 39–60.

Quillen D.
1. Homotopical algebra / Lect. Notes Math., Berlin – Heidelberg: Springer, 1967, v. 43.
2. Rational homotopy theory / Ann. Math., 1969, v. 90, p. 205–295.
3. On the (co)homology of commutative rings / Proc. Symp. Pure Math., 1970, v. 17, p. 65–87.
4. Higher algebraic K-theory, I / Lect. Notes Math., Berlin – Heidelberg: Springer, 1973, v. 341, p. 85–147.
5. Projective modules over polynomial rings / Invent. Math., 1976, v. 36, p. 167–171.

Reiman A.G., Semenov-Tyan-Shansky M.A., Faddeev L.D.
1. Quantum anomalies and cocycles on gauge groupes / Funkts. Anal. Prilozh., 1984, v. 18, no. 4, p. 64–72.

Roos J.-E.
1. Sur les foncteurs derivés de $\varprojlim$. Applications / C. R. Acad. Sci., Paris 1961, v. 252, no. 24, p. 3702–3704.
2. Sur les foncteurs derivés de produits infinis dans les catégories de Grothendieck. Exemples et contre-exemples / C. R. Acad. Sci., Paris 1966, v. 263, no. 25, p. 895–898.

Sato M., Kashiwara M., Kawai T.
1. Hyperfunctions and pseudodifferential equations / Lect. Notes Math., Berlin – Heidelberg: Springer, 1993, v. 287, p. 265–529.

Schapira P.
1. Théorie des hyperfonctions / Lect. Notes Math., Berlin – Heidelberg: Springer, 1970, v. 126.
2. Microdifferential systems in the complex domain / Berlin – Heidelberg: Springer, 1985.

Schechtman V.V.
1. Algebraic K-theory and characteristic classes / Usp. Mat. Nauk., 1978, v. 33, no. 6, p. 239–240.

Serre J.-P.
1. Homologie singulière des espaces fibrés. Applications / Ann. Math., 1951, v. 54, p. 425–505.
2. Groupes de homotopie et classes de groupes abéliennes / Ann. Math. 1953, v. 58, p. 258–294.
3. Faisceaux algébriques cohérents / Ann. Math., 1955, v. 61, p. 197-278.
4. Géométrie algébrique et géométrie analytique / Ann. Inst. Fourier, 1956, v. 6, p. 1–42.
5. Sur la dimension homologique des anneaux et de modules noethériens / Proc. Int. Sympos. algebraic number theory Tokyo – Nikko, 1956, p. 175–189.
6. Algèbre locale - multiplicités / Lect. Notes Math., Berlin – Heidelberg: Springer, 1965. v. 11; English edition: Local Algebra / Berlin – Heidelberg: Springer, 2000.
7. Cohomologie Galoisienne / Lect. Notes Math., Berlin – Heidelberg: Springer, 1965, v. 5; English edition: Galois Cohomology / Berlin – Heidelberg: Springer, 2002.
8. Cohomologie des groupes discrets / Ann. Math. Stud., 1971, v. 70, p. 77–169.

Spaltenstein N.
1. Resolutions of unbounded complexes / Comment. Math. Helv., 1988, v. 65, no. 2, p. 121–154.

Spanier E.
1. Algebraic topology / New York: Mc Graw-Hill, 1966.

Steenrod N., Eilenberg S.
1. Foundations of algebraic topology / Princeton: Princeton Univ. Press, 1952.

Sullivan D.
1. Infinitesimal computations in topology / Publ. Math. IHES, 1977, v. 47, p. 269–331.

Suslin A.A.
1. Projective modules over polynomial rings are free / Dokl. Akad. Nauk. SSSR, 1976, v. 229, no. 5, p. 1063–1066.
2. Algebraic K-theory / Itogi Nauki Tekh. Ser.: Algebra, Topologia, Geom., 1982, v. 20, p. 71–152.

Tanré D.
1. Homotopie rationnelle: modèles de Chen, Quillen, Sullivan / Lect. Notes Math., Berlin – Heidelberg: Springer, 1983, v. 1025.

Tate J.
1. Duality theorems in Galois cohomology over number fields, Proc. Int. Congress Math. 1962, Uppsala: Almquist, 1963, p. 288–295.

Verdier J.-L.
1. Théorème de dualité pour les cohomologies des espaces localement compacts / Sém. Heidelberg – Strasbourg 1966–1967, Strasbourg: Publ. Inst. Rech. Math. Avancée, 1969, no. 3, Exp. 4.
2. Théorème de dualité de Poincaré / C. R. Acad. Sci., Paris, 1963, v. 256, p. 2084–2086.
3. Catégories derivées, état 0 / Lect. Notes Math., Berlin – Heidelberg: Springer, 1977, v. 569, p. 262–311.
4. Extension of a perverse sheaf over a closed subspace / Astérisque, 1985, v. 130, p. 210–217.

Yoneda N.
1. On the homology theory of modules / J. Fac. Sci., Univ. Tokyo, Sec. IA, 1954, v. 7, p. 193–227.
2. On ext and exact sequences / J. Fac. Sci., Univ. Tokyo, Sec. IA, 1960, v. 8, p. 507–526.

Index

Abelian group filtered   117
-- simplicial   46
-- topological   118
Additivity of the derived category   162
Affine group scheme   83

Bialgebra   83
Bisimplicial set   18
Boundary   29

Canonical decomposition of a morphism (in an abelian category)   112
Cap lower   240
- upper   240
Cartier duality   85
Category   57
- abelian   113
- additive   113
- closed model   293
- cyclic   67
- derived   141
- dual   61
- exact   275
- examples of   58, 59
- Frobenius   276
- homotopic   159, 246, 332, 354
- of an ordered set   60
-- categories   67
-- coindices   196
-- complexes   58, 168
-- functors   63
-- indices   86
-- inductive limits   197
-- projective limits   197
- quotient   77
- simplicial   146
- stable   276
- triangulated   239
Chain   23
- degenerate   53
Class complementary   291

- of morphisms localizing   147
-- in a triangulated category   251
-- objects adapted to a functor   187
Classified space (of a group)   8, 61
Coboundary   24
Cochain   24
Coefficient system   28
Coequalizer   88
Cofibration   293, 331, 335, 350
Cohomology   47
- Čech   30, 47
- cyclic   54
- Hochschild   54
- of a cochain complex   25, 123
-- group   30
-- Lie algebra   48
-- sheaf   219
--- with compact support   227
Cokernel   111
Colimit   88
Complex
- acyclic   123
- bounded (from the left, from the right, from both sides)   147
- chain   25
- cochain   25, 123
- diagonal   208
- double   207, 208
- dualizing   234
- K-injective   199
- of Čech   46
-- Koszul   54, 108
-- de Rham   48
-- singular chains   47
- K-projective   199
- rigid   269
-- finite   269
- twisted   262
Cone   26, 154, 244
Convolution of a complex   258
Core of a triangulated category   278

Coskeleton 16, 93
Cycle 25
Cylinder of
– a morphism 154
– an object 332

Degeneration 14
Diagonal of a bisimplicial set 15
Differential graded algebra
  (DG-algebra) 338
– minimal 343
Dimension 227
– homological 171
– injective 172
– projective 172

Eilenberg–Zilber theorem 19
Element (of an object of an abelian
  category) 118
Epimorphism 114
Equalizer 87
Equivalence homotopic 301
– of categories 71
– – examples 71–76
– weak 293, 331, 334, 350
Exact pair 215
– sequence 40
– – of cohomology 40
– triple 40, 42
Extension 257
– and Ext^1 184
– by Yoneda 166

Face 14
Fibre of a sheaf at a point 34
Fibration 293, 331, 334, 350
– locally trivial 321
– minimal 317
– universal 325
Filtration
– and Ext^2 184
– of a complex 203
– – canonical 203
– – stupid 203
– regular 202
Five-lemma 120
Functor 60
– additive 122
– adjoint 89
– cohomological 243
– contravariant 61
– covariant 61
– derived 188
– – and Ext^i 194

– – – Tor_i 195
– – classical 194
– – of a composition 200
– – weak 197
– diagonal 86
– exact 125, 185, 285
– – from the left 125
– – right 125
– examples 60–62
– faithful 64
– forgetful 63, 91
– full 64
– representable 78
– – in a category of sheaves 229

Geometrical realization 6
– – of a simplicial topological space
  18
Germ of a section 34
Gluing 1, 288
– data 1
Grassmann algebra 267
Grothendieck topology 100
Group cosimplicial 46
– in a category 82
– simplicial 46

Hilbert theorem 175, 176
Homology 47
– cyclic 54
– Hochschild 54
– of a chain complex 25, 30
– of a group 30
– of a Lie algebra 48
– singular 47, 50
Homotopy 49, 53, 107, 301, 333, 351,
  353
– group 356
– – of a DG-algebra 353
– of morphisms of complexes 49,
  140
– of simplicial sets 301
– – of mappings 23
Horn 294
Hypercohomology 213
Hypersimplex 260

Image 113
– direct 133
– – higher 228
– – with compact support 224
– – – higher 227
– inverse 134–136, 223
– – with compact support 228

Injectivity diagram 125
Isomorphism of triangles 157
– – functors 69

Jordan–Holder theorem 122

Kan set 299
Kernel 111

Lie superalgebra 344
Limit
– inductive 88, 197
– projective 86
Localization 145

Manifolds 93
Massey product 266
Meier–Vietoris theorem 236
Minimal model of a
– de Rham algebra 352
– DG-algebra 347
Mittag–Leffler condition 198
Module flat 128, 217
Monomorphisms 114
Morita equivalence 77
Morphism 57
– diagonal 80
– flat 128, 224
– inner 105, 108
– of coefficient systems 44
– – complexes 41
– – functors 63
– – examples 63
– – sheaves 32
– – spectral sequences 202
– – triangles 157
– proper 224

Nerve of
– a category 104
– covering 7

Object 57
– $\mathcal{A}$-acyclic 133
– F-acyclic (for functor F) 195
– cofibrant 332
– fibrant 332
– final 65
– initial 65
– injective 125, 275
– projective 125, 275
– representing 78
– simple 121
– zero 110

Octahedron diagram 240

Poincaré duality 234
Polycategory 65
Postnikov system 257
Presheaf 32, 100
Product direct 80
– fibre 80
Projectivity diagram 128

Quadratic algebras 108
Quasi-isomorphism 143
Quiver 288
Quotient category 77
– (à la Serre) 121

Resolution
– Cartan–Eilenberg 207
– flat 217
– free 139
– injective 143, 179, 199
– projective 141, 199
Retraction 292
Ringed space 97
Roof 151

Scheme 98
– affine 95, 97
Section (of a sheaf) 32
– with compact support 225
– – support in a subset 234
Sheaf 32
– associated with a presheaf 35, 114
– constant 35
– fine 38
– flabby 37, 219
– flat 222
– injective 218
– locally constant 36
– of morphisms 38
– on a site 100
– quasicoherent 40
– soft 38, 226
Sieve 100
– covering 100
Simplex n-dimensional 2, 6
– nondegenerate 9
– singular 7
Simplices g-connected 316
Simplicial set 6
– N-truncated 16
– singular 60
Simplicial topological space 17
Site 100

Skeleton   3, 13, 92
Snake-lemma   120
Spectral sequence   201
- degenerate   202
- of a double complex   207
- of a filtered complex   203, 217
- Čech cohomology   214
- composition of functors   207
- Eilenberg–Moore   263
- hypercohomology   213
- Serre–Hochschild   214
Sphere simplicial   2, 295
Square cartesian   81
- cocartesian   81
t-structure   276, 286
Subcategory   64
- full   64, 76

- localizing   153
- thick   266
Sum amalgamated   81
- direct   138
Superspace   98
Suspension   276
Syzygies   139

Tensor product of algebras   82
Translation functor   157, 239
Triangle   156
- distinguished   157, 239
Triangulated space   2
Triangulation   2
- canonical of the product of two simplices   4
Truncation functor   279